THE AGES OF STARS

IAU SYMPOSIUM No. 258

COVER ILLUSTRATION: A star-spangled banner

Top left: $\sim 10^5$-10^6 year-old stars forming in dense molecular gas on the edge of an H II region illuminated by a $\sim 10^{6.5}$ year-old massive O-type star. *Top right:* $\sim 10^8$ year-old B-type stars in the Pleiades open cluster illuminating reflection nebulae. *Bottom right:* A $\sim 10^{10}$ year-old globular cluster – a metal-poor stellar city. *Bottom left:* Site of the 258th IAU symposium – "The Ages of Stars" – the Inner Harbor of Baltimore, Maryland, USA – illuminated by a $10^{9.66}$ year-old G2V star.

INTERNATIONAL ASTRONOMICAL UNION

UNION ASTRONOMIQUE INTERNATIONALE

THE AGES OF STARS

PROCEEDINGS OF THE 258th SYMPOSIUM OF THE INTERNATIONAL ASTRONOMICAL UNION HELD IN BALTIMORE, MARYLAND, USA OCTOBER 13–17, 2008

Edited by

ERIC E. MAMAJEK

Department of Physics and Astronomy, University of Rochester, Rochester, NY, USA

DAVID R. SODERBLOM

Space Telescope Science Institute, Baltimore, MD, USA

and

ROSEMARY F. G. WYSE

Department of Physics and Astronomy, The Johns Hopkins University, Baltimore, MD, USA

Shaftesbury Road, Cambridge CB2 8EA, United Kingdom

One Liberty Plaza, 20th Floor, New York, NY 10006, USA

477 Williamstown Road, Port Melbourne, VIC 3207, Australia

314–321, 3rd Floor, Plot 3, Splendor Forum, Jasola District Centre, New Delhi – 110025, India

103 Penang Road, #05–06/07, Visioncrest Commercial, Singapore 238467

Cambridge University Press is part of Cambridge University Press & Assessment, a department of the University of Cambridge.

We share the University's mission to contribute to society through the pursuit of education, learning and research at the highest international levels of excellence.

www.cambridge.org
Information on this title: www.cambridge.org/9780521889896

First published 2009

A catalogue record for this publication is available from the British Library

ISBN 978-0-521-88989-6 Hardback
ISSN 1743-9213

Table of Contents

Part 5. WHITE DWARFS

Chair: Robert Rood

Part 6. BROWN DWARFS

Chair: Michael Liu

Part 7. AGE-RELATED PROPERTIES OF SOLAR-TYPE STARS

Chair: Fred Walter

Part 8. ASTEROSEISMOLOGY AND THE SUN

Chair: Jeff Valenti

Part 9. NUCLEOCHRONOLOGY

Chair: John Stauffer

Preface

How old is that star?

That is one of the most difficult questions to answer in Galactic astrophysics. Of the fundamental properties that determine the state of a star (mass, composition and age, primarily), we can directly and accurately measure many stars' masses and then estimate the mass of a single star of the same spectral types. Similarly, stars directly reveal their compositions to us through their spectra. Many significant and interesting questions remain in studying the masses and compositions of stars, yet one can say that those subjects are well enough understood to know what the important questions are.

Age is another matter all together. We know a precise and accurate age for exactly one star: the Sun, and that age comes not from the Sun itself, but rather from studying solar system material in the laboratory, something we can do for no other star. For other stars, ages are inferred indirectly in one way or another. For example, in clusters we can assume a single composition and age and then fit an isochrone to the ensemble. Other age estimates can be derived from context, by drawing a connection among a group of stars or by applying some other constraint. For example, space motions can be used to estimate when in the past a young group of stars were in closest proximity, a likely time of formation. Or if we find a number of stars that share the same space motion, we presume a common origin so that an evolved star among them limits the age, just as for a bona fide cluster.

Stellar ages lie at the heart of much of astrophysics, and stellar evolution is all about time and how stars change with time. We wish to determine time-scales for physical processes such as angular momentum loss, nucleo-synthetic processing, changes in magnetic fields, and the like, or we wish to compare objects or groups of objects at different stages in their lives. Stellar and galactic evolution cannot be understood without some consideration of ages. If we could pin ages on individual stars we could determine the Galaxy's star formation history and we could understand the physics of low-mass stars much better. The well-studied spin-down of stars like the Sun and the concomitant decline of observed activity indices makes it possible to estimate rough ages for individual stars, but the scarcity and remoteness of older clusters makes calibrating and testing the activity-age (or rotation-age) relation problematic.

Age has been a slippery enough topic that it has been poorly examined in and of itself over the years. Only two international meetings have specifically addressed the ages of stars, in 1972 and 1989, a generation ago. The topic has come up in other contexts but has never been the focus of an IAU Symposium, which is truly amazing given the fundamental importance of age.

Yet now is an appropriate time to examine the problem of stellar ages in detail. We now understand models of stars and systems of stars much better than even a decade ago. Our understanding of the physics that goes into those models has greatly improved, including such aspects as opacities, nuclear reaction rates, diffusion, effects of rotation and gravity waves, and the influence of magnetic fields. At the same time, observations have tested the models, including now the detection of white dwarf cooling sequences in globulars and an open cluster, and evidence for multiple generations of stars within some clusters. At the other end of the age scale, we want to understand the mechanisms and duration of planet formation, but the stars around which we find circumstellar material have highly uncertain ages. We would like to be able to tell if the youngest star clusters and associations undergo multiple waves of formation, or if higher-mass stars form at different times from those of lower mass. So many aspects of stellar evolution beg for accurate ages!

This symposium brought together astronomers from the around the world to discuss the current state of the problem of estimating ages of individual stars and of populations, where the advances are now being made, and what the near future offers.

Some of the questions that we addressed included:

(*a*) What is the current state of our knowledge of stellar ages and how can we improve on that?

(*b*) How do we calibrate ages of systems and ensembles of stars, and what limits our ability to do so?

(*c*) What are current limits in the input physics for stellar models?

(*d*) How well can observations be matched to the models? How well can we model real spectra?

(*e*) How well do we, in fact, know the ages of ensembles and systems of stars? What limits the accuracy and precision of calculating and fitting isochrones for all kinds of stars at all stages of evolution? In other words, how well can we test and calibrate our models? To what extent does internal evolution within clusters limit us (e.g., blue stragglers)?

(*f*) What evidence is there for age spreads within systems and ensembles? How well does the lithium depletion boundary at the low end of the main sequence tell us ages?

(*g*) How good are ages of ensembles determined from tracing back kinematics?

(*h*) Can the morphology of the horizontal branch be used to determine ages?

(*i*) What do binary star systems tell us about ages and how can they help test models?

(*j*) Can we reach a point where we can reliably measure an accurate age for a single star in the field? How well does the nascent field of asteroseismology offer an independent means of determining ages?

These subjects then led to the essential question of the symposium: how well can we estimate the age of a single star or of groups of stars such as populations in nearby (resolved) stellar systems? Ages of stars can then be applied to some key astrophysical questions:

(*a*) How old are the primary components of our Galaxy; the thick disk and the thin disk; and over what span of time did they form? For instance, stars of the thick disk can be selected kinematically and from abundances of their alpha elements, but we need better and independent estimates of their ages.

(*b*) Over what span of time do stars within a cluster form?

(*c*) How old are the stars that have planets? In addition to the effect recently shown, in which metal- rich stars tend to be more likely to have planets, is there an age effect?

(*d*) How about the ages of stars with observed debris disks or other circumstellar material?

(*e*) Can we establish a well-defined comparison sample for the Sun, based on mass and age, so that we have a context for understanding long- and short-term changes in solar luminosity and activity?

(*f*) Are clusters, especially old ones, in fact representative of stars in the field? Or have they managed to survive as clusters because of different initial conditions?

(*g*) How well can we derive the star formation histories of composite systems such as the Local Group galaxies?

(*h*) How well can we extend our experience in our Galaxy to more distant realms, such as the halo of Andromeda, and what do we learn about the Milky Way as a result?

David R. Soderblom, chair SOC,
Baltimore, MD, USA

THE ORGANIZING COMMITTEE

Scientific

D. R. Soderblom (Chair, USA)
H. M. Antia (India)
N. Arimoto (Japan)
M. S. Bessell (Australia)
C. Charbonnel (Switzerland)
V. Hill (France)
L. Hillenbrand (USA)
B. Nordström (Denmark)
H. J. Rocha-Pinto (Brazil)
E. Tolstoy (the Netherlands)
D. A. VandenBerg (Canada)
R. Wyse (USA)
M. Zoccali (Chile)

Local

J. Valenti (chair)
J. Anderson
L. Bedin
T. Brown
K. Exter
A. Grocholski
R. de Jong
J. Kalirai
I.N. Reid
M. Robberto
E. Villaver

Acknowledgements

The symposium was sponsored and supported by IAU Division VII (Galactic System); and by IAU Commissions 26 (Binary and Multiple Stars), 29 (Stellar Spectra), and 33 (Structure and Dynamics of the Galactic System).

The Local Organizing Committee operated under the auspices of the Space Telescope Science Institute.

Financial support from the
International Astronomical Union
The U. S. National Science Foundation
Las Cumbres Observatory Global Telescope Network
and
the Space Telescope Science Institute
are gratefully acknowledged.

Row 1 (front): Kevin Covey, Benjamin Brown, Jørgen Christensen-Dalsgaard, Tim Brown, Jeff Cummings, Carla Cacciari, Christine Chen. Row 2 (center): Tracy Beck, Teresa Antoja, Chris Corbally, Chris Burke, Silvia Catalan, ?, Cameron Bell, Sydney Barnes. Row 3 (rear): Marcio Catelan, Adam Burgasser, Ann Marie Cody, Tabetha Boyajian, ?, Andrea Bellini, Michael Barker, Kelle Cruz.

Row 1 (front): Scott Engle, Thibaut Decressin, Aaron Dotter, Thomas de Boer, Trent Dupuy, Karl Gordon, David Fernández. Row 2 (center): Angela Bragaglia, Steven DeGennaro, Jonathan Fulbright, Sofia Feltzing, Andrea Dupree, Anna Frebel, Ane Garcés, Jacqueline Faherty, Jose Fernandez. Row 3 (rear): Pierre Demarque, Con Deliyannis.

Row 1 (front): Sandy Leggett, Yveline Lebreton, Krzysztof Helminiak, Adam Kraus, Ivan King, Robin Jefferies, Eric Jensen. Row 2 (center): Jonathan Irwin, ?, Leslie Hebb, Jan Lub, Michael Liu, Andreas Kaufer, Richard Jackson. Row 3 (rear): ?, Elizabeth Jeffery, Vera Kozhurina-Platais, Vanessa Hill, Lynne Hillenbrand.

Row 1 (front): Raquel Martinez-Arnaiz, David Montes, Gerhard Meurer, Steven Margheim, John Mackenty, Tim Naylor, Steven Pravdo. Row 2 (center): Antonela Monachesi, ?, Isa Oliviera, ?, Hashima Hasan, Birgitta Nordström, Søren Meibom. Row 3 (rear): Marc Pinsonneault, Jason Melbourne, Delbert McNamara, Walter Maciel, Eric Mamajek, Romas Mitalas, Russ Makidon, Robert Mathieu, Dougal Mackey, Georges Meynet.

Row 1 (front): Robert Rood, Maddelena Reggiani, Sofia Randich, Harvey Richer, Ricardo Schiavon, Maurizio Salaris, Ata Sarajedini. Row 2 (center): Evgenya Shkolnik, Simon Schuler, John Stauffer, Peter Stetson, Allen Sweigart, Ian Roederer, ?, Marian Doru Suran. Row 3 (rear): Massimo Roberto, David Soderblom, Gail Schaefer.

Row 1 (front): Patrick Young, Sukyoung Yi, Andrew West, Sylivie Vauclair, Jason Wright, Jeff Valenti. Row 2 (center): Bruce Weaver, Fred Walter, Ian Thompson, Aldo Valcarce, Gerard Vauclair. Row 3 (rear): Monica Tosi, Russel White, Kurtis Williams, Ed Guinan.

LOCAL AND SCIENTIFIC ORGANIZING COMMITTEE MEMBERS

Row 1 (front): Jay Anderson, Luigi Bedin, Jeff Valenti, David Soderblom, Andrea Bellini, Aaron Grocholski. Row 2 (rear): Massimo Robberto, Catherine Riggs, Darlene Spencer, Birgitta Nordström, Tom Brown.

Participants

Alessandra **Aloisi**, ESA/Space Telescope Science Institute aloisi@stsci.edu
Jay **Anderson**, Space Telescope Science Institute jayander@stsci.edu
Borja **Anguiano**, Astrophysikalisches Institut Potsdam baj@aip.de
Teresa **Antoja**, Universitat de Barcelona tantoja@am.ub.es
Mike **Barker**, Institute for Astronomy, Edinburgh University mkb@roe.ac.uk
Sydney **Barnes**, Lowell Observatory barnes@lowell.edu
Amelia **Bayo**, LAEFF-CAB, INTA-CSIC abayo@laeff.inta.es
Tracy **Beck**, Space Telescope Science Institute tbeck@stsci.edu
Luigi R. **Bedin**, Space Telescope Science Institute bedin@stsci.edu
Cameron **Bell**, University of Exeter bell@astro.ex.ac.uk
Andrea **Bellini**, UNIPD/Space Telescope Science Institute bellini@stsci.edu
Edouard **Bernard**, Instituto de Astrofisica de Canarias ebernard@iac.es
Tabetha **Boyajian**, Georgia State University tabetha@chara.gsu.edu
Angela **Bragaglia**, INAF-Osservatorio Astronomico di Bologna angela.bragaglia@oabo.inaf.it
Thomas **Brown**, Space Telescope Science Institute tbrown@stsci.edu
Benjamin **Brown**, University of Colorado, Boulder bpbrown@solarz.colorado.edu
Timothy **Brown**, LCOGT tbrown@lcogt.net
Eric **Bubar**, Clemson University ebubar@clemson.edu
Adam **Burgasser**, Massachusetts Institute of Technology ajb@mit.edu
Carla **Cacciari**, INAF - Osservatorio Astronomico Bologna carla.cacciari@oabo.inaf.it
Annalisa **Calamida**, ESO acalamid@eso.org
Silvia **Cataln**, Institut de Cincies de l'Espai (CSIC-IEEC) catalan@ieec.uab.es
Mrcio **Catelan**, Pontificia Universidad Catlica de Chile mcatelan@astro.puc.cl
Brian **Chaboyer**, Dartmouth College Brian.Chaboyer@Dartmouth.edu
Christine **Chen**, Space Telescope Science Institute cchen@stsci.edu
Joergen **Christensen-Dalsgaard**, University of Aarhus jcd@phys.au.dk
Chul **Chung**, Department of Astronomy Yonsei University mitchguy@gmail.com
Michele **Cignoni**, Department of Astronomy, Bologna University, Italy michele.cignoni@unibo.it
Christopher **Corbally**, Vatican Observatory corbally@as.arizona.edu
Kevin **Covey**, Harvard-Smithsonian Center for Astrophysics kcovey@cfa.harvard.edu
Jeff **Cummings**, Indiana University jdcummi@astro.indiana.edu
Valentina **D'Orazi**, Dipartimento di Astronomia, Universita di Firenze vdorazi@arcetri.astro.it
Thomas **de Boer**, Kapteyn Astronomical Institute deboer@astro.rug.nl
Roelof **de Jong**, Space Telescope Science Institute dejong@stsci.edu
Guido **De Marchi**, ESA gdemarchi@rssd.esa.int
Steven **DeGennaro**, University of Texas at Austin deg@astro.as.utexas.edu
Thibaut Argenlander **Decressin**, Institute for Astronomy, Bonn decressin@astro.uni-bonn.de
Constantine **Deliyannis**, Indiana University con@astro.indiana.edu
Pierre **Demarque**, Yale University pierre.demarque@yale.edu
Qader **Dorosti**, Institute for Studies in Theoretical Physics and Mathematics q_dorosti@yahoo.com
Aaron **Dotter**, University of Victoria, Dept. of Physics and Astronomy dotter@uvic.ca
Igor **Drozdovsky**, Instituto de Astrofisica de Canarias dio@iac.es
Andrea **Dupree**, Harvard-Smithsonian Center for Astrophysics dupree@cfa.harvard.edu
Trent **Dupuy**, Univ. of Hawaii tdupuy@ifa.hawaii.edu
Scott **Engle**, Villanova University scott.engle@villanova.edu
Katrina **Exter**, Space Telescope Science Institute kexter@stsci.edu
Jacqueline **Faherty**, Stony Brook University/AMNH jfaherty17@gmail.com
Sofia **Feltzing**, Lund Observatory sofia@astro.lu.se
Jose **Fenandez**, Harvard-Smithsonian Center for Astrophysics jfernand@cfa.harvard.edu
David **Fernández**, Universitat de Barcelona david.fernandez@am.ub.es
Giuliana **Fiorentino**, Kapteyn Astronomical Institute fiorentino@astro.rug.nl
Anna **Frebel**, McDonald Observatory, Univ. of Texas anna@astro.as.utexas.edu
Yves **Frmat**, Royal Observatory of Belgium yves.fremat@oma.be
Carme **Gallart**, Instituto de Astrofsica de Canarias carme@iac.es
Ane **Garcs**, Institut de Ciencies de l'Espai (CSIC-IEEC) garces@ieec.uab.es
Leo **Girardi**, Osservatorio Astronomico di Padova leo.girardi@oapd.inaf.it
George **Gontcharov**, Pulkovo Observatory georgegontcharov@yahoo.com
Karl **Gordon**, Space Telescope Science Institute kgordon@stsci.edu
Paul **Goudfrooij**, Space Telescope Science Institute goudfroo@stsci.edu
Claudia **Greco**, Observatoire de Genève claudia.greco@obs.unige.ch
Aaron **Grocholski**, Space Telescope Science Institute aarong@stsci.edu
Edward **Guinan**, Villanova University edward.guinan@villanova.edu
Hashima **Hasan**, NASA-Headquarters HHasan@nasa.gov
Leslie **Hebb**, University of St Andrews leslie.hebb@st-andrews.ac.uk
Krzysztof **Helminiak**, Nicolaus Copernicus Astronomical Center xysiek@ncac.torun.pl
Sebastian L. **Hidalgo**, Instituto de Astrofisica de Canarias shidalgo@iac.es
Vanessa **Hill**, GEPI, Observatoire de Paris and Cassiopee, OCA Vanessa.hill@obspm.fr
Lynne **Hillenbrand**, California Institute of Technology lah@astro.caltech.edu
Jonathan **Irwin**, Harvard-Smithsonian Center for Astrophysics jirwin@cfa.harvard.edu
Richard **Jackson**, Keele University UK richard.jackson01@btinternet.com
Kenneth **Janes**, Boston University janes@bu.edu
Elizabeth **Jeffery**, University of Texas at Austin ejeffery@astro.as.utexas.edu
Robin **Jeffries**, Keele University rdj@astro.keele.ac.uk
Eric **Jensen**, Swarthmore College ejensen1@swarthmore.edu
Prajwal **Kafle**, Tribhuwan University prrajkafle@gmail.com
Jason **Kalirai**, Space Telescope Science Institute jkalirai@stsci.edu
Janusz **Kaluzny**, Nicolaus Copernicus Astronomical Center jka@camk.edu.pl
Yongbeom **Kang**, Johns Hopkins University ybkang@pha.jhu.edu
Andreas **Kaufer**, European Southern Observatory akaufer@eso.org
Ivan **King**, Univ. of Washington king@astro.washington.edu
Jeremy **King**, Clemson University jking2@clemson.edu
Ali **Koohpaee**, IPM akoohpaee@gmail.com
Vera **Kozhurina-Platais**, Space Telescope Science Institute verap@stsci.edu
Adam **Kraus**, California Institute of Technology alk@astro.caltech.edu
Arunas **Kucinskas**, Institute of Theoretical Physics and Astronomy arunaskc@itpa.lt
Yveline **Lebreton**, Paris Observatory Yveline.Lebreton@obspm.fr
Young-Wook **Lee**, Yonsei University ywlee2@yonsei.ac.kr
Youngdae **Lee**, Chungnam National University hippo206@cnu.ac.kr
Sandy **Leggett**, Gemini Observatory sleggett@gemini.edu
Michael **Liu**, University of Hawaii mliu@ifa.hawaii.edu

Jan **Lub**, Leiden Observatory lub@strw.leidenuniv.nl
Walter **Maciel**, University of Sao Paulo maciel@astro.iag.usp.br
Dougal **Mackey**, University of Edinburgh dmy@roe.ac.uk
Eric **Mamajek**, University of Rochester emamajek@pas.rochester.edu
Steven **Margheim**, Gemini Observatory smargheim@gemini.edu
Raquel M. **Martinez-Arnaiz**, Universidad Complutense de Madrid (UCM) rma@astrax.fis.ucm.es
Robert **Mathieu**, University of Wisconsin - Madison mathieu@astro.wisc.edu
Delbert **McNamara**, Brigham Young University mcnamara@byu.edu
Soren **Meibom**, Harvard-Smithsonian Center for Astrophysics smeibom@cfa.harvard.edu
Jason **Melbourne**, California Institute of Technology jmel@caltech.edu
Erin **Mentuch**, University of Toronto mentuch@astro.utoronto.ca
Gerhardt **Meurer**, Johns Hopkins U. meurer@pha.jhu.edu
Michael **Meyer**, The University of Arizona mmeyer@as.arizona.edu
Georges **Meynet**, Observatory of Geneva University georges.meynet@obs.unige.ch
Amy **Mioduszewski**, NRAO amiodusz@nrao.edu
Romas **Mitalas**, University of Western Ontario rmitalas@rogers.com
Antonela **Monachesi**, Kapteyn Astronomical Institute monachesi@astro.rug.nl
Matteo **Monelli**, Instituto de Astrofsica de Canarias monelli@iac.es
David **Montes**, Universidad Complutense de Madrid, UCM dmg@astrax.fis.ucm.es
Saumitra **Mukherjee**, Jawaharlal Nehru University dr.saumitramukherjee@usa.net
Tim **Naylor**, University of Exeter timn@astro.ex.ac.uk
Andrzej **Niedzielski**, Torun Center for Astronomy, Nicolaus Copernicus University aniedzi@astri.uni.torun.pl
Birgitta **Nordstrom**, Niels Bohr Institute, Copenhagen University birgitta@astro.ku.dk
Grzegorz **Nowak**, Torun Centre for Astronomy, Nicolaus Copernicus University grzenow@astri.uni.torun.pl
Isa **Oliveira**, Leiden Observatory oliveira@strw.leidenuniv.nl
Nino **Panagia**, Space Telescope Science Institute panagia@stsci.edu
Marc **Pinsonneault**, Ohio State University pinsonneault.1@osu.edu
Giampaolo **Piotto**, Universita' di Padova giampaolo.piotto@unipd.it
Antonio **Pipino**, Oxford University/University of Southern California axp@astro.ox.ac.uk
Violet **Poole**, Washington State University vpoole@wsu.edu
Steven **Pravdo**, California Institute of Technology/JPL spravdo@jpl.nasa.gov
Sofia **Randich**, INAF-Osservatorio di Arcetri randich@arcetri.astro.it
Maddalena Maria **Reggiani**, Space Telescope Science Institute reggiani@stsci.edu
Soo-Chang **Rey**, Chungnam National University screy@cnu.ac.kr
Harvey **Richer**, University of British Columbia richer@astro.ubc.ca
Massimo **Robberto**, Space Telescope Science Institute robberto@stsci.edu
Ian **Roederer**, University of Texas at Austin iur@astro.as.utexas.edu
Robert **Rood**, University of Virginia rtr@virginia.edu
Kailash **Sahu**, Space Telescope Science Institute ksahu@stsci.edu
Maurizio **Salaris**, Liverpool John Moores University ms@astro.livjm.ac.uk
Ata **Sarajedini**, University of Florida ata@astro.ufl.edu
Gail **Schaefer**, Georgia State University (CHARA Array) schaefer@chara- array.org
Ricardo **Schiavon**, Gemini Observatory rschiavon@gemini.edu
Simon **Schuler**, NOAO/CTIO sschuler@ctio.noao.edu
Evgenya **Shkolnik**, DTM/CIW shkolnik@dtm.ciw.edu
Michal **Simon**, SUNY-Stony Brook michal.simon@sunysb.edu
Jan **Snigula**, MPE Garching snigula@mpe.mpg.de
David **Soderblom**, Space Telescope Science Institute drs@stsci.edu
Keivan **Stassun**, Vanderbilt University keivan.stassun@vanderbilt.edu
John **Stauffer**, IPAC stauffer@ipac.caltech.edu
Peter **Stetson**, Herzberg Institute of Astrophysics Peter.Stetson@nrc- cnrc.gc.ca
Laura **Sturch**, Carnegie Observatories lsturch@ociw.edu
Marian Doru **Suran**, Astronomical Institute of the Romanian Academy suran@aira.astro.ro
Allen **Sweigart**, NASA Goddard Space Flight Center Allen.V.Sweigart@nasa.gov
Yuhei **Takagi**, Kobe University takagi@stu.kobe-u.ac.jp
Ian **Thompson**, Carnegie Observatories ian@ociw.edu
Eline **Tolstoy**, Kapteyn Astronomical Institute etolstoy@astro.rug.nl
Monica **Tosi**, Osservatorio Astronomico di Bologna monica.tosi@oabo.inaf.it
Muneeb **ur Rahman**, Kohat University of Science & Technology muneebtj@gmail.com
Aldo **Valcarce**, Pontificia Universidad Catolica de Chile avalcarc@astro.puc.cl
Jeff **Valenti**, Space Telescope Science Institute valenti@stsci.edu
Sylvie **Vauclair**, LATT/OMP sylvie.vauclair@ast.obs-mip.fr
Gerard **Vauclair**, LATT/OMP gerardv@ast.obs-mip.fr
Paritosh **Verma**, Jaypee Institute of Engineering & Technology paritosh.dwarf05@gmail.com
Enrico **Vesperini**, Drexel University vesperin@physics.drexel.edu
Eva **Villaver**, Space Telescope Science Institute/ESA villaver@stsci.edu
Frederick **Walter**, Stony Brook University fwalter@mail.astro.sunysb.edu
Bruce **Weaver**, MIRA bw@mira.org
Andrew **West**, Massachusetts Institute of Technology awest@astro.berkeley.edu
Russel **White**, Georgia State University white@chara.gsu.edu
Kurtis **Williams**, Univ. of Texas at Austin kurtis@astro.as.utexas.edu
Jason **Wright**, Cornell University jtwright@astro.cornell.edu
Rosemary **Wyse**, Johns Hopkins University wyse@pha.jhu.edu
Sukyoung **Yi**, Yonsei University yi@yonsei.ac.kr
Mutlu **Yildiz**, Ege University mutlu.yildiz@ege.edu.tr
Patrick **Young**, Arizona State University patrick.young.1@asu.edu
David **Zurek**, American Museum of Natural History dzurek@amnh.org

Address by the SOC & LOC

Dear colleagues,

No meeting of this kind just happens without the assistance and support of many people and organizations. The Scientific Organizing Committee, listed on page xiii, provided critical guidance on the content of the symposium and potential speakers, which was invaluable. The Local Organizing Committee gave freely of their time to make the meeting the success it was, both in the program and the execution. Finally, many individuals at the Space Telescope Science Institute provided the support we can sometimes take for granted, and their willingness to assist in an unusual undertaking made all the difference. I particularly wish to thank Darlene Spencer and Catherine Riggs, who put in many hours and provided the keel to keep our ship steady.

Symposia also need external financial support to succeed. The following organizations provided such assistance, and all the attendees benefitted:

- Space Telescope Science Institute
- International Astronomical Union
- U.S. National Science Foundation
- Las Cumbres Observatory Global Telescope Network

D.R. Soderblom, SOC Chair, 15 October 2008

Thoughts on IAU 258 (with apologies to Edgar Allan Poe)

Sitting in a lecture dreary, pondering convective theory
From some vast and vacuous volume of astrophysics lore,
As I nodded, nearly napping, suddenly there came a rapping
As if the audience were clapping, begging of the speaker "More!"
Then I made a quick decision: all I had to do was listen,
Listen up and nothing more.

Lost amidst the storm and fury, delivered to that eager jury
Was the answer to this story, story that the stars foreswore.
Somewhere in that stellar history lay the answer to the mystery,
Mystery of the stellar ages, ages we'd not known before.
Could it be companion planet, orbiting in a field magnetic?
Or if not this, then how much more?

Elegantly he spun a theory connecting cluster's metallicity
Through the oscillating iron's penetrating to the core.
Utilizing new fiducials (calibration's ever crucial),
CMD are most unusual: take a look at Messier four
Sir, said I, unto the speaker, "Is it this or something deeper?"
Quoth the speaker, "Of such data - we need more."

Lithium in circulation's symptom of the star's rotation,
Rotation seen as modulation of the stellar spots before.
Does convective overshooting coupled with magnetic heating
Cause horizontal branch's splitting, splitting into three or four?
When we find a clear dependence relating ages to abundance
Then all is done in Baltimore.

Fred M. Walter
October 2008

The Ages of Stars
Proceedings IAU Symposium No. 258, 2008
E.E. Mamajek, D.R. Soderblom & R.F.G. Wyse, eds.

doi:10.1017/S1743921309031652

Some problems in studying the ages of stars

David R. Soderblom

Space Telescope Science Institute,
3700 San Martin Drive
Baltimore MD 21218 USA

Abstract. I list some questions and problems that have motivated this symposium, particularly with regard to single stars and low-mass stars.

Keywords. Galaxy: structure, Galaxy: disk, kinematics and dynamics, solar neighborhood, Hertzsprung-Russell diagram, stars: late-type

---

1. Motivation

The age of a star cannot be *measured*, not in the way we can measure mass or composition, the other key determinants of a star's physical state. I have always thought of the Vogt-Russell theorem as asserting that the state of a star is a function of its mass, composition, and *age*, but really it is just mass and composition. The composition of an individual star inexorably changes with age, due, for instance, to nuclear processes or diffusion, but age is not itself the direct agent of that change; age is not a force.

Yet a knowledge of age is essential, for age is how we place something on the time axis that runs through all of stellar evolution and nearly all of astrophysics. We start, of course, with the Sun – for which we can measure non-stellar material in the laboratory – and construct physical models that can reproduce all that we know (which is a lot, especially given what has come from helioseismology). We work from the Sun to other masses, and, especially, to ensembles of stars. With a star cluster, the precision with which we know vital parameters may be mediocre, but creating consistent models that can reproduce the entirety of the behavior of a large group with the same age and composition (we assume) allows those models to be tested in critical ways.

And so we work our way through a variety of means of estimating the age of a star or an ensemble. Each of these links in the chain has its own weaknesses, and in this short introduction to this symposium, I will list some of the questions and problems that come to mind in thinking about stellar ages.

2. Methods of age estimation for individual stars

Many of the scientific questions that motivated this symposium are centered on the ages of individual stars. For example: How old are the stars that we know to host planets? We may soon find planets around stars in clusters, but our major focus will remain on the nearest stars for obvious reasons. It should be seen as a challenge (and embarrassment) for us that our cosmologist colleagues can claim better precision for the age of the Universe than we can for the ages of the nearest stars. In order to understand the formation and evolution of our Galaxy, we need to be able to determine the ages of individual stars that may belong to the thick- or thin disk, or streams that have been captured from torn-apart dwarf galaxies. Any really unusual star begs for an age to be associated with it.

Determining the age of an individual, isolated star is a frustrating and thorny problem. Doing so for an ensemble – even a rich cluster – is not easy either. As is shown elsewhere in this volume, some cluster ages are claimed to have uncertainties of $\sim$10%, but there are not independent tests to verify that accuracy, and it is systematic effects (reddening, metallicity, opacities, and so on) that dominate.

Through decades of painstaking effort, models have been constructed that reproduce the current state of the Sun and which also fit the considerable information we now possess on its interior properties (and yet questions can still remain on so basic a matter as the solar oxygen abundance). Given that confidence in our knowledge of the physics of the Sun, we can then understand the state and behavior of stars with different masses and abundances, especially when we have additional constraints, such as needing to match all the stars in a cluster at once. Doing this gives us confidence that we understand the essential physics of stellar structure and evolution.

Of particular interest are lower-mass stars, near and below 1 $M_\odot$, and in particular those found in the field or in small groupings. These objects pose their own problems. The main sequence lifetime at 1 $M_\odot$ is about 10 Gyr, and so such stars have ages spanning the entire age range of the Galactic disk; this makes them useful as a population to study that disk. The Sun itself, according to models, is slightly brighter and slightly warmer than when it first arrived on the Zero-Age Main Sequence (ZAMS) 4.5 Gyr ago. Much of that evolution in the H-R diagram has been nearly parallel to the ZAMS. In addition, main sequences for different metallicities lie on top of one another, and so there are several kinds of degeneracy in trying to determine the age of a solar-type star solely from photometry and parallax.

The uncertainties inherent in almost all of the age-dating techniques are significant. Sometimes, however, the goal is not so much to arrive at precise ages for individual targets as it is to be able to reliably order and bin the targets in age (τ), or, more appropriately, in $\log \tau$. Many of the relationships that depend on age are power laws or exponentials, and so lend themselves to estimating $\log \tau$ more consistently than τ itself. Complicating the problem is the fact that the various indicators available to us all too often yield inconsistent results, or only limits in some cases, and combining the information into a single best judgment of age is not straightforward. The ages we will consider we place into four types: fundamental, model-based, empirical, and statistical. This order is from most reliable to least.

2.1. *Fundamental ages*

I regard an age as *fundamental* if the underlying physics is completely understood and well characterized. There is only one fundamental age, that of the Sun, and it is based on radioactive decay of meteoritic material. There remain some uncertainties in the chronology of the solar system (Chaussidon 2007), but in comparison to the problems faced with astrophysical ages they are minuscule. We take the Sun to be $4,567 \pm 5$ Myr old (Chaussidon 2007). This is the one and only stellar age that is both precise and accurate.

2.2. *Model-dependent ages*

We may think of the ages of clusters determined from their color-magnitude diagrams as being reliable, and they are probably the best we have, yet they depend inherently on our detailed knowledge of stellar physics. The next tier of age-estimation techniques all depend to some extent on models or very basic assumptions to work.

2.2.1. *Isochrone ages*

Ages determined from a star's position in an H-R diagram (HRD) are model-based. They are largely self-consistent, at least, and on the whole are probably reliable, but there are many steps involved in applying our knowledge of physics to the problem, adding uncertainty and model dependency. Not all models give the same answers, and this is especially true for pre-main sequence (PMS) stars because of our poor knowledge of how to treat convection.

For low-mass stars, the difficulties of placing stars in an HRD primarily arise from their very slow evolution. Just placing a star on a given set of isochrones implies both precise and accurate knowledge of a star's temperature, luminosity, and metallicity. Despite decades of effort, our ability to determine the T_{eff} of a star is still limited to about 50–100 K (Clem *et al.* 2004; Ramírez & Meléndez 2005; Masana *et al.* 2006), which is a substantial uncertainty (with the Sun again being a notable exception). To derive luminosities, we need good parallaxes, which are now available for the nearer stars, and also bolometric corrections. Our ability to determine metallicities has improved substantially, but they still remain somewhat model-dependent, uncertain, and inconsistent. We do have the advantage, sometimes, of working on stars like the Sun, so that spectrum features are abundant and narrow, and we can work differentially relative to the Sun and so reduce some systematic uncertainties. Gustafsson & Mizuno-Wiedner (2001) have noted potential problems with isochrone ages, based on uncertain knowledge of stellar interiors, but they are concerned with thick-disk and halo stars that can have overall abundances and abundance patterns that are very different from solar. For most field stars, the differences from solar conditions are fairly minor.

For clusters, a number of presentations at this symposium were included to address concerns such as the adequacy of current models. In addition, these questions arise:

- Can we determine key cluster parameters well- and reliably enough to further reduce uncertainties? Gaia should certainly contribute significantly.
- Can we establish T_{eff} scales and bolometric corrections for main sequence stars reliably and derive accurate luminosities? This is critical for understanding the scatter we see in CMDs and interpreting it as age spreads.
- Can we understand the atmospheres of ZAMS stars well enough to determine T_{eff} in the presence of high levels of activity?

The ages of PMS stars are also estimated from HRDs, but with some different problems. PMS evolution is rapid, and so isochrones are well separated, but the physics of PMS stars is still incompletely understood and different models can yield substantially different isochrones. The observed quantities – temperature and luminosity – are also harder to determine accurately for PMS stars. The accuracy of T_{eff} values is limited by the inherent variability of PMS stars and by their conspicuously inhomogeneous atmospheres; in other words, it is not straightforward to convert an observed color index into T_{eff}. The same problems limit our ability to determine the luminosity, and, in addition, nearly all PMS stars are far enough away for the parallaxes to be not quite good enough, although that is being remedied. PMS isochrones are also metallicity-dependent, just as for the main sequence, but the same problems that inhibit our determining T_{eff} and luminosity (excess continuum emission and non-standard atmospheric structures that lead to line emission, among other things) also make accurate abundances problematic. Finally, precise and accurate masses for PMS stars are badly needed so that we can calibrate the evolutionary tracks. For the most part the masses of PMS stars are estimated from their position in an HRD, but few stars have measured masses. The few PMS stars in binaries that have measured orbits are critical tests of the models.

For PMS stars, these questions arise:

• Can we test PMS models and isochrones well enough to have confidence in them? Perhaps the inconsistencies observed are in part due to the different models being applicable in different mass ranges? In other words, maybe all the models are partly right and partly wrong?

• Can we establish T_{eff} scales for PMS stars reliably and derive accurate luminosities? This is critical for understanding the scatter we see in CMDs and interpreting it as age spreads.

• PMS ages in particular are confused by the inherent uncertainty in the zero point. Can that be reduced or resolved?

2.2.2. *The lithium depletion boundary*

In recent years it has been possible to detect Li in the lowest mass members of some young groups and clusters and to then compare the location of the Li depletion boundary (LDB) to models. This method was proposed by Rebolo *et al.* (1992) and has now been used for several clusters. It promises to provide a sensitive indicator of cluster age (Bildsten *et al.* 1997) that is independent from that from the main sequence turnoff. However, the LDB ages for the three clusters studied by Barrado y Navascués *et al.* (2004), for example, are significantly higher than the turn-off ages (by about 50%), indicating a possible systematic effect.

At the present, LDB ages are attractive in that they involve many fewer assumptions than isochrone ages, but the age range for which the LDB can be used is small and so it has been difficult to test this method. The difficulties are worsened by the very-low-mass stars being so faint, even in the nearer clusters. More detections of the LDB in young clusters are needed.

2.2.3. *Ages from isotope decay (nucleochronology)*

Some ages determined from radioactive decay are model-dependent because they are for distant stars and there is a need to estimate the initial abundance of a species. Some of the Galaxy's older stars have had ages estimated from the decay of Th or U (Cayrel *et al.* 2001; Sneden *et al.* 1996; Gustafsson & Mizuno-Wiedner 2001; Kratz *et al.* 2004; Dauphas 2005; del Peloso *et al.* 2005a,b,c).

• Isotope-decay ages offer one of the only checks on the ages of old stars that is independent of isochrones and CMDs. Can we improve their accuracy?

2.2.4. *Asteroseismology*

In the past few years we have seen ages determined by matching stellar interior models to observed asteroseismological oscillation frequencies (Floranes *et al.* 2005). The underlying concept is that the lowest-frequency modes one can see in spatially-unresolved observations penetrate the core of the star, and the sound speed there is sensitive to the density, which is to say the helium fraction, which directly results from the star getting old. The results can be very precise, and they are accurate as well to the extent that the models are only modestly different from the solar models. For example, Eggenberger & Carrier (2006) determined the age of β Vir (an F8V star slightly more massive than the Sun) by this method, with age uncertainties of about 3 to 8%, although two separate solutions gave equally good fits to the observations. This asteroseismological method offers great potential for determining ages in ways independent of current techniques. It is particularly good for deriving the ages of older stars, which are also those most difficult to age-date in other ways. The observations to be obtained by the *Corot* and *Kepler* missions should be very important for this, and those asteroseismic ages can, in turn, help to calibrate better empirical age relations. At present there are not enough asteroseismic ages published to draw conclusions about them.

• The ages derived from asteroseismology depend on essentially the same physics as those from isochrones, but asteroseismologic ages will likely work best for older stars. Will the new generation of very large telescopes (30m) allow us to detect oscillations in solar-type stars in old clusters, so we can compare ages directly?

2.2.5. *Kinematic traceback ages*

On time scales of 10^8 to 10^9 years, stars and clusters in the Galactic disk encounter massive objects that disrupt their Galactic orbits; this leads to disk "heating" (see the reviews by Wyse and Nordström in this volume). This effect erases some of the past kinematic history of a star, but at very young ages these tidal encounters have not yet occurred and it is possible to trace back the Galactic motion of a star. When we find stars with common Galactic space motions, we can do this for them as a group and see when in the past they were in closest proximity. This is a simple application of mechanics but it involves assumptions about the Galactic potential. Blaauw (1978) appears to have been first to determine such an expansion age, and Brown *et al.* (1997) and Fernández *et al.* (2008) provide an analysis for a number of nearby OB associations that takes advantage of *Hipparcos* observations. The analysis of Brown *et al.* (1997) shows that although ages determined from kinematic traceback may avoid some model dependency, they are subject to significant systematic errors from a number of effects. These errors all conspire to make the association appear to be kinematically younger than its true age. As D. Fernández reports in this volume, these ages from kinematic traceback turn out to be unreliable.

2.3. *Empirical ages*

For the non-cluster stars, especially older ones, it is necessary to use predominantly *empirical* age indicators. In these cases we observe a relationship between the quantity and age that appears to be monotonic, and there is a reasonable scenario to account for much of what is observed. However, not enough is known about the underlying physics to calculate the way in which we believe the observed quantity ought to change with age. One of these indicators is the surface Li abundance. The others are all variations on the theme of the rotation-activity relation that has been so well studied during recent years.

There are three types of empirical age methods applicable to low-mass stars:

(*a*) The decline in surface lithium abundance.

(*b*) The loss of angular momentum and spindown of stars.

(*c*) The decline of magnetically-related activity with age, seen in such indicators as Ca II H and K, Hα, or x-rays.

All of these are discussed in detail in this volume, and each has its advantages, disadvantages, and useful range of ages and stellar masses to which it can be applied.

• Can we turn some empirical techniques into model-dependent ones? That would mean reaching an understanding of, say, rotational spindown in late-type stars, well enough to create models that predict stellar behavior. Despite our detailed knowledge of the Sun, we have not even been able to explain or predict the solar activity cycle, so the prospects for this seem poor.

• Will we ever really understand Li depletion? For every trend that is seen there always seems to be at least one exception. Recent work on stellar models (see Deliyannis in this volume) is at least encouraging.

2.4. *Statistical Ages*

Several properties of stars correlate with age, but there is not a one-to-one relationship that can be used to derive an age. Instead, only broad limits on the age can be set. For example, Galactic disk heating leads to older stars tending to have greater net space motions than younger stars. Indeed, the older populations of our Galaxy – the thick disk and halo – are defined by their large space motions. But disk heating is only a tendency, and it is easy to point out counterexamples to the trend: both the Sun and α Centauri are 4–5 Gyr old yet have low net space motions. Thus kinematics is at best suggestive of an age.

The so-called age-metallicity relation is even less useful in any practical way. It is not clear if an actual relationship exists between age and overall metallicity for the Galactic thin disk; what may appear as such may really be a relation between metallicity and the Galactocentric radius at which a star forms. For the stars of interest in this paper, none are old enough to even age-date very roughly from their metallicity.

- Is there really an age-metallicity relation or does it just seem that way because of other underlying Galactic trends?

3. Open clusters

Open clusters (OCs) would seem to present a best-case situation for estimating an age. Their overall abundances are generally close to solar, so the models applied can use well-tested physics. Several are close enough to have distances determined from trigonometric parallaxes. Reddening is generally low. Some are reasonably well-populated.

As an example, consider the Pleiades. It is nearby, making its members accessible to high-resolution spectroscopy. The precise distance to the Pleiades remains contentious despite efforts to measure the distance in a number of independent ways. All of those methods are consistent to within the stated uncertainties (Pinsonneault *et al.* 1998; Narayanan & Gould 1999; Gatewood *et al.* 2000; Stello & Nissen 2001; Makarov 2002; Munari *et al.* 2004; Pan *et al.* 2004; Zwahlen *et al.* 2004; Johns-Krull & Anderson 2005; Soderblom *et al.* 2005; Southworth *et al.* 2005) with the notable exception of the result from the *Hipparcos* mission, although the *Hipparcos* value (van Leeuwen 2007) has gradually approached the distances determined by other studies as corrections have been applied. The distances and metallicities of the nearby OCs are closely interrelated (An *et al.* 2007), so having independent measures of those quantities is vital.

Most of the Pleiades has only slight reddening ($E(B-V) = 0.03$), although there are some patches with high reddening. It is fairly populous and photometry of excellent quality is available over the full range of stellar types from B to brown dwarfs.

Despite the importance and accessibility of the Pleiades, there have been few determinations of the cluster's metallicity: [Fe/H] $= -0.034 \pm 0.024$ (Boesgaard & Friel 1990); $+0.06 \pm 0.05$ (King *et al.* 2000); $+0.06$ (Groenewegen *et al.* 2007); $+0.06 \pm 0.02$ (Gebran & Monier 2008). Some other studies have looked at chemical peculiarities among the A stars, which we do not consider here. An *et al.* (2007) reanalyzed the data of Boesgaard & Friel (1990) and eliminated cluster non-members to get [Fe/H] $= +0.03 \pm 0.02$. Thus the average measured metallicity for the Pleiades appears to be $\sim 10\%$ supra-solar.

Quoted turn-off ages include: 78 Myr (Mermilliod 1981); 100 Myr (Meynet *et al.* 1993); 120 Myr (Kharchenko *et al.* 2005); 135 Myr (Webda database†); and 79 ± 52 Myr (Paunzen & Netopil 2006). The lithium depletion boundary age is given as 120–130 Myr (Stauffer *et al.* 1998) and 130 ± 20 Myr (Barrado y Navascués *et al.* 2004). We adopt

† http://obswww.unige.ch/webda

$\tau = 120 \pm 20$ Myr as an average, making the Pleiades an exemplar of a Zero-Age Main sequence cluster for intermediate-mass (about 0.5 to 2.0 $M_\odot$) stars. In other words, for one of the best-studied OCs available to us, the uncertainty in age is $\sim 20\%$. We should bear that uncertainty clearly in mind when we seek to estimate the ages of less-well-studied associations or young clusters.

OCs are critical in attempting to estimate the ages of individual stars because we rely on them as calibrators. This leads to a very basic problem. The majority of all OCs are no more than ~ 100 Myr old, and this is due to Galactic processes that rend them and strew their members into the field. This is where field stars come from and it means that there are few OCs at greater ages to test and calibrate age-estimation methods. Also, the rarity of old OCs means that they tend to be fairly distant and not so easily studied. Finally, the rarity of old OCs makes me wonder if the few that are left are truly representative of field stars of the same age. The old OCs that survive must have started out being rich and dense, which is unlike the star-forming regions we see in our part of the Galaxy. I can imagine, for instance, that a rich and dense OC might partition its internal angular momentum differently than a sparse cluster, leading to different distributions of apparent rotation. That is pure speculation, but we are often relying on rotation or a related quantity (activity) to estimate the age of an older star, and we use clusters such as M67 or NGC 188 to calibrate.

That leads to these questions about OCs and their ages:

• How can we improve OC ages, both in precision and accuracy? Eclipsing binaries can be very helpful if we are lucky enough to find a detached system near the cluster's turn-off so that we have well-determined masses, for instance.

• Can we at least rank-order clusters by age more reliably?

• Can we tell if there are real cluster-to-cluster differences in helium?

• Can we test the uniformity of composition for OCs?

• How can we tie field stars and calibrate empirical age indicators better?

4. Globular clusters

That basic assumption of uniform composition and age for a star cluster is now being shown to be problematic, especially for globulars. Questions about uniform composition in globulars have been raised for years. Among the most interesting and challenging astrophysical breakthroughs in recent years has been the discovery of multiple populations on the main sequences and turn-offs in old clusters (Bedin *et al.* 2004), both globulars and open clusters. The same considerations apply in studying nearby (resolved) galaxies and their populations, as discussed in several reviews herein:

• Can we accurately disentangle the various effects that have been put forth to explain multiple main sequences, such as different ages, different metallicities, or different helium abundances? Is there unambiguous evidence for multiple populations within globulars (i.e., stars formed separately and then merged, as opposed to various effects taking place within a globular after it forms).

• Are we considering all the options in trying to explain what's seen, or are there new classes of physical effects responsible?

• Can we reconcile the ages determined from a cluster turn-off with that from its white dwarf cooling sequence?

References

An, D., Terndrup, D. M., Pinsonneault, M. H., Paulson, D. B., Hanson, R. B., & Stauffer, J. R. 2007, *ApJ*, 655, 233

Barrado y Navascués, D., Stauffer, J. R., & Jayawardhana, R. 2004, *ApJ*, 614, 386

Bedin, L. R., Piotto, G., Anderson, J., Cassisi, S., King, I. R., Momany, Y., & Carraro, G. 2004, *ApJLett*, 605, L125

Bildsten, L., Brown, E. F., Matzner, C. D., & Ushomirsky, G. 1997, *ApJ*, 482, 442

Blaauw, A. 1978, in *Problems of Physics and Evolution of the Universe*, ed. L. V. Mirzoyan (Yerevan: Armenian Acad. Sci), p. 101

Boesgaard, A. M. & Friel, E. D. 1990, *ApJ*, 351, 467

Brown, A. G. A., Dekker, G., & de Zeeuw, P. T. 1997, *MNRAS*, 285 479

Cayrel, R., *et al.* 2001, *Nature*, 409, 691

Chaussidon, M. 2007, in Lectures in Astrobiology II, Adv. Astrobiol. Biogeophys., ed. M. Gargaud *et al.*, (Berlin: Springer), p. 45

Clem, J. L., VandenBerg, D. A., Grundahl, F., & Bell, R. A. 2004, *AJ*, 127, 1227

Dauphas, N. 2005, *Nature*, 435, 1203

del Peloso, E. F., da Silva, L., & Arany-Prado, L. I. 2005a, *A&A*, 434, 301

del Peloso, E. F., da Silva, L., & Porto de Mello, G. F. 2005b, *A&A*, 434, 275

del Peloso, E. F., da Silva, L., Porto de Mello, G. F., & Arany-Prado, L. I. 2005c, *A&A*, 440, 1153

Eggenberger, P. & Carrier, F. 2006, *A&A*, 449, 293

Fernández, D., Figueras, F., & Torra, J. 2008, *A&A*, 480, 735

Floranes, H. O., Christensen-Dalsgaard, J., & Thompson, M. J. 2005, *MNRAS*, 356, 671

Gatewood, G., de Jonge, J. K., & Han, I. 2000, *ApJ*, 533, 938

Gebran, M. & Monier, R. 2008, *A&A*, 483, 567

Groenewegen, M. A. T., Decin, L., Salaris, M., & De Cat, P. 2007, *A&A*, 463, 579

Gustafsson, B. & Mizuno-Wiedner, M. 2001, in *Astrophys. Ages and Time Scales*, eds. T. von Hippel, C. Simpson, & N. Manset, ASP Conf. Ser., 245, 271

Johns-Krull, C. M. & Anderson, J. 2005, in Proc. 13th Cool Stars Workshop, Cool Stars, Stellar Systems, and the Sun, ed. F. Favata, G. Hussain, & B. Battrick (Nordwijk: ESA), 683

Kharchenko, N. V., Piskunov, A. E., Röser, S., Schilbach, E., & Scholz, R.-D. 2005, *A&A*, 438, 1163

King, J. R., Soderblom, D. R., Fischer, D., & Jones, B. F. 2000, *ApJ*, 533, 944

Kratz, K.-L., Pfeiffer, B., Cowan, J., & Sneden, C. 2004, New. Astr. Rev., 48, 105

Makarov, V. V. 2002, *AJ*, 124, 3299

Masana, E., Jordi, C., & Ribas, I. 2006, *A&A*, 450, 735

Mermilliod, J. C. 1981, *A&A*, 97, 235

Meynet, G., Mermilliod, J.-C., & Maeder, A. 1993, *A&A Supp.*, 98, 477

Munari, U., Dallaporta, S., Siviero, A., Soubiran, C., Fiorucci, M., & Girard, P. 2004, *A&A*, 418, 31

Narayanan, V. K. & Gould, A. 1999, *ApJ*, 523, 328

Pan, X., Shao, M., & Kulkarni, S. 2004, *Nature*, 427, 326

Paunzen, E. & Netopil, M. 2006, *MNRAS*, 371, 1641

Pinsonneault, M. H., Stauffer, J., Soderblom, D. R., King, J. R., & Hanson, R. B. 1998, *ApJ*, 504, 170

Ramírez, I. & Meléndez, J. 2005, *ApJ*, 626, 465

Rebolo, R., Martín, E. L., & Magazzù, A. 1992, *ApJLett*, 389, L83

Sneden, C., McWilliam, A., Preston, G. W., Cowan, J. J., Burris, D. L., & Armosky, B. J. 1996, *ApJ*, 467, 819

Soderblom, D. R., Nelan, E., Benedict, G. F., McArthur, B., Ramirez, I., Spiesman, W., & Jones, B. F. 2005, *AJ*, 129, 1616

Southworth, J., Maxted, P. F. L., & Smalley, B. 2005, *A&A*, 429, 645

Stauffer, J. R., Schultz, G., & Kirkpatrick, J. D. 1998, *ApJLett*, 499, L199

Stello, D. & Nissen, P. E. 2001, *A&A*, 374, 105

van Leeuwen, F. 2007, *A&A*, 474, 653

Zwahlen, N., North, P., Debernardi, Y., Eyer, L., Galland, F., Groenewegen, M. A. T., & Hummel, C. A. 2004, *A&A*, 425, L45

At the symposium banquet in the historic Tremont Grand ballroom, Dave Soderblom insists that dinner not be served and the exits blocked until the situation with stellar ages is improved markedly.

William Noel, curator of manuscripts at the Walters Art Museum in Baltimore, shows the Archimedes Palimpsest to astronomers attending the "Ages of Stars" symposium.

The Ages of Stars
Proceedings IAU Symposium No. 258, 2008
E.E. Mamajek, D.R. Soderblom, & R.F.G. Wyse, eds.

doi:10.1017/S1743921309031664

The star-formation history of the Milky Way Galaxy

Rosemary F. G. Wyse
Department of Physics & Astronomy, Johns Hopkins University,
Baltimore, MD 21218, USA
email: wyse@pha.jhu.edu

Abstract. The star-formation histories of the main stellar components of the Milky Way constrain critical aspects of galaxy formation and evolution. I discuss recent determinations of such histories, together with their interpretation in terms of theories of disk galaxy evolution.

Keywords. Galaxy: disk, (Galaxy:) evolution, (Galaxy:) stellar content

1. Context: Disk galaxy formation and evolution

The Milky Way appears to be a typical disk Galaxy, albeit the one for which we can obtain the most detailed information and for the largest samples of tracer objects (I will mostly discuss samples of stars). The star-formation histories of the main stellar components of the Milky Way constrain critical aspects of disk galaxy formation and evolution. These include aspects of the merger history, such as what merged with the Milky Way and when did it merge, and also the epoch at which extended disks started to form. These in turn depend upon the nature of dark matter, possible 'feedback' mechanisms once stars start to form, and the amplitudes, onset, and duration, and signs, of gas flows.

Much observational evidence, in particular from large-scale structure such as the spectrum of fluctuations in the cosmic microwave background, and the statistics of massive galaxy clusters, has led to a 'concordance' cosmological model, wherein the recent (since a redshift of ~ 1) rate of overall expansion of the Universe is driven by Dark Energy, and the matter content of the Universe is predominantly non-baryonic Cold Dark Matter (e.g., Spergel *et al.* 2007). The primordial power spectrum of Cold Dark Matter (CDM) has most power on small scales, resulting in a hierarchical sequence of structure formation, whereby small scales form first, then subsequently merge to form larger systems. This concordance model will be referred to below as ΛCDM.

While baryons are a minor constituent of such a ΛCDM Universe, contributing less than 3% of the energy density at the present day, they are of course how we trace most of the mass in the Universe. The baryonic physics of gaseous dissipational cooling and subsequent star formation is much more difficult to model than is the (Newtonian) gravity that is the only force of significance for dissipationless CDM. The most detailed simulations of the formation and evolution of an analog of the present-day Milky Way have therefore been purely N-body, with gravity the only force and all matter being CDM. The state-of-the-art in October 2008 is illustrated by the Via Lactae II simulation (Diemand *et al.* 2008). This simulation follows the formation of the dark halo of a model Milky Way Galaxy, with initial conditions selected from a larger simulation to ensure that there will be no major merger after a redshift of 1.7 (a look-back time of somewhat less than 10 Gyr – as we discuss below, this restriction on the merger history is to prevent destruction of the thin disk within the lifetime of old stars presently observed in the local disk).

A striking feature of this very-high-resolution simulation is the persistent substructure, even near the analog of the solar circle: 97% of subhaloes that are initially identified at redshift unity still leave a bound remnant at the present epoch. Depending on the details of their mass and density distributions, these subhaloes are expected to lead to heating and fattening of the thin stellar disk (e.g., Hayashi & Chiba 2006).

The overall trends of stellar ages in such hierarchical clustering models are illustrated by the predictions of Abadi *et al.* (2003), who simulated the formation of a disk galaxy using a hybrid N-body and Smoothed Particle Hydrodynamics code, with simple gas-cooling and star-formation criteria.† The galaxy that forms is more bulge-dominated than is the Milky Way, reflecting the typical, active merger history in the ΛCDM cosmogony (and hence the need to pre-select initial conditions to form a Milky Way analog, as done by Diemand *et al.*). The model galaxy can be decomposed into spheroid (bulge/halo), thick disk and thin disk, with these components being distinct in terms of both surface-brightness profiles and kinematics. The spheroid is old, and was created by major mergers that disperse the pre-existing disk, plus tidal debris from more minor mergers (in this simulation, about half by each mechanism). The thin disk seen today is mostly stars that formed *in situ* after the cessation of merger activity from gas that was accreted smoothly into the disk plane, as it cooled in the dark halo. The inner thin-disk formed faster, reflecting the shorter accretion times of lower-angular-momentum gas, destined to dissipate into circular orbits in the central regions (Mo, Mao & White 1998). There is therefore a gradient in mean stellar age at the present epoch, declining from a mass-weighted age of 6 Gyr close to the center to 3 Gyr in the outer parts. The active merging, typical for the ΛCDM cosmogony, means that the overall galactic potential is not steady enough for formation of an extended thin disk until more recently than a look-back time of ~ 8 Gyr (redshifts less than unity). A disk can form early, then be destroyed, but the early disks are compact, due to the low angular-momentum content of their progenitor gas, reflected in the 'inside-out' formation of disks (e.g., Scannapieco *et al.* 2009).

In the Abadi *et al.* realization, the thick disk is predominantly old, and the bulk of it consists not of stars formed in the potential well of the final galaxy, but instead of stars formed in satellite galaxies that were later accreted, with half due to a satellite that merged ~ 6 Gyr ago, while the stars are older than ~ 10 Gyr old. The old stars (older than ~ 8 Gyr) in the thin disk are also predominantly accreted. This of course requires that the satellite hosts be on high-angular-momentum orbits, with periGalacticons within the disk, before their stars are assimilated into the disk. Given that the typical initial orbit of a satellite dark-halo is far from circular, this in turn requires that the satellite halo be rather massive, so that dynamical friction can operate to circularize and shrink the orbit.

1.1. *Tracing the merger history*

In a merger, orbital energy is absorbed into the internal degrees of freedom of the merging systems, heating them. A major merger, one with approximately equal mass ratio, essentially destroys pre-existing disks, transforming them into a pressure-supported spheroid or stellar halo (with $\sim 1:3$ being the limiting mass ratio; Cretton *et al.* 2001). The effects of the accretion of satellites (galaxies plus pure dark matter) depends on the time of accretion, their initial orbits, masses and density profiles, since these dictate how easily they are tidally disrupted, and to which component their stellar debris will contribute.

Minor mergers (mass ratios of less than $\sim 1:5$) of a fairly robust satellite should puff up an existing thin stellar disk into a thick disk (e.g., Quinn & Goodman 1986; Kazantzidis

† Watch the movie at `http://www.aip.de/People/MSteinmetz/Movies.html`

et al. 2008), plus deposit tidal debris from the satellite along its orbit. Some fraction of the stellar mass of the thin disk is also often assumed, in semi-analytic models, to be directly added to the bulge after a minor merger (Kauffmann 1996; de Lucia & Blaizot 2007). Gravitational torques during mergers cause transport of angular momentum, resulting in gas being taken into the central regions, where, after an induced star-burst, it can contribute to the bulge (Mihos & Hernquist 1996). Gas flows driven by mergers can possibly also build-up the bulge by triggering disk instabilities, in a fusion of dynamical and secular mechanisms (Bower *et al.* 2006). Dense, inner regions of massive satellites (the higher mass meaning dynamical friction timescales are shorter) could also contribute stars to the bulge (Ostriker & Tremaine 1975).

The stellar age distributions of thin disk, thick disk, stellar halo and bulge populations therefore depend on the merger history, and observational determination of these distributions can constrain the mergers.

2. The star-formation history of the thin disk

Unfortunately, the star-formation history (SFH) of the thin disk is poorly known far from the solar neighborhood. This lack of data needs to be rectified; hopefully the next generation of imaging surveys, such as Pan-STARRS, will provide data for both the inner disk and the outer disk. Theoretical expectation in a wide range of models is that star formation should proceed on faster timescales in the denser regions, due to the shorter dynamical times there, so that inner (denser) regions of disks are expected to have an older mean age, even if the time of the onset of star formation is fixed. Chemical evolution models (see Pipino & Matteucci's contribution to this volume) also favor slower star-formation in the outer parts, to match metallicity gradients. In hierarchical-clustering models, as noted above, the onset of star formation in the thin disk is later for the outer parts, due to the later accretion of the higher-angular momentum gas to form the outer disk.

2.1. *The onset of star formation in the local disk*

The star-formation history of the local disk has been derived using a variety of techniques, the details of which are discussed (with limitations and advantages) elsewhere in this volume. Ages of the oldest stars have been estimated from isochrone fitting, with the result that the oldest stars are less than 2 Gyr younger than the metal-poor globular clusters, i.e., ages of $\sim$ 11 Gyr (e.g., analyses of the Hipparcos dataset by Binney *et al.* 2000; analyses of local stars with Strömgren photometry by Nordström *et al.* 2004; Nordström, this volume; Holmberg, Nordström & Andersen 2008). White-dwarf cooling ages are model-dependent, as discussed at length in the contributions by Salaris and by Kalirai in this volume, and oldest ages of $\sim$ 12 Gyr are compatible with the luminosity function data (e.g., Fig 8 of Salaris, this volume, but note that the models of Hansen *et al.* 2002, utilized in Kalirai's paper in this volume, favor a younger oldest age). These old ages are consistent with an early onset of star formation in the local disk (assuming that the stars found locally were born locally), the lookback time of the onset corresponding to redshift $z \gtrsim 2$.

The exponential scale-length of low-mass stars (of spectral type like the Sun and later) in the disk is $\sim 2-3$ kpc (e.g., Jurić *et al.* 2008 who used M-dwarfs as tracers). Thus if the old stars in the solar neighborhood were formed close to their present location in the disk, star formation was initiated at $\sim 3-4$ scalelengths at $z > 2$. This would then imply that the formation of extended disks was *not* delayed until after a redshift of unity – the typical epoch of the last major merger for a $10^{12}\ M_\odot$ halo, in ΛCDM – as has been proposed in CDM-models with feedback (e.g., Weil *et al.* 1998; Thacker & Couchman 2001; Governato *et al.* 2007).

Alternatively, the old stars in the local thin disk could have formed elsewhere and more recently arrived in the solar neighborhood – two such scenarios have been proposed, with these old stars forming in either (a) satellite galaxies that are assimilated later, on circular orbits (Abadi *et al.* 2003), or (b) the inner disk and migrating outwards due to the influence of transient spiral arms† (Roškar *et al.* 2008a,b). In the first scenario, the satellites that could provide stars on near-circular orbits at the solar neighborhood would most probably have to be massive, so that dynamical friction (operating on a timescale proportional to the inverse of the satellite mass) can be effective in damping the satellite's orbit prior to its member stars being accreted. These satellites are the most capable of self-enrichment, and may be expected to contribute not just old stars, but also younger stars. If the pattern of elemental abundances produced were anything like those found in the surviving satellites (bearing in mind that these 'old disk' satellites are proposed to be accreted relatively recently, after a redshift of unity) then these should give a distinct signature, in particular low values of $[\alpha/\mathrm{Fe}]$ at low $[\mathrm{Fe/H}]$. At least two groups – Ruchti, Fulbright, Wyse *et al.* (2009, in prep.), using the RAVE survey (e.g., Steinmetz *et al.* 2006) to select their sample, and Reddy & Lambert (2008) – are obtaining and analyzing elemental abundance data for metal-poor (thick) disk stars, and have found nothing distinctive.

In the second scenario, one again expects a signature in the elemental abundance pattern of local stars, since efficient radial mixing of stars from distant regions with different star-formation histories gives increased scatter (e.g., François & Matteucci 1993; Schoenrich & Binney 2009). Migration has been proposed to explain in particular metal-poor disk stars locally in the disk (Haywood 2008) with a large fraction of these stars to have come inwards to the solar neighborhood, from the outer disk. These metal-poor disk stars are again of all ages (except younger that ~ 2 Gyr, plausibly the travel time of the migration), including the oldest ages, so this migration would imply an early onset for the outer disk beyond the solar circle, even more difficult to explain in ΛCDM than early star formation at the solar circle. Haywood (2008) proposes that migrated stars do indeed complicate the interpretation of trends in the elemental abundance patterns, but as we discuss below (section 3.2, cf. Reddy, Lambert & Allende Prieto 2006), the scatter is small and the uncertainties in the kinematic basis for the assignment of stars to the different components of the Galaxy provides an alternative explanation for outliers.

2.2. *Variation over time*

Again, various techniques have been used to estimate the temporal variation of the star-formation rate in the local disk. Several find evidence for 'bursts' in star-formation activity, of amplitude $2-3$, superposed on an underlying slow variation. Examples include the isochrone-based analysis of the Color Magnitude Diagram (CMD) of the Hipparcos dataset by Hernandez, Gilmore & Valls-Gabaud (2000), providing a temporal resolution of 50 Myr, albeit with a sample selection that limited the analysis to only the last ~ 3 Gyr (and adopting a fixed metallicity, which is a reasonable simplification for this narrow age range). These authors found a quasi-periodic variation with period of ~ 0.5 Gyr, which they suggested could be due to the passage of spiral arms.

Cignoni *et al.* (2006) developed their own approach to the analysis of the Hipparcos CMD and derived the star-formation history back to ~ 12 Gyr (adopting an age-metallicity relation). They found good agreement with the earlier results of Hernandez

† I thank Roelf de Jong for his question after my talk.

et al. for the younger stars, after rebinning them into 1 Gyr bins to match time resolutions. As is well-known, older disk stars have higher-amplitude random motions, with resultant wider epicyclic excursions about their orbital guiding-center. The derived star-formation rates at older ages then trace a larger portion of the disk, and stars formed locally a long time ago may have been lost from the sample. Cignoni *et al.* investigated the possible kinematic dependence of their derived SFH by excluding stars more than $n\sigma$ (n = 1, 2) away from the canonical local thin-disk 3D-space motion distribution, and found that the resulting distributions, focusing on ages less than $\sim$ 6 Gyr, were indistinguishable, and thus unaffected by dynamical diffusion.

The major result is a broad peak at ages $\sim$ 2–6 Gyr, with the star formation rate at $\sim$ 3 Gyr being a factor of 2–3 higher than either of the present-day rate or the rate at ages greater than 6 Gyr. This increase is perhaps attributable to triggering of star-formation activity by an accretion event. This overall SFH is consistent with the analysis of low-mass M-stars (using Hα activity as an age indicator) by Fuchs, Jahreiß & Flynn (2009; see their Fig. 5 for comparisons with others); the chromospheric-activity based analysis of G-stars by Rocha-Pinto *et al.* (2000) shows more high-frequency variations and less of a consistent increase in star-formation activity back to $\sim$ 5 Gyr.

Fuchs *et al.* (2009) argue that the Milky Way follows the 'Schmidt-Kennicutt' star-formation law found for external disk galaxies, which would imply that an increase in gas supply (inflow? accretion?) accompanied the increased level of star-formation rate $\sim$ 5 Gyr ago. In any case it is reassuring to find that the Milky Way is typical.

The possibility of significant re-distribution of stars via 'radial migration,' in addition to the radial epicyclic excursions considered already in the above papers, needs to be considered, as it raises the issue of to which region(s) of the disk does the derived SFH apply? Roškar *et al.* (2008a) appeal to radial migration in particular to build-up the outer disk, defined by a break in the gas surface density (and correspondingly, following the Schmidt-Kennicutt law, in the stellar surface density). The outer H I disk of the Milky Way does indeed show such a break, at Galactocentric distance of 12–13 kpc, well beyond the solar circle (see Fig. 5 of Levine, Blitz & Heiles 2006), and similar in location to the 'edge' of the stellar disk at 12–14 kpc (Reylé *et al.* 2008). In the model of Roškar *et al.* (2008a), stars currently beyond the break are those that are most likely to have migrated several kpc outwards (ΔR $\sim$ 4 kpc, see their Fig. 2), a significantly larger distance than their typical epicyclic excursions of $\sim$ 2 kpc. The root-mean-square change in radius across the scale, and lifetime, of all the disk stars in this model is $\sim$ 2.4 kpc, more comparable to the $\sim \pm 1.5$ kpc epicyclic excursions (estimated from observed kinematics) of the $\sim$ 3–5 Gyr-old stars that dominate locally. Looking more closely at the model 'solar circle', Roškar *et al.* (2008b) find that as many as half of the stars could have been born outside 7–9 kpc, with a bias to metal-rich stars from interior regions. As discussed above, the mixing of stars from regions of different star-formation histories should lead to scatter in the elemental abundances, and the evidence is that only a small fraction of stars do not follow well-defined trends, within observational uncertainty (Bensby *et al.* 2005; Feltzing & Bensby, this volume; Haywood 2008; Reddy, Lambert & Allende Prieto 2006; Reddy & Lambert 2008). The implication is that radial excursions are limited to mixing predominantly regions of very similar star-formation histories and chemical enrichments – either because there is little radial variation in star-formation history over much of the disk, or there is little radial migration. It will be very interesting to analyze the significantly larger samples of fainter stars with elemental abundances that will be feasible with planned instruments such as HERMES on the Anglo-Australian telescope, and WFMOS (a Gemini instrument) on the Subaru telescope.

3. The star-formation history of the thick disk

3.1. *Formation and early evolution*

The Galactic thick disk was defined 25 years ago through star counts at the South Galactic Pole in which two vertical exponential components were manifest (Gilmore & Reid 1983), with the general consensus now that it is separate and distinct from the Galactic thin disk. Analysis (Jurić *et al.* 2008) of the deep, uniform, wide-field imaging data from the Sloan Digital Sky Survey (SDSS) confirmed the necessity for two disk components, and the best-fit 'global' thick-disk parameters are an exponential scale-length of 3.6 kpc and exponential scale-height of 900 pc, with 12% of the local stellar mass-density in the thick disk. These combine to give a total mass equal to 10–20% of the stellar mass of the thin disk, or $\sim 10^{10}\, M_{\odot}$. Stars in the thick disk have distinct kinematics, elemental abundances (and ratios) and age distributions, when compared to the thin disk or stellar halo. Similar structures have been identified in the resolved stellar populations of external disk galaxies (e.g., Mould 2005; Yoachim & Dalcanton 2006).

There are several mechanisms by which a thick stellar disk could result (see, e.g., Gilmore, Wyse & Kuijken 1989; Majewski 1993). Thick disks seems inevitable in ΛCDM due to the heating inherent during the expected late merging and assimilation of satellites into pre-existing thin stellar disks. Simulations with a cosmological distribution of subhalos, in terms of both their mass function and their orbital characteristics, confirm this, and also show that the most massive satellite dominates the heating (e.g., Hayashi & Chiba 2006; Kazantzidis *et al.* 2008; see Hopkins et al. 2008 for a dissenting view, based however on an analysis using the same satellite orbital distribution as Hayashi & Chiba). In agreement with earlier simulations that focused on the accretion of one satellite in isolation, a satellite that is dense enough to survive to influence the disk, and of total dissipationless mass ratio 10–20% of that of the stellar disk, will produce a thick stellar disk.

The heating is achieved via a mix of local deposition of energy plus excitation of resonances (Sellwood, Nelson & Tremaine 1998), and a thin-disk component can persist (Kazantzidis *et al.* 2008). Of course, subsequent accretion of gas, and perhaps stars, can (re)-form a thin disk. Dissipation naturally leads to a thin disk after accretion of gas, while accretion of stars into a thin-disk component requires circular orbits for the parent system of those stars.

The stars in the thick disk in this scenario (creation by heating from a pre-existing thin disk) would have an age distribution that reflected that of the thin disk that was heated into the thick disk. With the continuous, fairly smooth, derived star-formation history of the (local) thin disk, starting at the earliest epochs, (lookback times $\sim 10-12$ Gyr), recent accretion, merging and associated heating would then produce a thick disk that at the present time would have stars of age equal to the look-back time of the accretion, unless the accreted satellite were easily destroyed; i.e., it was of low (relative) density. Turning this around, if the thick disk originated through merger-induced heating of the thin stellar disk, the last significant merger (defined as $> 20\%$ mass ratio to the disk, robust, dense satellite of stars and dark matter) can be dated by the young limit of the age distribution of stars in the thick disk: an age distribution of stars in the thick disk that goes down to, say, 5 Gyr would allow a merger and heating at a redshift of $z_{\rm last} \sim 0.5$, when the lookback time equals 5 Gyr, but if all thick disk stars are old, then the last significant merger was long ago. The age of the oldest thick disk stars, in this scenario, also further constrains the epoch at which an extended, thin stellar disk was in place, available to be heated.

3.2. *Ages of the oldest stars*

Analyses of the turn-off age of the thick disk stellar population within a few kpc of the solar neighborhood agree that the bulk of the thick disk stars are *old*. The (well-defined) turn-off color, for the spectroscopically derived typical metallicity of a star in the thick disk of ~ -0.6 dex, is equal to that of Galactic globular clusters of similar metallicity, e.g., 47 Tuc, corresponding to an age of 10–12 Gyr (e.g. Gilmore & Wyse 1985; Carney, Latham & Laird 1989; Gilmore, Wyse & Jones 1995; Ivezic *et al.* 2008, their appendix). This is equal to the estimated age of the oldest *thin* disk stars locally, as discussed above.

Ages for *individual*, slightly evolved thick-disk stars can be estimated from Strömgren photometry. These analyses show that the mean age is certainly older than that of the local thin disk, but there are disagreements in the fraction of 'thick disk' stars that are younger than the globular cluster ages, plausibly largely due to the difficulties in the assignment of an individual nearby star to a specific component, either the thin or thick disk. This probabilistic assignment is based on kinematics, and the standard assumption is that the space motions of each of the local thin disk and thick disk are adequately modeled by three one-dimensional Gaussians (e.g., see Feltzing & Bensby's contribution to this volume; Bensby *et al.* 2007a, 2004a,b; Reddy, Lambert & Allende Prieto 2006). This is clearly an over-simplification for the thin disk, given that moving groups have been robustly identified in the local disk (e.g., Dehnen 1998; Dehnen & Binney 1998; Famaey *et al.* 2005; Bensby *et al.* 2007b), with properties consistent with being dynamically induced by a combination of the Galactic bar and transient spiral arms (e.g. Dehnen 2000; de Simone, Wu & Tremaine 2004). As noted above, radial migration within the thin disk may occur and may also be a complicating issue for the kinematics-based population assignments, and this has yet to be analysed in detail. The likely production of high-velocity outliers in the thin disk kinematics by such mechanisms as three-body interactions can also complicate the population assignments.

Indeed, Reddy *et al.* (2006; their Fig. 24) suggest that a significant fraction of 'younger' thick disk stars are in fact contaminants from the thin disk. Evidence for an age-metallicity trend within the thick disk does remain in their sample, although once these thin disk contaminants are removed, the thick disk shows no evidence for the incorporation of iron from Type Ia supernovae, in the elemental abundance pattern (their Fig. 20). An age spread of several Gyr could be consistent with this constant 'Type II plateau' in the $[\alpha/\mathrm{Fe}]$ ratios, plus a typical timescale for Type Ia chemical enrichment of ~ 1 Gyr, if the thick disk consisted of stars from several independent star-formation regions, each of which had a short ($\lesssim 1$ Gyr) duration, but different onset times. However, with age uncertainties of ± 2 Gyr, the data are also consistent with a narrow age range, with mean ~ 12 Gyr (on their age-scale), and a simpler interpretation of the elemental abundance pattern as reflecting a global short duration of star formation, ~ 1 Gyr.

3.3. *Implications for merger history*

The old age of thick-disk stars, $\gtrsim 10$ Gyr, limits the last significant minor merger to have occurred at a redshift $\gtrsim 2$. While the last *major* merger in ΛCDM typically is at this epoch, it is minor mergers that are constrained by heating of the thin disk into a thick disk, and these are expected to continue to lower redshifts (e.g., Stewart *et al.* 2008). The inferred quiescent merger history of the Milky Way is atypical in ΛCDM. As we will now discuss, consistent limits on the minor-merger history are obtained from the derived star-formation history of the central bulge.

4. The star-formation history of the central bulge

During mergers, the expectation is that existing disk stars, and gas to fuel star formation, will be added to the bulge (e.g., Kauffmann 1996), perhaps through an intermediate stage of build-up of a massive inner disk that subsequently becomes unstable (e.g., Bower *et al.* 2006). The dense, inner regions of satellites can also survive to be added to the bulge, if dynamical friction is efficient enough (and the satellite massive and dense enough). Gravitational torques due to the (mildly) triaxial inner bulge/bar will also drive modest gas inflows at the present day (e.g., Englmaier & Gerhard 1999) in the plane of the disk.

While there are younger stars in the central regions, these are confined to the disk plane, and the central bulge is dominated by old (10–12 Gyr), metal-rich stars, with enhanced α-element ratios. These properties point to bulge formation in an intense short-lived burst of star formation, *in situ* (a deep potential-well being required to reach the observed high metallicities), a long time ago (e.g., Elmegreen 1999; Ferreras, Wyse & Silk 2003; and the contribution by Pipino & Matteucci in this volume). The inferred star-formation rate is reasonable, of order 10 $M_\odot$ yr^{-1} for a total mass of $\sim 10^{10}$ $M_\odot$ and an age range of ~ 1 Gyr.

Thus, unless the inner disk is composed of uniformly old stars, there can have been no recent disk instability to form the bar/bulge. There is little room for significant build-up of the central bulge by recent mergers, with the old age again limiting significant merger activity to redshifts $\gtrsim 2$. This matches the constraints from the thick disk, and leads to the suggestion (Wyse 2001) that perhaps both bulge and thick disk were formed by the same last significant minor merger.

An alternative suggestion, consistent with the low value of the angular momentum content of the bulge, and the similarity of the specific angular momentum distributions of the bulge and stellar halo (see Fig. 1), is that the bulge formed from gas ejected from the early star-formation regions in the halo (Carney, Latham & Laird 1990; Wyse & Gilmore 1992). Gas must have been lost from the stellar halo since all indications are that the stellar Initial Mass Function (IMF) was normal, at both high and low masses (e.g., Wyse 1998), but the mean metallicity is far below the yield for that IMF (see Hartwick 1976 for the basic model).

5. The star-formation history of the halo

The bulk of the stellar halo by mass, interior to Galactocentric distances of ~ 20 kpc, is rather uniform in its properties: the stars – and globular clusters – are old and metal-poor, and show enhanced elemental abundances ([α/Fe] in particular) that are indicative of short duration(s) of star formation, in low-mass star-forming regions, with 'normal' Type II-progenitor stellar IMF. These properties are unlike those of most stars in satellite galaxies now. Particular attention recently has been paid to the differences in elemental abundance patterns (cf., Venn *et al.* 2004; Geisler *et al.* 2007). The differences can be understood in terms of the different star-formation histories of the field halo and the surviving satellite galaxies – the former having a narrow stellar age-range and rapid chemical enrichment, and the latter having a wide stellar age-range and slow, inefficient enrichment, allowing incorporation of iron from Type Ia supernovae (and hence low [α/Fe]) even at low levels of overall enrichment (e.g., Gilmore & Wyse 1991; Unavane, Wyse & Gilmore 1996; Lanfranchi & Matteucci 2003).

Comparison of the age distributions alone leads to the conclusion (Unavane *et al.* 1996) that accretion into the field halo from stellar satellites with a typical extended SFH has

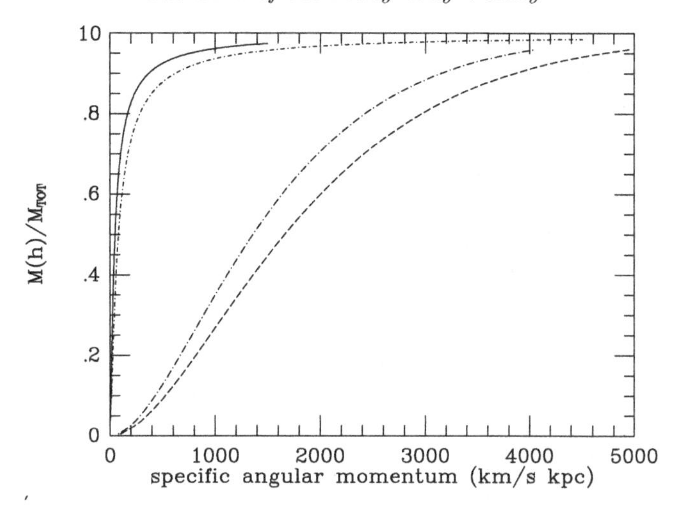

Figure 1. Adapted from Wyse & Gilmore 1992, their Figure 1. Specific angular momentum distributions of the bulge (solid curve), the stellar halo (short-dashed/dotted curve), the thick disk (long-dashed/dotted curve) and the thin disk (long-dashed curve). The bulge and stellar halo have similar distributions, with most of the mass at low angular momentum (curves passing through upper left of the figure). The distributions of thick and thin disks are also similar to each other, with significant mass fraction at high angular momentum.

not been important for the last $\sim$ 8 Gyr, this time-scale coming from the estimated lower limit to the ages of stars in the field halo. Their analysis of the distribution of the field halo stars in the color-metallicity plane allows perhaps $\sim$ 10% by mass to have been accreted later, predominantly in the metal-rich tail of the halo. Of course satellites like the Ursa Minor dSph, composed of only old, metal-poor stars, could be accreted and assimilated into the field halo at any time and would not be distinguishable on the basis of age or metallicity. However, such satellites are rare at the present time. Early disruption of a few massive satellites could form the stellar halo, of total stellar mass $\sim 10^9 M_\odot$ (e.g., Robertson *et al.* 2005), but we need to understand what causes the necessary cessation in star formation, when, for example, the LMC has not been so affected.

A significant fraction of stars in the outer halo could have been accreted from the Sgr dSph (Ibata, Gilmore & Irwin 1994), which is on an orbit with periGalacticon of $\sim$ 25 kpc and apoGalacticon $\sim$ 50 kpc (e.g., Ibata *et al.* 1997). Tidal arms from this system are seen across the Galaxy (e.g., Majewski *et al.* 2003; Fellhauer *et al.* 2006).

6. Concluding remarks

The old mean stellar ages and short durations of star formation of the thick disk and bulge argue for little late accretion and merging into the Milky Way. Accretion after a redshift of $\sim$ 2 should have been predominantly smooth, of gas-dominated, low-density systems. The relatively high mean metallicity, plus inferred rapid star formation (from the ages and elemental abundance patterns), of both components argues for star formation within deep potential wells. This favors *in situ* star formation of each of the bulge and thick disk, and rapid mass assembly of the overall Milky Way.

The presence of very old stars, ages $\gtrsim$ 10 Gyr, in the local thin and thick disks argues for the existence of an extended disk at redshift $\sim$ 2.

The old mean stellar age and inferred short duration of star formation of the bulk of the stellar halo implies that the field halo cannot have formed from systems like the existing satellites, which typically have had much more extended star-formation histories. The low mean metallicity, plus curtailed star formation, argues for star formation within shallow potential wells, so that early mass loss is facilitated. Instead of such low-mass systems, the ΛCDM-based models favor a few massive satellites, accreted early, as the source of the field halo (Robertson *et al.* 2005; Sales *et al.* 2007); rapid gas loss is assumed to occur by ram pressure stripping, but this assumption needs to be modelled.

The quiescent merging history, and rapid mass assembly, of the Milky Way, is unusual in ΛCDM. However, surveys of galaxies at redshift $z \sim 2$, find rapid star formation and chemical enrichment, and extended disks, similar to the inferences for the Milky Way at that look-back time (e.g., Maiolino *et al.* 2008; Daddi *et al.* 2008; Genzel *et al.* 2008).

However, our understanding of the global star-formation history of the Milky Way remains incomplete. For this we need to determine the detailed age distributions, spatial distributions, space motions, and elemental abundances for large samples of Galactic stars, both locally and more globally. Several large imaging surveys are planned in the near future, and these need to be matched by large spectroscopic surveys, at both high and low spectral resolution. These will be challenging, but worth the investment.

References

Abadi, M., Navarro, J., Steinmetz, M., & Eke, V. 2003, *ApJ*, 597, 21
Bensby, T., Feltzing, S., & Lundström, I. 2004a, *A&A*, 415, 155
Bensby, T., Feltzing, S., & Lundström, I. 2004b, *A&A*, 421, 969
Bensby, T., Feltzing, S., Lundström, I., & Ilyin, I. 2005, *A&A*, 433, 185
Bensby, T., Zenn, A., Oey, S., & Feltzing, S. 2007a, *ApJ*, 663, L13
Bensby, T., Oey, S., Feltzing, S., & Gustaffson, B. 2007b, *ApJ*, 655, L89
Binney, J., Dehnen, W., & Bertelli, G. 2000, *MNRAS*, 318, 658
Bower, R., *et al.* 2006, *MNRAS*, 370, 645
Carney, B., Latham, D., & Laird, J. 1989, *AJ*, 97, 423
Carney, B., Latham, D., & Laird, J. 1990, *AJ*, 99, 572
Cignoni, M., Degl'Innocenti, S., Prada Moroni, P., & Shore, S., 2006, *A&A*, 459, 783
Cretton, N., Naab, T., Rix, H-W., & Burkert, A. 2001, *ApJ*, 554, 291
Daddi, E. *et al.* 2008, *ApJ*, 673, L21
Dehnen, W. 1998, *AJ*, 115, 2384
Dehnen, W. 2000, *AJ*, 119, 800
Dehnen, W. & Binney, J. 1998, *MNRAS*, 298, 387
Diemand, J. *et al.* 2008, *Nature*, 454, 735
Elmegreen, B. 1999, *ApJ*, 517, 103
Englmaier, P. & Gerhard, O. 1999, *MNRAS*, 304, 512
Famaey, B. *et al.* 2005, *A&A*, 430, 165
Fellhauer, M. *et al.* 2006, *ApJ*, 651, 167
Ferreras, I., Wyse, R. F. G., & Silk, J. 2003, *MNRAS*, 345, 1381
François, P. & Matteucci, F. 1993, *A&A*, 280, 136
Fuchs, B., Jahreiß, H., & Flynn, C. 2009, *AJ*, 137, 266
Geisler, D., Wallerstein, G., Smith, V., & Casetti-Dinescu, D. 2007, *PASP*, 119, 939
Genzel, R. *et al.* 2008, *ApJ*, 687, 59
Gilmore, G. & Reid, I. N. 1983, *MNRAS*, 202, 1025
Gilmore, G. & Wyse, R. F. G. 1985, *AJ*, 90, 2015
Gilmore, G. & Wyse, R. F. G. 1991, *ApJ*, 367, L55
Gilmore, G., Wyse, R. F. G., & Jones, J. B. 1995, *AJ*, 109, 1095
Gilmore, G., Wyse, R. F. G., & Kuijken, K. 1989, *ARAA*, 27, 555

Governato, F. *et al.* 2007, *MNRAS*, 374, 1479
Hansen, B. *et al.* 2002, *ApJ*, 574, L115
Hartwick, F. D. A. 1976, *ApJ*, 209, 418
Hayashi, H. & Chiba, M. 2006, *PASJ*, 58, 835
Haywood, M. 2008, *MNRAS*, 388, 1175
Hernandez, X., Gilmore, G., & Valls-Gabaud, D. 2000, *MNRAS*, 317, 831
Holmberg, J., Nordström, B., & Andersen, J. 2008, *A&A*, submitted (arXiv:0811.3982)
Hopkins, P. *et al.* 2008, *ApJ*, 688, 757
Ibata, R., Gilmore, G., & Irwin, M. 1994, *Nature*, 370, 194
Ibata, R., Wyse, R. F. G., Gilmore, G., Irwin, M., & Suntzeff. N. 1997, *AJ*, 113, 634
Ivezic, Z. *et al.* 2008, *ApJ*, 684, 287
Jurić, M. *et al.* 2008, *ApJ*, 673, 864
Kauffmann, G. 1996, *MNRAS*, 281, 487
Kazantzidis, S. *et al.* 2008, *ApJ*, 688, 254
Lanfranchi, G. & Matteucci, F. 2003, *MNRAS*, 345, 71
Levine, E. S., Blitz, L., & Heiles, C. 2006, *ApJ*, 643, 881
de Lucia, G. & Blaizot, J. 2007, *MNRAS*, 375, 2
Maiolino, R. *et al.* 2008, *A&A*, 488, 463
Majewski, S. 1993, *ARAA*, 31, 575
Majewski, S., Skrutskie, M., Weinberg, M., & Ostheimer, J. 2003, *ApJ*, 599, 1082
Mihos, J. C. & Hernquist, L. 1996, *ApJ*, 464, 641
Mo, H., Mao, S., & White, S. D. M. 1998, *MNRAS*, 295, 319
Mould, J. 2005, *AJ*, 129, 698
Nordström, B. *et al.* 2004, *A&A*, 418, 989
Ostriker, J. & Tremaine, S. 1975, *ApJ*, 256, L113
Quinn, P. & Goodman, J. 1986, *ApJ*, 309, 472
Reddy, B. & Lambert, D. 2008, *MNRAS*, 391, 95
Reddy, B., Lambert, D., & Allende Prieto, C. 2006, *MNRAS*, 367, 1329
Reylé, C., Marshall, D. J., Robin, A. C., & Schultheis, M. 2008, *A&A*, submitted (arXiv:0812.3739)
Robertson, B. *et al.* 2005, *ApJ*, 632, 872
Rocha-Pinto, H. J., Scalo, J., Maciel, W., & Flynn, C. 2000, *A&A*, 358, 869
Roškar, R. *et al.* 2008a, *ApJ*, 675, L65
Roškar, R. *et al.* 2008b, *ApJ*, 684, L79
Sales, L., Navarro, J., Abadi, M., & Steinmetz, M. 2007, *MNRAS*, 379, 1464
Scannapieco, C., White, S. D. M., Springel, V., & Tissera, P. B. 2009, *MNRAS*, submitted (arXiv:0812.0976)
Schroenrich, R. & Binney, J. 2009, *MNRAS*, submitted (arXiv:0809.3006)
Sellwood, J. A., Nelson, R. W., & Tremaine, S., 1998, *ApJ*, 506, 590
de Simone, R., Wu, X., & Tremaine, S. 2004, *MNRAS*, 350, 627
Spergel, D. N. *et al.* (WMAP Team) 2007, *ApJS*, 170, 337
Steinmetz, M., *et al.* (the RAVE collaboration) 2006, *AJ*, 132, 1645
Stewart, K., *et al.* 2008, *ApJ*, 683, 597
Thacker, R. J. & Couchman, H. 2001, *ApJ*, 555, L17
Unavane, M., Wyse, R. F. G., & Gilmore, G. 1996, *MNRAS*, 278, 727
Venn, K., Irwin, M., Shetrone, M., Tout, C., Hill, V., & Tolstoy, E. 2004, AJ, 128, 1177
Weil, M. L., Eke, V., & Efstathiou, G. 1998, *MNRAS*, 300, 773
Wyse, R. F. G. 1998, in: G. Gilmore & D. Howell (eds.), ASP Conf. Ser. 142, *The Stellar Initial Mass Function*, (San Francisco: ASP), p. 89
Wyse, R. F. G. 2001, in: J. G. Funes & E.M. Corsini (eds.), ASP Conf. Ser. 230, *Galaxy Disks and Disk Galaxies*, (San Francisco: ASP), p. 71
Wyse, R. F. G.& Gilmore, G. 1992, *AJ*, 104, 144
Yoachim, P. & Dalcanton, J. 2006, *AJ*, 131, 226

Discussion

R. DE JONG: Recently it has, for instance, been argued by Roškar *et al.* (2008) that radial migration of stars is much larger than previously estimated, meaning that stars in the solar neighbourhood do not reflect the local star formation history, but instead a combination of central Galaxy star formation and migration history. If this is the case, how can we disentangle these effects?

R. WYSE: The thin disk at the solar neighborhood shows little scatter in elemental abundance ratios, and this is hard to reconcile with stellar migration over many kpc, mixing regions of different star-formation history. If migration is limited to $\lesssim 1$ kpc then I do not believe it will affect the derived local SFH significantly since normal stellar orbits sample that range.

J. MELBOURNE: What percentage of stars in the thin disk are old? Are they coeval with the thick disk? Would they be thick disk stars?

R. WYSE: The analyses of the derived local star-formation histories do not have a very good handle on the oldest stars (since they are faint) but estimates (e.g. Cignoni *et al.* 2006) suggest $\sim$10% in the 10-12 Gyr range. These have thin-disk kinematics so they are probably not thick-disk members (modulo uncertainties in population assignment). It is difficult to distinguish ages 10-12 Gyr, so although the oldest thin disk stars may be ~ 1 Gyr younger than the thick disk, it is best to say both are 'old'.

KING: You've said little or nothing about the bar. Is it merely the manifestation of a dynamical instability, or should we be able to learn something from it about the formation history of the Milky Way?

WYSE: We still don't know much about the evolutionary state of the bar. It is clear that the Milky Way bulge shows some characteristics of 'pseudo-bulges', for example an exponential surface-density profile, but the uniform old age argues against a recent disk-bar instability to form the bulge, given the evidence for recent star formation in the inner disk. But we do need to get more data on the stellar populations in the inner disk.

The Ages of Stars
Proceedings IAU Symposium No. 258, 2008
E.E. Mamajek, D.R. Soderblom & R.F.G. Wyse, eds.

doi:10.1017/S1743921309031676

The age of the Galaxy's thick disk

Sofia Feltzing[1]† **and Thomas Bensby**[2]

[1]Lund Observatory,
Box 43, SE-22100, Lund, Sweden
email: sofia@astro.lu.se

[2]European Southern Observatory,
Alonso de Cordova 3107, Vitacura, Casilla 19001, Santiago, Chile
email: tbensby@eso.org

Abstract. We discuss the age of the stellar disks in the solar neighborhood. After reviewing the various methods for age dating, we discuss current estimates of the ages of both the thin- and the thick disks. We present preliminary results for kinematically-selected stars that belong to the thin- as well as the thick disk. All of these dwarf and sub-giant stars have been studied spectroscopically and we have derived both elemental abundances as well as ages for them. A general conclusion is that in the solar neighborhood, on average, the thick disk is older than the thin disk. However, we caution that the exclusion of stars with effective temperatures around 6500 K might result in a biased view of the full age distribution for the stars in the thick disk.

Keywords. Galaxy: structure, Galaxy: disk, kinematics and dynamics, solar neighborhood, Hertzsprung-Russell diagram, stars: late-type

---

1. Introduction

The age of a stellar population can be determined in several ways. For groups of stars, isochrones may be fitted to the stellar sequence in the Hertzsprung-Russell diagram (HRD; see, e.g., Schuster *et al.* 2006), or the luminosity function for the white dwarfs can be fitted with cooling tracks (e.g., Leggett *et al.* 1998). Ages for individual stars can be determined from the HRD (if the star is a turn-off or sub-giant star) or by utilizing relations that relate the rotation or atmospheric activity of a star to its age (examples are given by Barnes 2007 and Mamajek & Hillenbrand 2008). Asteroseismology provides the possibility to constrain the stellar ages very finely. A recent example of the age determination for a young star is given in Vauclair *et al.* (2008). Finally, the age of a star can be estimated by studying the amount of various elements present in the photosphere of the star. In particular the amount of elements such as U and Th that decay radioactively can be used to estimate the age. Examples of this are given by del Peloso *et al.* (2005). Estimating the age from the decay of radioactive isotopes is sometimes called *nucleocosmochronology.*

All but one of these methods, nucleocosmochronology, relies on our understanding of stellar evolution. Some of the methods work well for young stars. This is especially true for rotation and stellar activity (see Mamajek 2009 and Barnes 2009) while the determination of stellar ages using isochrones is limited in various ways depending on the type of star under study.

The isochrones give the best results for turn-off and sub-giant stars, but with very poor power to differentiate between different ages on the red giant branch. In fact, the stars on the sub-giant branch are the most desirable tracers of the age of a particular

† SF is a Royal Swedish Academy of Sciences Research Fellow supported by a grant from the Knut and Alice Wallenberg Foundation.

stellar population (Sandage *et al.* 2003). In particular, it does not matter if the stellar temperature is well determined or not (Bernkopf & Fuhrmann 2006).

However, we would argue that the power of isochrone ages mainly lies in the *relative* ages – i.e., being able to say "star A is older than star B and it is about this big an age difference between star A and star B." Such statements and determinations are, of course, less desirable if we want to determine the absolute age of a star or stellar population, but they are very powerful if we want to know in which order the stars formed and what time-scales were involved, i.e., the study of galaxy formation and evolution. The good thing with the isochrone method is that it is reasonably straightforward to derive the ages also for large samples of stars (but see Jørgensen & Lindegren 2005) as well as for old stars. The less useful aspect is that we are mainly limited to using the turn-off stars. For an older population this implies the inherently faint, but numerous, F- and G-type dwarf and sub-giant stars. In order to construct the HRD, we need to know the distances to the stars. This is difficult to do for large numbers of stars once we are outside the volume covered by Hipparcos. However, it is possible to derive the distance if the star is assumed to be a dwarf or if the star can be determined to be a dwarf star. Strömgren photometry and some other photometric systems are able to determine the evolutionary state of a star. Some examples of how the Strömgren photometry can be used to this end are given in Schuster *et al.* (2006), von Hippel & Bothun (1993), and Jønch-Sørensen (1995). So far these studies have mainly been limited to the solar neighbourhood due to the observational equipment available. Recent studies are trying to remedy this situation by using CCD images obtained with wide-field cameras. An early example is given in Árnadottír *et al.* (2008).

2. The ages of the stellar disks

The main tracers for age-dating the thin disk are open clusters, the luminosity function of white dwarfs, and, recently, nucleocosmochronology. Generally, estimates of the age of the thin disk using the luminosity function of white dwarfs find a lower limit for the age of around 9 Gyr (e.g Oswalt *et al.* 1995; Leggett *et al.* 1998; Knox *et al.* 1999).

Open clusters indicate a similar lower age for the thin disk. It is interesting to note the existence of open clusters that are both old as well as metal-rich. NGC 6791 has a metallicity of +0.35 dex and an age between 8 and 9 Gyr (Grundahl *et al.* 2008). Such old stars are normally not considered to be able to be that metal-rich. In our new, local sample of stars we seem to pick up a few metal-rich and old stars that have thin disk kinematics and thin disk abundance patterns.

The thin disk hosts the majority of the younger stars. In general young stars rotate more rapidly than older stars and they have more chromospheric activity. As they grow older they rotate more slowly and their outer atmospheres become less active. These characteristics can be utilized to estimate the age of a star (Mamajek & Hillenbrand 2008; Barnes 2007). However, none of these measures are particularly straightforward. A recent example of how they could be combined in order to give better age estimates is given by Mamajek & Hillenbrand (2008). They show that with their new measure of stellar ages, combined from rotation and activity measures, the star formation history of the thin disk has been less variable than previously thought.

The thick disk is in general found to be exclusively old. The age estimates for the thick disk have been done either by studying local, kinematically-selected samples, or by studying the turn-off color for stars well above the Galactic plane where the thick disk dominates (typically about 1 kpc and higher; Gilmore *et al.* 1995).

Recent studies of kinematically-selected thick disk samples in the solar neighborhood appear to agree that the thick disk is old and, essentially, all older than kinematically-defined thin disk samples (see, e.g., Bensby *et al.* 2005; Reddy *et al.* 2006). It is clear that the kinematic definitions are only statistical and that we will never be able to create a sample that is completely free from thin-disk stars. It is especially important to keep in mind that the young stars in the stellar disk (thin or thick) have a rather lumpy distribution in velocity space. This enables the identification of stars that potentially have a common origin but it complicates the division of stars into thin- and thick disk (for a recent discussion see Holmberg *et al.* (2007)). As shown in Holmberg *et al.* (2007), as we progress to older stars the kinematics change and the velocity distributions get smoother. This should not be surprising as any older sample will be more dominated by the thick disk, for which not much lumpiness has been observed so far (but see Gilmore *et al.* (2002), Schuster *et al.* (2006), and Wyse *et al.* (2006) for discussions of the last merger and how that has influenced the local as well as not-so-local stellar kinematics). Not only the velocity dispersions are important to consider but also how large a portion the thick disk contributes in the solar neighborhood (the normalization of the stellar number density). In Bensby *et al.* (2005), we show that our selection criteria are rather robust against changes in this normalization. It remains to be fully investigated how sensitive the selection criteria are to the presence of lumpy velocity distributions.

In this context the study of volume-limited samples becomes increasingly important. Fuhrmann (2008) studied a volume-limited sample of stars within 25 pc from the Sun. He identifies the stars that are enhanced in [Mg/Fe] with the thick disk. All of these stars are found to be older than the stars he associates with the thin disk, but no specific ages are given. In the next section we will revisit the volume-limited samples in comparison to the samples selected based on kinematics.

3. A new local sample of late-F and early-G dwarf stars – the local disk(s) revisited

We have obtained high-resolution, high S/N spectra for about 900 dwarf stars. The data have been obtained with several spectrographs but in general S/N$>$ 250 and $R \geqslant$ 65,000 (apart from the subset of stars originally observed with FEROS [Bensby *et al.* (2003)] which have $R = 48,000$). In Feltzing & Bensby (2008) we presented the kinematic properties and some elemental abundances for a subsample of about 550 F- and G dwarf stars.

The ages for these stars have been derived using Yonsei-Yale isochrones (see Bensby *et al.* 2005), where we also allow for enhancement in α-elements. Taking the α-enhancement into account is important because for a given star the age will be lower should it be enhanced in these elements as opposed to if it is not. The effect of taking the α-enhancement into account is thus that any age-gap between the thick- and the thin disk decreases (the thin disk stars are not at all or only moderately enhanced in α-elements and thus there is only a small or no effect on their ages when α-enhancement is included in the age estimates).

In Fig. 1a and Fig. 1c we are attempting a comparison with results from Fuhrmann (2008) and show the stars within 25 and 50 pc, respectively. In the plots, we have also coded the age of the stars such that an older star has a bigger symbol. It is clear from Fig. 1a and Fig. 1c that stars that are enhanced in an α-element, here Si, are also older than stars that are not enhanced. On average, our volume-limited samples appear to show the same sort of trends that Fuhrmann (2008) found. Although the sample within

25 pc is very small and incomplete, still stars enhanced in Si are old and the young stars are not enhanced and also show a tight trend of [Si/Fe] vs. [Fe/H]. Fig. 1c further shows that there appears to be a real separation between the two trends, i.e., one for younger and one for older stars. It should be kept in mind that our sample is not volume-complete and also that Fuhrmann (2008) imposes some further criteria on the stellar parameters for those stars that he includes in his final plots. For now, we are showing all stars, covering the full parameter space sampled within our program (compare, e.g., Feltzing & Bensby (2008) and Bensby *et al.* (2009, in preparation)).

Figures 1b and d then show the volume-limited samples but with a kinematic selection imposed as well such that we select stars that are ten times more likely to be thick disk than thin disk to represent the thick disk and vice versa for the thin disk. It is intriguing to see that the kinematically-selected thin disk stars mimic the trend found for the younger stars and the thick disk mimics the trend found for the older stars.

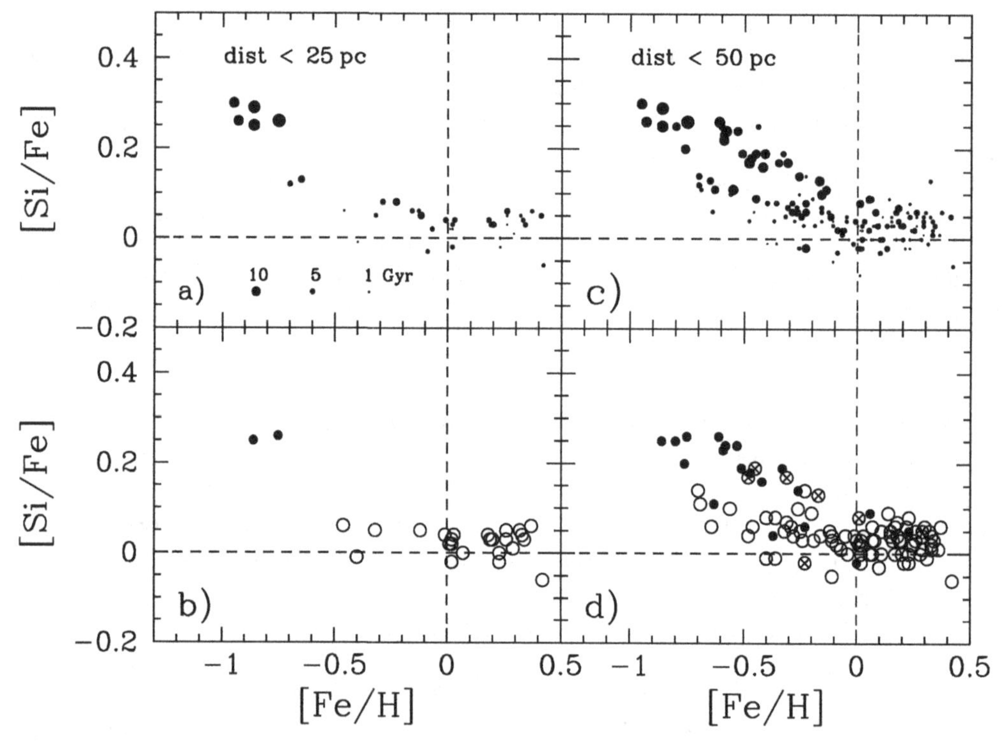

Figure 1. [Si/Fe] vs. [Fe/H] for four subsamples drawn from the full sample of Bensby *et al.* (2009, in preparation). All stars shown have age determinations with relative errors less than 2 Gyr. Note that these are not volume-complete samples, only volume-limited. **a.** All stars from Bensby *et al.* within 25 pc. The size of the symbols indicate their ages with larger symbols representing higher ages. The scale for the ages is indicated at the lower part of the panel. **b.** A kinematically-selected subset of the stars in **a.** A • marks stars that are ten times more likely to be thick- than thin disk members and a ∘ marks stars that are ten times more likely to be thin- than thick disk members. **c.** All stars from Bensby *et al.* within 50 pc. The size of the symbols indicate their ages with larger symbols representing higher ages. Same sizes are used as in panel a. **d.** A kinematically-selected subset of the stars in **c.** A • marks stars that are ten times more likely to be thick- than thin disk members and a ∘ marks stars that are ten times more likely to be thin- than thick disk members. Stars marked with an additional × are stars that are ten times more likely to be thin disk members than thick disk but also have an age larger than 8 Gyr.

In Feltzing & Bensby (2008) we identified a small number of stars on typical thin-disk orbits but with enhanced abundances for the α-elements. These stars were found to be old (older than about 8 Gyr in our determination). In Fig. 1d these stars are explicitly marked. For further discussion about plausible origins for these stars we refer to Feltzing & Bensby (2008). It is worth noting, however, that it is essentially these stars that make the downward trend of [Si/Fe] in the kinematically-selected thick disk sample blend in with the thin-disk sample.

Figure 2 shows the age-metallicity plot for a first selection of stars with kinematics that make them very likely thick-disk candidates. All of these stars are ten times more likely to belong to the thick- than to the thin disk. As can be seen, the bulk of these stars are older than the Sun and they have a mean age of around 10 Gyr. They cover that whole metallicity range from −1 dex to solar. For this first attempt at establishing if there is an age-metallicity relation present in our kinematically defined sample, we have only included stars for which we could determine the ages to better than 2 Gyr. As our stars originally are essentially selected only based on their kinematic properties and a metallicity estimated from photometry, we cover a reasonably large range of $T_{\rm eff}$. In Fig. 2 we have chosen to show the stars with $T_{\rm eff} > 6000$ K with a separate symbol. Not surprisingly, these stars are in general young. If they really belong to the thick disk then that would be rather challenging for any of the models put forward for the formation of the thick disk. However, our method to determine the stellar ages is "simple" and as these apparently young stars are in regions of the HRD where the stellar tracks show various "kinks" such a simple age estimate might go wrong in the estimate of the error. We will therefore redo all our ages using the method developed by Jørgensen & Lindegren (2005). This method provides a better and more realistic estimate of the error in the age determination. For now we would, however, like to caution against over-interpreting apparent young ages present in kinematically-defined thick-disk samples.

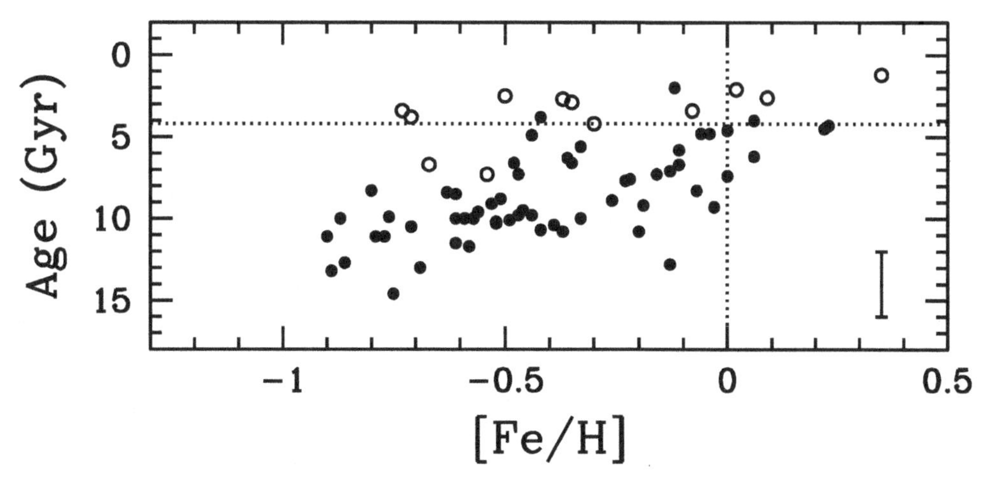

Figure 2. Ages and metallicities for a sample of thick disk stars from our new study. The stars shown all are ten times more likely to belong to the thick as opposed to the thin disk. The estimated errors in the derived ages are less than 2 Gyr. The error-bar in the lower right hand corner shows a 2 Gyr error. Metallicities are based on spectroscopy. The position of the Sun is marked by two dotted lines. α-enhancement has been taken into account when determining the ages (see Sect. 3). The filled circles show stars with $T_{\rm eff} < 6000$ K and the open circles the 9 thick disk stars that have $T_{\rm eff} > 6000$ K.

4. Summary

Age determinations of the stellar disk(s) of the Galaxy are inherently complicated. There are several factors that make it hard to define the age of either disk, not least the mixture of stellar populations in the solar neighborhood. The absolute ages of individual stars may be obtained through, e.g., asteroseismology and nucleocosmochronology. However, for the study of the stellar populations as such, fitting of isochrones to well-defined samples and the fitting of the white dwarf luminosity function using cooling tracks might be more appropriate.

In most current studies, stars with kinematics typical of the thick disk are, on average, found to be older than stars with kinematics typical of the thin disk. There appears to be an age-metallicity relation present in the thick disk. This is found in studies using various techniques. However, the exact definition of the thick disk in relation to the thin disk in terms of stellar kinematics is not straightforward and will need more work.

References

Árnadottír, A. S., Feltzing, S., & Lundström, I. 2008, *ArXiv*:0807.1665
Barnes, S. A. 2009, these proceedings
Barnes, S. A. 2007, *ApJ*, 669, 1167
Bensby T., Feltzing, S., & Lundström, I. 2003, *A&A*, 410, 527
Bensby T., Feltzing, S., Lundström, I., & Ilyin, I. 2005, *A&A*, 433, 185
Bernkopf, J. & Fuhrmann, K. 2006, *MNRAS*, 369, 673
del Peloso, E. F., da Silva, L., Porto de Mello, G. F., & Arany-prado, L. I. 2005, *A&A*, 440, 1153
Feltzing, S. & Bensby, T. 2008, *arXiv:* 0811.1777
Fuhrmann, K. 2008, *MNRAS*, 384, 173
Gilmore, G., Wyse, R. F. G., & Jones, J. B. 1995, *AJ*, 109, 1095
Gilmore, G., Wyse, R. F. G., & Norris, J. E. 2002, *ApJ*, 574, L39
Grundahl, F., Clausen, J. V., Hardis, S., & Frandsen, S. 2008, *arXiv*:0810.2407G
Holmberg, J., Nordström, B., & Andersen, J. 2007, *A&A*, 475, 519
Jønch-Sørensen, H. 1995, *A&A*, 298, 799
Jørgensen, B. R. & Lindegren, L. 2005, *A&A*, 436, 127
Knox, R. A., Hawkins, M. R. S., & Hambly, N. C. 1999, *MNRAS*, 306, 736
Leggett, S. K., Ruiz, M. T., & Bergeron, P. 1998, *ApJ*, 497, 294
Mamajek, E. E. 2009, these proceedings
Mamajek, E. E. & Hillenbrand, L. A. 2008, *ApJ*, 687, 1264
Oswalt, T. D., Smith, J. A., Wood, M. A., & Hintzem, P. 1995, *Nature*, 382, 692
Reddy, B. E., Lambert, D. L., & Allende Prieto, C. 2006, *MNRAS*, 367, 1329
Sandage, A., Lubin, L. M., & VandenBerg, D. A. 2003, *PASP*, 115, 1187
Schuster, W. J., Moitinho, A., Mrquez, A., Parrao, L., & Covarrubias, E. 2006, *A&A*, 445, 939
Vauclair, S., Laymand, M., Bouchy, F., Vauclair, G., Bon Hoa, A. H., Charpinet, S., & Bazot, M. 2008, *A&A*, 482, L5
von Hippel, T. & Bothun, G. D. 1993, *ApJ*, 407, 115
Wyse, R. F. G., Gilmore, G., Norris, J. E., Wilkinson, M. I., Kleyna, J. T., Koch, A., Evans, N. W., & Grebel, E. 2006, *ApJ*, 639, L13

Discussion

R. WYSE: The elemental abundance pattern (enhanced $[\alpha/\mathrm{Fe}]$, type II dominated) of the thick disk implies a short duration, less than the time for iron from Type Ia, ~ 1 Gyr. How is this consistent with a 5 Gyr or so spread in ages? Also the thin-disk kinematics are not well modeled by Gaussians so I would caution against population assignment based on Gaussians.

S. FELTZING: I agree that at least the thin disk is not well modeled by a Gaussian. It is obvious that the distribution of the velocities for nearby F- and G dwarf stars is lumpy. About abundances and ages: Taking a second look at our sample and also looking at the enhancement of α-elements, in this case Ti, I would say that the data might still be compatible with the picture you sketch in that the majority of the thick disk stars are enhanced in α elements, and their mean age is ~ 10 Gyr with some scatter, and there is a younger group with α about solar.

A. FREBEL: What are the abundance trends for stars in the z-selected boxes? Similar trends; different trends?

S. FELTZING: The abundance trends for our sample for the different boxes differ. For example, the box with high eccentricity and high $z_{\rm max}$ shows a very tight trend, e.g., for Ti, while stars with low eccentricity and high $z_{\rm max}$ show a much less tight trend and also potentially different levels of α enhancement.

J. MELBOURNE: I was interested in the younger stars identified in the thick disk. Does this argue for a more extended process of thick disk formation than a single monolithic event?

S. FELTZING: When a kinematic decomposition of the stars into a thin- and thick disk is done (e.g., Bensby *et al.* 2003) there are essentially no stars younger than about solar age. The thin disk stars selected in the same way appear to show a younger mean age than the thick disk stars. I am not sure we are in a position yet to interpret the few apparently young thick disk stars in terms of monolithic collapse or not. I still think a formation of a thickened disk through a merger event is favored by most of the data. This means today's thick disk stars formed in a thin disk.

Sofia Feltzing

Rosie Wyse

George Meynet and Chris Corbally

The Ages of Stars
Proceedings IAU Symposium No. 258, 2008
E.E. Mamajek, D.R. Soderblom & R.F.G Wyse, eds.

doi:10.1017/S1743921309031688

Signatures of heating processes in the Galactic thin disk

Birgitta Nordström

The Niels Bohr Institute, University of Copenhagen
Juliane Maries Vej 30, DK-2100 Copenhagen, Denmark
E-mail: birgitta@astro.ku.dk

Abstract. The term "heating" is used loosely to refer to a range of processes that result in an increase in velocity dispersion with age for subgroups of disk stars. We briefly summarise the observational basis for studies of disk heating and show that qualitative differences exist between the evolution of the in-plane and vertical motions. Ways to discriminate between various heating scenarios are discussed; the most recent galaxy merger simulations may in fact suggest that discrimination on purely kinematic grounds might be unfeasible, even with large samples of stars with excellent ages.

Keywords. stars: kinematics, Galaxy: solar neighborhood, Galaxy: disk, Galaxy: evolution, galaxies: kinematics and dynamics

---

1. Introduction

Observations demonstrate that old stellar populations in the Galactic disk show larger random motions (velocity dispersions) than groups of younger stars; see Wielen (1977) for a classic reference. This observation may reflect *(i):* initial conditions in a monolithic collapse scenario for the disk; *(ii):* gradual scattering of the initially circular orbits of newborn stars by spiral arms or massive objects within the disk itself, such as giant molecular clouds (GMCs) or ; *(iii):* the effect of the gradual accretion of dwarf galaxies into the disk in a hierarchical merger scenario. Thus, the velocities (and, by inference, Galactic orbits) of disk stars contain vital information on the formation and evolution of the disk.

The term "disk heating" is often applied loosely to the sum of the effects that may cause larger velocity dispersions in old stars. Strictly speaking, however, "heating" should only be used to refer to processes that, over time, inject kinetic energy into the random component of the stellar motions. Fossilised initial conditions for, e.g., halo, perhaps thick disk stars are not the result of "heating", nor is the combination of ordered motions of subgroups of stars such as moving groups or differential effects of Galactic rotation. Disentangling these various effects from the observations and placing them in proper historical perspective is not a simple matter, however.

In order to understand the origin of the present assemblage of disk stars, it is therefore necessary to quantify the kinematic properties of subpopulations in the disk and characterise the origin of their stars as accurately as possible. This contribution provides a brief review of the observational material available for such studies and what can be learned from the data at present.

2. Available data sets

The obvious requirement for detailed studies of the kinematic evolution of disk stars with time is a set of complete kinematic and astrophysical data for a large sample of stars:

Even after subdividing the stars by such parameters as age, metallicity, or type of Galactic orbit, the resulting subsamples must be large enough to reduce the sampling noise in the resulting velocity dispersions to a level where observed differences are statistically significant. E.g., the often-quoted paper by Quillen & Garnett (2001) was based on less than 200 stars – clearly insufficient for their conclusions to be robust. Freedom from kinematic selection biases is an equally important condition.

The Hipparcos mission yielded large numbers of stars with accurate parallaxes, proper motions, and hence tangential velocities. A kinematically unbiased subsample of ~12,000 main-sequence stars was analysed in considerable detail by Dehnen & Binney (1998). However, radial velocities, and thus true three-dimensional velocity vectors, were missing for most stars, and the *B-V* colour index was used as a crude proxy for age. The lack of multiple radial velocities also precluded the detection of close (spectroscopic) binaries in the sample.

The lack of radial velocities was mitigated in part by Famaey *et al.* (2005), who provided accurate radial velocities for ~6,000 northern K and M giants observed by Hipparcos. They computed the full Galactic velocity components *U, V, W* and analysed the systematic patterns in the data with sophisticated maximum likelihood techniques. Rough ages were estimated from isochrones for the subgroupings they defined. The strength of this data set is the large sample of accurate velocities and freedom from kinematic bias; the weakness is the lack of full sky coverage and of reliable age estimates for the individual stars.

Most recently, the RAVE project (Zwitter *et al.* 2008) has started producing radial velocities of large numbers of faint southern stars, selected without kinematic bias. Again, the lack of complete sky coverage and accurate stellar parameters for most stars, including accurate parallaxes, proper motions, and accurate ages, impedes the use of these data for the purposes discussed here.

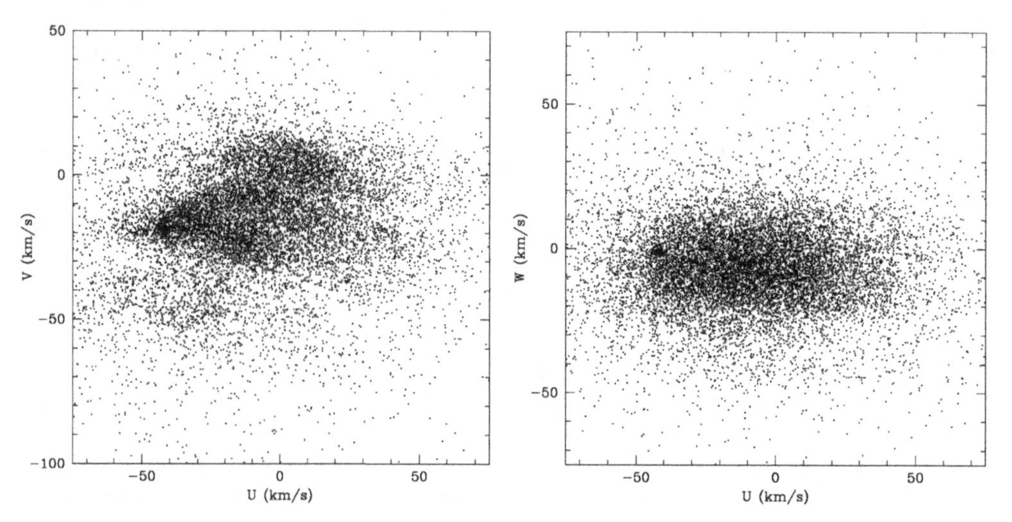

Figure 1. The observed *UV* (left) and *UW* planes (right) for the GCS stars. W is the velocity perpendicular to the Galactic plane.

The currently largest and, by design, most complete data set from which disk heating can be studied is the Geneva-Copenhagen Survey of the Solar Neighbourhood by Nordström *et al.* (2004; GCS I). The GCS contains a magnitude-limited, kinematically unbiased sample of ~14,000 F and G dwarf stars with Hipparcos parallaxes and Tycho-2 proper motions, accurate, multiple radial-velocity observations (important for binary detection), and derived *U, V, W* velocity components and Galactic orbital parameters. Metallicities and other astrophysical parameters were determined from Strömgren $uvby\beta$ photometry, and ages and their uncertainties were computed by the method of Jørgensen & Lindegren (2005). The key calibrations were revisited and updated by Holmberg *et al.* (2007; GCS II), and the revised Hipparcos parallaxes by van Leeuwen (2007) in Holmberg *et al.* (2009; GCS III), which contains the latest version of the data set. Most stars are within 150 pc of the Sun, and the sample is essentially volume complete out to ~40 pc.

In the following, we illustrate the use of the data in studies of the kinematic history of the Galactic disk. There are qualitative differences between the in-plane velocities and those perpendicular to the plane, so the two cases will be discussed separately.

3. Motions in the Galactic plane

The qualitative difference between the in-plane and vertical velocity distributions is obvious from Fig. 1. The *UV* plane shows anything but the sum of smooth, Gaussian distributions that would correspond to a simple model of thin and thick disk plus a sprinkling of halo stars. Apart from the Hyades (prominent at left), the stars are distributed in a number of bands that must correspond to distinct dynamical features such as spiral arms, moving groups, or other phenomena. These bands are not visible in the *UW* plane, however, indicating that different mechanisms dominate the in-plane and vertical dynamical histories.

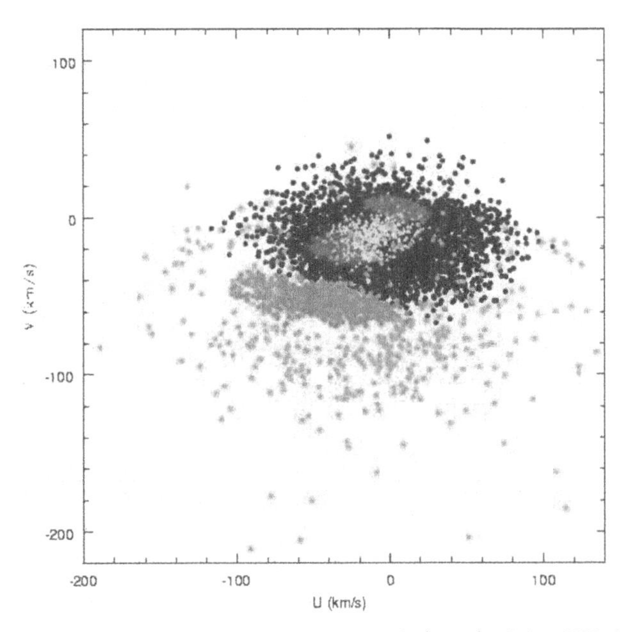

Figure 2. The kinematic decomposition by Famaey et al. (2005) of the *UV* plane for their stars. The smooth background of black and outer blue points shows the general thin and thick disk (plus halo) population. The central (yellow) group consists of the youngest stars, with the Sirius (magenta) and Hyades-Pleiades (red) groups directly above and below. The dense band of green points below the main disk concentration is the Hercules stream.

The in-plane velocity distribution was analysed in detail by Famaey *et al.* (2005), who used a maximum likelihood technique to identify a number of distinct dynamical groups in their *UV* plane; see Fig. 2. Already their limited age and metallicity data enabled them to conclude that these groups we not dominated by stars of a common origin, i.e. with the same age and chemical composition, a conclusion that has been refined later with better data, e.g. high resolution spectroscopic data for the Hercules stream by Bensby *et al.* (2007).

Thus, these features appear to arise from some mechanism(s) in the disk capable of dynamically focusing stars into similar orbits. E.g., the Hercules stream has been ascribed to the Outer Lindblad Resonance with the Galactic bar (Dehnen 2000, Fux 2001), but other mechanisms are possible, such as stationary or transient spiral arms. As pointed out by Seabroke & Gilmore (2007), a simple velocity dispersion is then inadequate to characterise the dynamical state of disk stars of a given age, and more detailed dynamical simulations are needed to clarify the origin of the observed in-plane motions. "Disk heating" then becomes an integral chapter of the origin and evolution of the Milky Way Galaxy itself.

4. The vertical kinematics

As suggested by Fig. 1 (right), the vertical (W) motions of disk stars do not reflect the complicated substructure seen in the *UV* plane. Instead, as noted by Seabroke & Gilmore (2007), the W velocities show an essentially pure Gaussian distribution at all ages (see Fig. 3), suggesting that "heating" is indeed an appropriate term for the gradual increase of random motions seen in the vertical direction.

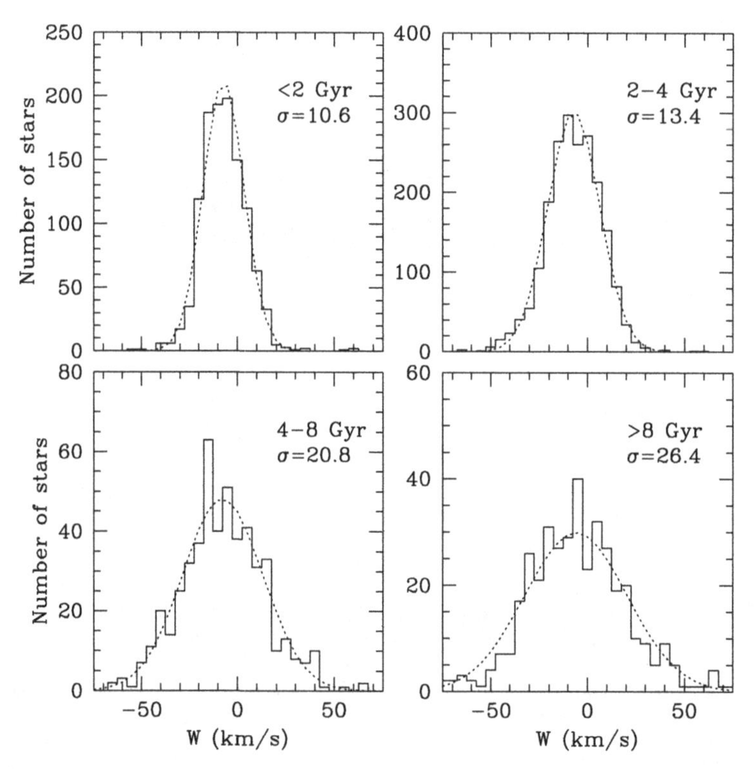

Figure 3. Distribution of W velocities and fitted Gaussians for four age groups of the GCS (single stars with $\sigma(Age) < 25\%$; GCS II, Fig. 32).

This does not, however, by itself mean that identifying the mechanism(s) responsible for the observed heating is a simple matter. Standard local gravity enhancement candidates are stationary spiral arms or density waves and giant molecular clouds (GMCs). However, such spiral arms do not affect the stellar motions in the vertical direction significantly, and GMCs are unable to reproduce the observed heating rate (Hänninen & Flynn 2002).

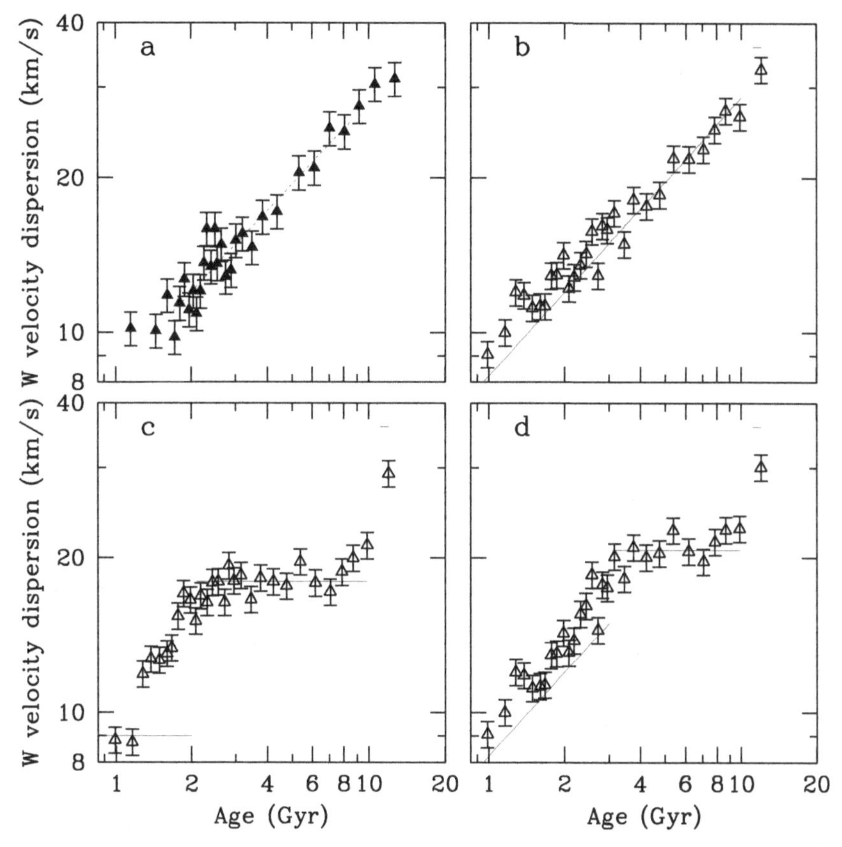

Figure 4. *a:* Observed AVR in W from GCS III (single stars with $\sigma_{Age} < 25\%$) with a fitted power law. *b-d:* Simulated AVRs for three disk heating scenarios: Continuous heating, early saturation as in Quillen & Garnett (2001), and a late minor merger. *Open symbols in b, c, d:* Rederived ages and velocity dispersions for the simulated stars.

Transient spiral arms appear to be able to match the data (De Simone *et al.* 2004), but their properties remain to be fully explored. More sophisticated calculations involve large-scale redistribution of stars within the disk over its lifetime without major net deformation of their initially circular orbits. Such models also seem able to reproduce the observed relations (see Roškar *et al.* 2008 and references therein), but offer few options for verification on kinematic grounds alone.

In the hierarchical merger scenario for galaxy formation, a galaxy like the Milky Way should be built up by gradual accretion of dwarf galaxies, such as the Sagittarius dwarf. Such mergers are expected to inject kinetic energy into the pre-existing (thin) disk stars, puffing up the disk and producing the observed heating. Satellites of appreciable mass are expected to produce a sharp jump in velocity dispersion or even disrupt the disk entirely; thus, the details of the W velocity dispersion as a function of time – the age-velocity relation (AVR) – should provide significant clues to the actual history (see Fig. 4).

For this to be possible, accurate stellar ages and reliable error estimates are obviously indispensable.

Fig. 4 suggests that the detailed shape of the AVR in W should allow to discriminate between heating mechanisms internal to the disk and the effects of mergers. However, we note that the very recent calculations by Hopkins *et al.* (2008) indicate that, in more realistic merger scenarios, minor mergers would primarily heat the bulge of a spiral galaxy with little effect on the disk itself. If so, the two scenarios may indeed be indistinguishable from kinematic data alone, even with significantly larger samples and better ages.

5. Conclusions

Recent observational work has greatly improved the empirical data with which to test models of the dynamical heating of the disk. Further improvements in stellar age determination may allow us to refine the dynamical history of the disk in even greater detail than today, but this will require very significant improvements in the observational determination of effective temperatures and chemical compositions, and in the stellar models and their transformation to the observational plane. It is perhaps ironic that, at the same time, state-of-the-art dynamical simulations may be telling us that the ages and velocities of stars cannot tell us the full history of the dynamical history of the disk; very detailed chemical analyses ("chemical tagging") may be required as well.

Acknowledgements: Financial support from the Carlsberg Foundation, the Danish Natural Science Foundation and the International Astronomical Union is gratefully acknowledged. This research was also supported in part by the National Science Foundation under Grant No. PHY05-51164. I thank the Geneva-Copenhagen Survey Team for our pleasant collaboration on the whole project.

References

Bensby, T., Oey, M.S., Feltzing, S., & Gustafsson, B. 2007, *ApJ*, 655, L89
Dehnen, W. 2000, *AJ*, 119, 800
Dehnen, W. & Binney, J. J. 1998, *MNRAS* 298, 387
De Simone, R. S., Wu, X., & Tremaine, S. 2004, *MNRAS* 350, 627
Famaey, B., Jorissen, A., Luri, X., *et al.* 2005, *A&A* 430, 165
Holmberg, J., Nordström, B., & Andersen, J. 2007, *A&A* 475, 519 (GCS II)
Holmberg, J., Nordström, B., & Andersen, J. 2009, *A&A*, in press (GCS III); also arXiv:0811.3982
Hopkins P. F., Hernquist, L., Cox, T. J., Younger, J. D., & Besla, G. 2008, *ApJ*, in press; also arXiv:0806.2861v2
Hänninen, J. & Flynn, C. 2002, *MNRAS* 337, 731
Jørgensen, B. R. & Lindegren, L. 2005, *A&A* 436, 127
Nordström, B., Mayor, M., Andersen, J., Holmberg, J., *et al.* 2004, *A&A* 418, 989 (GCS I)
Quillen, A. C. & Garnett, D. 2001, in Galaxy Disks and Disk Galaxies, eds. J. G. Funes, S. J. & E. M. Corsini. *ASP Conf. Ser.* 230, 87
Roškar, R., Debattista, V. P., Quinn, T. R., Stinson, G. S., & Wadsley, J. 2008, *ApJ* 684, L79
Seabroke, G. M. & Gilmore, G. 2007, *MNRAS* 380, 1348
van Leeuwen, F. 2007, *A&A* 474, 653
Wielen, R. 1977, *A&A* 60, 263
Zwitter, T. Siebert, A. Munari, U., *et al.* 2008, *AJ* 136, 421

Discussion

J. MELBOURNE: I was again struck by the large scatter in metallicity at a range of ages from 5 Gyr on. Does this argue for a more extended formation age for the thick disk?

B. NORDSTRÖM: The spread in metallicities is in fact large (and the same) for all ages. This could point to dynamical effects. New models by Roskar *et al.* (2008) and Schonrich & Binney (2008) point to a kinematic origin of the spread.

I. KING: How good are your 3-D velocities? That is, how sure are you of assignment of each individual star to a particular moving group? In the past this was poor; studies of metallicities by Strömgren photometry in earlier days (e.g., Breger, PASP 1971) showed that only half the stars assigned to a group were really members.

B. NORDSTRÖM: The 3-D velocities are accurate to 1.5 $\mathrm{km\,s^{-1}}$. This makes the verification of membership of the moving groups more certain than before. There might not be sharp borders to field stars. The groups seem to be an effect of dynamical mechanisms. The metallicities and ages of the "moving groups" do not point to a common birthplace.

D. SODERBLOM: You stated that studying a volume- limited sample of the solar neighborhood does not work for understanding some Galactic properties because models show the stars come from a large range of locations in the Galaxy. What would work?

B. NORDSTRÖM: We should still look at a volume- limited sample and try to understand the large scatter in the diagnostic diagrams (age-metallicity and age-velocity). The scatter is much larger than the observational errors and I think that there is information there that tells us something. Many of the stars in the solar neighborhood follow the Sun, and others that are leaving might be replaced by similar stars.

Birgitta Nordström

The Ages of Stars
Proceedings IAU Symposium No. 258, 2008
E.E. Mamajek, D.R. Soderblom & R.F.G. Wyse, eds.

doi:10.1017/S174392130903169X

The timescales of chemical enrichment in the Galaxy

Antonio Pipino[1] **and Francesca Matteucci**[2]

[1]Physics & Astronomy, University of Southern California, Los Angeles 90089-0484, USA
email: pipino@usc.edu

[2]Dipartimento di Astronomia, Universita'di Trieste, via GB Tiepolo 11, 34100 Trieste, Italy

Abstract. The time-scales of chemical enrichment are fundamental to understand the evolution of abundances and abundance ratios in galaxies. In particular, the time-scales for the enrichment by SNe II and SNe Ia are crucial in interpreting the evolution of abundance ratios such as [α/Fe]. In fact, the α-elements are produced mainly by SNe II on time-scales of the order of 3 to 30 Myr, whereas the Fe is mainly produced by SNe Ia on a larger range of time-scales, going from 30 Myr to a Hubble time. This produces differences in the [α/Fe] ratios at high and low redshift and it is known as "time-delay" model. In this talk we review the most common progenitor models for SNe Ia and the derived rates together with the effect of the star formation history on the [α/Fe] versus [Fe/H] diagram in the Galaxy. From these diagrams we can derive the timescale for the formation of the inner halo (roughly 2 Gyr), the timescale for the formation of the local disk (roughly 7-8 Gyr) as well the time-scales for the formation of the whole disk. These are functions of the galactocentric distance and vary from 2-3 Gyr in the inner disk up to a Hubble time in the outer disk (inside-out formation). Finally, the timescale for the formation of the bulge is found to be no longer than 0.3 Gyr, similar to the timescale for the formation of larger spheroids such as elliptical galaxies. We show the time-delay model applied to galaxies of different morphological type, identified by different star formation histories, and how it constrains differing galaxy formation models.

Keywords. Galaxy: abundances, Galaxy: bulge, Galaxy: disk, Galaxy: evolution, Galaxy: formation, galaxies: elliptical and lenticular, cD, galaxies: evolution, galaxies: formation

1. How to model galactic chemical evolution: role of SNIa vs. SNII

Before going into the detailed chemical evolution history of the Milky Way, it is necessary to understand how to model, in general, galactic chemical evolution. The basic ingredients to build a model of galactic chemical evolution can be summarized as: i) Initial conditions; ii) Stellar birthrate function (the rate at which stars are formed from the gas and their mass spectrum); iii) Stellar yields (how elements are produced in stars and restored into the interstellar medium); iv) Gas flows (infall, outflow, radial flow). We refer the reader to Matteucci (2008) for a thorough review of their relative roles. Here we just briefly focus on some properties of both Type II and Ia SNe and recall the mass range of their progenitors:

1.0.1. *Massive stars ($8 < M/M_{\odot} \leqslant 40$)*

In the mass range 10-40 $M_{\odot}$, available calculations are from Woosley & Weaver (1995), Thielemann *et al.* (1996), Meynet & Maeder (2002), Nomoto *et al.* (2006), among others. These stars end their life as Type II SNe and explode bycore-collapse; they producemainly α-elements (O, Ne, Mg, Si, S, Ca), some Fe-peak elements, s-process elements ($A < 90$) and r-process elements. Lifetimes are below 30 Myr.

1.0.2. *Type Ia SN progenitors*

There is a general consensus about the fact that SNeIa originate from C-deflagration in C-O white dwarfs (WD) in binary systems, but several evolutionary paths can lead to such an event. The C-deflagration produces $\sim 0.6 - 0.7 M_{\odot}$ of Fe plus traces of other elements from C to Si, as observed in the spectra of Type Ia SNe. Two main evolutionary scenarios for the progenitors of Type Ia SNe have been proposed:

Single Degenerate (SD) scenario: the classical scenario of Whelan and Iben (1973), recently revised by Han & Podsiadlowsky (2004), namely C-deflagration in a C-O WD reaching the Chandrasekhar mass $M_{Ch} \sim 1.44 M_{\odot}$ after accreting material from a red giant companion. One of the limitations of this scenario is that the accretion rate should be defined in a quite narrow range of values.

The clock to the explosion is given by the lifetime of the secondary star in the binary system, where the WD is the primary (the originally more massive one). Therefore, the largest mass for a secondary is $8 M_{\odot}$, which is the maximum mass for the formation of a C-O WD. As a consequence, the minimum timescale for the occurrence of Type Ia SNe is $\sim$ 30 Myr (i.e. the lifetime of a $8 M_{\odot}$) after the beginning of star formation. Recent observations in radio-galaxies by Mannucci *et al.* (2005, 2006) seem to confirm the existence of such prompt Type Ia SNe. The minimum mass for the secondary is $0.8 M_{\odot}$, which is the star with lifetime equal to the age of the Universe. Stars with masses below this limit are obviously not considered.

Double Degenerate (DD) scenario: the merging of two C-O white dwarfs, due to loss of angular momentum caused by gravitational wave radiation, which explode by C-deflagration when M_{Ch} is reached (Iben and Tutukov 1984). In this scenario, the two C-O WDs should be of $\sim 0.7 M_{\odot}$ in order to give rise to a Chandrasekhar mass after they merge, therefore their progenitors should be in the range (5-8)$M_{\odot}$. The clock to the explosion here is given by the lifetime of the secondary star plus the gravitational time delay which depends on the original separation of the two WDs. The minimum timescale for the appearance of the first Type Ia SNe in this scenario is one million years more than in the SD scenario (e.g. Greggio 2005 and references therein). At the same time, the maximum gravitational time delay can be as long as more than a Hubble time.

A way of defining the typical Type Ia SN timescale is to assume it as the time when the maximum in the Type Ia SN rate is reached (Matteucci & Recchi, 2001). *This timescale varies according to the chosen progenitor model and to the assumed star formation history, which varies from galaxy to galaxy.* For the solar vicinity, this timescale is at least 1 Gyr, if the SD scenario is assumed, whereas for elliptical galaxies, where the stars formed much more quickly, this timescale is only 0.5 Gyr (Matteucci & Greggio, 1986; Matteucci & Recchi 2001).

2. Chemical evolution time-scales and the formation of the Milky Way

The kinematical and chemical properties of the different Galactic stellar populations can be interpreted in terms of the Galaxy formation mechanism. Here we focus on the results of the two-infall model of Chiappini, Matteucci & Gratton (1997) to which we refer the reader for details. Such a scenario predicts two main episodes of gas accretion: during the first one, the halo the bulge and most of the thick disk formed, while the second gave rise to the thin disk. We first witness the sequence of the formation of the stellar halo, in particular the inner halo, following a monolithic-like collapse of gas (first infall episode) but with a longer timescale than originally suggested by Eggen *et al.* (1962): here the time scale is 0.8-1.0 Gyr. During the halo formation also the bulge is formed on a very short timescale in the range 0.1-0.5 Gyr (see below). During this phase also the thick disk assembles or at least part of it, since part of the thick disk, like the outer halo, could have been accreted. Then the thin disk formation, namely the assembly of the innermost disk regions just around the bulge, begins; this is due to the second infall episode. The thin-disk assembles inside-out, in the sense that the outermost regions take a much longer time to form. It is clear that the early phases of the halo and bulge formation are dominated by Type II SNe (and also by Type Ib/c SNe) producing mostly α-elements such as O and Mg and part of Fe. On the other hand, Type Ia SNe start to be non negligible only after 1Gyr and they pollute the gas during the thick and thin disk phases.

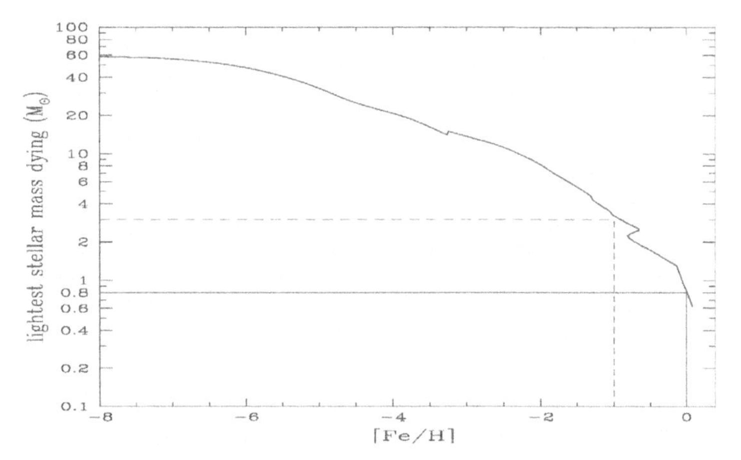

Figure 1. In this figure we show the smallest stellar mass which dies at any given [Fe/H] achieved by the ISM as a consequence of chemical evolution (see text).

More in detail, the main assumptions of the two-infall model are: the IMF is that of Scalo (1986) normalized over a mass range of 0.1-100$M_{\odot}$. The infall law gives the rate at which the total surface mass density changes because of the infalling gas. It is normalised to reproduce the total present time surface mass density in the solar vicinity (σ_{tot}= 51 $\pm$ 6 $M_{\odot}$ pc^{-2} , see Boissier & Prantzos 1999), 1 Gyr is the time for the maximum infall on the thin disk, 0.8 Gyr is the e-folding time for the formation of the halo thick-disk (which means a total duration of 2 Gyr for the complete halo-thick disk formation) and $\tau_D(r)$ is the timescale for the formation of the thin disk and it is a function of the galactocentric distance (formation inside-out, Matteucci & François 1989; Chiappini *et al.* 2001). In particular, it is assumed that $\tau_D = 1.033r(Kpc) - 1.267\ (Gyr)$ where r is the galactocentric distance.

The SFR is the Kennicutt law with a dependence on the surface gas density and also on the total surface mass density (see Dopita & Ryder 1994). The exponent of the surface gas density is set equal to 1.5, similar to what suggested by Kennicutt (1998a). These choices for the parameters allow the model to fit very well the observational constraints, in particular in the solar vicinity. We recall that below a critical threshold for the surface gas density ($7M_{\odot}pc^{-2}$ for the thin disk and $4M_{\odot}pc^{-2}$ for the halo phase) we assume that the star formation is halted. The existence of a threshold for the star formation has been suggested by Kennicutt (1998a,b).

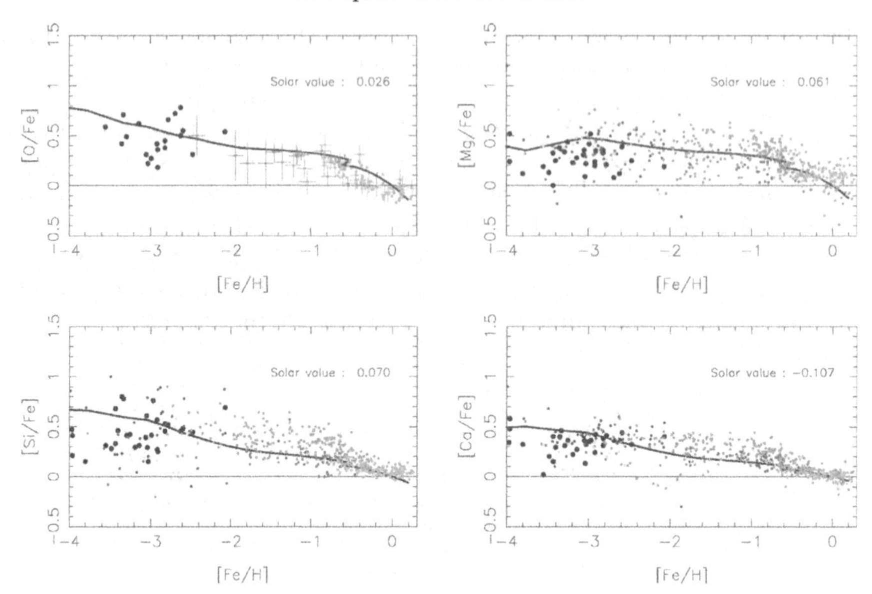

Figure 2. Predicted and observed [α/Fe] vs. [Fe/H] in the solar neighborhood. The models and the data are from François *et al.* (2004). The models are normalized to the predicted solar abundances. The predicted abundance ratios at the time of the Sun formation (Solar value) are shown in each panel and indicate a good fit (all the values are close to zero).

2.1. *The chemical enrichment history of the solar vicinity*

In Fig. 1 we show the smallest mass dying at any cosmic time corresponding to a given predicted abundance of [Fe/H] in the ISM. This is because there is an age-metallicity relation and the [Fe/H] abundance increases with time. Thus, it is clear that in the early phases of the halo only massive stars are dying and contributing to the chemical enrichment process. Clearly this graph depends upon the assumed stellar lifetimes and upon the age-[Fe/H] relation. It is worth noting that the Fe production from Type Ia SNe appears before the gas has reached [Fe/H] =-1.0, therefore during the halo and thick disk phase. This clearly depends upon the assumed Type Ia SN progenitors (in this case the single degenerate model).

2.1.1. *The time-delay model*

The time-delay refers to the delay with which Fe is ejected into the ISM by SNe Ia relative to the fast production of α-elements by core-collapse SNe. Tinsley (1979) first suggested that this time delay would have produced a typical signature in the [α/Fe] vs. [Fe/H] diagram. Matteucci & Greggio (1986) included for the first time the Type Ia SN rate formulated by Greggio & Renzini (1983) in a detailed numerical model for the chemical evolution of the Milky Way. The effect of the delayed Fe production is to create an overabundance of O relative to Fe ([O/Fe]$>$ 0) at low [Fe/H] values, and a continuous decline of the [O/Fe] ratio until the solar value ($[O/Fe]_\odot = 0.0$) is reached for [Fe/H]> -1.0 dex. This is what is observed and indicates that during the halo phase the [O/Fe] ratio is due only to the production of O and Fe by SNe II. However, since the bulk of Fe is produced by Type Ia SNe, when these latter start to be important then the [O/Fe] ratio begins to decline. This effect was predicted by Matteucci & Greggio (1986) to occur also for other α-elements (e.g. Mg, Si). At the present time, a great amount of stellar abundances is available and the trend of the α-elements has been since

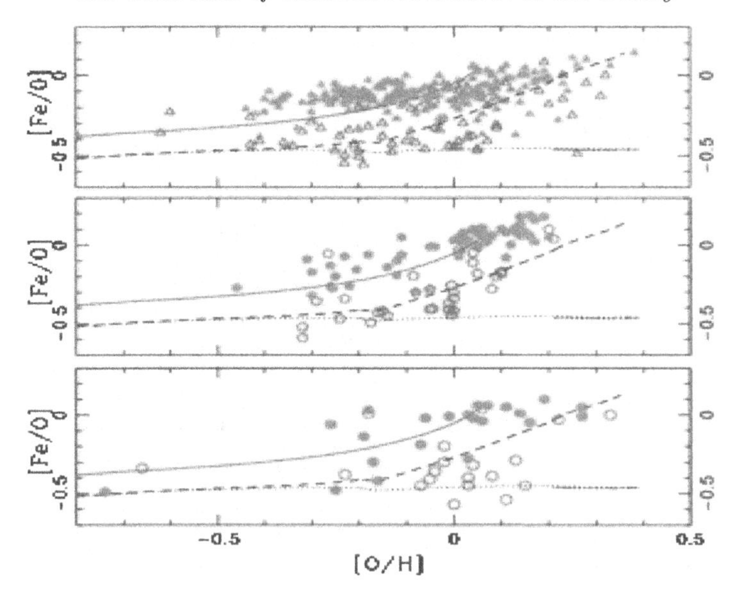

Figure 3. Thick vs thin disk [Fe/O]-[O/H] diagrams. Theoretical predictions for the thick and the thin disk are shown by dashed and solid lines, respectively. The dotted line is a model for the thick disk without pollution by SNIa. Data for the thick and the thin disk are represented by open and full symbols, respectively. For details refer to Chiappini (2008).

long confirmed. In Fig. 2 shows one among our most recent models applied to the latest compilation of data (Franqis *et al.* 2004). A good fit of the [O/Fe] ratio as a function of [Fe/H] is obtained only if the α-elements are mainly produced by Type II SNe and the Fe by Type Ia SNe. If one assumes that only SNe Ia produce Fe as well as if one assumes that only Type II SNe produce Fe, the agreement with observations is lost. Therefore, the conclusion is that both Types of SNe should produce Fe in the proportions of 1/3 for Type II SNe and 2/3 for Type Ia SNe. The model in Fig. 2 does not distinguish between the thin and the thick disk. Recently, Chiappini (2008) presented a model where the evolutions of the thin and thick disks were considered separately. The main hypothesis was that the thick disk formed and evolved faster than the thin disk (see Fig. 3), and the conclusion was that a good agreement with the observed abundances in thick disk stars can be obtained if the thick disk assembled by gas accretion on a timescale no longer than 2 Gyr. This conclusion favors the formation of the thick disk stars in situ instead than by accretion of extant stellar satellites.

2.1.2. *The G-dwarf metallicity distribution and constraints on the thin disk formation*

The G-dwarf metallicity distribution is quite an important constraint for the chemical evolution of the solar vicinity: it is the fossil record of the star formation history of the thin disk. Originally, there was the "G-dwarf problem" which means that the Simple Model of galactic chemical evolution could not reproduce the distribution of the G-dwarfs. It has been since long demonstrated that relaxing the closed-box assumption and allowing for the solar region to form gradually by accretion of gas can solve the problem (Tinsley, 1980). Assuming that the disk forms from pre-enriched gas can also solve the problem but still the gas infall is necessary to have a realistic picture of the disk formation. The two-infall model can reproduce very well the G-dwarf distribution and also that of K-dwarfs (see Fig. 4), as long as a timescale for the formation of the disk in the solar vicinity

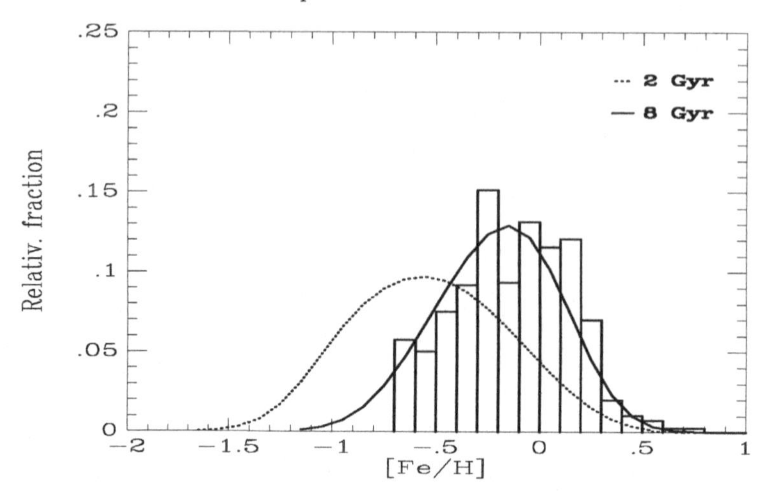

Figure 4. The figure is from Kotoneva *et al.* (2002) and shows the comparison between a sample of K-dwarfs and model predictions in the solar neighborhood. The dotted curve refers to the two-infall model with a timescale $\tau = 2$ Gyr, whereas the continuous line refers to $\tau = 8$ Gyr.

of 7-8 Gyr is assumed. This conclusion is shared by other authors (Alibés *et al.* 2001; Prantzos & Boissier 1999)

2.2. *The Galactic disk*

The chemical abundances measured along the disk of the Galaxy suggest that the metal content decreases from the innermost to the outermost regions, in other words there is a negative gradient in metals.

In Fig. 5 we show theoretical predictions of abundance gradients along the disk of the Milky Way compared with data from HII regions, B stars and PNe. The adopted model is from Chiappini *et al.* (2001) and is based on an inside-out formation of the thin disk. The model does not allow for exchange of gas between different regions of the disk. The disk is, in fact, divided in several concentric shells 2 Kpc wide with no interaction between them.

The reason for the steepening is that in the model of Chiappini et al. there is included a threshold density for SF, which induces the SF to stop when the density decreases below the threshold. This effect is particularly strong in the external regions of the Galactic disk, thus contributing to a slower evolution and therefore to a steepening of the gradients with time. Other authors (e.g. Boissier & Prantzos, 1999) found a flattening of the gradient with time in absence of a star formation threshold.

2.3. *The Galactic bulge*

In the context of chemical evolution, the Galactic bulge was first modeled by Matteucci & Brocato (1990) who predicted that the [α/Fe] ratio for some elements (O, Si and Mg) should be super-solar over almost the whole metallicity range, in analogy with the halo stars, as a consequence of assuming a fast bulge evolution which involved rapid gas enrichment in Fe mainly by Type II SNe. Recent data concerning medium- and high-resolution spectroscopy of bulge stars (Rich & McWilliam, 2000; Fulbright *et al.*, 2006, 2007; Zoccali *et al.*, 2006; Lecureur *et al.* 2007) and more recent models (e.g. Ballero *et al.* 2007) have confirmed the previous predictions and suggested that the bulge must have formed on a time scale of ~ 0.3 Gyr and in any case no longer than 0.5 Gyr. This is in agreement with Elmegreen (1999) and Ferreras *et al.* (2003). A similar model, although

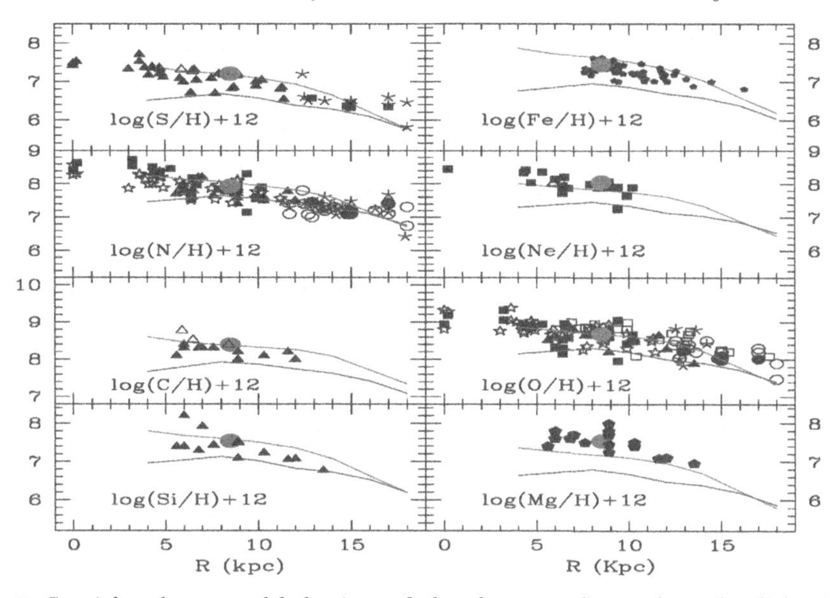

Figure 5. Spatial and temporal behaviour of abundance gradients along the Galactic disk as predicted by the best model of Chiappini *et al.* (2001). The upper lines in each panel represent the present time gradient, whereas the lower ones represent the gradient a few Gyr ago (see Chiappini *et al.* 2001).

updated with the inclusion of the development of a galactic wind and more recent stellar yields, has been presented by Pipino *et al.* (2008b): it shows how a model with intense star formation and rapid assembly of gas can best reproduce the most recent accurate data on abundance ratios (both in single stars and in the integrated spectra) as well as the metallicity distribution .

3. From the Bulge to external galaxies: Ellipticals

In Fig. 6 we present the predictions by Matteucci (2003) of the [α/Fe] ratios as functions of [Fe/H] in galaxies of different morphological type. In particular, for the Galactic bulge or an elliptical galaxy of the same mass, for the solar vicinity region and for an irregular Magellanic galaxy (LMC and SMC). The underlying assumption is that different objects undergo different histories of star formation, being very fast in the spheroids (bulges and ellipticals), moderate in spiral disks and slow and perhaps gasping in irregular gas rich galaxies. The effect of different star formation histories is evident in Fig. 6 where the predicted [α/Fe] ratios in the bulge and ellipticals remain high and almost constant for a large interval of [Fe/H]. This is due to the fact that, since star formation is very intense, the bulge reaches very soon a solar metallicity thanks only to the SNe II; then, when SNe Ia start exploding and ejecting Fe into the ISM, the change in the slope occurs at larger [Fe/H] than in the solar vicinity. In the extreme case of irregular galaxies the situation is opposite: here the star formation is slow and when the SNe Ia start exploding the gas is still very metal poor. This scheme is quite useful since it can be used to identify galaxies only by looking at their abundance ratios.

The above mentioned models for the bulge can be extended to study elliptical galaxies. In fact, similarly to bulges, elliptical galaxies show over-solar [α/Fe] ratios. Moreover, these ratios are found to increase with galactic mass. In the framework of the

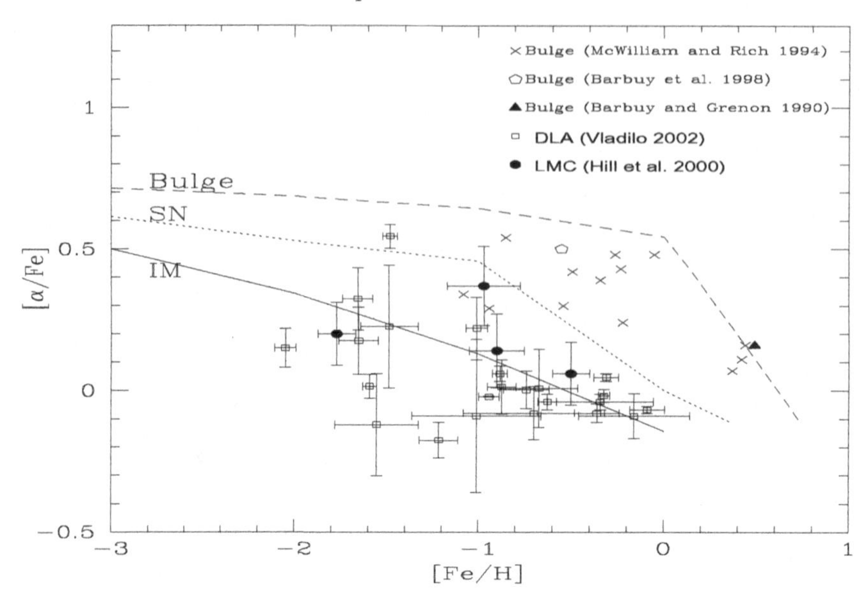

Figure 6. The predicted [α/Fe] vs. [Fe/H] relations for the Galactic bulge (upper curve), the solar vicinity (median curve) and irregular galaxies (low curve). Data for the bulge are reported for comparison. Data for the LMC and DLA systems are also shown for comparison, indicating that DLAs are probably irregular galaxies. Figure and references are from Matteucci (2003).

time-delay model, we can explain such a behaviour as an increase of the SF efficiency as a function of galactic mass (Matteucci 1994, Pipino & Matteucci, 2004). To explain the higher star formation efficiency in the most massive galaxies in the framework of the *monolithic collapse*, Pipino *et al.* (2008c) appeal to massive black holes-triggered SF: a short ($10^6 - 10^7$ yr) super-Eddington phase can provide the accelerated triggering of associated star formation. According to Pipino *et al.* (2008c) models, the galaxy is fully assembled on a time-scale of 0.3-0.5 Gyr.

On the other hand, when the same detailed treatment for the chemical evolution is implemented in a semi-analytic model for galaxy formation based on the *hierarchical clustering scenario* (e.g. White & Rees, 1978), it does not produce the observed mass-[α/Fe] relation (Pipino *et al.* 2008a). In particular, the slope is too shallow and the scatter too large (see Fig. 7), in particular in the low and intermediate mass range. The model shows significant improvement at the highest masses and velocity dispersions, where the predicted [α/Fe] ratios are now marginally consistent with observed values. Moreover, an excess of low-mass ellipticals with too high a [α/Fe] ratio is predicted.

A thorough exploration of the parameter space shows that the failure in reproducing the mass- and σ-[α/Fe] relations can partly be attributed to the way in which star formation and feedback are currently modelled. The merger process is responsible for a part of the scatter. We suggest that the next generation of semi-analytical model should feature feedback (either stellar of from AGN) mechanisms linked to single galaxies and not only to the halo, especially in the low and intermediate mass range. The scatter is also intrinsic to the merger history, thus calling for further modification of the baryons behaviour with respect to the CDM. In other words we envisage a lack of a self-regulating mechanisms which acts on a galactic scale and counterbalances to some extent the random nature of the merger trees.

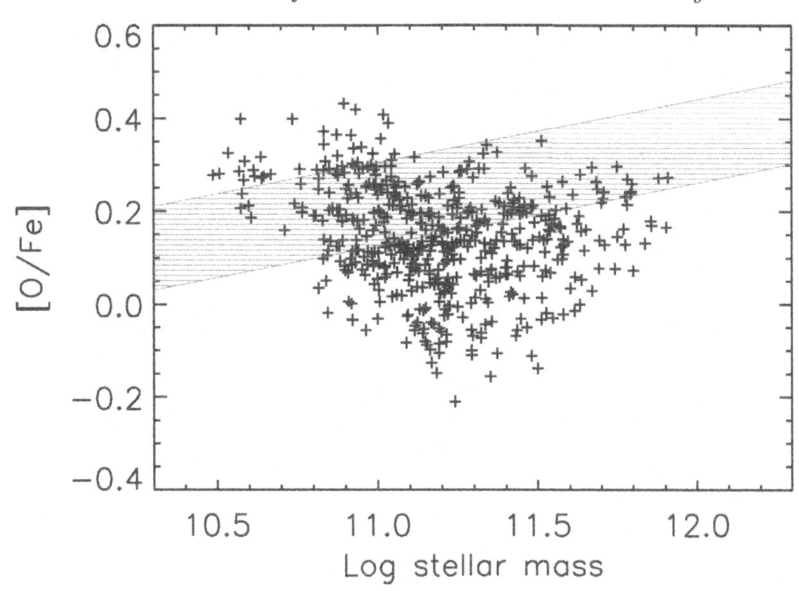

Figure 7. The α/Fe-mass and relations as predicted by GalICS (Pipino *et al.* 2008a) for the whole sample of ellipticals (black points). The thick solid lines encompass the 1σ-region (hatched area) around the mean trend reported by Thomas *et al.* (2008, in prep.).

4. Conclusions

From the discussions of the previous sections we can extract some important conclusions on the formation and evolution of the Milky Way, derived from chemical abundances. In particular, the inner halo formed on a timescale of 1-2 Gyr at maximum, the outer halo formed on longer time-scales perhaps from accretion of satellites or gas. The disk at the solar ring formed on a timescale not shorter than 7 Gyr. The whole disk formed inside out with time-scales of the order of 2 Gyr or less in the inner regions and 10 Gyr or more in the outermost regions. The bulge is very old and formed very quickly on a timescale smaller than even the inner halo and not larger than 0.5 Gyr. The abundance gradients arise naturally from the assumption of the inside-out formation of the disk. A threshold density for the star formation helps in steepening the gradients in the outer disk regions. The IMF seems to be different in the bulge and the disk, being flatter in the bulge, although more abundance data are necessary before drawing firm conclusions. Elliptical galaxies and bulges share similar chemical properties (at a given mass), therefore a very short and intense formation is predicted for them. While this can be accommodated in the *monolithic framework* by means of, e.g., AGN-triggered star formation, it is still hard for models based on *hierarchical clustering* to account for the observed chemical properties.

References

Alibés, A., Labay, J., & Canal, R., 2001, *A&A*,, 370, 1103

Ballero, S., Matteucci, F., Origlia, L., & Rich, R. M. 2007, *A&A*, 467, 123

Boissier, S. & Prantzos, N., 1999, *MNRAS*, 307, 857

Chiappini, C., 2008, in: J. Funes, S. J., Corsini, E. M. (eds.), *Formation and Evolution of Galaxy Disks*, ASP Conference Series, Vol. 396, (San Francisco: ASP), p.113

Chiappini, C., Matteucci F., & Gratton R. 1997, *ApJ*, 477, 765

Chiappini, C., Matteucci, F., & Romano, D., 2001, *ApJ*, 554, 1044

Dopita, M. A. & Ryder, S. D., 1994, *ApJ*, 430, 163
Eggen, O. J., Lynden-Bell, D., & Sandage, A. R., 1962, *ApJ*,, 136, 748
Elmegreen, B. G., 1999, *ApJ*, 517, 103
Ferreras, I., Wyse, F. G., & Silk, J., 2003, *MNRAS*, 345, 1381
François, P., Matteucci, F. Cayrel, R., Spite, M., Spite, F., & Chiappini, C., 2004, *A&A*, 421, 613
Fulbright, J. P., McWilliam, A., & Rich, R. M., 2006, *ApJ*, 636, 831
Fulbright, J. P., McWilliam, A., & Rich, R. M., 2007, *ApJ*, 661, 1152
Greggio, L., 2005, *A&A*, 441, 1055
Greggio, L. & Renzini, A., 1983, *A&A*, 118, 217
Han, Z. & Podsiadlowski, Ph., 2004, *MNRAS*, 350, 1301
Iben, I. Jr. & Tutukov, A., 1984, *ApJ*, 284, 719
Kennicutt, R. C. Jr., 1998a, *ApJ*, 498, 541
Kennicutt, R. C. Jr., 1998b, *ARA&A*, 36, 189
Kotoneva, E., Flynn, C., Chiappini, C., & Matteucci, F. 2002, *MNRAS*, 336, 879
Lecureur, A., Hill, V., Zoccali, M., Barbuy, B., Gomez, A., Minniti, D., Ortolani, S., & Renzini, A., 2007, *A&A*, 465, 799
Mannucci, F., Della Valle, M., Panagia, N., Cappellaro, E., Cresci, G., Maiolino, R., Petrosian, A., & Turatto, M., 2005, *A&A*, 433, 807
Mannucci, F., Della Valle, M., & Panagia, N., 2006, *MNRAS*, 370, 773
Matteucci, F., 2001, *The Chemical Evolution of the Galaxy*, ASSL, Kluwer Academic Publisher
Matteucci, F. 2008, in: E. Grebel and B. Moore (eds.), *The Origin of the Galaxy and the Local Group*, astro-ph/0804.1492
Matteucci, F. & Brocato, E., 1990, *ApJ*, 365, 539
Matteucci, F. & François, P., 1989, *MNRAS*, 239, 885
Matteucci, F. & Greggio, L., 1986, *A&A*, 154, 279
Matteucci, F. & Recchi, S., 2001, *ApJ*, 558, 351
Matteucci, F., Romano, D., & Molaro, P., 1999, *A&A*, 341, 458
Meynet, G. & Maeder, A., 2002, *A&A*, 390, 561
Nomoto, K., Tominaga, N., Umeda, H., Kobayashi, C., & Maeda, K., 2006, *Nucl. Phys. A*, 777, 424
Pipino, A., Devriendt, J. E. G., Thomas, D., Silk, J., & Kaviraj, S., 2008a, submitted, arXiv0810.5753
Pipino, A. & Matteucci, F. 2004, *MNRAS*,, 347, 968
Pipino, A., Matteucci, F., & D'Ercole, A., 2008b, *IAUS*, 245, 19
Pipino, A., Silk, J., & Matteucci, F., 2008c, *MNRAS*, in press, arXiv0810.2045
Prantzos, N., 2003, *A&A*, 404, 211
Prantzos, N. & Boissier, S., 2000, *MNRAS*, 313, 338
Rich, R. M. & McWilliam, A., 2000, in: Jacqueline Bergeron (ed.), *Discoveries and Research Prospects from 8- to 10-Meter-Class Telescopes*, Proc. SPIE Vol. 4005, p. 150
Salpeter, E. E., 1955, *ApJ*, 121, 161
Scalo, J. M., 1986, *Fund. Cosmic Phys.*, 11, 1
Schmidt, M., 1959, *ApJ*, 129, 243
Thielemann, F. K., Nomoto, K., & Hashimoto, M., 1996, *ApJ*, 460, 408
Tinsley, B. M., 1980, Fund. Cosmic Phys. Vol. 5, 287
Tinsley, B. M., 1979, *ApJ*, 229, 1046
Whelan, J. & Iben, I. Jr., 1973, *ApJ*, 186, 1007
White, S. D. M. & Rees, M. J., 1978, *MNRAS*, 183, 341
Woosley, S. E. & Weaver, T. A., 1995, *ApJS*, 101, 181
Zoccali, M., Lecureur, A., Barbuy, B., Hill, V., Renzini, A., Minniti, D., Momany, Y., Gómez, A., & Ortolani, S., 2006, *A&A*, 457, L1

Discussion

E. MAMAJEK: 1. What is assumed about the composition of the gas being accreted by the Milky Way? 2. Given your chemical evolution model and its assumptions, what is the inferred mass accretion rate for the Milky Way?

A. PIPINO: 1. The composition of the gas is primordial; the same applies to our models for ellipticals/bulges. However in a multi-zone model (for spheroids) the outer parts pollute the pristine gas sinking toward the galactic center. 2. See comments by M. Tosi.

J. KALIRAI: How significantly are the chemical evolution models affected by changes to the threshold mass that distinguishes Type II SNe production from white dwarf formation? For example, initial masses of 6 versus 10 solar masses?

A. PIPINO: The answer is two-fold: 1. High-mass regime : since Mg and O yields are mass-dependent (increasing with stellar mass), forming SNe II from either 8, 10, or 12 $M_{\odot}$ will produce different results. However the difference is within 0.1 to 0.2 dex. In averaged abundance ratios such as those inferred from Lick indices, the above difference will translate into a small offset in the predicted [Mg/Fe] vs. σ relation. 2. Low-mass regime: Chemical evolution models are tuned to reproduce the present-day SN Ia rate. A change (reduction) in the upper mass for WD would probably require a retuning of the model so that the number of SNe Ia stays nearly the same. I therefore expect the changes to be small. Of course the minimum delay needed in order to have a SN Ia will be larger than the standard 30 Myr (lifetime of a 8 $M_{\odot}$ star).

B. NORDSTRÖM: What is the reason for "observational spread"? And could that be explained by dynamical effects? Comment: Transient spiral waves (Roskar 2008) can also explain the "G-dwarf problem."

A. PIPINO: The spread can be accounted for with small-scale inhomogeneities such as slight differences and gas flows between adjacent shells (if one models the galaxy with a spherically symmetrical geometry). For instance, Pipino, Matteucci & Dercole show it for the galactic bulge by plotting the contours of the 2-D *G-dwarf-like* distribution in [Fe/H] and [α/H] plane instead of the standard average [O/Fe] vs. [Fe/H] behavior of the ISM. Stellar dynamical effects might play a role although if stars move only for < 2 kpc there is not much room for improving upon the G-dwarf diagram predicted by dynamical evolution models.

M. TOSI: This is an answer to the previous question on infall rate and metallicity. From the point of view of chemical evolution models we need infall rates an average of 1 $M_{\odot}$ per year. From the observational point of view, the high velocity clouds for which the rate has been estimated provide $\sim 0.2 M_{\odot}$ per year, but by extrapolating to the other clouds a rate of about 1 $M_{\odot}$ per year can be reached. The metallicity required by the models to allow for the diluting infall effect is lower than $0.2 Z_{\odot}$, and indeed the high-velocity clouds, where metallicity has been measured have this value.

Tabetha Boyajian

Tim Brown

The Ages of Stars
Proceedings IAU Symposium No. 258, 2008
E.E. Mamajek, D.R. Soderblom & R.F.G. Wyse, eds.

doi:10.1017/S1743921309031706

The star formation history of the Magellanic Clouds

Carme Gallart[1], Ingrid Meschin[1], Noelia E. D. Noël[1,3], Antonio Aparicio[1], Sebastián L. Hidalgo[1] & Peter B. Stetson[2]

[1]Instituto de Astrofísica de Canarias and Departamento de Astrofísica, Universidad de La Laguna. 38200 La Laguna. Tenerife, Spain.
email:carme,imeschin,noelia,antapaj,shidalgo@iac.es

[2]Herzberg Institute of Astrophysics, National Research Council, Victoria, BC, Canada V9E 2E7 email: Peter.Stetson@nrc-cnrc.gc.ca

[3]Current address: Institute for Astronomy, Royal Observatory, University of Edinburgh

Abstract. The star formation history of the Magellanic Clouds, including the old and intermediate-age star formation events, can be studied reliably and in detail through color-magnitude diagrams reaching the oldest main sequence turnoffs. This paper reviews our current understanding of the Magellanic Clouds' star formation histories and discusses the impact of this information on general studies of galaxy formation and evolution.

Keywords. galaxies: formation and evolution; galaxies: individual (LMC, SMC); Magellanic Clouds

---

1. Introduction

The Magellanic Clouds (MCs) are close enough for the oldest main sequence turnoffs (MSTO) in the color-magnitude diagram (CMD)s of their resolved stars to be easily reachable from the ground, or using the HST in the case of the most crowded fields. This allows us to obtain accurate and detailed determinations of their spatially resolved star formation histories (SFH) which, in turn, will provide important information about some important aspects of galaxy formation and evolution. For example:

(i) The MCs are intermediate in size between the large spirals such as M31 or the Milky Way, and dwarf galaxies. In addition, they may have originated independently of the Milky Way system (D'Onghia & Lake 2008; Besla *et al.* 2007). Therefore, they can offer insight into the important question: does the onset of star formation in a galaxy depend on galaxy mass, type or environment?

(ii) The MCs appear to be an interacting system, both with each other and with the Milky Way. The combination of their detailed SFHs with knowledge of their past orbits can provide key information on how interactions affect the SFH of a galaxy.

(iii) Stellar clusters are often used as a proxy for the SFH of a galaxy. The MCs, where very accurate cluster ages and the field SFH can be obtained, are ideal places to test this common assumption.

(iv) Stellar population gradients are often inferred from the integrated light of distant galaxies, but their nature is difficult to ascertain (e.g. Taylor *et al.* 2005). The stellar population gradients in the MCs (e.g. Gallart *et al.* 2008) can be understood once the SFH is determined in representative portions of each galaxy, and this can provide insight into the nature of population gradients in galaxies in general.

The determination of accurate SFHs in representative portions of both MCs is necessary to properly approach the above issues. However, the amount of information published so far is actually quite limited, particularly as compared with that available about other aspects of the MCs. The huge size in the sky of both galaxies and the relatively (in comparison) small field of view of current optical detectors have certainly played a role in this situation. Most published MC CMDs reaching the oldest MSTO have been obtained using the WFPC2 or the ACS on board the HST, which implies a tiny field of view. In very crowded areas, the use of the HST is mandatory, but for less crowded regions excellent CMDs, comparable to the HST ones, can be obtained from the ground using 2–4m class telescopes (e.g. Gallart *et al.* 2008; Noël *et al.* 2007; see Figure 1). This produces well populated CMDs which allow us to overcome small number statistics problems at all ages. In this paper, we will review the existing literature on the SFH of the MCs, and present some ongoing work by our group on the same matter. We will preferentially discuss SFH determinations based on CMDs reaching the oldest MSTOs.

2. The LMC

2.1. *The cluster formation history*

Deep CMDs obtained with the WFPC2 confirmed the true old nature of a sample of candidate old clusters in the LMC (Olsen *et al.* 1998; Johnson *et al.* 1999; see also: Brocato *et al.* 1996 for earlier, high quality ground based CMDs of some of these clusters; Mackey & Gilmore 2004 for ACS data of three more additional old clusters). These works concluded that clusters as old as the old Milky Way globulars exist in the LMC, and that the age spread among them is actually very small. In contrast with the Milky Way, a type of populous young and intermediate-age clusters exists in the LMC in large numbers (Da Costa 1991), but careful searches have found only one cluster (ESO 121-SC03; Mackey *et al.* 2006) in the so-called LMC "age gap" (Geisler *et al.* 1997), which extends from $\simeq$ 4 to 13 Gyr ago.

2.2. *The field star formation history*

The published LMC field SFHs obtained from CMDs reaching the oldest MSTO are based on WFPC2 observations. In the case of bar fields, the CMDs are relatively well populated (e.g. Holtzman *et al.* 1999; Olsen 1999; Smecker-Hane *et al.* 2002), and yield relatively consistent SFHs, with star formation rate depressed from $\simeq$ 10 to 6 Gyr ago and somewhat increased or bursty in the last couple Gyr. Due to the small field of view, the WFPC2 disk CMDs are very sparsely populated (Holtzman *et al.* 1999; Javiel *et al.* 2005; Smecker-Hane *et al.* 2002 tried to overcome this problem by mosaicking 10 WFPC2 fields in a disk field located $\simeq 1.7^\circ$ southwest of the LMC center). While the corresponding SFHs present field-to-field variations, particularly at the young side, the general trend is, in contrast to the cluster age distribution, a relatively flat star formation rate as a function of time, with mild enhancements in the last few Gyr, but star formation not always continuing to the present time. While the scatter in the field SFH results could be related to small number statistics in the CMDs, the variations at young ages may be related to an actual gradient in the stellar populations present across the LMC disk, as discussed in Gallart *et al.* (2008).

We have undertaken a major project aimed at deriving the SFH of the LMC and its gradients using CMDs obtained with ground-based wide field imagers such as MOSAIC II at the CTIO 4m and WFI at the ESO 2.2 m. These CMDs are well populated and reach the oldest MSTO with good photometric precision. In this paper, we will discuss results for four fields located from 2.3° to 7.1° from the LMC center.

Figure 1 shows the $[(V-I)_0, M_I]$ CMDs of the four LMC fields. The number of stars observed with good quality photometry down to $M_I \lesssim 4$ in each field—in order

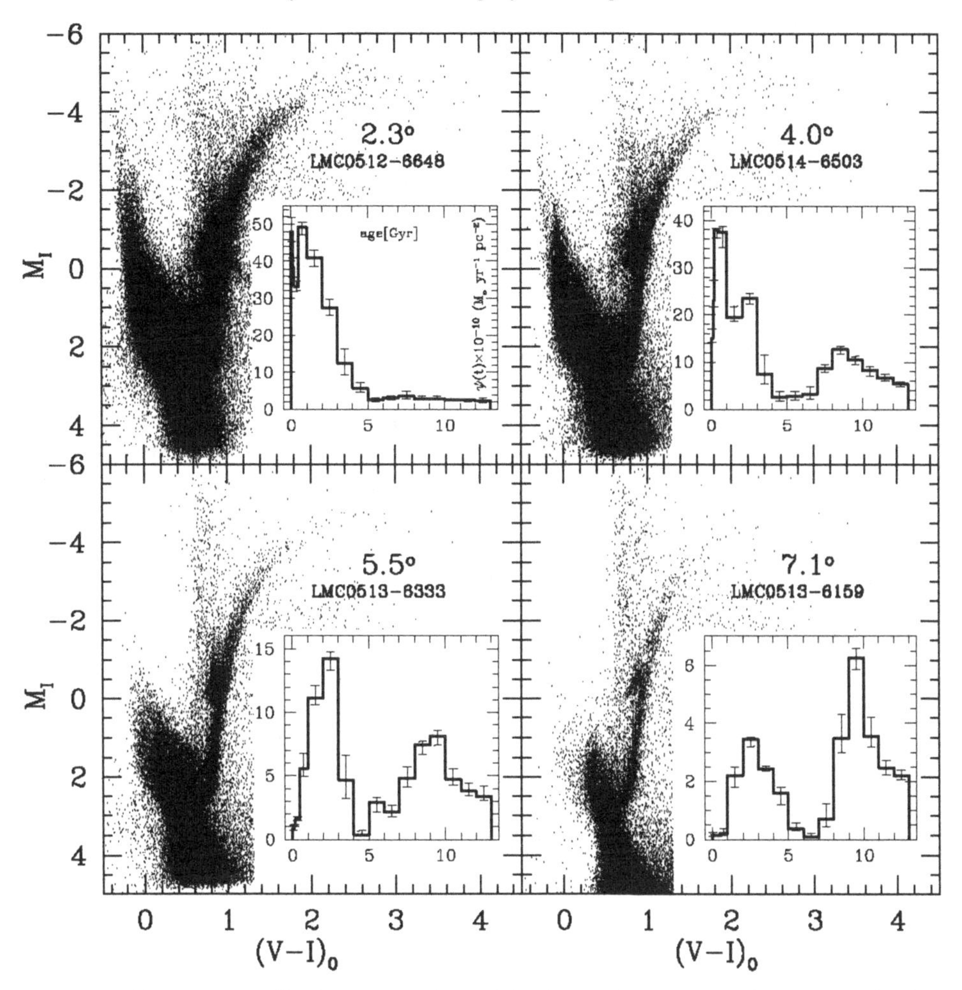

Figure 1. $[(V-I)_0, M_I]$ CMDs for the four LMC fields discussed in this paper. A distance modulus $(m-M)_0 = 18.5$ and E(B-V)=0.10, 0.05, 0.037 and 0.026 magnitudes, for the innermost to the outermost field respectively, have been assumed. The insets show the SFH projected on the $(\psi(t), t)$ plane (the age in Gyr refers to stellar ages or look-back time). We have assumed the Kroupa (2002) IMF, a 40% or binary stars with mass ratios $q \geqslant 0.5$. The BaSTI stellar evolution library (Pietrinferni *et al.* 2004) has been used as input to IAC-STAR to compute the synthetic CMDs.

of increasing galactocentric distances—are 300 000, 214 000, 86 000 and 39 000 respectively. All the CMDs reach the oldest MSTO ($M_I \simeq 3.0$) with good photometric precision and completeness fractions over 75% (except for the innermost field, in which crowding is severe). The two innermost fields show CMDs with a prominent, bright main sequence and a well populated red clump, typical of a population which has had ongoing star formation from $\simeq$13 Gyr ago to the present time. The two outermost fields clearly show a fainter main sequence termination, indicating a SFH recently truncated or sharply decreasing. No extended horizontal branch is observed in any of the fields, but all fields host a number of stars redder than the RGB tip, which are candidate AGB stars.

To quantitatively derive the SFH, we have used the IAC-STAR, IAC-POP and MINNIAC codes (Aparicio & Gallart 2004; Aparicio & Hidalgo 2008; Hidalgo *et al.* 2008, in preparation) to compute synthetic CMDs and compare the distribution of stars in the observed and synthetic CMDs, obtaining the best solution through χ^2 minimization (see the references above and Meschin *et al.* (2008) for details on the synthetic CMD technique and our particular implementation of it). Preliminary solutions for each field are shown in Figure 1. We are in the process of testing their robustness and dependence on variations of some of the input parameters, such as the binary fraction, the IMF, or the distance modulus and reddening adopted. We believe, however, that the present solutions provide a reasonable description of the main features characterizing the LMC SFH at different galactocentric radii.

Some coherent features, together with systematic variations of the SFH among fields, are noticeable. In all fields, local maxima of the star formation rate as a function of time, $\psi(t)$, are found around $\simeq$ 3-4 and 8-10 Gyr ago, with relatively low star formation activity in the approximate age interval $\simeq$ 4-7 Gyr ago. The ratio between the amount of star formation younger and older than the age of the minimum $\psi(t)$ decreases toward the outer part of the galaxy, thus the population is on average older there. Star formation in all fields seems to have started around 13 Gyr ago. The youngest age in each field showing a substantial amount of star formation gradually increases with galactocentric radius, from 2.3° to 7.1°. In particular, in the fields at 2.3°and 4.0°, star formation goes on to ages younger than 1 Gyr; however, only in the innermost field the star formation continues vigorously to the present time.

The age gradient in the youngest LMC population is correlated with the HI column density as measured by Staveley-Smith *et al.* (2003): the two innermost fields are located at $\simeq$ 0.7 Kpc on either side of $R_{H\alpha}$, LMC0512-6648 on the local maximum of the azimuthally averaged HI column density with $\simeq 1.63\times10^{21}\mathrm{cm}^{-2}$ (close to the HI threshold for star formation; Skillman 1987) and LMC0514-6503 where the azimuthally averaged HI column density is only $\simeq 5\times10^{20}\mathrm{cm}^{-2}$. Finally, the two outermost fields are close to the HI radius considered by Staveley-Smith *et al.* to be at an HI density of $10^{20}\mathrm{cm}^{-2}$. The outermost field, LMC0513-6159, is approximately halfway to the tidal radius (van der Marel *et al.* 2002). If the youngest stars in each field were formed *in situ*, we are observing an outside-in quenching of the star formation at recent times ($\simeq$ 1.5 Gyr), possibly implying a decrease in size of the HI disk able to form stars. Alternatively, star formation may have been confined to the central $\simeq$ 3–4 Kpc, where gas resides (or is accreted to), and stars then migrate outwards (e.g. Roškar *et al.* 2008). In fact, it is expected that both star-formation sites and stars migrate across the LMC disk due to tidal interactions with the Milky Way and the SMC (e.g. Bekki & Chiba 2005). Of course, a combination of the two scenarios is also possible.

3. The SMC

3.1. *The cluster formation history*

No clusters as old as the Milky Way globulars are known in the SMC. Its oldest cluster, NGC121, has been shown to be $\simeq$ 2-3 Gyr younger than the oldest galactic globular clusters (Glatt *et al.* 2008a). Unlike the LMC, the SMC contains a fair amount of intermediate-age populous clusters (Glatt *et al.* 2008b; Piatti *et al.* 2008) and young clusters (Pietrzynski & Udalski 1999; Chiosi *et al.* 2006).

3.2. *The field star formation history*

Three studies have presented SFHs derived from CMDs reaching the oldest MSTOs: McCumber *et al.* (2005) and Chiosi & Vallenari (2007) studied small HST fields located toward the east near the SMC center and find a SFH with a conspicuous increase in the rate of star formation at ages younger than $\simeq$ 1 Gyr ago, together with star formation ongoing, with varying intensities, for the rest of the galaxy's lifetime. Dolphin *et al.* (2001) derived the SFH for a field located $\simeq 2°$ northwest from the SMC center and found a broadly peaked SFH, with the largest star formation rate occurring between 5 and 8 Gyr ago, and some small amount of star formation going on since a very early epoch and down to $\simeq$ 2 Gyr ago.

Harris & Zaritsky (2004) derived a global SFH for the SMC based on the Magellanic Clouds Photometric Survey (Zaritsky *et al.* 1997) UBVI catalog that includes over 6 million SMC stars. They concluded that there was a significant epoch of star formation up to 8.4 Gyr ago when $\simeq$ 50% of the SMC stars were formed, followed by a long quiescent period in the range 3–8.4 Gyr ago, and a approximately continuous period of star formation (somewhat peaked at 2-3 Gyr, 400 Myr and 60 Myr ago) starting 3 Gyr ago and extending to the present time. However, this study, which is often taken as a reference on the SMC SFH, is based on CMDs that are not deep enough to derive the full SFH from the information on the main sequence (the CMDs reach B$\simeq$ 22 mag., which corresponds to main sequence stars younger that $\simeq$ 3 Gyr old). Therefore, while this dataset and many studies derived from it are invaluable for the study of the central region of the SMC, we believe that deeper data are necessary to derive a detailed and reliable old and intermediate-age SFH.

We have also undertaken a project (see Noël *et al.* 2007) to derive the SMC SFH, using observations obtained with the LCO 2.5m telescope. These yield CMDs reaching the oldest MSTO, of similar quality to the LMC ones shown in Figure 1, in twelve $8.8' \times 8.8'$ fields. They range from 1.3° to 4.0° from the center of the SMC and have different position angles: three fields are located to the west of the SMC center, three to the east and six to the south. Noël *et al.* (2008, in preparation; Noël 2008) have derived the SFHs for these fields, which are reproduced in Figure 2.

In each panel showing the SFH of a given field, three solutions obtained with three different age binnings have been represented, together with a spline fit to them, which we will adopt as the final SFH (see the Figure caption for additional details on underlying parameters such stellar evolution models used, IMF and binary stars adopted in these solutions). It can be seen that common patterns, which vary smoothly with position, appear in most fields: there are four episodes of enhanced star formation rate: i) one at young ages, only present in the eastern fields and in the most central one located to the south, peaked at 0.2–0.5 Gyr ago; ii) two at intermediate ages: a conspicuous one peaked at 4–5 Gyr ago in all fields, and a less significant one peaked at 1.5–2.5 Gyr ago. Finally, iii) an enhancement at old ages, between $\simeq$ 8 and 12 Gyr old (the exact time varying among fields).

This spatially resolved SFH shows, for the first time, that the underlying spheroidally distributed population in the SMC (Zaritsky *et al.* 2000; Cioni, Habing & Israel 2000) is actually quite homogeneously composed of stars of a wide range of ages, from $\simeq$ 2-13 Gyr ago. In fact, the intermediate age star formation rate is relatively high out to large galactocentric distances (at least 4.5 kpc from the center, in field smc0053), in agreement with Noël & Gallart (2007). They also show that the young star formation event that is responsible for the irregular appearance of the SMC and the formation of the Magellanic Bridge (Harris 2007), does not represent, at least at galactocentric radius over $\simeq 1.5°$, an exceptional increase over the mean star formation rate.

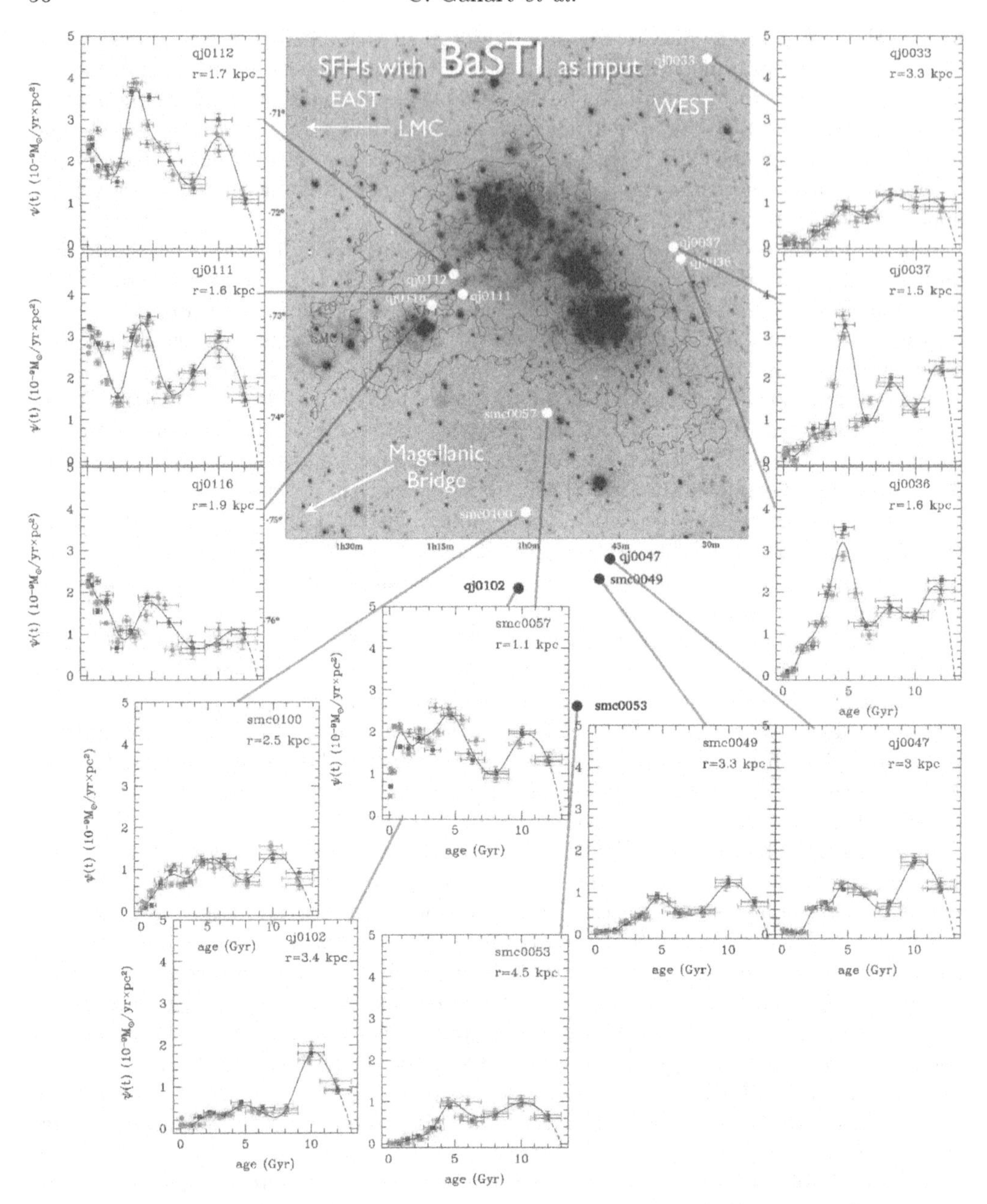

Figure 2. The derived SFHs of our SMC fields, projected on the $(\psi(t), t)$ plane (the age in Gyr refers to stellar ages or look-back time). The BaSTI stellar evolution library (Pietrinferni *et al.* 2004), with the bolometric corrections from Castelli & Kurucz (2003), was used as input of IAC-star. The Kroupa (2002) IMF was used. We assumed that 30% of the stars in the galaxy are binaries with q$\geqslant$0.7. A distance modulus $(m - M)_0 = 18.9$ and a reddening according to Schlegel *et al.* (1998; except for the innermost fields, where the reddening was estimated from the CMD itself, see Noël et a. 2007 for details) were adopted. Each solution shows the SFH obtained for three age binnings. Each point carries a vertical error bar that is the formal error from IAC-pop, calculated as the dispersion of 20 solutions with $\chi^2_\nu = \chi^2_{\nu,min} + 1$. Horizontal tracks are not error bars, but show the age interval associated to each point. The solid line shows the results of a cubic spline fit to the results. We do not have a constraint on the $\psi(t)$ at 13 Gyr old and the end point of our spline fit was chosen to be zero arbitrarily (dashed lines between 12 and 13 Gyr ago in the spline fit).

4. Discussion

As discussed in the introduction, the knowledge of detailed SFHs in representative portions of both the LMC and the SMC will provide important insight into general questions related to the galaxy formation and evolution process. We will discuss here some of the evidence provided by the MC SFHs reviewed in this paper on the questions posed in the introduction.

Both the LMC and the SMC seem to have begun forming stars as early as the Milky Way and basically all its satellites. This statement is particularly secure in the case of the LMC, which hosts a number of old clusters with relative ages measured with respect to *bona fide* old Milky Way globulars. Also the field star formation in both galaxies seems to have started very early on, even though a precise dating (to within $\simeq$ 1-2 Gyr) is difficult. If the recent claims about a late accretion of the MCs into the Milky Way system are confirmed (e.g. Besla *et al.* 2007; Costa *et al.* 2008, in preparation), the MCs would be added to the list of *external* galaxies with a very old stellar population, thus adding further evidence for an independence on the time of onset of star formation with respect to host galaxy type, mass or environment.

It is interesting, however, that the main SFH features differ substantially between the LMC and the SMC: the most ubiquitous SMC episode of enhanced star formation occurs at the time of lowest activity in the LMC ($\simeq$ 4-7 Gyr ago). It will be important to relate these SFHs with more secure orbits of both galaxies, which will provide information about to what extent, and in which conditions, a galaxy's SFH is affected by galaxy interactions.

In relation to the question of whether the cluster formation history can be considered representative of the field SFH, we note that the features in the LMC SFH are somewhat reminiscent of the age distribution of star clusters in this galaxy, particularly regarding the existence of the age gap which could be related to the epoch of decreased field star formation at intermediate age. The old field star formation event, however, seems to span a quite wide age range, in contrast with the narrow age range of the old LMC clusters. The SMC, on the other hand, has a continuous cluster age distribution which is consistent with the fact that no important epochs of quiescence are observed in its SFH.

Finally, stellar population gradients are observed both in the LMC and in the SMC, in the sense that the strength of the intermediate-age and young star formation in relation to that of the old population decreases toward the outer part of both galaxies, i.e. the average age of the stellar population becomes older as galactocentric radius increases. Even though the reasons for this behavior (e.g. changes in the star formation activity as a function of radius and time, or migration of the stars from the central part of the galaxy where star formation preferentially takes place) will still need a fair amount of research, this information provides some insight into the nature of stellar population gradients observed in more distant late type galaxies.

Acknowledgements

C.G., A.A., I.M., N.E.D.N. & S.L.H. acknowledge the support from the IAC and the Spanish MEC (AYA2004-06343 and AYA2007-67913).

References

Aparicio, A. & Gallart, C. 2004, *AJ*, 128, 1465
Aparicio, A. & Hidalgo, S. L. 2008, *AJ*, in press
Bekki, K. & Chiba, M. 2005, *MNRAS*, 356, 680

Besla, G., Kallivayalil, N., Hernquist, L., Robertson, B., Cox, T. J., van der Marel, R. P., & Alcock, C. 2007, *ApJ*, 668, 969

Brocato, E., Castellani, V., Ferraro, F. R., Piersimoni, A. M., & Testa, V. 1996, *MNRAS*, 282, 614

Castelli, F. & Kurucz, R. L. 2003, in: N. Piskunov, W. W. Weiss & D. F. Gray (eds), *Modelling of Stellar Atmospheres*, Proc. IAU Symposium No. 210 (ASP), p. A20

Chiosi, E. & Vallenari, A. 2007, *A&A*, 466, 165

Chiosi, E., Vallenari, A., Held, E. V., Rizzi, L., & Moretti, A. 2006, *A&A*, 452, 179

Cioni, M.-R. L., Habing, H. J., & Israel, F. P. 2000, *A&A*, 358, L9

Da Costa, G. 1991, in: R. Haynes & D. Milne (eds), *The Magellanic Clouds*, Proc. IAU Symposium No. 148 (Kluwer Academic Publishers, Dordrecht), p. 183

Dolphin, A. E., Walker, A. R., Hodge, P. W., Mateo, M., Olszewski, E. W., Schommer, R. A., & Suntzeff, N. B. 2001, *ApJ*, 562, 303

D'Onghia, E. & Lake, G. 2008, *ApJ* (Letters), 686, 61

Gallart, C., Stetson, P. B., Meschin, I. P., Pont, F., & Hardy, E. 2008, *ApJ* (Letters), 628, L89

Geisler, D., Bica, E., Dottori, H., Claria, J. J., Piatti, A. E., & Santos, J. F. C., Jr. 1997, *AJ*, 114, 1920

Glatt *et al.* 2008b, *AJ*, 136, 1703

Glatt *et al.* 2008a, *AJ*, 135, 1106

Harris, J. 2007, *ApJ*, 658, 345

Harris, J. & Zaritsky, D. 2004, *AJ*, 127, 1531

Holtzman, J. A. *et al.* 1999, *AJ*, 118, 2262

Javiel, S. C., Santiago, B. X., & Kerber, L. O. 2005, *A&A*, 431, 73

Johnson, J. A., Bolte, M., Stetson, P. B., Hesser, J. E., & Somerville, R. S. 1999, *ApJ*, 527, 199

Kroupa, P. 2002, *Science*, 295, 82

Mackey, A. D. & Gilmore, G. F. 2004, *MNRAS*, 352, 153

Mackey, A. D., Payne, M. J., & Gilmore, G. F. 2006, *MNRAS*, 369, 921

McCumber, M. P., Garnett, D, R., & Dufour, R. J. 2005, *AJ*, 130, 1083

Meschin, I., Gallart, C., Aparicio, A., Carrera, R., Monelli, M., Hidalgo, S., & Stetson, P. B. 2008, in: J. Th. van Loon & J. M. Oliveira (eds), *The Magellanic System: Stars, Gas and Galaxies*, Proc. IAU Symposium No. 256, in press

Noël, N. E. D. 2008, PASP, 120, in press

Noël, N. E. D. & Gallart, C. 2007, *ApJ* (Letters), 665, L23

Noël, N., Gallart, C., Costa, E., & Méndez, R. 2007, *AJ*, 133, 2037

Olsen, K. A. G. 1999, *AJ*, 117, 2244

Olsen, K. A. G., Hodge, P. W., Mateo, M., Olszewski, E. W., Schommer, R. A., Suntzeff, N. B., & Walker, A. R. 1998, *MNRAS*, 300, 665

Piatti, A. E., Geisler, D., Sarajedini, A., Gallart, C., & Wischnjewsky, M. 2008, *MNRAS*, 389, 429

Pietrinferni, A., Cassisi, S., Salaris, M., & Castelli, F. 2004, *ApJ*, 612, 168

Pietrzynski, G. & Udalski, A. 1999, *AcA*, 49, 157

Roškar, R., Debattista, V. P., Stinson, G. S., Quinn, T. R., Kaufmann, T., & Wadsley, J. 2008, *ApJ*, 675, L65

Schlegel, D. J., Finkbeiner, D. P., & Davis, M. 1998, *ApJ*, 500, 525

Skillman, E. 1987, in C. J. Lonsdale Persson (ed), *Star Formation in Galaxies* (NASA CP-2466; Washington: NASA), p. 263

Smecker-Hane, T. A., Cole, A. A., Gallagher, J. S., III, & Stetson, P. B. 2002, *ApJ*, 566, 239

Staveley-Smith, L., Kim, S., Calabretta, M. R., Haynes, R. F., & Kesteven, M. J. 2003, *MNRAS*, 339, 87

Taylor, V. A., Jansen, R. A., Windhorst, R. A., Odewahn, S. C., & Hibbard, J. E., 2005, *ApJ*, 630, 784

van der Marel, R. P., Alves, D. R., Hardy, E., & Suntzeff, N. B. 2002, *AJ*, 124, 2639

Zaritsky, D., Harris, J., Grebel, E. K., & Thompson, I. B. 2000, *ApJ* (Letters), 534, L53

Zaritsky, D., Harris, J., & Thompson, I. 1997, *AJ*, 114, 1002

Discussion

A. SARAJEDINI: Do the star formation bursts seem in the field stars correlate with those of the clusters? Do they correlate with close encounters of the LMC/SMC/MW system?

C. GALLART: The relation between cluster formation epochs and enhancements of star formation in field stars is not tight, but I think a loose correlation is observed in both the LMC and the SMC: the two main events of field star formation in the LMC are separated by a decrease in the star formation activity around $\sim$ 7 Gyr ago, and this might be related to the LMC cluster age gap. In contrast, in the SMC the SFH appears more continuous in the field (the difference between the amount of star formation in the peaks and the valleys seems to be smaller), and so is the cluster formation history. In any case, I think that much more work is needed both in increasing the area for which the field SFH is determined and the number of cluster for which accurate ages are known. About the correlation with close encounters of the LMC/SMC/MW system, we have identified some correlation between SMC/MW pericenter passages as predicted by the new orbit proposed by Kallivayalil *et al.* (2006, ApJ, 638, 772). However, I think that more precise orbits are necessary, and we will certainly have these in the next few years, thus opening a very interesting prospect of studying in detail the relationship between interactions and star formation history.

G. DEMARCHI: Your data show a nice radial trend between age and distance from the LMC. Could it be that you see very few young stars moving away from the LMC because massive stars do not form there? Could it be that low-mass stars of younger ages are still there, hidden in the photometry?

C. GALLART: Under the reasonable, I think, assumption, of an IMF constant with time, the scenario you propose can be excluded.

V. HILL: I presume in the case of the SMC the line-of-sight effect is non-negligible. How does this affect the ages and star formation histories derived from the various fields?

C. GALLART: It would tend to smooth the SFH features, for example, widening the time-span of episodes of increased star formation. The depth effect can be modeled in the synthetic CMDs used to obtain the SFH to constrain the possible effect on the derived SFH.

Carme Gallart

The Ages of Stars
Proceedings IAU Symposium No. 258, 2008
E.E. Mamajek, D.R. Soderblom & R.F.G. Wyse, eds.

doi:10.1017/S1743921309031718

Star formation histories of resolved galaxies

Monica Tosi

INAF - Osservatorio Astronomico di Bologna
Via Ranzani 1, I-40127, Bologna, Italy
email: monica.tosi@oabo.inaf.it

Abstract. The colour-magnitude diagrams of resolved stellar populations are the best tool to study the star formation histories of the host galactic regions. In this review the method to derive star formation histories by means of synthetic colour-magnitude diagrams is briefly outlined, and the results of its application to resolved galaxies of various morphological types are summarized. It is shown that all the galaxies studied so far were already forming stars at the lookback time reached by the observational data, independently of morphological type and metallicity. Early-type galaxies have formed stars predominantly, but in several cases not exclusively, at the earliest epochs. All the other galaxies appear to have experienced rather continuous star formation activities throughout their lifetimes, although with significant rate variations and, sometimes, short quiescent phases.

Keywords. Hertzsprung-Russell diagram, galaxies: stellar content, galaxies: individual (M31, M33, LMC, SMC, Leo A), Local Group, Magellanic Clouds, galaxies: evolution

---

1. Introduction

Two complementary approaches are necessary to understand galaxy evolution: on the one hand, we need to develop theoretical models for galaxy formation, chemical and dynamical evolution, and on the other hand, we need to collect as many and as accurate as possible observational data to constrain such models. In particular, we need to know the masses, chemical abundances and kinematics of the various galactic components, namely gas, stars and dark matter; we need to know the star formation history (SFH), the initial mass function (IMF), etc. In this review our current knowledge of the SFHs, as derived from the colour-magnitude diagrams (CMDs) of their resolved stellar populations, is summarized.

Resolved stellar populations are the best tracers of the SFH of a galactic region, and their CMD the best tool to exploit the tracers. This is due to the well known circumstance that the location of any individual star in a CMD is uniquely related to its mass, age and chemical composition. From the CMD we can thus disentangle directly these evolution parameters. In the case of simple stellar populations, i.e. coeval stars with the same chemical composition, isochrone fitting is the most frequently used method to infer the system age. In the case of galaxies, with rather complicated mixtures of different stellar generations, the age determination is less straightforward, but their CMDs remain the best means to derive the SFH.

2. CMD synthesis and star formation histories

To visualize how the SFH affects the CMD morphology, a few representative cases are displayed in Fig. 1. The six panels of the Figure show the effect of different SFHs on the synthetic CMD of a hypothetical galactic region with number of resolved individual stars, photometric errors, blending and incompleteness factors typical of a region in the

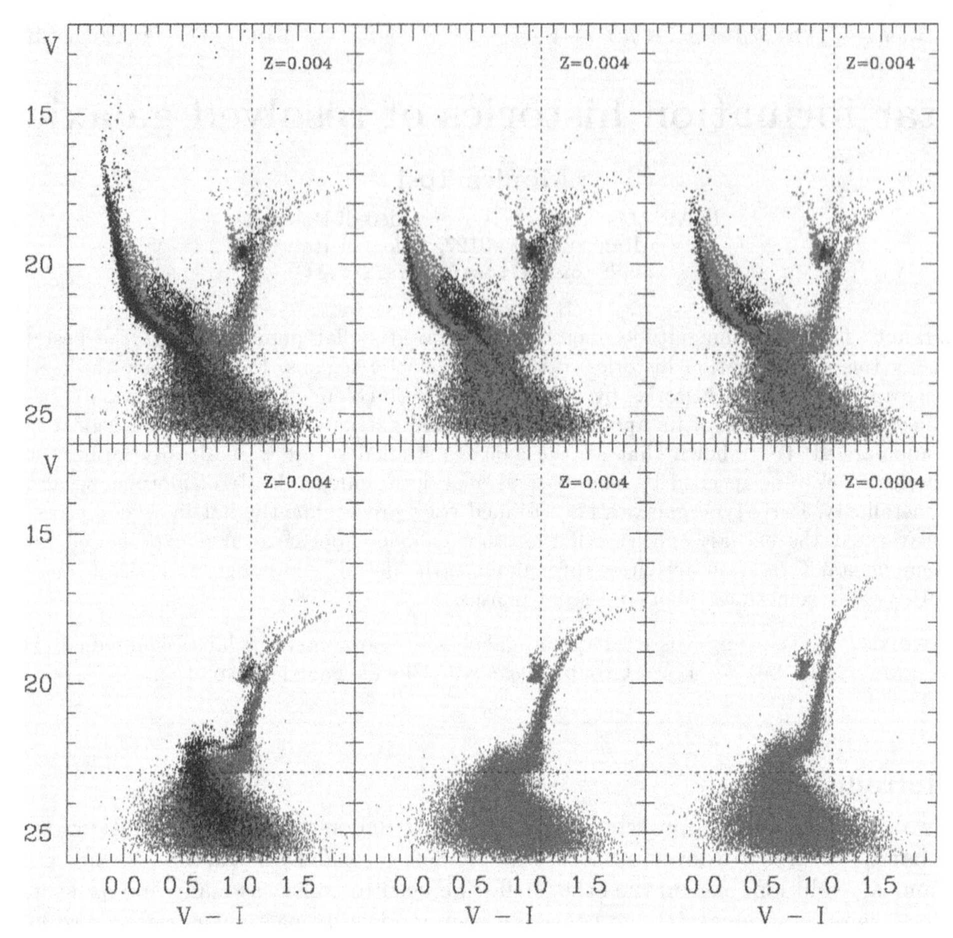

Figure 1. The effect of the SFH on the theoretical CMD of a hypothetical galactic region with $(m\text{-}M)_0$=19, E(B-V)=0.08, and with the photometric errors and incompleteness typical of HST/WFPC2 photometry. All the shown synthetic CMDs contain 50000 stars and are based on the Padova models (Fagotto *et al.* 1994a, Fagotto *et al.* 1994b) with the labeled metallicities. Top-central panel: the case of a SFR constant from 13 Gyr ago to the present epoch. Top-left panel: the effect of adding a burst 10 times stronger in the last 20 Myr to the constant SFR. The CMD has a much brighter and thicker blue plume. Top-right panel: same constant SFR as in the first case, but with a quiescence interval between 3 and 2 Gyrs ago; a gap appears in the CMD region corresponding to stars 2-3 Gyr old, which are completely missing. Bottom-central panel: SF activity only between 13 and 10 Gyr ago with Z=0.004. Bottom-right panel: SF activity only between 13 and 10 Gyr ago with Z=0.0004: notice how colour and luminosity of turnoff, subgiant and red giant branches differ from the previous case. Bottom-left panel: SF activity between 13 and 11 Gyr ago, followed by a second episode of activity between 5 and 4 Gyr ago: a gap separates the two populations in the CMD, but less evident than in the top-right panel case, when the quiescent interval was more recent.

SMC imaged with HST/WFPC2. If the SFH of the studied region has been one of the following six cases, then, according to stellar evolution models, the CMD of its stars is one of those shown in Fig. 1. The top three panels show examples of CMDs typical of late-type galaxies, with ongoing or recent star formation activity. If the star formation rate (SFR) has been constant for all the galaxy lifetime, the CMD of the region is expected to have the morphology of the top-central panel, with a prominent blue plume mostly populated by main-sequence (MS) stars and an equally prominent red plume resulting from the overposition of increasingly bright and massive stars in the red giant branch (RGB), asymptotic giant branch (AGB) and red supergiant phases. At intermediate colours, for decreasing brightness, stars in the blue loops and subgiant phases are visible, as well stars at the oldest MS turnoff (MSTO) and on the faint MS of low mass stars. Stars of all ages are present, from those as old as the Hubble time to the brightest ones a few tens Myr old.

If we leave the SFH unchanged except for the addition of a burst ten times stronger concentrated in the last 20 Myr, the CMD (top-left panel) has a much brighter and more populated blue plume, now containing also stars a few Myr old. In the top-right panel the same constant SFR as in the first case is assumed, but with a quiescent interval between 3 and 2 Gyrs ago: a gap is clearly visible in the CMD region corresponding to the age of the missing stars.

The three bottom panels of Fig. 1 show CMDs typical of early-type galaxies, whose SF activity is concentrated at the earliest epochs. If only one SF episode has occurred from 13 to 10 Gyr ago, with a constant metallicity Z=0.004 as in the top panel cases, the resulting CMD is shown in the bottom-central panel. If the SF has occurred at the same epoch, but with a metallicity ten times lower, the evolutionary phases in the resulting CMD (bottom-right panel) have colours and luminosities quite different from the previous case. Finally, the bottom-left panel shows the case of two bursts, the first from 13 to 11 Gyr ago and the second from 5 to 4 Gyr ago. The gap corresponding to the quiescent interval is evident in the CMD, although not as much as the more recent gap of the top-right panel.

The tight dependence of the CMD morphology on the SFH is the cornerstone of the synthetic CMD method, which consists in comparing the observational CMD of a galactic region with synthetic CMDs, such as those of Fig. 1, created via Monte Carlo extractions on stellar evolution tracks or isochrones for a variety of SFHs, IMFs, binary fractions and age-metallicity relations (see e.g. Tosi *et al.* 1991, Tolstoy 1996, Greggio *et al.* 1998, Aparicio & Gallart 2004 for detailed descriptions of different procedures). The synthetic CMDs take into account the number of stars, photometric errors, incompleteness and blending factors of the observational CMD (or portions of it). Hence, a combination of assumed parameters is acceptable only if the resulting synthetic CMD reproduces all the features of the observational one: morphology, colours, luminosity functions, number of stars in specific evolutionary phases. The method does not provide unique solutions, but significantly reduces the possible SFH scenarios.

At its first applications to photometric data from ground-based, moderate size telescopes the synthetic CMD method demonstrated its power, showing that even in tiny galaxies such as Local Group dwarf irregulars (dIrrs) the SFH varies from one region to the other and that their star formation regime is rather continuous, with long episodes of moderate activity, separated by short quiescent intervals (the so-called gasping regime, Ferraro *et al.* 1989, Tosi *et al.* 1991, Marconi *et al.* 1995, Gallart *et al.* 1996, Tolstoy 1996) and not the bursting regime (short episodes of strong SF activity separated by long quiescent phases) that most people attributed to late-type dwarfs at the time.

When the first non-aberrated images were acquired with HST, the impressive improvement in the achievable photometric resolution and depth, and the corresponding quantum leap in the quality of the CMDs, triggered a worldwide burst of interest in the derivation of the SFHs of nearby galaxies and in the synthetic CMD method. Many people developed their own procedures and to date a large fraction of Local Group galaxies have had the SFH of at least some of their regions derived with the synthetic CMD method. Nowadays, in LG galaxies it is possible to resolve individual stars down to faint/old objects in all galactic regions and we can thus infer the SFHs over long lookback times τ (up to the Hubble time), with an average time resolution around $(0.1\text{–}0.2)\tau$.

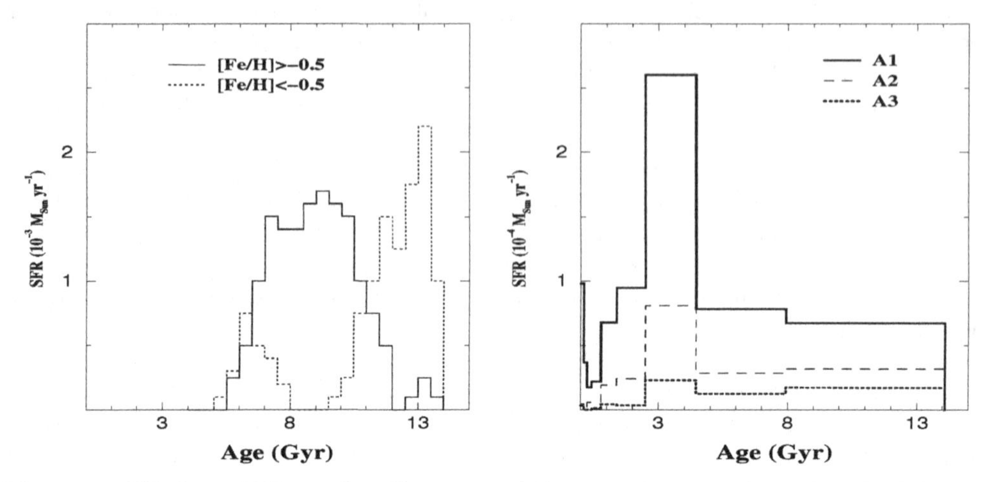

Figure 2. SFH from CMDs in (small) regions of the two LG external spirals. The left-hand panel shows the SFH of first HST/ACS field of M31 studied by Brown (2006), who divided the stars according to their metallicity. The right-hand panel shows the SFHs of the three HST/ACS fields of M33, A1, A2 and A3, studied by Barker *et al.* (2007).

3. Star formation histories of Local Group galaxies

The current census of LG galaxies with SFH derived in some of their regions with the synthetic CMD method is impressive. SFHs have been inferred from CMDs of both the two external spiral galaxies, M31 and M33, the two Magellanic Clouds, LMC and SMC, a dozen dIrrs, 5 transition type dwarfs and about 20 early-type dwarfs (dwarf spheroidals, dSphs, and dwarf ellipticals, dEs).

In M31, long HST/ACS exposures have allowed Brown *et al.*(2008) to resolve stars fainter than the oldest MSTO in three regions and derive their SFH back to the earliest epochs. They find a fairly continuous activity through the whole lifetime of Andromeda. The SFH in the first M31 field studied by Brown is shown in the left-hand panel of Fig. 2: if the SFH resulting from both metal poor and rich stars is considered, it turns out to have been rather constant. In M33, HST/ACS imaging has allowed Barker *et al.* (2007) to study three different regions, again resolving their oldest stars. The resulting SFH (right-hand panel of Fig. 2) clearly differs from one region to the other and shows significant bumps and gasps over a rather continuous mode. In all the three regions was the SF activity already in place a Hubble time ago. It is apparent that the SF activity in the M31 field has been both stronger and more constant than in M33.

The SFHs of several regions of the Magellanic Clouds have been studied by a number of authors, both from space and from ground (e.g. Holtzman *et al.* 1999, Dolphin *et al.* 2001, Smecker-Hane *et al.* 2002, Harris & Zaritsky 2004, Chiosi *et al.* 2006, Noel *et al.* 2007, Cignoni *et al.* 2009). Their proximity makes the oldest stars visible also from ground, with the advantage of fields of view larger than those of the HST cameras. Harris & Zaritsky (2004) even covered the whole SMC. On the other hand, the exquisite spatial resolution of HST is necessary to resolve and study the fainter stars in crowded regions, such as those of the star forming clusters. While stars at the oldest MSTOs and subgiant branches are the unique means to firmly establish the SFH at the earliest epochs, pre-MS

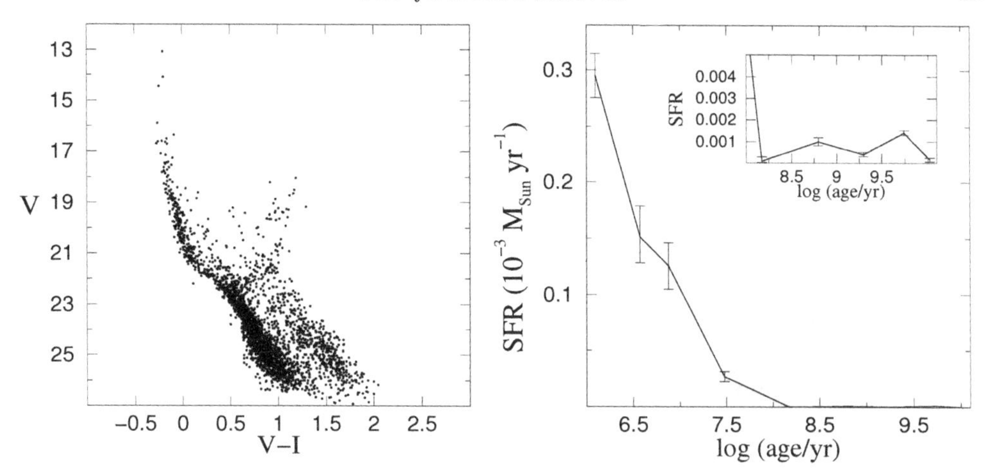

Figure 3. Left-hand panel: CMD of the HST/ACS field around the young cluster NGC 602 in the SMC. The red sequence of pre-MS stars is easily recognizable parallel to the lower MS. The bright blue plume contains the young cluster stars while the lower MS is only populated by field stars, since the cluster stars with mass below $\sim$1 $M_{\odot}$ haven't yet had time to reach it. Right-hand panel: corresponding SFH as derived with the synthetic CMD method (Cignoni *et al.* 2009). The oldest part of the SFH is zoomed-in in the upper right inset.

stars are precious tools to study the details of the most recent SFH in terms of time and space behaviour (Cignoni *et al.* 2009). The SMC regions of intense recent star formation can provide key information on the star formation mechanisms in environments with metallicity much lower than in any Galactic star forming region. Fig. 3 shows the CMD of the young cluster NGC 602 in the Wing of the SMC, observed with HST/ACS. Both very young stars (either on the upper MS or still on the pre-MS) and old stars are found. The SFH of the cluster and the surrounding field is also shown, revealing that the cluster has formed most of its stars around 2.5 Myr ago, while the surrounding field has formed stars continuously since the earliest epochs. All the studies on the MC fields have found that the SFHs of their different regions differ from one another in the details (e.g. epoch of activity peaks, enrichment history, etc.) but are always characterized by a gasping regime, i.e. a rather continuous activity since the earliest epochs, but with significant peaks and gasps. In the LMC a clear difference has been found between the SFH of field stars and of star clusters, the latter showing a quiescence phase, several Gyr long, absent in the field.

Dwarf irregulars were the first systems to which synthetic CMD analyses were applied. HST has had a large impact on studies of these systems. The high spatial resolution of its cameras have allowed Dohm-Palmer *et al.* (1998) and Dohm-Palmer *et al.* (2002) to spatially resolve and measure the SF activity over the last 0.5 Gyr in all the sub-regions of the dIrrs Gr8 and Sextans A, close to the borders of the LG. The resulting space and time distribution of the SF, with lightening and fading of adjacent cells, is intriguingly reminiscent of the predictions of the stochastic self-propagating SF theory proposed by Seiden, Schulman, & Gerola(1979) 30 years ago. The HST/ACS is currently providing the deepest and tighter CMDs of dIrrs ever obtained, likely to remain unequaled for a very long time. These spectacular CMDs reach well below the oldest MSTO and allow the derivation of the SFH back to a Hubble time ago. The first of such impressive studies is that of Leo A (Cole et al. 2007), whose CMD and SFH are plotted in Fig. 4. In Leo A the star formation activity was present, although quite low, at the earliest epochs, and

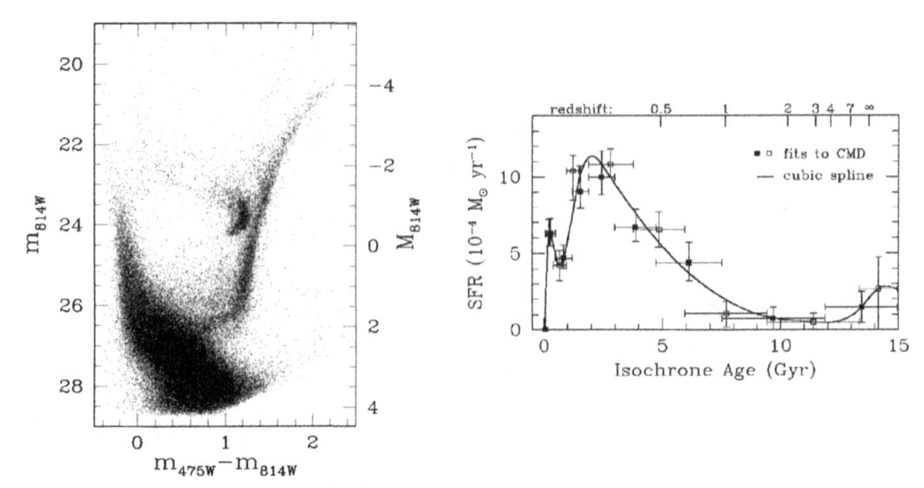

Figure 4. CMD and SFH of Leo A as derived by Cole *et al.* (2007) from HST/ACS data. Notice the impressive depth and tightness of the CMD, allowing to infer the SFH even at the earliest epochs.

90% of the activity occurred in the last 8 Gyr, with the main peak around 2 Gyr ago and a secondary peak a few hundreds Myr ago. Once again, the results obtained so far show that dIrrs experience a rather continuous star formation since the earliest epochs, but with significant peaks and gasps.

To find SFHs peaked at earlier epochs one needs to look at early-type dwarfs: dEs, dSphs and even transition-type dwarfs clearly underwent their major activity around or beyond 10 Gyr ago. The latter also have significant activity at recent epochs (e.g. Young *et al.* 2007). The former have few (or no) episodes of moderate activity in the last several Gyrs (e.g. Smecker-Hane *et al.* 1996, Hurley-Keller *et al.* 1998, Hernandez, Gilmore & Valls-Gabaud 2000, Dolphin 2002, Dolphin *et al.* 2005).

The beautiful CMDs from Carme Gallart's L-CID HST program on 6 dwarfs of different type (two dIrrs, two dSphs and two transition type, see Hidalgo *et al.* this volume) promise to provide SFHs of unprecedented time resolution for external galaxies. Another interesting project is trying to treat homogeneously all the LG galaxies observed with the HST/WFPC2, deriving the CMDs of their resolved populations in a self-consistent way (Holtzman, Afonso & Dolphin 2006) and the corresponding SFH with the same technique and assumptions (Dolphin *et al.* in preparation, see also Dolphin *et al.* 2005). Homogeneous data sets and analyses are valuable to obtain a uniform overview of the properties of the different galaxies in the LG.

4. Star formation histories of galaxies outside the Local Group

In galaxies beyond the LG, distance makes crowding more severe, and even HST cannot resolve stars as faint as the MSTO of old populations. The higher the distance, the worse the crowding conditions, and the shorter the lookback time τ reachable even with the deepest, highest resolution photometry. Depending on distance and intrinsic crowding, the reachable τ in galaxies more than 1 Mpc away ranges from several Gyrs (in the best cases, when the RGB or even the HB are clearly identified), to several hundreds Myr (when AGB stars are recognized), to a few tens Myr (when only the brightest supergiants are resolved).

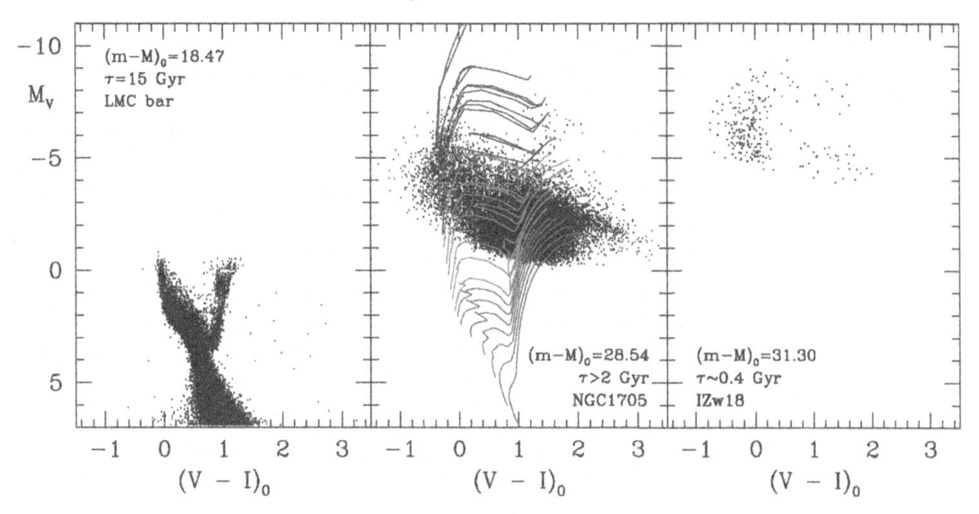

Figure 5. Effect of distance on the resolution of individual stars and on the corresponding lookback time τ for the SFH. CMD in absolute magnitude and colour of systems observed with the HST/WFPC2 and analysed with the same techniques, but at different distances; from left to right: 50 Kpc (LMC bar), 5.1 Mpc (NGC1705) and 18 Mpc (IZw18). The central panel also shows stellar evolution tracks from Fagotto *et al.* (1994b) for reference: red lines refer to low-mass stars, green lines to intermediate mass stars, and blue lines to massive stars.

The effect of distance on the possibility of resolving individual stars, and therefore on the reachable τ, is shown in Fig. 5, where the CMDs obtained from WFPC2 photometry of three late-type galaxies are shown: the LMC bar (Smecker-Hane *et al.* 2002), with a distance modulus of 18.47 (50 kpc) and a CMD reaching several mags below the old MSTO; NGC1705 (Tosi *et al.* 2001), with distance modulus 28.54 (5.1 Mpc) and a CMD reaching a few mags below the tip of the RGB; and IZw18 (Aloisi, Tosi & Greggio 1999), with the new distance modulus 31.3 (18 Mpc) derived by Aloisi *et al.* (2007). Notice that the latter modulus is inferred from the periods and luminosities of a few classical Cepheids measured from HST/ACS data which also allowed us to reach the RGB, but the WFPC2 data shown in Fig. 5 allow to reach only the AGB. The CMD obtained from the ACS is shown in Fig. 6.

Since the Local Group doesn't host all types of galaxies, with the notable and unfortunate absence of both the most and the least evolved ones (ellipticals and Blue Compact Dwarfs, BCDs, respectively), a few people have tackled the challenging task of deriving the SFH of more distant galaxies. In spite of the larger uncertainties and the shorter lookback time, these studies have led to quite interesting results, which wouldn't have been possible without HST.

First of all, all the galaxies, including BCDs, where individual stars have been resolved by HST, and the SFH has been derived with the synthetic CMD method, have turned out to be already active at the lookback time reached by the photometry (see e.g. Lynds *et al.* 1998, Aloisi, *et al.* 1999, Schulte-Ladbeck *et al.* 2000, Schulte-Ladbeck *et al.* 2001, Annibali *et al.* 2003, Rejkuba, Greggio & Zoccali 2004, Vallenari Schmidtobreik & Bomans 2005). None of them appears to be experiencing now its first star formation activity, including the most metal poor ones, such as SBS1415 and IZw18 (see Aloisi *et al.* 2005 and Aloisi et al. 2007). Fig. 7 sketches the SFHs derived by various authors for some of the starburst dwarfs studied so far and one low surface brightness dwarf, UGC 5889. The lookback time is indicated and in all cases stars with that age were detected.

All the late-type dwarfs of Fig. 7 present a recent SF burst, which is what let people discover them in spite of the distance, and none of them exhibits long quiescent phases within the reached τ. It is interesting to notice that the SFH of the low surface brightness dwarf UGC 5889 (Vallenari *et al.* 2005) is also qualitatively similar to that of starburst dwarfs, except that the SFR is definitely moderate. In all the shown galaxies the strongest SF episodes are overimposed over a rather continuous, moderate SF, already in place at the τ reached by the photometry. Indeed, no one has ever found yet a galaxy without stars as old as τ from the CMDs of the resolved populations.

The SF rate differs significantly from one galaxy to the other. The two most powerful bursts measured so far are the recent ones in NGC 1705 and NGC 1569, with SFR per unit area a factor 10-100 higher than in the other starbursting dwarfs studied through their CMDs. Intriguingly, the strongest of all is not a BCD, but the dwarf irregular NGC 1569, suggesting that the morphological classification of these faint small galaxies was possibly affected by their distance and the capability of resolving their shape with ground-based small telescopes, at the time of their discovery. Had it been at 20 Mpc, NGC 1569 would have probably been classified as a BCD.

5. Discussion

From the comparison of the SFHs of starburst dwarfs with those of Local Group dwarf irregulars, one can see that in both cases the SF regime is rather continuous (gasping), with two main differences: starburst dwarfs always have the strongest SF episode at recent epochs, while the current SF activity of local irregulars is not necessarily the highest peak. Leo A, with the main SF peak a few Gyr ago (Fig. 4) is quite typical. On the other hand, it is interesting to notice that the SF of the SMC region around the very young cluster NGC 602 (Fig. 3), host of HII regions, shows time distribution and current rate per unit area similar to those (Fig. 7) of starburst dwarfs (once called extragalactic HII regions). The former however involves a small area, corresponding to a tiny fraction of the SMC, while the latter are global behaviours, referring to the whole galaxy.

By comparing with each other the SFHs derived from the CMDs of (few, small) regions of the LG spirals (Fig. 2), one is tempted to speculate over a possible dependence of the SFHs on their morphological type and luminosity class. M31 (SA b I-II) seems to have had very continuous, almost constant SF, since the earliest epochs. The solar neighbourhood

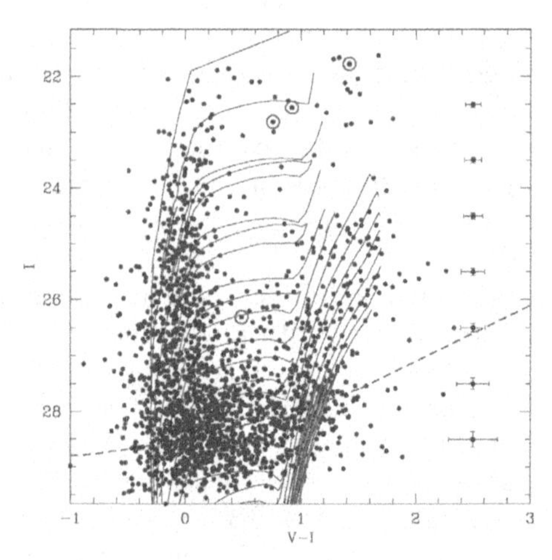

Figure 6. CMD of IZw18, obtained from HST/ACS imaging (Aloisi *et al.* 2007). Overimposed are the Z=0.0004 isochrones by Bertelli *et al.* (1994) with the RGB in red. Also shown is the average position of the 4 classical Cepheids with reliable light-curves obtained from these data.

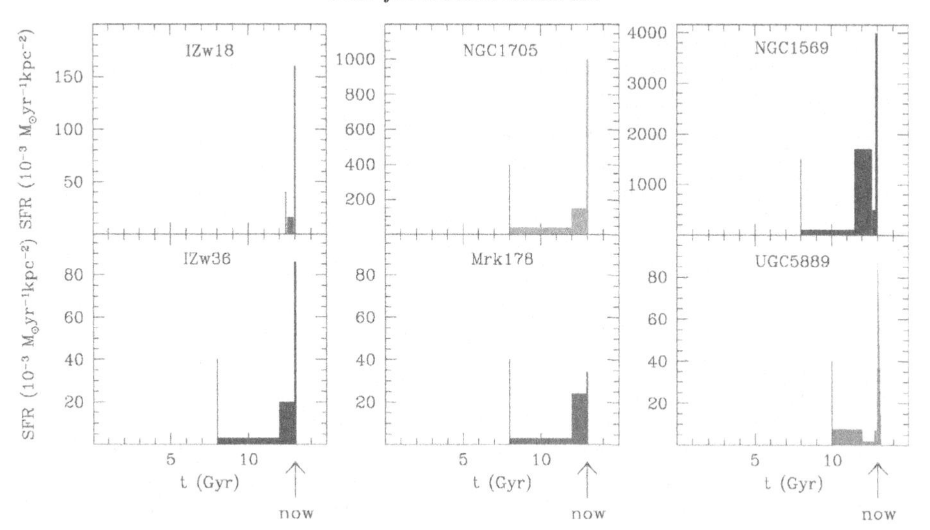

Figure 7. SFHs in late-type galaxies derived with the synthetic CMD method. In all panels the SF rate per unit area as a function of time is plotted. The thin vertical line in each panel indicates the reached lookback time. References: NGC 1569, Greggio *et al.* (1998), Angeretti *et al.* (2005); NGC 1705, Annibali *et al.* (2003); IZw18, Aloisi *et al.* (1999); IZw36, Schulte-Ladbeck *et al.* (2001); Mrk178, Schulte-Ladbeck *et al.* (2000); UGC 5889, Vallenari *et al.* (2005).

of the MW (SAB bc II-III) also shows a rather continuous SF regime, but with larger differences between the rate of the SF peaks and dips, see e.g. Rocha-Pinto & Maciel (1997), Hernandez, Valls-Gabaud & Gilmore (2000) and Cignoni *et al.* (2006). M33 (SA c III) definitely has significant bumps and gasps over its SFH, with a distribution of SFR with time almost indistinguishable from that derived for the late-type dwarf Leo A. One could then argue that the later the morphological type and the lower the luminosity class of the spirals, the more similar their SFH to those of late-type dwarfs.

Aside from speculations, the general results drawn from all the SFHs derived so far for galaxies with CMDs studies can be summarized as follows:

- Evidence of long interruptions in the SF activity is found only in early-type galaxies;
- Few early-type dwarfs have experienced only one episode of SF activity concentrated at the earliest epochs: many show instead extended or recurrent SF activity;
- No galaxy currently at its first SF episode has been found yet;
- No frequent evidence of strong SF bursts is found in late-type dwarfs;
- There is no significant difference in the SFH of dIrrs and BCDs, except for the current SFR.

Acknowledgements The Symposium organizers and IAU are gratefully acknowledged for partial financial support. I thank M. Cignoni and A. Cole for preparing figures *ad hoc* for this paper, and A. Dolphin and C. Gallart for the SFH plots shown in advance of publication. Some of the results described here have been obtained thanks to fruitful, recurrent and pleasant collaborations with A. Aloisi, L. Angeretti, F. Annibali, M. Cignoni, L. Greggio, A. Nota, and E. Sabbi.

References

Aloisi, A., Tosi, M., & Greggio, L. 1999, *AJ*, 118, 302

Aloisi, A., Clementini, G., Tosi, M., Annibali, F., Contreras, R. *et al.* 2007, *ApJ*, 667, L151

Aloisi, A., van der Marel, R. P., Mack, J., Leitherer, C., Sirianni, M., & Tosi, M. 2005, *AJ*, 631, L45

Angeretti, L., Tosi, M., Greggio, L., Sabbi, E., Aloisi, A., & Leitherer, C. 2005, *AJ*, 129, 2203

Annibali, F., Greggio, L., Tosi, M., Aloisi, A., & Leitherer, C. 2003, *AJ*, 126, 2752

Aparicio, A. & Gallart, C. 2004, *AJ*, 128, 1465

Barker, M. K., Sarajedini, A., Geisler, D., Harding, P., & Schommer, R. 2007, *AJ* 133, 1138

Bertelli, G., Bressan, S., Chiosi, C., & Nasi, E. 1994, *A&AS*, 106, 275

Bertelli, G. & Nasi, E. 2001, *AJ*, 121, 101

Brown, T. M. 2006, in *The Local Group as Astrophysical Laboratory*, M. Livio, T. M. Brown eds, STScI Symp. Ser. 17 (CUP), p.111

Brown, T. M., Beaton, R., Chiba, M., Ferguson, H. C., Gilbert, K. M. *et al.* 2008, *ApJ*, 658, L121

Chiosi, E., Vallenari, A., Held, E. V., Rizzi, L., & Moretti, A. 2006, *A&A*, 452, 179

Cignoni, M., Degl'Innocenti, S., Prada Moroni, P. G., & Shore, S. N. 2006, *A&A*, 459, 783

Cignoni, M., Sabbi, E., Nota, A., Tosi, M., Degl'Innocenti, S., Prada Moroni, P., Angeretti, L., Carlson, L., Gallagher, J., Meixner, M., Sirianni, M., & Smith, L. J. 2009, *AJ*, in press

Cole, A. A., Skillman, E. D., Tolstoy, E., Gallagher, J. S., Aparicio, A., *et al.* 2007, *ApJ*, 659, L17

Dohm-Palmer, R. C., Skillman, E. D., Gallagher, J. S., Tolstoy, E., Mateo, M., Dufour, R. J., Saha, A., Hoessel, J., & Chiosi, C. 1998, *AJ*, 116, 1227

Dohm-Palmer, R. C., Skillman, E. D., Mateo, M., Saha, A., Dolphin, A., Tolstoy, E., Gallagher, J. S., & Cole, A. A. 2002, *AJ*, 123, 813

Dolphin, A. E. 2002, *MNRAS*, 332, 91

Dolphin, A. E., Weisz, D. R., Skillman, E. D., & Holtzman, J. A. 2005, *astro-ph/0506430*

Dolphin, A. E., Walker, A. R., Hodge, P. W., Mateo, M., Olszewski, W. W., Schommer, R. A., & Suntzeff, N. B. 2001, *ApJ*, 562, 303

Fagotto, F., Bressan, A., Bertelli, G., & Chiosi, C. 1994a, *A&AS*, 104, 365

Fagotto, F., Bressan, A., Bertelli, G., & Chiosi, C. 1994b, *A&AS*, 105, 29

Ferraro, F. R., Fusi Pecci, F., Tosi, M., & Buonanno, R. 1989, *MNRAS*, 241, 433

Gallart, C., Aparicio, A., Bertelli, G., & Chiosi, C. 1996, *AJ*, 112, 1950

Greggio, L., Tosi, M., Clampin, M., De Marchi, G., Leitherer, C., Nota, A., & Sirianni, M. 1998, *ApJ*, 504, 725

Harris, J. & Zaritsky, D. 2004, *ApJ*, 127, 1531

Hernandez, X., Gilmore, G., & Valls-Gabaud, D. 2000, *MNRAS*, 317, 831

Hernandez, X., Valls-Gabaud, D., & Gilmore, G. 2000, *MNRAS*, 316, 605

Holtzman, J. A., Gallagher, J. S., Cole, A. A., Mould, J. R., *et al.* 1999, *AJ*, 118, 2262

Holtzman, J. A., Afonso, C., & Dolphin, A. E. 2006, *ApJS*, 166, 534

Hurley-Keller, D., Mateo, M., & Nemec, J. 1998, *AJ*, 115, 1840

Lynds, R., Tolstoy, E., O'Neil., E. J. Jr., & Hunter, D. A. 1998, *AJ*, 116, 146

Marconi, G., Tosi, M., Greggio, L., & Focardi, P. 1995, *AJ*, 109, 173

Noel, N. E. D., Gallart, C., Costa, E., & Mendez, R. A. 2007, *AJ*, 133, 2037

Rejkuba, M., Greggio, L., & Zoccali, M. 2004, *A&AS*, 415, 915

Rocha-Pinto, H. J. & Maciel, W. J. 1997, *MNRAS*, 289, 882

Seiden, P. E., Schulman, L. S., & Gerola, H. 1979, *ApJ*, 232, 709

Schulte-Ladbeck, R. E., Hopp, U., Greggio, L., & Crone, M. M. 2000, *AJ*, 120, 1713

Schulte-Ladbeck, R. E., Hopp, U., Greggio, L., Crone, M. M., & Drozdovsky, I. O. 2001, *AJ*, 121, 3007

Smecker-Hane, T. A., Cole, A. A., Gallagher, J. S. III., & Stetson, P. B. 2002, *ApJ*, 566, 239

Smecker-Hane, T. A., Stetson, P. B., Hesser, J. E., & Vandenberg, D. A. 1996, in *From stars to galaxies*, *PASP Conf.Ser.*, 98, 328

Tolstoy, E. 1996, *ApJ*, 462, 684

Tosi, M., Greggio, L., Marconi, G., & Focardi, P. 1991, *AJ*, 102, 951

Tosi, M., Sabbi, E., Bellazzini, M., Aloisi, A., Greggio, L., Leitherer, C., & Montegriffo, P. 2001, *AJ*, 122, 127

Vallenari, A., Schmidtobreick, L., & Bomans, D. J. 2005, *A&Ap*, 435, 821

Young, L. M., Skillman, E. D., Weisz, D. R., & Dolphin, A. E. 2007, *ApJ*, 659, 331

Discussion

G. Meurer: Concerning the recent "gasp" in star formation seen in low-surface-brightness dwarf galaxies, I don't think these can be attributed to the star formation history, but rather are induced by an under-populated upper end of the IMF. This is all explained in my poster (#42).

M. Tosi: I have not myself worked on the LSBs. Could your results be attributed to stochastic effects on the IMF in very low-density environments, rather than to actual IMF variations?

R. Wyse: I was very interested in your result that you always see stars as old as your limits. Could you say more about the prospects of pushing deeper, to see if it's like the Local Group, where all galaxies contain 10-12 Gyr old stars?

M. Tosi: It will not be easy to reach CMD phases (e.g., HBs) typically that old outside the Local Group in star-forming galaxies, because their young stars will always tend to hide the much fainter oldest stars. The best instrument for such a goal remains a space-based telescope.

J. Melbourne: In response to the question on Adaptive Optics, I do AO at Keck and I would agree that you will not do well with the very young stars but you might do very well on the AGB.

M. Tosi: Yes, AGBs will be very well measured with AO, but won't necessarily trace the oldest stellar populations.

Monica Tosi

The Ages of Stars
Proceedings IAU Symposium No. 258, 2008
E.E. Mamajek, D.R. Soderblom & R.F.G. Wyse, eds.

doi:10.1017/S174392130903172X

The ACS LCID project: Variable stars as tracers of population gradients

Edouard J. Bernard[1], **for the LCID Team**†

[1]Instituto de Astrofísica de Canarias, E-38205 La Laguna, Tenerife, Spain
email: ebernard@iac.es

Abstract. We present a few highlights concerning the search for short-period variable stars in four galaxies, namely IC 1613, LGS 3, Cetus and Tucana, based on very deep, multi-epoch HST/ACS photometry. These are discussed in the context of the star formation histories obtained from our very deep color-magnitude diagrams. In particular, we show how the pulsational properties of the RR Lyrae stars, which represent the vast majority of the observed variables, can trace subtle differences in the age and metallicity of the old population. For example, in the dwarf spheroidal galaxy Tucana we find that the fainter RR Lyrae stars, having a shorter period, are more centrally concentrated than the more luminous, longer period RR Lyrae variables. Through comparison with the predictions of theoretical models of stellar evolution and stellar pulsation, we interpret the fainter RR Lyrae stars as a more metal-rich subsample. In addition, we show that they must be older than about 10 Gyr, indicating that the metallicity gradient must have appeared very early on in the history of this galaxy. We also compare the populations of Cepheids in the galaxies of our sample based on their period-Wesenheit diagram. We tentatively classify them as classical short-period Cepheids in the two gas-rich galaxies (IC 1613 & LGS 3), and as anomalous Cepheids in the dwarf spheroidals.

Keywords. stars: horizontal-branch, stars: oscillations, stars: Population II, stars: variables: other, galaxies: dwarf, galaxies: evolution, Local Group

1. Introduction

Pulsating variable stars play a major role in the study of stellar populations and in cosmology, as their pulsational properties are traditionally used to determine distances and to put constraints on stellar physical properties. Because the pulsations occur at a particular phase of their evolution depending on the star mass, variable stars trace the spatial distribution of stellar populations of given ages, therefore highlighting the eventual radial trends across the studied galaxy (e.g., Gallart *et al.* 2004).

This, in turn, provides important clues to the formation mechanisms and the star formation history (SFH) of the host galaxy. Therefore, by providing information about the properties of the underlying population, variable star research procures a way to study the histories of these galaxies, independent and complementary to the color-magnitude diagram (CMD) analysis.

With the goal of understanding these processes, we are carrying out a large project (LCID) aiming at reconstructing the full SFH of a sample of isolated dwarf galaxies of the Local Group (LG), based on very deep, multi-epoch *Hubble Space Telescope* (*HST*) ACS data (see Monelli *et al.*, these proceedings). The sample includes representatives of the three main dwarf galaxy morphological types—irregular (dIrr), spheroidal (dSph) and so-called transition dIrr/dSph—located further than about two virial radii from both

† Local Cosmology from Isolated Dwarfs: http://www.iac.es/project/LCID/.

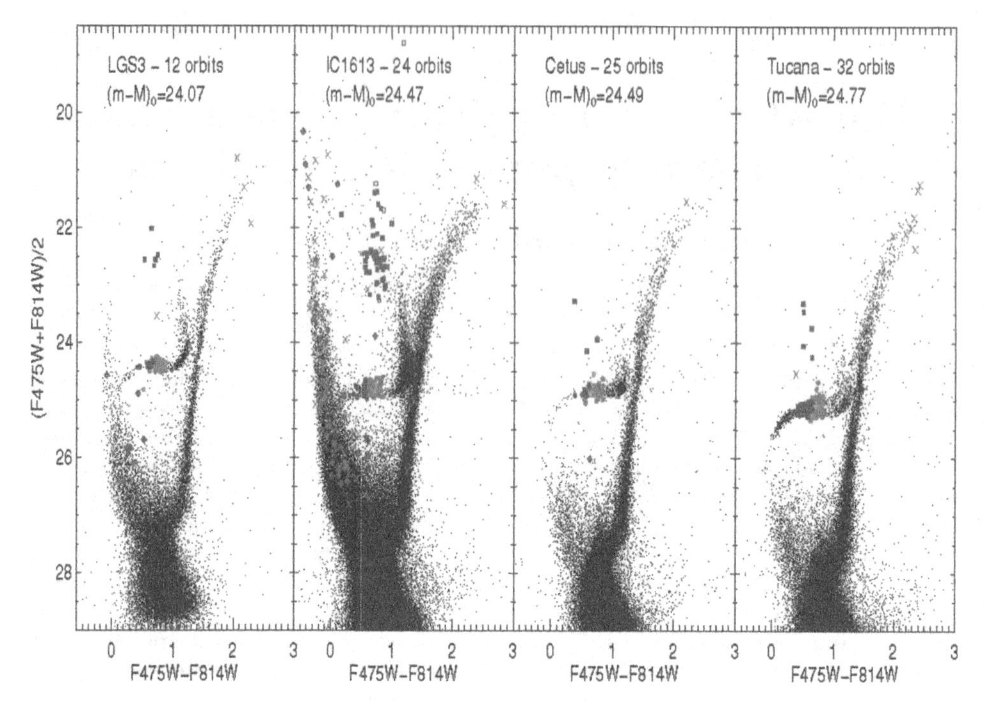

Figure 1. Color-magnitude diagrams of four of the galaxies of our sample, where the variables have been overplotted: Cepheids (*squares*), RR Lyrae stars (*circles*), eclipsing binaries (*diamonds*), and other candidates (*crosses*).

the Milky Way (MW) and M31. The project will be described in detail in a forthcoming paper (Gallart *et al.* 2009, in preparation), and the first results concerning the SFH of Leo A were presented in Cole *et al.* (2007).

Figure 1 shows the CMDs of the four galaxies that have been searched for variable stars to date: LGS3, IC1613, Cetus and Tucana. The number of orbits devoted to each galaxy, which also corresponds to the number of datapoints per band (F475W and F814W), is also indicated in the panels. In total, we found more than 900 variables in these galaxies. The vast majority (~700) are of the RR Lyrae type. The other classified variables include about 60 Cepheids—both classical and anomalous (see below)—and 50 eclipsing binaries. For LGS3, Cetus and Tucana, these are also the first confirmed variable stars. The remaining candidates are mainly located on the main sequence of IC1613 and, while most of them are probably eclipsing binaries, some of the brightest stars exhibit low-amplitude, pulsating-like variation. The variables in the dSphs Cetus and Tucana are presented in Bernard *et al.* (2009a). We will describe the variables in IC1613 in a forthcoming paper (E. Bernard *et al.* 2009b, in preparation). Here, we present a few highlights of the project, namely the use of RR Lyrae stars as tracers of old population gradients, and the relationship between the SFH and the type of Cepheid present in a galaxy.

2. RR Lyrae stars in Tucana

In Fig. 1, one can see that Tucana harbors a rather complex horizontal-branch (HB), which is well populated on both sides of the IS. The red side also presents a small gap in magnitude, suggesting the combination of two HB of different luminosities. This is supported by the unusual width in luminosity of the HB inside the IS (~0.3 mag at

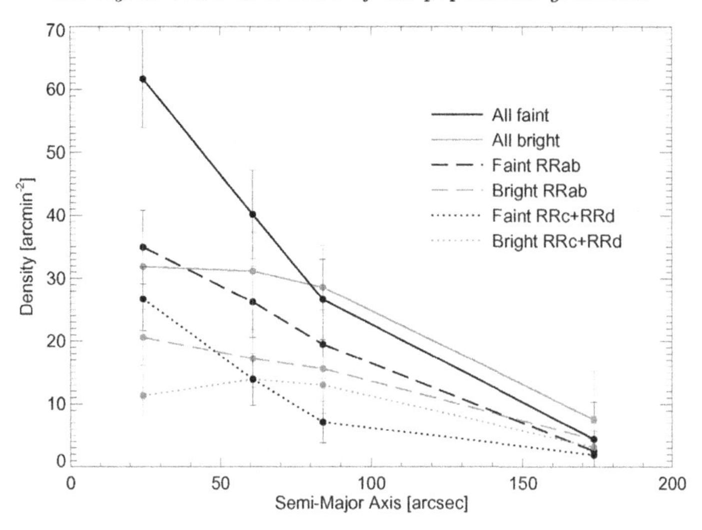

Figure 2. Radial distribution of the different subsamples of RR Lyrae stars in Tucana, showing how the fainter RR Lyrae stars are systematically more concentrated in the center of the galaxy than the brighter variables.

F475W-F814W=0.7, excluding the few bright and faint outliers), which is larger than expected from the evolutionary tracks, even taking into account the evolution off the zero-age HB (ZAHB).

We arbitrarily used the mean magnitude of the variables to split the RR*ab*, RR*c* and RR*d* variables into bright and faint subsamples, each type of variable having approximately the same number of stars in each subsample.

On a period-amplitude (PA) diagram, two distinct RR*ab* sequences are clearly identified, characterized by different mean periods (0.574 vs. 0.640 days) and dispersions around the fit. As predicted by nonlinear pulsation models (Bono *et al.* 1997b), the mean period is a function of luminosity, in the sense that the more luminous variables tend to have a larger period. The period difference also shows up in the RR*c* and RR*d* subtypes, and is of the order of 0.02 days.

The difference in the mean period of the RR Lyrae stars of each sample could be attributed to both a difference in metallicity (from the period-metallicity correlation) or a difference in the evolutionary status of the individual stars, the stars in the redward evolution off the ZAHB having a longer period (Bono *et al.* 1997b, their Fig. 16-17). However, stellar evolution models indicate that evolution off the ZAHB alone cannot account for the range of luminosity spanned by the RR Lyrae stars. Hence, it is necessary to invoke a range of metallicity to reproduce the distribution of stars within the IS, the more metal-rich stars being fainter. The hypothesis of a bimodal metallicity distribution is strengthened by the double "bell-shape" of the RR*c* in the PA diagram (Bono et al. 1997a) and the presence of two RGB bumps separated by ~0.2 mag in F814W.

The presence of different populations in a galaxy, whether due to the details of its SFH or to the accretion of an external stellar system, generally leads to gradients in the observable properties of its stars. Figure 2 presents the radial profile for each subsample of RR Lyrae stars. The radii were chosen so that each concentric region contains the same number of variables. It shows that, for each type of RR Lyrae stars, the fainter, more metal-rich variables are systematically more concentrated near the center of the galaxy,

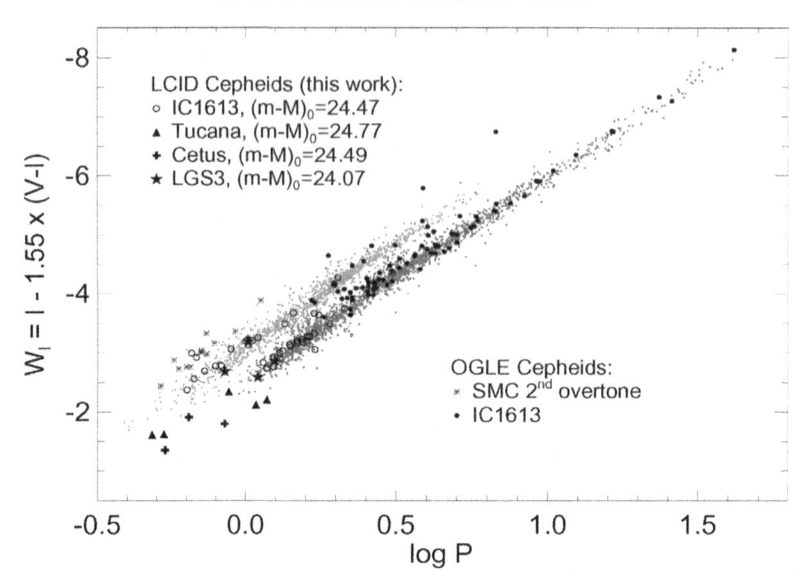

Figure 3. Period-Wesenheit diagram of the Cepheids in the galaxies of our sample (*see inset*), overplotted on the OGLE Cepheids (F: *dark gray*, FO: *light gray*).

while the brighter, metal-poor RR Lyrae stars are spatially extended. The combination of different intrinsic properties of the individual stars with the different spatial distribution supports the hypothesis that they represent separate populations.

In addition, our artificial HB computations indicate that the faintest, highest metallicity ($Z>0.0006$) RR Lyrae variables of Tucana must be >10 Gyr old. Therefore, under the reasonable assumption that chemical enrichment follows age in star forming galaxies, the presence of gradients in the RR Lyrae populations shows that these metallicity gradients appeared very early on in the history of this galaxy.

3. Cepheids

As shown in Fig. 1, Cepheids were detected in all four galaxies. However, their mean luminosities are very different between the gas-rich galaxies, LGS3 and IC1613, and the dSphs. This is also apparent in Fig. 3, where we plot the Cepheids found in the four galaxies of our sample in the period-Wesenheit diagram. The use of the Wesenheit, or reddening-free, index instead of the luminosity reduces the scatter due to interstellar reddening. The Johnson magnitudes for our Cepheids were obtained as described in Bernard et al. (2009a). They are shown overplotted on the classical Cepheids of the Large and Small Magellanic Clouds (LMC & SMC; fundamental mode: *dark gray*, first overtone: *light gray*) and of IC 1613 from the OGLE collaboration (Udalski *et al.* 2001, and references therein). The apparent magnitudes were converted to absolute magnitude assuming a distance modulus of 18.54 to the LMC (which is the value we adopted to calculate the distance of the LCID galaxies), and a distance offset of 0.51 of the SMC relative to the LMC (Udalski *et al.* 1999). The distances for the galaxies of our sample were calculated from the properties of the RR Lyrae stars (see Bernard *et al.* 2009a) and are shown in the inset.

It shows that the Cepheids in the gas-rich galaxies fit very well on the PL relations, while those in the dSphs fall significantly below. Note that the uncertainty on the

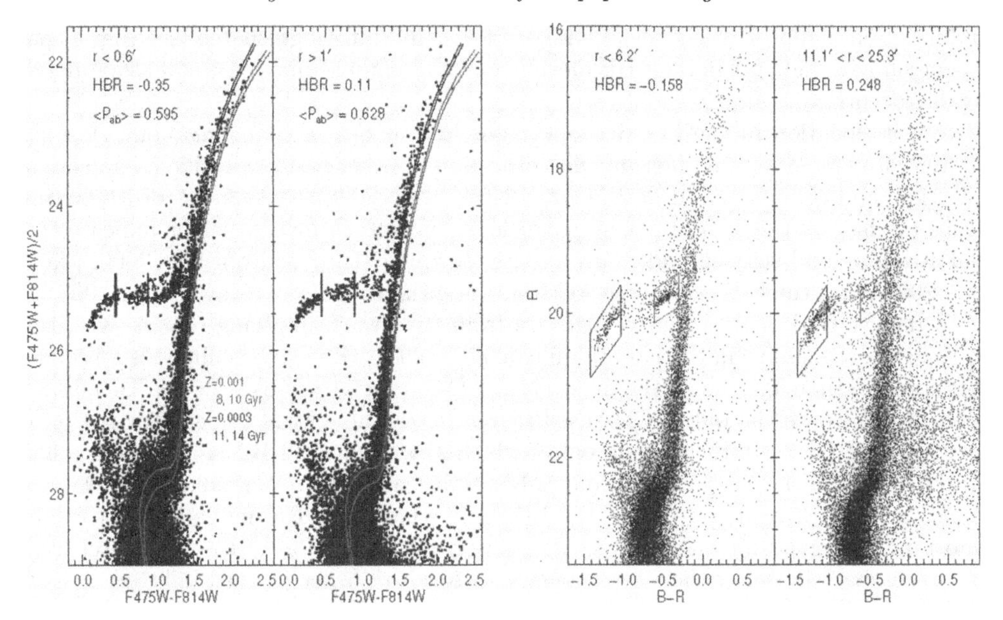

Figure 4. CMDs of Tucana (*left*, from Bernard *et al.* 2008) and Sculptor (*right*, in instrumental magnitude, from WFI@ESO 2.2-m data; Giuffrida *et al.* 2006) for two galactocentric distances. Note the similarity between the two galaxies in terms of slope of the upper RGB, horizontal-branch morphology and turnoff stars, both in the inner and outer regions.

distance estimates for the LCID Cepheids is about 0.1, which is insufficient to explain the magnitude shift. Their position with respect to the PL relations is similar to that of the anomalous Cepheids found in the dwarf galaxy Phoenix (Gallart *et al.* 2004). The difference between anomalous and classical Cepheids, from a theoretical point of view, is that the former initiated the core He burning under degenerate conditions while the latter are massive enough to burn He in non-degenerate conditions. Given that the dSphs did not form stars in the past few Gyrs, the only stars that can cross the instability strip above the HB have relatively low mass. On the other hand, the gas-rich galaxies have current star formation, at least residual in the case of LGS3, and therefore still have massive stars producing luminous, short-period classical Cepheids.

4. Isolated vs. Satellite Dwarf Spheroidals

As stated in the previous section, we found that the pulsational properties of the RR Lyrae stars in Tucana trace metallicity gradients, and that these gradients must have appeared very early on. This was the first time spatial variations of these properties were observed in a dwarf galaxy, thanks to the large spatial coverage and number of discovered variables. Is Tucana unique in this respect, or can we expect the same to occur in other dSphs?

In the Local Group (LG), the gas-deficient dSph galaxies are preferentially found as satellites to the MW or M31, while the gas-rich dwarfs are usually isolated. This suggests that environmental factors must play a fundamental role in the formation and evolution of these dwarf galaxies. However, our very deep HST photometry of Tucana, which is one of only two *really isolated* dSphs in the LG—at 870 kpc from the MW and 1340 kpc from M31—showed that its stellar population is disturbingly similar to that of the nearby satellite dSph Sculptor: their deep CMD are basically indistinguishable (see Fig. 4) and present similar gradients in their HB morphology and main-sequence turnoff as a function of galactocentric distance.

Given its location in the LG, Tucana might have experienced *at most* one close encounter with the MW in its lifetime. Sculptor, on the other hand, with its apogalacticon of 122 kpc and orbital period of 2.2 Gyr (Piatek *et al.* 2006), spent most of its lifetime within the halo of the MW. In these conditions, theoretical investigations indicate that tidal stripping and stirring and ram pressure stripping (Blitz & Robishaw 2000, Mayer *et al.* 2006), and the local UV radiation from the primary galaxy (Mayer *et al.* 2007) all act to remove dark matter and/or baryons from the dwarf, implying that it was likely ten times more massive in the past (Kravtsov *et al.* 2004). Even so, Sculptor is still much more massive than Tucana at the present time ($M_V = -11.1$ versus -9.6 for Tucana).

However, to date the only CCD-based search for variable stars in Sculptor is that of Kaluzny *et al.* (1995, in the V-band only), which identified 226 RR Lyrae stars in a field of only 15'x15', while the galaxy's tidal radius is at least of 80' (Westfall *et al.* 2006) and $\sim$1050 variables are expected (van Agt 1978). The fact that Tucana and Sculptor have important differences in mass and in interaction history makes of them an interesting pair to compare in detail their evolutionary histories in order to shed light on possible mechanisms shaping up the early evolution of galaxies. For these reasons, as a parallel project we recently obtained deep, multi-epoch photometry of a 1.7 square degree field centered on Sculptor using the MOSAIC-II camera mounted on the CTIO-4m Blanco telescope, specifically to study the gradients in the properties of the variable stars and the star formation histories, and to compare the results with those of Tucana.

Acknowledgements

I am grateful to the co-investigators of the projects discussed in these proceedings for allowing me to show results in advance of publication, and to Giuseppe Bono for providing the ESO-2.2m photometry of Sculptor. Support for this work was provided by a Marie Curie Early Stage Research Training Fellowship of the European Community's Sixth Framework Programme (contract MEST-CT-2004-504604), the IAC (grant 310394) and the Education and Science Ministry of Spain (grants AYA2004-06343 and AYA2007-3E3507).

References

Bernard, E. J., *et al.* 2008, *ApJ*, 678, L21
Bernard, E. J., *et al.* 2009a, *ApJ*, *submitted*
Blitz, L. & Robishaw, T. 2000, *ApJ*, 541, 675
Bono, G., Caputo, F., Cassisi, S., Incerpi, R., & Marconi, M. 1997a, *ApJ*, 483, 811
Bono, G., Caputo, F., Castellani, V., & Marconi, M. 1997b, *A&AS*, 121, 327
Cole, A. A., *et al.* 2007, *ApJ*, 659, L17
Gallart, C., *et al.* 2004, *AJ*, 127, 1486
Giuffrida, G., *et al.* 2006, *MemSAI*, 77, 125
Mayer, L., *et al.* 2006, *MNRAS*, 369, 1021
Mayer, L., *et al.* 2007, *Nature*, 445, 738
Ngeow, C., *et al.* 2008, *arXiv:astroph/0811.2000*
Piatek, S., *et al.* 2006, *AJ*, 131, 1445
Udalski, A., *et al.* 1999, *Acta Astronomica*, 49, 201
Udalski, A., *et al.* 2001, *Acta Astronomica*, 51, 221
van Agt, S. 1978, *Publications of the David Dunlap Observatory*, Vol. 3, No. 7, p. 205
Westfall, K., *et al.* 2006, *AJ*, 131, 375

Discussion

A. PIPINO: Can you give an estimate of the metallicity gradient on the basis of the RR Lyrae analysis alone? Is the method accurate enough?

E.J. BERNARD: We did not try to quantify the gradients from the RR Lyrae stars since their location in the CMD is also a function of their evolutionary status, and the populations are mixed in the center of the galaxy.

A rough estimate based on the difference of average RR Lyrae luminosity between the inner—where the populations are mixed—and outer—mostly metal-poor—fields (~0.05 mag) and the slope of the luminosity-metallicity relation (ΔM_V(RR)/Δ[Fe/H]=0.214; Clementini *et al.* 2003, AJ, 125, 1309) gives a 0.2 dex difference over the sampled radius. Assuming the bright and faint subsamples are *bona fide* metal-poor and metal-rich populations, that is, ignoring the evolutionary effects on the luminosity of individual stars and the fact that the HB is not precisely horizontal, their difference in luminosity of 0.1 mag implies a difference in metallicity of ~0.5 dex.

Edouard Bernard

Adam Kraus and Eric Jensen

Jeff Cummings and David Montes

The Ages of Stars
Proceedings IAU Symposium No. 258, 2008
E.E. Mamajek, D.R. Soderblom & R.F.G. Wyse, eds.

doi:10.1017/S1743921309031731

Age-related observations of low mass pre-main and young main sequence stars

Lynne A. Hillenbrand
California Institute of Technology
MC 105-24, Pasadena, CA 91125 (USA)
email: `lah@astro.caltech.edu`

Abstract. This overview summarizes the age dating methods available for young sub-solar mass stars. Pre-main sequence age diagnostics include the Hertzsprung-Russell (HR) diagram, spectroscopic surface gravity indicators, and lithium depletion; asteroseismology is also showing recent promise. Near and beyond the zero-age main sequence, rotation period or vsin*i* and activity (coronal and chromospheric) diagnostics along with lithium depletion serve as age proxies. Other authors in this volume present more detail in each of the aforementioned areas. Herein, I focus on pre-main sequence HR diagrams and address the questions: Do empirical young cluster isochrones match theoretical isochrones? Do isochrones predict stellar ages consistent with those derived via other independent techniques? Do the observed apparent luminosity spreads at constant effective temperature correspond to true age spreads? While definitive answers to these questions are not provided, some methods of progression are outlined.

Keywords. stars: pre–main-sequence, (stars:) Hertzsprung-Russell diagram (Galaxy:) open clusters and associations: general

---

1. Techniques for assessing young star ages

Standard stellar age dating techniques can be divided in to those which are purely empirical in nature and those which are more theoretically grounded. The former include measurements related to stellar kinematics and cluster membership, stellar rotation as derived from periodic photometric modulation or spectroscopic absorption line broadening, stellar chromospheric activity as measured by e.g. fractional Ca II H&K or Hα line luminosity, stellar coronal activity as measured by soft x-rays, lithium depletion trends, and age-metallicity as well as age-velocity dispersion relations. Most of the empirical correlations have dependencies on the stellar mass in addition to the age effect which is sought, adding necessary complication to any analysis. The latter methods, those referenced more directly to theory, include location in the Hertzsprung-Russell (HR) diagram relative to calculated isochrones, theoretical nuclear burning as traced though e.g. lithium abundances, and Asteroseismological constraints.

Each of the above diagnostics is in principle also applicable in the pre-main sequence phase of stellar evolution, as well as to the zero-age and young main sequence phases. However, many of them are significantly diminished in value as quantitative pre-main sequence age indicators due to "saturation" or "degeneracy" effects. Specifically, the coronal and chromospheric activity indicators which generally decline in strength with advancing main sequence age (e.g. Mamajek & Hillenbrand, 2008) are either very shallowly dependent on, or constant with, age for solar-type pre-main sequence stars. Rotational behavior likewise deviates from the monotonic spindown characteristic of main sequence angular momentum evolution, with significantly higher dispersion in rotation properties observed at ages younger than ~200 Myr. This is explained as remnant behavior related to earlier interaction of the stellar magnetosphere with the primordial circumstellar disk, specifically star-to-star variation in the time scale for star-disk coupling. Kinematics and cluster membership still apply as age diagnostics in the pre-main sequence phase, and are employed in the same "guilt by association" manner as utilized for main sequence clusters. However, the absolute age dating is more difficult.

Another method that is sometimes used as a relative age dating technique for the very youngest stars is the fractional infrared excess luminosity, or the shape of the mid-infrared spectral energy distribution. While it is true that the vast majority of stars with remaining detectable primordial circumstellar dust are younger than ~10 Myr, and that the vast majority of stars with so much dust that they are seen via scattered light or are still partially or totally self-embedded are younger than ~1-2 Myr, there is no evidence for a monotonic relationship between circumstellar dust characteristics and absolute stellar age. On the contrary, there are strong arguments for significant *dispersion* in the amount of circumstellar dust (and gas) among stars aged less than ~10 Myr – even those located in the same cluster or association. Thus, while stellar youth is certainly indicated by the presence of circumstellar material, the quantitative use of circumstellar properties as stellar chronometers is not recommended and will not be discussed further here.

We are thus left with three stellar age dating methods that are both applicable and increasingly well-calibrated at young – pre-main sequence and zero-age main sequence – ages: (1) the theoretical HR diagram or extinction-correction color-magnitude diagram, (2) inferences of $\log g$ vs. $\log T_{eff}$ from spectra, and (3) lithium abundance measurements and depletion trends. There is also some promise from (4) asteroseismology but this method has not yet proved itself. In what follows I discuss each of these four techniques and recent results.

2. Stellar age dating in regions of recent star formation

Stars form within giant molecular clouds that become unstable to fragmentation and subsequent gravitational collapse to produce: stellar clusters, multiple star systems, and individual stars. The time for an individual protostar to collapse is related to the local sound speed, and is expected to be 0.1-0.2 Myr (Shu, *et al.* 1987). On larger spatial scales, two main theories of star formation suggest different regulating phenomena and therefore time scales for the start-to-finish process of star formation in a molecular cloud. Regulation by quasi-static ambipolar diffusion processes takes ~3-10 Myr (Shu, 1977, Mouschovias, 1976) while regulation by turbulence dissipation occurs on the dynamical time scale of only ~0.5-few Myr (Ballesteros-Paredes *et al.* 1999, Elmegreen 2000).

We can hope to probe the relative importance of these physical processes by studying the mean ages and detailed age distributions in regions of current and recent star formation. Our main tools are those mentioned above: stellar bolometric luminosities and the HR diagram, spectral diagnostics of stellar surface gravity, and measurements of Li I 6707 Å abundances. The bulk of my discussion concerns HR diagrams.

2.1. *HR diagrams*

A good case study that informs our understanding of stellar ages and age spreads in star forming regions is the Orion Nebula Cluster (ONC). Hillenbrand (1997) published a synthesis of existing and new photometry and spectroscopy in this region, enabling the location of over 900 stars on the theoretical HR diagram. Now, new and better photometry from HST/ACS and ESO/WFI as well as over 600 new optical spectral types from WIYN/Hydra, Palomar/Norris, and Keck/LRIS are available. Also, recent estimates of the ONC distance place the cluster ~15% closer than previously accepted values. Further, we now have a better understanding of the photometric variability trends and amplitude ranges of individual ONC stars, enabling use of median photometry. The improved data along with revisions in our understanding of the temperature and bolometric correction scales appropriate for young pre-main sequence stars makes it worth revisiting the finding of a substantial luminosity spread in the Hillenbrand (1997) study.

First attempts at revision are presented by Da Rio *et al.* (2009), considering very carefully the subtleties of young star de-reddening and the effects of accretion, and Reggiani *et al.* (poster at this meeting), considering only the least photometrically variable stars. These authors find essentially no reduction in the ~1.5 dex luminosity spread [or $\sigma(\log \mathrm{L}/\mathrm{L}_{\odot}) \approx 0.55$ dex at fixed $\log \mathrm{T}_{eff}$] characteristic of the lower quality and single epoch Hillenbrand (1997) data. Thus, observational errors and biases, and known causes of scatter do not appear to be the main culprit in creating the large luminosity spreads that are observed in the ONC HR diagram.

Indeed, such apparent luminosity spreads have been seen for decades in young cluster and association HR diagrams. Literature from the 1960s and 1970s, e.g. the venerable Iben & Talbot (1966) and Ezer & Cameron (1967) showed them. Such early comparisons between data and theoretical pre-main sequence isochrones are primarily responsible for long standing paradigms such as "molecular clouds form stars for about 10 Myr" and "circumstellar disks last about 10 Myr." Although the above statements have been modified with better data and modern interpretation, the evidence for cluster luminosity spreads has persisted. Thus the questions remain: Are the apparent luminosity spreads real? Do they indicate true age spreads? Can we use them to infer star formation histories?

2.1.1. *HR diagram methodology*

Before embarking on these questions of HR diagram interpretation, it is important to review how stars are located in the HR diagram based on observational data and available techniques, as well as the accompanying complications to such procedures.

In practice, a spectral type determined at blue optical (BV), red optical (RI), or near-infrared (YJHK) wavelengths is used along with a spectral-type- to-effective-temperature conversion to set the abscissa in the HR diagram. Photometry within some subset of optical or near-infrared bands is used along with the spectral type to calculate and correct for reddening, and then a bolometric correction appropriate to the spectral type is adapted in order to calculate the ordinate of bolometric luminosity.

Complications to this standard process that are unique to young stars include effects related to the ubiquitous presence of circumstellar disks for some portion of the early pre-main sequence (see Meyer, this volume). Accretion from the disk on to the star creates a hot excess which makes blue photometry "too blue," while thermal plus accretion emission from the inner disk makes red photometry "too red." Both phenomena confuse de-reddening procedures. The potential existence of both blue and red excess means that, in fact, there may be no truly photospheric wave band at which to apply bolometric corrections to the reddening corrected photometry. Furthermore, some young sources are not seen directly, but via light scattered through circumstellar disks or envelopes which leads to significant luminosity underestimation. For example, all Taurus scattered light sources sit on or below the zero-age main sequence. The extent to which scattered light affects other systems, in which it is not known from spatially resolved images, is unknown. Luminosity effects resulting from typical parameter distributions star plus disk systems were modeled by Kenyon & Hartmann (1990) who found induced luminosity deviations relative to non-disked stars of $\sigma(\log \mathrm{L}/\mathrm{L}_{\odot}) < 0.2$ dex. Other concerns for HR diagram construction at young ages include generally large values of visual extinction (>1-10 mag), uncertainty regarding the appropriate extinction law, and significant photometric variability at typically <0.2 mag levels though >1 mag in more extreme cases.

Additionally present for the young stars are the usual complications affecting all HR diagram determinations. These include random errors due to spectral type and photometric uncertainties, and systematic errors deriving from unresolved multiplicity that result in luminosity overestimates (e.g. Simon *et al.* 1993).

2.1.2. *Checks on methodology*

Checks on our ability to locate young stars in the HR diagram are provided by binary and higher order multiple systems, whose components we expect to be coeval. Previous work in this area includes that by Hartigan *et al.* (1994), White *et al.* (1999) Prato *et al.* (2003), and Ammler *et al.* (2005). Recently, Kraus & Hillenbrand (2009) have used more modern temperature and luminosity pairs based on improved photometry and spectral types from high spatial resolution data to determine that, indeed, the binaries and higher order multiples in the Taurus-Auriga region are more coeval than random pairings of member stars. However, while some multiple systems lie on theoretical isochrones (within the errors), others are significantly mismatched. It is unclear at present whether the observed effects can be attributed to random or systematic errors, or if they indicate true non-coevality, but the result should be kept in mind as we proceed to discuss luminosity spreads in clusters as a whole.

Totally independent checks on $\log \mathrm{L}/\mathrm{L}_{\odot}$ and $\log \mathrm{T}_{eff}$ conversions via the theoretical HR diagram to stellar *masses* come from comparison of such predictions to dynamical mass measurements. However, similarly independent checks of $\log \mathrm{L}/\mathrm{L}_{\odot}$, $\log \mathrm{T}_{eff}$ to stellar *age* predictions are more difficult to develop. At best, we can demand that the theoretical isochrones are parallel to observed cluster sequences, similar to the expectations for binaries and higher order multiples. We can also hope for consistency with other techniques, such as e.g. turn-off ages for the higher mass stars in the same cluster, surface gravity measurements, lithium abundance determinations, etc.

2.1.3. *HR diagram theory*

As we aim to use HR diagrams to gain knowledge about absolute stellar ages, in addition to investigating the evidence for or against spreads in age, it is important to discuss in brief the calibrating theory for the stellar age determinations.

As detailed in Hillenbrand *et al.* (2008), there are significant systematic effects between available theoretical predictions of pre-main sequence evolution. Specifically, the trend at sub-solar masses is that for a given location in the HR diagram, the youngest ages are those inferred from the D'Antona & Mazzitelli (1994, 1997 and 1998 update) theory with increasingly older ages predicted by Yi *et al.* (2001, 2003, 2004), Swenson *et al.* (1993), Palla & Stahler (1993, 1999), Siess *et al.* (2000), and finally the Baraffe *et al.* (1995, 1998) theory predicting the oldest ages. Age differences between these various track sets for the same $\log \mathrm{L}/\mathrm{L}_{\odot}$, $\log \mathrm{T}_{eff}$ pair rise to $\sim$0.75 dex at the youngest ages!

Furthermore, with all isochrone sets, trends of stellar age with stellar mass are present, as widely reported in observational papers. Along any empirical isochrone $<$50-80 Myr, the higher mass stars generally appear older than the lower mass stars. This suggests either that we are still missing physics associated with the initial appearance of stars in the HR diagram, or that the mass-dependent physics of stellar interiors is still not adequately understood. As detailed by Palla (this volume), additional complications to pre-main sequence evolutionary theory such as the effects of initial conditions (i.e. the "birthline"), disk accretion, stellar magnetic fields, and stellar rotation may explain some of the observed dispersion between observationally derived and theoretically predicted effective temperatures and luminosities.

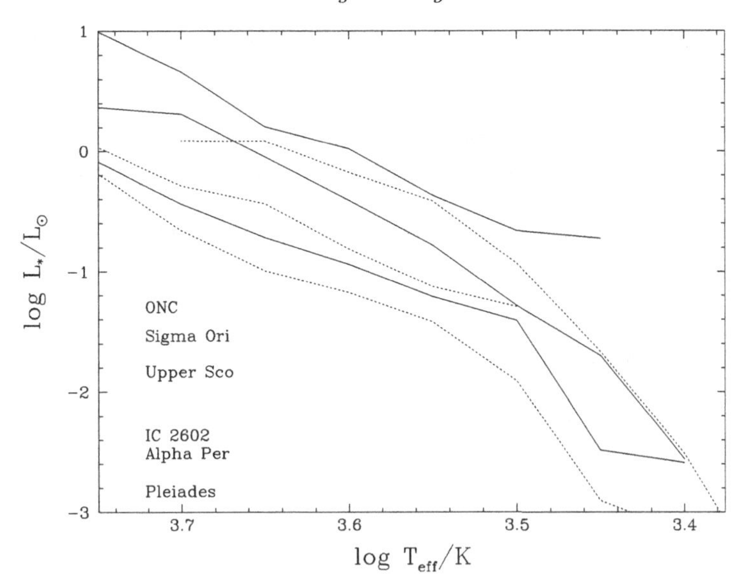

Figure 1. A sequence of empirical isochrones for representative young clusters and associations in the solar neighborhood. Shown is the median stellar bolometric luminosity as a function of stellar effective temperature for member star and brown dwarfs. Ordering in the legend corresponds to relative luminosity at $\log \mathrm{T}_{eff}$=3.6.

2.1.4. *Empirical results for young clusters*

I apply the methods outlined above and return now to the question of how well pre-main sequence clusters compare to theoretical isochrones. Shown in Hillenbrand *et al.* (2008) are the $\log \mathrm{L}/\mathrm{L}_{\odot}$ and $\log \mathrm{T}_{eff}$ distributions – represented as the mean and 1-σ luminosity vs. binned effective temperature – for over 25 young clusters, associations, and currently active star forming regions within 500 pc of the Sun. I take the mean luminosity with effective temperature sequence in each region as an empirical isochrone. As illustrated comparatively in Figure 1, there is good representation among this sample of stellar populations having ages from $<$ 1 Myr to just over 100 Myr. The median luminosities at fixed effective temperature span approximately 1.5 dex.

I quantify the luminosity *spreads* in each region by calculating the distribution of $\Delta \log \mathrm{L}/\mathrm{L}_{\odot}$= [$\log L_{observed}(i)/L_{\odot}-$ $\log L_{median}(\log T_{eff})/L_{\odot}$], or the deviation of individual luminosities from the median value appropriate for the temperature. These distributions, also, are illustrated in Hillenbrand *et al.* (2008). In most cases, Gaussian fits appear to describe adequately the luminosity deviations, suggesting that random processes are the dominant contributor to the luminosity spreads.

In detail, the fitted dispersions to the $\Delta \log \mathrm{L}/\mathrm{L}_{\odot}$ distributions for somewhat older (>30 Myr) near-main sequence young clusters, such as the Pleiades, α Per, IC 2602, IC 2391, and the Tucanae / Horologium Association, are low with $\sigma(\Delta \log \mathrm{L}/\mathrm{L}_{\odot})$ = 0.10-0.15 dex. We can take this as the typical luminosity dispersion that may be expected from the HR diagram placement methods described above. Towards younger ages, however, empirical dispersion increases substantially with $\sigma(\Delta \log \mathrm{L}/\mathrm{L}_{\odot})$ = 0.2-0.6 dex for clusters younger than 3-10 Myr. These spreads may be compared to the 0.15-0.25 dex dispersions estimated as plausible by Hartmann (2001) for young stellar populations in which significant accretion luminosity is present, and the even smaller spreads discussed by Burningham *et al.* (2005) as characteristic of young variables. For pre-main sequence contraction going roughly as $L \propto \tau^{-2/3}$, the *implied* age dispersion from literal interpretation of observed luminosity dispersion is then $\sigma(\log \tau) \propto 1.5\ \sigma(\log L)$.

However, it is not only the Gaussian width that is important to assess in considering cluster luminosity dispersion. Rather, it may be the subtle deviations from pure Gaussianity that convey the important information apropos, e.g. star formation history of a region, or other factors such as binary properties of the sample. Monte Carlo simulation of the luminosity distributions that accounts for these various details can help illuminate the important effects, and is discussed in a later section.

At this point I would like to (re-)emphasize that before any apparent luminosity dispersion is considered real, that observational fidelity must be verified so as to minimize any contaminating effects to the already complex interpretation of the luminosity spread phenomenon. First, we should ensure that we are considering only certain cluster or association members and regions that are not confused by superposed episodes of star formation. Next, we should strive to obtain exquisite photometry and high quality spectroscopy so as to reduce the influence of random observational errors. We should account for possible scattered light (causing luminosity underestimates) in young regions and multiplicity (causing luminosity overestimates) in all regions; although these both are systematic effects, they apply to only some portion of the population and therefore contribute to apparent luminosity spreads.

In summary, only pristine samples and the best data should be used in probing luminosity distributions. I turn now to discussion of potential correlates with $\Delta \log L/L_{\odot}$.

2.1.5. *Independent observational checks*

We can test the reality of the observed apparent luminosity spreads via their confirmation by independent observational means. Specifically, we can look for correlations between $\Delta \log L/L_{\odot}$ and surface gravity indicators or lithium abundance trends. Further, we can take advantage of the asteroseismological checks that have recently come to fruition for pre-main sequence stars in certain mass regimes. In addition to the discussion below, I refer the reader to clever techniques pioneered by Jeffries (this volume) and Naylor (this volume) which also provide checks on the observed apparent luminosity spreads in young clusters.

2.2. *Surface gravity diagnostics*

Low mass stars have a number of surface gravity sensitive spectral features in the red optical wavelength range – that most often used to classify such objects in modern studies – with others available at near-infrared wavelengths but not discussed here. For stars with spectral types later than ~M2, the CaH 6975 Å band and the Na I 8183,8195 Å doublet lines are surface gravity sensitive at ages younger than ~20-30 Myr (Schiavon *et al.* 1995, Slesnick *et al.* 2006). Towards later spectral types, beyond ~M6 and extending well into the L types, the K I 7665,7699 doublet lines, and several VO bands are additionally useful surface gravity diagnostics at ages younger than ~ 100 Myr (Steele & Jameson 1995, Kirkpatrick *et al.* 2008).

One test of the reality of the observed apparent luminosity spreads in young clusters is whether there is any correlation between $\Delta \log L/L_{\odot}$, the deviation from the median cluster luminosity normalized for effective temperature, and a quantitative index of surface gravity. With $L \propto R^2 T_{eff}^4$ and $g \propto M/R^2$, we expect for low mass stars of constant mass contracting along Hayashi (roughly constant temperature) tracks that $\log L/L_{\odot}$ and $\log g$ should be inversely correlated. Slesnick *et al.* (2008) demonstrate for stars in the Upper Sco region that objects with the same or similar measures of the surface gravity sensitive Na I 8190 Å spectral index can have a broad range of luminosity. Although the

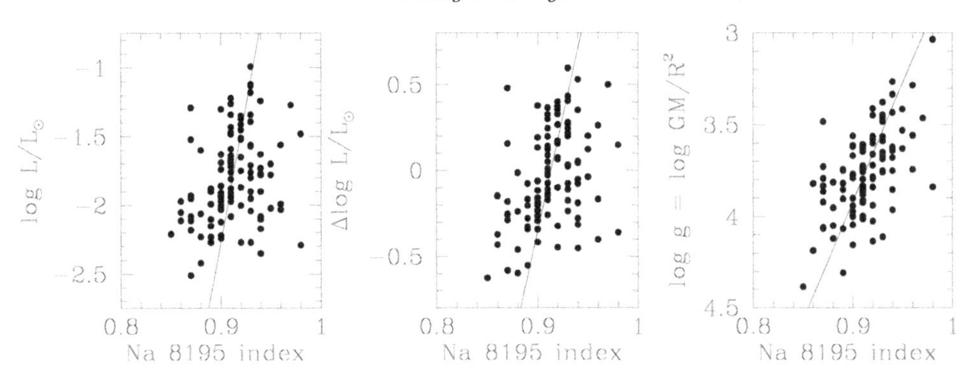

Figure 2. Based on data presented in Slesnick *et al.* (2008) for M4-M7 stars in the Upper Sco region. The left, middle, and right panels correlate $\log L/L_\odot$ (stellar bolometric luminosity), $\Delta \log L/L_\odot$ (deviation from mean luminosity normalized to effective temperature), and log GM/R^2 (surface gravity) computed from the pre-main sequence $\log L/L_\odot$ and $\log T_{eff}$ location in the HR diagram – all with the surface gravity sensitive Na I 8190 Å spectroscopic index defined by Slesnick *et al.* Correlation coefficients and the linear least squares fits are poor for the left and middle panels, but -0.6 (inversely correlated) in the right panel with 0.27 dex rms for the displayed fit of [$\log g = (-12.9 \pm 0.7) \times NaI + (15.5 \pm 0.6)$].

median age of the cluster is ~5 Myr, individual stars with the same Na I index have ages predicted based on their luminosities from <3 to >14 Myr. If these luminosity-based ages are to be believed, we would expect corresponding differences in the Na I index.

What is observed is shown in Figure 2. Although there is significant scatter, it does appear that the surface gravities implied from HR diagram location do correlate in the expected way with a completely independent (spectroscopic) indicator of surface gravity. However, neither the straight $\log L/L_\odot$ nor the $\Delta \log L/L_\odot$ values exhibit similar correlation. Luminosity based ages for individual stars thus still warrant considerable skepticism, and cautions remain against uncritical assessment of observed apparent luminosity dispersion as true age dispersion. However, the quantitative results shown here do imply non-zero spread in both luminosity-based surface gravity and spectroscopic surface gravity, and thus by implication perhaps age at the several Myr level.

2.3. *Lithium 6707 Å measurements*

Low mass stars burn both deuterium and lithium during the pre-main sequence evolutionary stages, essentially early steps in the hydrogen burning set of reactions that take place later on the main sequence. Contracting objects between $\sim 1 - 2.5 M_\odot$ undergo lithium burning processes for only a few tens of Myr to <1 Myr, while those below $\sim 1 M_\odot$ and down to the hydrogen burning limit deplete their lithium essentially forever, and brown dwarfs burn only deuterium but never lithium (e.g. D'Antona & Mazzitelli, 1994; Nelson *et al.* 1993). Lithium depletion trends in young pre-main sequence and main sequence populations have been used at sub-solar masses to estimate stellar ages, as discussed elsewhere in this volume. There is considerable scatter at constant age (e.g. within clusters) in both the observed equivalent widths and the derived abundances at constant mass or spectral type for stars younger than a few hundred Myr. Physically, this dispersion in surface abundance is likely related to the dispersion in rotation speeds over the same age range.

For any given star, lithium depletion is monotonically related to stellar age in the sense that lithium is never created via nuclear reactions, only destroyed. A test, therefore, of the reality of observed apparent luminosity spreads in young clusters is whether there is any correlation between $\Delta \log L/L_\odot$, the deviation from the median cluster luminosity normalized for effective temperature, and lithium abundance. Palla *et al.* (2005, 2007) have argued in the case of a small sample in the ONC that this is indeed the case, with isochronal and lithium depletion ages agreeing to within 5% in most cases based on the models of Siess *et al.* (2000). The agreement is particularly noted for those objects which sit low in the HR diagram relative to the main locus. However, it is just these stars which are suspected of being slightly foreground interlopers, part of the Orion Ic association which is indeed older and envelops the Orion Id (ONC) region. Thus, although intriguing and certainly an excellent way to test the conundrum of large luminosity spreads, in the particular case of the ONC there may be other complications which overshadow the main effect of this comparison.

An interesting case is that of St 34 in Taurus (White & Hillenbrand, 2005), a near-equal mass binary with both components sitting low in the HR diagram relative to other Taurus members (isochronal age ~8 Myr) and also near-fully lithium depleted (depletion age >25 Myr). Otherwise, the star has all the characteristics typical of classical T Tauri stars: strong Hα, He I, other accretion/outflow spectroscopic diagnostics, infrared excess, etc. It is thus either an unusually long-lived accretion disk system, or has had a somewhat unusual radial contraction and very unusual lithium depletion history. Another mysterious young object with apparent lithium depletion age much older than its isochronal age is Beta Pic group member HIP 112312 A (Song *et al.* 2002).

The existence of a few potentially anomalous objects like the examples above not withstanding, the correlation between lithium depletion and $\log L/L_\odot$ should be investigated more broadly. Recent studies of lithium in Taurus by Sestito *et al.* (2008) and in older nearby associations by Mentuch *et al.* (2008) provide some of the needed data.

2.4. *Asteroseismology*

Pulsational behavior in the Sun and other stars has offered important checks on our stellar interior models. Typically, e.g. in the case of the Sun as well as near the classical instability strip on the HR diagram, the pulsation mode is driven by opacity sources (the κ mechanism). Some pre-main sequence stars of intermediate mass lie near this strip (Marconi & Palla, 1998) and are being monitored for pulsations with a good number of detections to date (e.g. Zwintz *et al.* 2008 and references therein). Additionally, there is a prediction (Palla & Baraffe, 2005) at the lowest masses that stellar/sub-stellar interior instability can be driven by deuterium burning (ϵ mechanism) and also result in observable pulsational behavior. A *narrow* instability strip that is nearly parallel to the isochrones offers for these very low mass stars and brown dwarfs strong age constraints that are totally independent of the HR diagram – if pulsators can be found.

Many of the known brown dwarfs in star forming regions such as Chamaeleon I/II, Lupus, Ophiuchus, Upper Scorpius, IC 348, Sigma Ori, and the ONC lie near the predicted instability strip. Candidate objects are being monitored in thesis work by A.M. Cody (poster presented at this meeting) for photometric variability to determine which might be pulsators. Thus far, interesting variability at the right amplitudes (<0.02 mag) and on the right time scales (several hours) has been detected through periodogram analysis.

Significant work of a very detailed nature still needs to be conducted on both the observational side and the theoretical side of pre-main sequence pulsations. In principle, however, asteroseismology is a powerful technique for assessing independently the stellar ages inferred from HR diagrams.

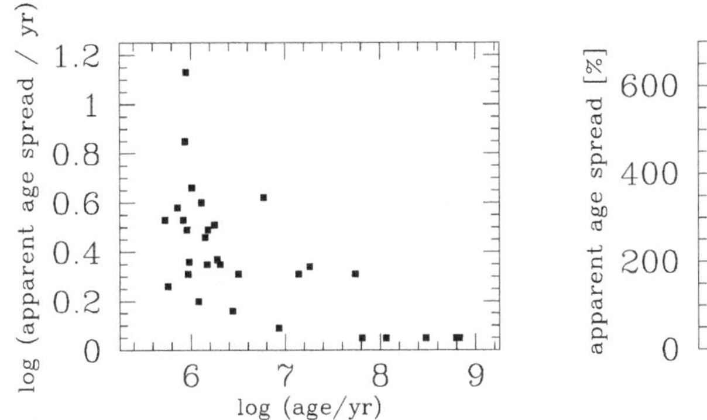

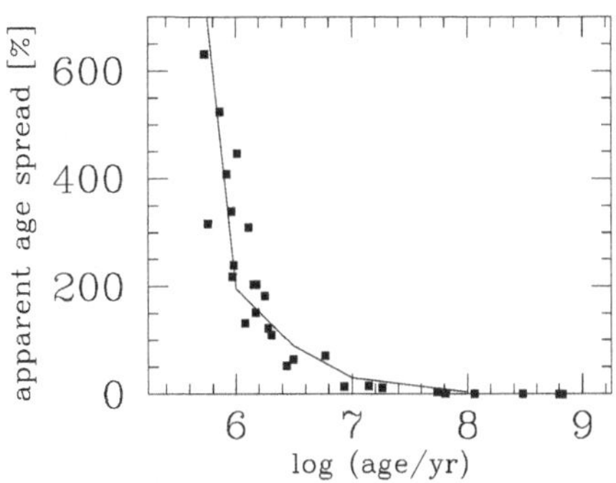

Figure 3. Left panel: 1-sigma log age dispersion versus mean log age as predicted from the $\log L/L_\odot$ and $\log T_{eff}$ data by D'Antona & Mazzitelli (1997/1998). Other tracks generally yield older ages and even larger age dispersions. Right panel: corresponding percentage (linear) age error. The line indicates the expected age error introduced by an imposed luminosity scatter of 0.2 dex, and reproduces the observed "age scatter" for the youngest clusters though overpredicts the "age scatter" for older clusters where the actual luminosity spread is indeed lower. We can conclude that the observed luminosity dispersion is comparable to the luminosity spread expected from the random and systematic errors suffered during HR diagram placement, rather than being dominated by true age spread.

3. Simple HR diagram simulations

Having discussed young cluster HR diagrams and several independent checks on the observed apparent luminosity spreads, we turn briefly now to simulation of those luminosity spreads.

We consider as a first simplistic look, a situation in which the uncertainty (σ) in individual values of $\log L/L_\odot$ which empirically characterizes the luminosity dispersion in older (main sequence) clusters, ~0.1-0.2 dex. We can then propagate such $\sigma(\log L)$ values into $\sigma(\log \tau)$ values, and compare to empirically inferred (from the observed $\Delta \log L/L_\odot$ distribution, or the apparent luminosity spread) values of $\Delta \log \tau$, or apparent age spread. I show in Figure 3 the predicted trend of $\sigma(\log \tau)$ vs $\log \tau$ compared to the trend actually inferred from young cluster HR diagrams. As can be seen, the assumption of luminosity errors typical of those on the main sequence leads to the expectation that age errors should rise towards younger ages to >100-500% at ages <1-3 Myr.

Slesnick *et al.* (2008) performed a more realistic Monte Carlo simulation that projected on to the HR diagram the combined effects of various errors appropriate for late type pre-main sequence stars. An underlying 5 Myr coeval population was masked by: photometric error of 0.025 mag in each observation band (typical of observations), spectral type error of 1/2 spectral subclasses, (typical of M-types), distance spread in the population (rendering the model cluster as deep as it is wide), and stellar multiplicity with 1/3 of the systems being equal mass binaries (consistent with the mass ratios typically observed for low mass stars and brown dwarfs). The simulation resulted in a mean age and age dispersion totally consistent both visually and statistically with that found for the low mass population in the Upper Scorpius region. In other words, despite the apparent luminosity spread, no true age spread was needed in order to model the data.

Even more sophisticated Monte Carlo simulations were performed by A. Bauermeister in undergraduate thesis work. The models consider different possibilities for both evolutionary tracks and star formation histories, along with realistic input error and binary distributions in order to simulate cluster age spreads. The simulations can be analyzed in the same way as empirical data, e.g. calculating for the resulting HR diagrams the median luminosity as a function of effective temperature, the dispersion and the detailed shape of the $\Delta \log L/L_\odot$ distribution, and the slope of $\log L/L_\odot$ vs $\log T_{eff}$ fit over a limited spectral type range.

Results thus far (as reported previously in Hillenbrand *et al.* 2008) indicate that the main effects of random errors are on the widths of the Gaussian core in the $\Delta \log L/L_\odot$ distributions, and the main effects of binaries are on the shape of the high side of the $\Delta \log L/L_\odot$ distribution. True age spread, if present, may be detectable as additional spread in $\Delta \log L/L_\odot$ present on both the high and low luminosity sides. We have found from extensive K-S testing that when observational errors are modest (~10%) and binarity properties of the underlying population are well understood, age spreads larger than ~15-20% can be distinguished from no age spread or a "burst" star formation scenario. We continue to test the various parameter spaces, including variation of the star formation history (e.g. burst, constant, gaussian, linearly/exponentially increasing or decreasing). We also find that the fitted slope in $\log L/L_\odot$ vs $\log T_{eff}$ can inform the choice of evolutionary tracks, modulo the binarity properties.

4. Findings and implications

Pre-main sequence evolutionary tracks: (1) vary significantly and systematically between theory groups; (2) under-predict stellar masses by 30-50%; (3) under-predict likely low-mass stellar ages by 30-100%; and (4) over-predict likely high-mass stellar ages by 20-100%. The above imply large and systematic uncertainties in both mass and age distributions for young low mass populations, and hence: star formation histories in molecular clouds, disk evolutionary time scales, and angular momentum evolutionary time scales.

The reality of the observed apparent luminosity spreads in recently star forming regions can be tested via detailed correlation of the $\Delta \log L/L_\odot$ distributions with surface gravity indicators, lithium abundance measurements, and perhaps soon seismology checks in certain mass regimes. HR diagram simulations that account for plausible error, binary and other astrophysical effects are needed in order to determine the relevant luminosity spread or $\sigma(L) = \sqrt{\sigma^2_{observed}(L) - \sigma^2_{understood}(L)}$ that might then be assessed as a real luminosity spread for inference of any extended star formation history.

At present, there is only marginal or no strong evidence for moderate age spreads in young clusters. However, this conclusion does not preclude the "popcorn" effect for cluster star formation history, in which a few stars form first, preceding the main event, and a few stars lag, forming last – just like an episode of popcorn production relative to the interval between popcorn events in a typical household microwave oven or other popcorn nursery.

Acknowledgement

This presentation has included results established by my collaborators on various projects: Massimo Robberto and the HST Orion Treasury team plus Aaron Hoffer on the ONC, Adam Kraus on multiplicity and Taurus scattered light sources, Catherine Slesnick plus Davy Kirkpatrick on young star surface gravities, Russel White plus Amber Bauermeister on HR diagram simulations, and Ann Marie Cody on brown dwarf pulsations.

References

Ammler, M., Joergens, V., & Neuhauser, R., 2005, A&A, 440, 1127

Ballesteros-Paredes, J., Hartmann, L., & Vázquez-Semadeni, E., 1999, *ApJ*, 527, 285
Baraffe, I., Chabrier, G., Allard, F., & Hauschildt, P. H. 1995, ApJL, 446, 35
Baraffe, I., Chabrier, G., Allard, F., & Hauschildt, P. H. 1998, AA, 337, 403
Burningham, B., Naylor, T., Littlefair, S. P., & Jeffries, R. D., 2005, *MNRAS*, 363, 1389
D'Antona, F. & Mazzitelli, I. 1994, *ApJS*, 90, 467
D'Antona, F. & Mazzitelli, I. 1997, in Cool stars in Clusters and Associations, ed. R. Pallavicini, & G. Micela, Mem. S. A. It., 68, 807
Da Rio, N., Robberto, M., Soderblom, D. R., Panagia, N., Hillenbrand, L. A., Palla, F., & Stassun, K., 2009, *ApJ*, submitted
Elmegreen, B. G., 2000, *ApJ*, 530, 277
Ezer, D. & Cameron, A. G. W., 1967, Can. J. Phys., 45, 3429
Hartigan, P., Strom, K. M., & Strom, S. E., 1994, *ApJ*, 427, 961
Hartmann, L. W., 2001, *AJ*, 121, 1030
Hillenbrand, L. A., 1997, *AJ*, 113, 1733
Hillenbrand, L. A., Bauermeister, A., & White, R. J., 2008, ASPC, 384, 200
Iben, I. & Talbot, R., 1966, *ApJ*, 144, 968
Kenyon, S. J. & Hartmann, L. W., 1990, *ApJ*, 349, 197
Kirkpatrick, J. D., Cruz, K. L., Barman, T. S., Burgasser, A. J., *et al.*, 2008, *ApJ*, 689, 1295
Kraus, A. L. & Hillenbrand, L. A., 2009, *ApJ*, submitted.
Mamajek, E. E. & Hillenbrand, L. A., 2008, *ApJ*, 687, 1264
Marconi, M. & Palla, F., 1998, *ApJ*, 507, 141
Mentuch, E., Brandeker, A., van Kerkwijk, M. H., Jayawardhana, R., & Hauschildt, P. H., 2008, *ApJ*, 689, 1127
Mouschovias, T. C. 1976, *ApJ*, 207, 141
Nelson, L. A., Rappaport, S., & Chiang, E., 1993, *ApJ*, 413, 364
Palla, F. & Barraffe, I., 2005
Palla, F., & Stahler, S. W. 1993, *ApJ*, 418, 414
Palla, F., & Stahler, S. W. 1999, *ApJ*, 525, 772
Palla, F., Randich, S., Pavlenko, Y. V., Flaccomio, E., & Pallavicini, R., 2007, *ApJ*, 659, L41
Palla, F., Randich, S., Flaccomio, E., & Pallavicini, R., 2005, *ApJ*, 626, L49
Prato, L., Greene, T. P., & Simon, M., 2003, *ApJ*, 584, 853
Schiavon, R. P., Batalha, C., & Barbuy, B., 1995, A&A 301, 840
Sestito, P., Palla, F., & Randich, S, 2008, *A&A*, 487, 965.
Shu, F. H., 1977, *ApJ*, 214, 488
Shu, F. H., Adams, F. C., & Lizano, S. 1987, *ARAA*, 25, 23
Siess, L., Dufour, E., & Forestini, M. 2000, *A&A*, 358, 593
Simon, M., Ghez, A. M., Leinert, Ch., 1993, *ApJ*, 408, L33
Song, I., Bessell, M. S., & Zuckerman, B., 2002, *ApJ*, 581, L43
Slesnick, C. L., 2008, PhD thesis, California Institute of Technology
Slesnick, C. L., Carpenter, J. M., & Hillenbrand, L. A., 2006, *AJ*, 131, 3016
Slesnick, C. L., Hillenbrand, L. A., & Carpenter, J. M., 2008, *ApJ*, 688, 377
Steele, I. A. & Jameson, R. F., 1995, *MNRAS*, 272, 630
Swenson, F. J., Faulkner, J., Rogers, F. J., & Iglesias 1994, *ApJ*, 425, 286
White, R. J. & Hillenbrand, L. A., 2005, *ApJ*, 621, L65
White, R. J., Ghez, A. M., Reid, I. N., & Schultz, G., 1999, *ApJ*, 520, 811
Yi, S., Kim, Y.-C., & Demarque, P. 2003, *ApJS*, 144, 259
Yi, S., Demarque, P., & Kim, Y.-C., 2004, *Ap&SS*, 291, 261
Yi, S., Demarque, P., Kim, Y. -C., Lee, Y.-W., Ree, C. H., Lejeune, T., & Barnes, S. 2001, *ApJS*, 136, 417
Zwintz, K., 2008, *ApJ*, 673, 1088

Discussion

R. MATHIEU: This is the session where we need to address a very important topic for this symposium: What is the meaning of $t = 0$? (Follow up comment: This issue underlies any attempt to define or test coevality in very young binaries or associations. Reversing the question, a coeval ONC population may constrain formation processes and correlations of formation across clouds?)

L. HILLENBRAND: Hayashi would say $t = 0$ corresponds to when the boundary is crossed establishing hydrostatic equilibrium. However, a further concept is that of the stellar birthline or the maximum $R(M)$ that establishes the L and T at which the star first "appears" in the HRD and begins its contraction vertically downward. Now, even if we agree on this, a further problem is the interaction of a young pre-main sequence star with its disk. These stars are accreting so they are increasing in mass by maybe 10% over the pre-MS phase, which causes motion in the HRD towards higher T and L. Further, the clock can be "re-set" by outburst events that increase the luminosity significantly; several of these may occur in the first Myr or more, perhaps contributing to the luminosity dispersion in clusters.

F. WALTER: You have done an extraordinary job compressing a very broad topic into a short period. Consequently, you had to gloss over many details. Two aspects of the environment: 1. The Orion Nebula is viewed through the Orion OBIc association – OBIc is older and surely contaminates the sample to some extent. 2. There may be a difference in age spreads between OB associations and T associations. In the former, feedback from the massive stars may terminate low mass SF by dispersing the gas and dust; no such mechanism exists in the T associations.

L. HILLENBRAND: 1. I very much agree about the foreground contamination from the Orion Ic population which veils the Orion Id population I discussed. Some of these stars are likely present as the low luminosity outliers in the HRD. 2. One of Francesco Palla's slides I did not show included an argument based on the multiplicity of the massive ONC trapezium stars. Francesco demonstrated the locations of these individual components in the HRD and wanted us to appreciate that they are right on the birthline. Therefore, he says, the massive stars indeed formed last and may have terminated star formation as you say.

B. WEAVER: Answer to the question of where is $t = 0$? Choose a unique time which is independent of idiosyncratic starting conditions; therefore: use the ZAMS and use negative time for the PMS For Francesco: why not run models backwards in time to handle starting condition problems?

L. HILLENBRAND: Other than reminding us of the meaning of "quasistatic contraction" I think I should defer this to a theorist (see Pinsonneault comment #1).

M. PINSONNEAULT: 1) Comment: One would expect stars to converge to a unique state on a thermal timescale, so the range in true ages should be comparable to the range in assembly timescales. One does have difficulty in interpreting stars at the birthline. 2) Question: We know from ZAMS eclipsing binaries that spots can affect stellar radii. What information do we have about temperature variations in protostars?

L. HILLENBRAND: 1. See Bruce Weaver's comment above. 2. Surely the stars are spotted, both in the traditional sense of cool spots related to the rotation-driven chromospheric and coronal activity, and because of hot spots due to accretion within rings at high latitudes and having filling factors of a few to maybe 10%. The temperature inferred from low-resolution spectra of pre-MS stars do not account for any such effects.

E. JENSEN: You mentioned cleaning up the Orion Nebula cluster analysis by removing photometrically variable stars. While this will improve the HR diagram, will it bias you toward or against particular types of stars?

L. HILLENBRAND: I don't see any evidence that the median luminosities or effective temperatures are biased, although the dispersion in luminosity is reduced. However, if it is true that so-called CTTS sit higher in the HRD than WTTS as Palla and Bouvier argue for Taurus and if the large-amplitude variable sample is dominated by CTTS as we would expect from accretion effects, your concern is valid.

I. KING: I understand that your beautiful CMD of Orion came from a mosaic of considerable size. What are the chances of getting a second epoch of observation, to get proper motions and cleanup the CMD?

L. HILLENBRAND: There is both archival data and new observations planned with HST over smaller portions of the area covered in the Robberto *et al.* program that are geared towards astrometric studies of the ONC. I recall the beautiful demonstration of the power of such techniques by you and Jay Anderson in NGC 6397.

M. MEYER: Comment: If one constrains an age spread to be coeval it seems to me that would constrain the distribution of infall rates (defining a plausible range of birthlines, cf. Hartmann et al. 1997. Question: In addition to the upper limits to the age spreads in OB associations and Taurus have you constrained them to be below some age in other low-mass T associations?

L. HILLENBRAND: 1. Yes, in addition to all of the other effects already mentioned, any remaining luminosity spread will reflect some combination of spread in initial conditions (one of Palla's main points) plus any intrinsic age spread. 2. I do not yet have a rigorous analysis of each of the regions I discussed observationally. However, what you suggest is exactly the goal of this work.

R. MATHIEU: We have been working on this issue for decades, and the issue of coeval formation remains uncertain. Is there a "killer" observation (or observations) that would/will resolve this?

L. HILLENBRAND: I think membership constraints from kinematics are most crucial, so Gaia and other astrometric studies will be very important, as will large radial velocity surveys.

Lynne Hillenbrand

Ann Marie Cody, Kelle Cruz, and Anna Frebel

The Ages of Stars
Proceedings IAU Symposium No. 258, 2008
E.E. Mamajek, D.R. Soderblom & R.F.G. Wyse, eds.

doi:10.1017/S1743921309031743

Age spreads in star forming regions?

R. D. Jeffries

Astrophysics Group, Keele University, Keele, Staffordshire, ST5 5BG, UK
email: rdj@astro.keele.ac.uk

Abstract. Rotation periods and projected equatorial velocities of pre-main-sequence (PMS) stars in star forming regions can be combined to give projected stellar radii. Assuming random axial orientation, a Monte-Carlo model is used to illustrate that distributions of projected stellar radii are very sensitive to ages and age dispersions between 1 and 10 Myr which, unlike age estimates from conventional Hertzsprung-Russell diagrams, are relatively immune to uncertainties due to extinction, variability, distance etc. Application of the technique to the Orion Nebula cluster reveals radius spreads of a factor of 2–3 (FWHM) at a given effective temperature. Modelling this dispersion as an age spread suggests that PMS stars in the ONC have an age range larger than the mean cluster age, that could be reasonably described by the age distribution deduced from the Hertzsprung-Russell diagram. These radius/age spreads are certainly large enough to invalidate the assumption of coevality when considering the evolution of PMS properties (rotation, disks etc.) from one young cluster to another.

Keywords. stars: pre–main-sequence, stars: rotation, stars: formation

---

1. Introduction

Does star formation take a long time, or is it all over on a dynamical free-fall timescale? This is a keenly debated question in star formation theory, with implications spanning topics as diverse as investigating early star/disk/planet evolution using populations in young star formation regions (SFRs) which are often *assumed to be coeval*, through to assessing overall star formation efficiency and the build up of galactic populations.

According to one paradigm, the collapse of molecular clouds is a quasi-static process slowed by magnetic pressure. The timescale for star formation is governed by ambipolar diffusion and could be $\simeq$10 Myr (e.g. Tan, Krumholz & McKee 2006). Alternatively, on the basis of short deduced molecular cloud lifetimes, others argue that star formation is a rapid process, taking place in compressed filamentary structures on free-fall timescales $\leqslant$ 1 Myr (e.g. Elmegreen 2007).

A crucial piece of evidence for star formation timescales is the presence (or not) of age spreads among stars in young SFRs. Low-mass pre-main-sequence (PMS) stars can be assigned model-dependent ages from their position in Hertzsprung-Russell (H-R) diagrams as they contract along Hayashi tracks. Using this technique several authors (e.g. Palla & Stahler 2000; Huff & Stahler 2006) claim star formation "accelerates" exponentially up to the present day, on timescales of $\simeq$10 Myr. These apparent age spreads favour quasi-static, "slow" star formation. However, conventional H-R diagrams are severely affected by (i) intrinsic variability, (ii) extinction uncertainties, (iii) accretion luminosity, (iv) binarity, (v) distance dispersion – all of which can mimic age spreads where none exist (e.g. Hartmann 2001; Hillenbrand, Bauermeister & White 2008).

In this contribution I illustrate a technique to circumvent these difficulties using the rotational properties of PMS stars. This produces an alternative H-R diagram (radius versus temperature) that can be modelled to reconstruct a star formation history free from the problems above (e.g. see Jeffries 2007a, b).

2. Projected stellar radii

New wide-field surveys are finding rotation periods (P in days) for hundreds of magnetically spotted PMS stars in SFRs. At the same time, it is now possible to obtain projected equatorial velocities ($v \sin i$ in km s^{-1}) for these stars from rotational line broadening using multi-object spectrographs such as FLAMES at the VLT and Hectoechelle at the MMT. Combining these measurements gives geometric estimates of radii, $R \sin i = 0.02\, P\, v \sin i$ (in solar radii). The inclination angle, i, is unknown, but if it is assumed random (for which there is some evidence – e.g. Jackson & Jeffries in these proceedings – and no counter evidence) and the measurement uncertainties are understood, then distributions of $R \sin i$ can be Monte-Carlo modelled to estimate the true R for any group of stars.

As an example of the technique's power, in Fig. 1 I show a *simulation* of what could be achieved by observing projected equatorial velocities for 458 PMS objects with rotation periods in the young SFR NGC 2264 (from Lamm *et al.* 2004 and Makidon *et al.* 2004), with a 10% precision and a threshold for detection of $v \sin i \geqslant 15$ km s^{-1} – which is routinely possible. The simulation assumes that rotation axes are randomly oriented but that objects with $i < 30°$ do not show rotational modulation. The left hand panels show the recovered $R \sin i$ values versus $V - I$ for coeval populations at several ages, where the Siess, Dufour & Forestini (2000, S00) isochrones are used to assign the intrinsic stellar radii. The right hand panels collapse this distribution to 1-dimensional form by normalising $R \sin i$ at each colour by the value of R at 3 Myr.

With typical measurements, a set of 20 $R \sin i$ values can give R to $\pm 5\%$. But at a given colour or $T_{\rm eff}$, R is expected to change by a factor of three between 1 and 10 Myr! Hence the $R \sin i$ distribution is *very sensitive* to age differences and age dispersions in this range (see right panels of Fig. 1), but becomes less so at older ages. Any inferred ages and age spreads are of course model-dependent, but the radii are absolute. The technique is almost immune to problems associated with variability, binarity, extinction uncertainty and accretion luminosity. It is also distance-independent to boot!

3. Results for the Orion Nebula Cluster

The first attempts to use this technique were made in the Orion Nebula Cluster (ONC). The results were described in detail by Jeffries (2007b) and are summarised here. The ONC is a young and populous SFR with a sample of 95 K- and M-stars that have measured rotation periods (Herbst *et al.* 2002), effective temperatures (Hillenbrand 1997) and $v \sin i$ (Rhode, Herbst & Mathieu 2001; Sicilia-Aguilar *et al.* 2005).

I calculated $R \sin i$ for these stars and then simulated the normalised (to $R_{\rm 3Myr}$) distributions using the Monte Carlo model which produced the simulations in Fig. 1. The models were tested against the observed data using the Kolmogorov-Smirnoff statistic on the cumulative distributions. The simulations are insensitive to the threshold i below which it is assumed no rotational periodicity would be found, but *are* sensitive to the choice of radius isochrones. I ran models using the S00 and D'Antona & Mazzitelli (1997, DAM97) isochrones. Uncertainties in periods, $v \sin i$ and $T_{\rm eff}$ were taken from the sources cited above.

The first models I tried were coeval with the age as a free parameter. The best-fitting ages were 1.78 Myr and 0.76 Myr for the S00 and DAM97 isochrones, but both were rejected as good models at the >95% level (see Fig. 2). New models were generated by allowing the radius to spread around a single coeval value. The spread was characterised by a Gaussian σ_r in $\log_{10} R$. These generated good fits with central ages very similar to

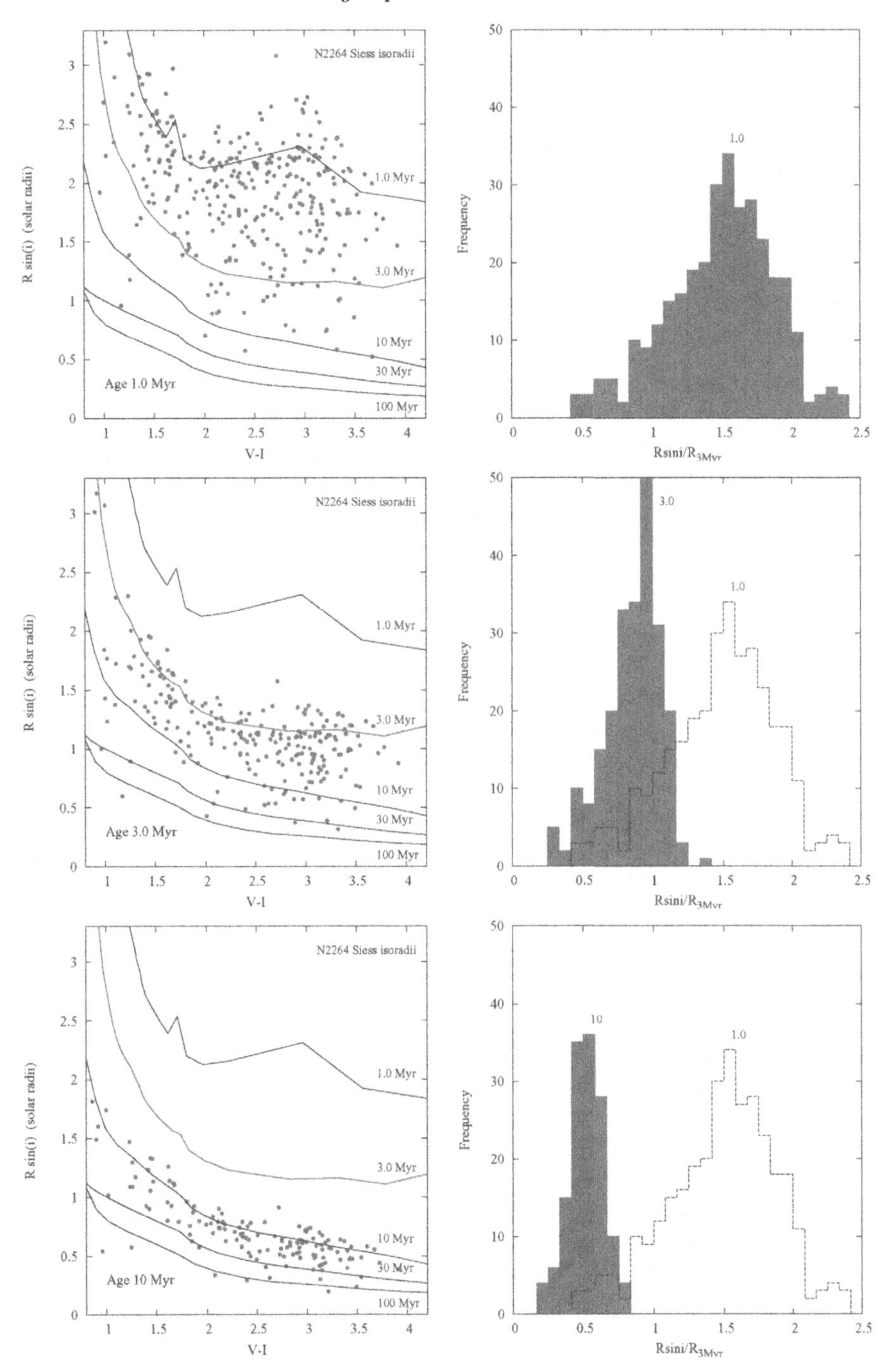

Figure 1. A simulation of the expected $R\sin i$ values that would be obtained from a sample of 458 periodic PMS stars in the young SFR NGC 2264. The left hand panels show $R\sin i$ values versus colour assuming random rotation axis orientation and that only $v\sin i$ values $\geqslant 15\,\mathrm{km\,s^{-1}}$ are detectable. The right hand panels show the 1-dimensional collapsed distribution obtained by normalising by the expected radius at an age of 3 Myr. The simulations include typical measurement uncertainties in colour, $v\sin i$ and period. Each row shows how the distribution would look if the NGC 2264 stars were coeval and at ages of 1, 3 or 10 Myr. The solid lines are radius isochrones from Siess *et al.* (2000).

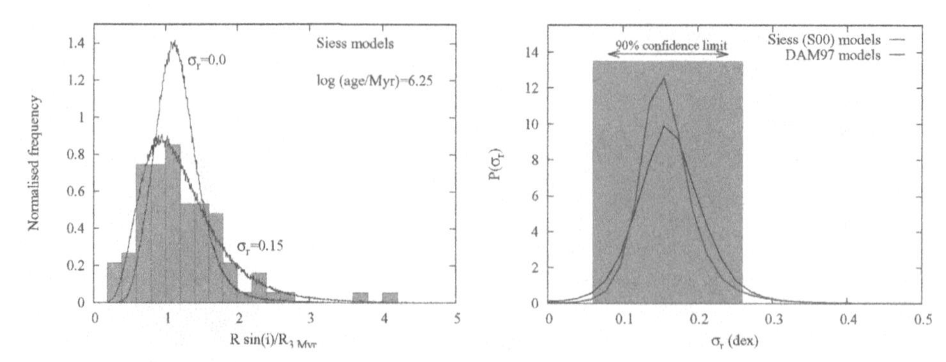

Figure 2. (Left) The measured normalised $R\sin i$ distribution for the ONC compared with a coeval model ($\sigma_r = 0.0$). This model distribution is too narrow. Also shown is a model with a Gaussian spread ($\sigma_r = 0.15$ dex) in $\log_{10} R$ which provides a much better fit. In both cases, the central age for the distribution is 1.78 Myr. (Right) The probability distribution for σ_r using either the S00 or DAM97 isochrones. The 90% confidence region for the S00 isochrones is shown, but the result is almost identical for the DAM97 isochrones. This modelling implies a spread of a factor 2–3 (FWHM) in radius at a given $T_{\rm eff}$.

the previous coeval model, but with $\sigma_r \simeq (0.15 \pm 0.08)$ dex (90% confidence interval) for both sets of isochrones (see Fig. 2). This implies linear radius spreads of a factor of 2–3 (FWHM) at a given $T_{\rm eff}$.

Rather than a simple radius spread it is natural to interpret the results in terms of an age spread. I fitted two types of analytic age spread: a Gaussian spread (σ_a) in $\log_{10}$ age about a central value

$$f(\log_{10}\text{age}) = N_0 \exp\left(\frac{-(\log_{10}\text{age} - \log_{10}\text{central age})^2}{2\sigma_a^2}\right);$$

or an exponentially accelerating star forming rate with timescale λ_a and an abrupt cut-off (or zero-point) age

$$f(\text{age}) = N_0 \exp\left(-\frac{\text{age}}{\lambda_a}\right) \qquad \text{for age} > \text{zeropoint age}.$$

Finally, I modelled the $R\sin i$ distribution by assuming that the stars had the age distribution implied by their positions in the H-R diagram (assuming an ONC distance of 392 pc – Jeffries 2007a).

There are three main results, summarised below.

(*a*) Both classes of model require an age spread ($\sigma_a > 0$, $\lambda_a > 0$ – see Fig. 3). For the Gaussian model the best fitting dispersion $\sigma_a \simeq 0.4$ dex is independent of isochrone choice, but with model-dependent central ages similar to those given by the coeval models. The exponential model has a best-fitting $\lambda_a \simeq 1.1$ Myr for the DAM97 isochrones and $\lambda_a \simeq 1.9$ Myr for the S00 isochrones.

(*b*) The data are incapable of distinguishing between the exponentially accelerating model or the Gaussian spread in $\log_{10}$age.

(*c*) Modelling the $R\sin i$ distribution using the ages derived from the H-R diagram (see Fig. 4) gives a reasonable fit for both sets of isochrones.

4. Discussion

Although the absolute ages and age dispersions derived with this technique are to some extent model-dependent, the absolute radii and radius dispersion are geometric

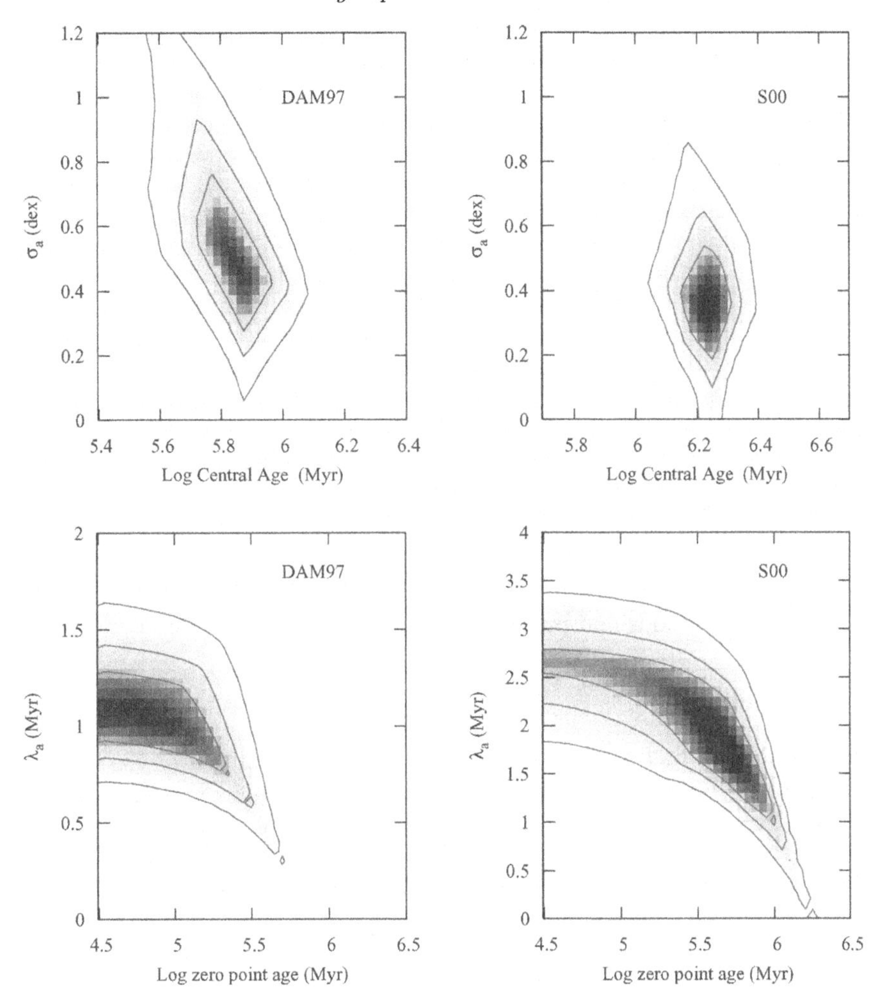

Figure 3. Relative probability distributions of a good fit in the cases of a Gaussian distribution in $\log_{10}$age (top row, with free parameters of a central age and dispersion σ_a in dex) and an exponentially accelerating star formation rate (bottom row, with free parameters of a timescale λ_a and a zero-point cut-off age. For each model we show the results using either the S00 or DAM97 isochrones. The contours enclose 68%, 90% and 99% of the probability.

estimates. We conclude that there is very strong evidence for spreads amounting to factors of 2–3 (FWHM) in radius at a given $T_{\rm eff}$ in PMS stars of the ONC. As PMS tracks are close-to-vertical in the H-R diagram for low-mass stars, this implies order-of-magnitude spreads in moment of inertia – a fact that cannot be ignored when considering the angular momentum evolution of PMS stars in SFRs.

Whether these radius spreads represent real age spreads is a moot point. It is possible that differing accretion histories could lead to luminosity/radius differences for coeval stars of similar present-day $T_{\rm eff}$. However, according to current, non-accreting models, the data imply age spreads in the ONC that are larger than its mean age (>2 Myr for the S00 models), consistent with age spreads judged from its conventional H-R diagram, and certainly large enough to compromise any coeval assumption. In addition, the spreads we have found may actually be underestimates. The rotation sample in the ONC is clearly biased against the faintest (possibly oldest?) stars (see Fig. 4). We cannot comment on

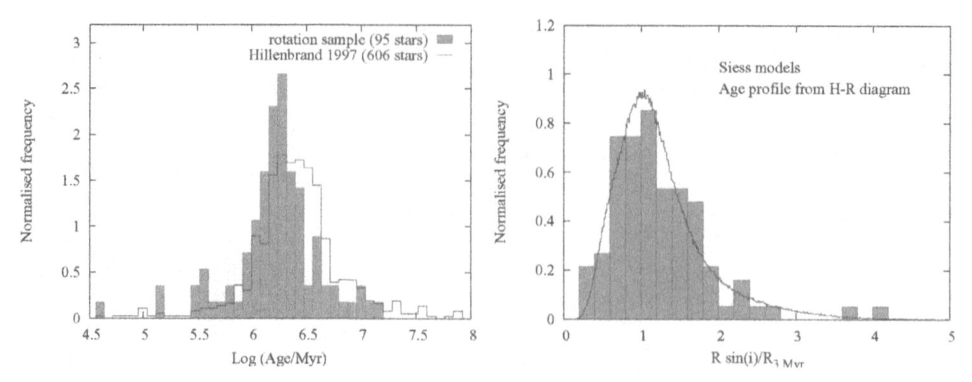

Figure 4. (Left) Normalised age distribution of ONC PMS stars estimated from the H-R diagram and S00 isochrones. I show normalised distributions for the rotation sample and the full sample of Hillenbrand (1997). The rotation sample is missing some of the "oldest" stars. (Right) The $R \sin i$ distribution modelled using the age distribution for the rotation sample in the left-hand panel is a reasonable fit to the observed $R \sin i$ distribution.

whether age spreads as large as 10 Myr are likely until these low-luminosity outliers in the ONC have their periods and projected rotation velocities measured.

References

D'Antona, F. & Mazzitelli, I. 1997, *MemSAI*, 68, 807

Elmegreen, B. G. 2007, *ApJ*, 668, 1064

Hartmann, L. W. 2001, *AJ*, 121, 1030

Herbst, W., Bailer-Jones, C. A. L., Mundt, R., Meisenheimer, K. & Wackermann, R. 2002, *A&A*, 396, 513

Hillenbrand, L. A. 1997, *AJ*, 113, 1733

Hillenbrand, L. A., Bauermeister, A. & White, R. J. 2008, in G. van Belle (ed.), *14th Cambridge Workshop on Cool Stars, Stellar Systems and the Sun*, (San Francisco: Astronomical Society of the Pacific), p. 200

Huff, E. M. & Stahler, S. W. 2006, *ApJ*, 644, 355

Jeffries, R. D. 2007a, *MNRAS*, 376, 1109

Jeffries, R. D. 2007b, *MNRAS*, 381, 1169

Lamm, M., Bailer-Jones, C. A. L., Mundt, R. & Herbst, W. 2005, *A&A*, 417, 557

Makidon, R. B., Rebull, L. M., Strom, S. E., Adams, M. T. & Patten, B. M. 2004, *AJ*, 127, 2228

Palla, F. & Stahler, S. W. 2000, *ApJ* 540, 255

Rhode, K. L., Herbst, W. & Mathieu, R. D. 2001, *AJ* 122, 3258

Sicilia-Aguilar, A., Hartmann, L. W., Hernández, J., Briceño, C. & Calvet, N. 2005, *AJ* 130, 188

Siess, L., Dufour, E. & Forestini, M. 2000, *A&A* 358, 593

Tan, J. C., Krumholz, M. R. & McKee, C. F. 2006, *ApJ*, 641, L121

Discussion

L. HILLENBRAND: Is there an effect related to the absolute value of the rotation period, i.e., does consideration of only rapid rotators or only slow rotators show a systematic effect in the resulting radius distributions?

R. JEFFRIES: We haven't tested that but what I can say is that the slow- and rapid rotators do not occupy distinct locations in the HRD, so I don't expect any differences.

K. COVEY: Lynne Hillenbrand's talk demonstrated that model isochrones and empirical ones have different slopes in L vs. $T_{\rm eff}$ space. If this means that models have mass-dependent systematic errors in radius, you could get an increased spread when you normalize by model values of $R_{\rm 3Myr}$. Do you see any systematic shifts that depend on mass in $R \sin i/R_{\rm 3Myr}$? Marc Pinsonneault asked this more coherently!

R. JEFFRIES: The average value of $R \sin i/R_{\rm 3Myr}$ does not show any major $T_{\rm eff}$ dependence for either of the models considered. Thats because we use stars with (mainly) $T_{\rm eff} < 5500$ K, where the R-color vs. $T_{\rm eff}$ relation is quite flat.

E. JENSEN: If there is a real age spread, does your model include a different distribution of rotational velocities at different ages? If older stars are more slowly- or quickly rotating, more or fewer of them could fall below the observational $v \sin i$ limit.

R. JEFFRIES: No, we use a single true $v_{\rm eq}$ distribution. It is chosen so that in the simulation, the observed $v \sin i$ distribution is well-matched by the simulated $v \sin i$ distribution.

R. MATHIEU: Building on your geometric model, if any of your stars are double-lined spectroscopic binaries, then you get a direct measure of stellar density (assuming alignment of spin and orbital axes). Density and effective temperature give you an age (and mass) from isochrones; several of these give a direct test of coevality. The key for your error budget is the $v \sin i$ measurement.

R. JEFFRIES: Yes, good idea. Uncertainties in $v \sin i$ are typically 10%. We currently exclude SB2s from the analysis, because it may not be clear which star the period "belongs" to. Also, I think the uncertainty in total system mass may be more important than $v \sin i$ errors.

I. KING: Without knowing anything about the underlying astronomy, I am struck by something in the statistics: You see a log-normal distribution of ages. The central limit theorem will produce this if the distribution is the product of a large number of distributions, no matter what their shapes (i.e., the distribution of logs the sum of a large number of logs). Have you any idea of what the physical nature of each of these distributions might be?

R. JEFFRIES: I certainly cannot discern between the exponentially "accelerating" star-forming scenario and a log-normal distribution.

M. ROBBERTO: The rotation period of T Tauri stars is controlled by the accreting disk. In environments like the Trapezium cluster disks seem to have a shorter lifetime being photo-evaporated by the OB stars. Could the rotation period vs. age relation be dependent on the environment and not "universal"?

R. JEFFRIES: Yes it could, but all I have done is estimate the distribution of radii in the ONC, and not made any assumption about how the period evolves. In fact stars of all periods appear at essentially random positions in the ONC HR diagram.

M. Pinsonneault: Do you see difference between hot and cool stars, which might reflect errors in the color dependence of $R \sin i$?

R. Jeffries: See answers to Covey's question.

Rob Jeffries

The Ages of Stars
Proceedings IAU Symposium No. 258, 2008
E.E. Mamajek, D.R. Soderblom & R.F.G. Wyse, eds.

doi:10.1017/S1743921309031755

New methods for determining the ages of PMS stars

Tim Naylor[1], N.J. Mayne[1], R.D. Jeffries[2], S.P. Littlefair[3] & Eric S. Saunders[4]

[1]School of Physics, University of Exeter Stocker Road, Exeter, EX4 4QL, UK.

[2]Astrophysics Group, School of Physical and Geographical Sciences, Keele University, Keele, Staffordshire ST5 5BG UK.

[3]Department of Physics and Astronomy, University of Sheffield, Sheffield S3 7RH, UK.

[4]Las Cumbres Observatory Global Telescope network, 6740 CortonaDrive, Suite 102, Goleta, CA93117, USA

Abstract. We present three new methods for determining the age of groups of pre-main-sequence stars. The first, creating empirical isochrones allows us to create a robust age ordering, but not to derive actual ages. The second, using the width of the gap in colour-magnitude space between the pre-main-sequence and main-sequence (the radiative convective gap) has promise as a distance and extinction independent measure of age, but is as yet uncalibrated. Finally we discuss τ^2 fitting of the main sequence as the stars approach the terminus of the main sequence. This method suggests that there is a factor two difference between these "nuclear" ages, and more conventional pre-main-sequence contraction ages.

Keywords. methods: statistical, methods: data analysis, stars: pre–main-sequence

1. Introduction

Good age determinations for pre-main-sequence (PMS) clusters and associations are crucial for our understanding of this phase of stellar evolution. For example, modelling the interaction of young stars with their (presumably) planet forming discs requires observational measurements of the disc dissipation and stellar spin-up timescales, both of which require accurate ages for the clusters and associations studied. Equally, any determination of the mass functions in young groups is strongly dependent on the assumed age. In this contribution we review three methods we are developing for measuring the ages of PMS clusters and associations using colour-magnitude diagrams (CMDs).

2. Empirical isochrones

Figure 1 shows CMDs for members of a selection of young clusters and associations. The majority of the stars lie on the PMS, which is elevated in the diagram with respect to the main sequence (MS). For older groups the stars lie closer to the MS, and this decline in luminosity with time is an age indicator. If the isochronal models for PMS stars were good fits to the data we could simply use the best-fitting ages. However, in practice the data deviate systematically from the isochrone (e.g. Bonatto 2004; Pinsonneault *et al.*, 2004). Furthermore the derived ages can depend on which colour is fitted (e.g. Naylor *et al.*, 2002). An obvious alternative is to abandon attempts to allocate absolute ages, but develop an age order (or ladder) based on the luminosity of the PMS. Simply plotting the data for different groups in the same diagram does not lead to useful results because the spread in each sequence is large. We therefore fit splines through each sequence, and place

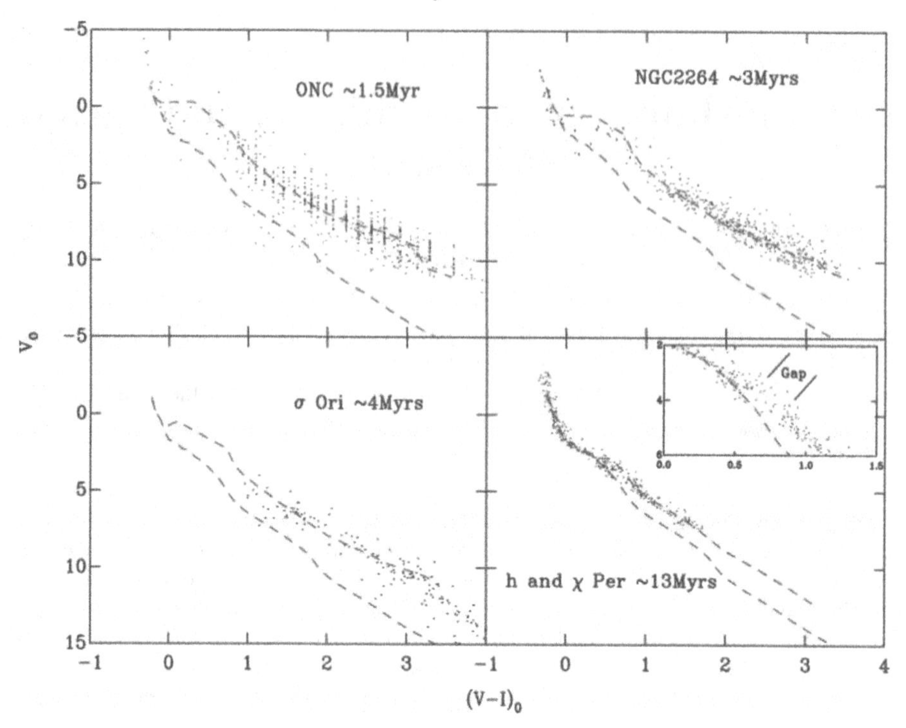

Figure 1. The CMDs for a selection of young groups in absolute magnitude and intrinsic colour. In each case the lower red dotted line is the position of the MS, the upper an appropriate Siess *et al.* (2000) isochrone.

Table 1. The Empirical Isochrone Age Ladder

Age	Groups
1 Myr	IC 5146
2 Myr	ONC, NGC 6530
3 Myr	λ Ori, σ Ori, NGC 2264
4-5 Myr	IC 348, Cep OB3b[1], NGC 2362
5-10 Myr	γ Vel[2]
10 Myr	NGC 7160
13 Myr	h and & χ Per
40 Myr	NGC 2547

Notes: From Mayne & Naylor (2008), except for: [1]Littlefair (in prep); [2]Jeffries *et al.* (2008)

the splines in absolute-magnitude, intrinsic-colour space. The result of such a procedure is shown in Figure 2 (left), where we can see that NGC2264 is older than the ONC, but younger than NGC1960 and about the same age as σ Ori.

We used the above method in Mayne *et al.* (2007) to obtain an age order for a set of well known clusters and associations, but realised that the limiting factor was the determination of the distance (with which the age is degenerate). So, in Mayne & Naylor (2008) we measured the distances to these clusters and associations in a consistent way. As can be seen in Figure 1 the most massive stars are actually on the MS. We therefore fitted these stars to a MS model using τ^2 fitting (see Naylor & Jeffries, 2006 and Section 4) to derive distances. In Table 1 we present the resulting age ordering, including two more associations from more recent work. Although we give ages in Table 1, it is worth emphasising that strictly speaking we only derive an order. The ages represent an informed average of literature PMS ages for these groups.

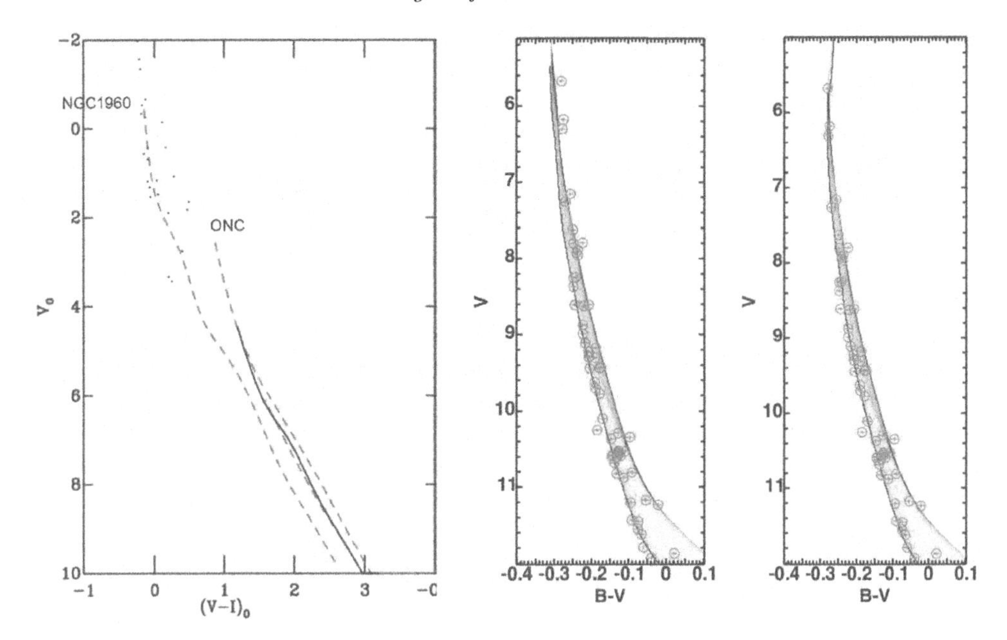

Figure 2. Left: The empirical isochrone for NGC2264 (black curve), compared with other clusters and associations (the red dashed curves). The unmarked red dashed isochrone close to that for NGC2264 is σ Ori. **Right**: Geneva-Bessell isochrones fitted to the NGC6530 data of Walker (1957). Each star has been dereddened using its position in a $U-B/B-V$ diagram. The fit in the left hand panel had the age fixed at 0.25 Myr-old, yielding $Pr(\tau^2) = 0.03$. On the right the age was a free parameter, found to be 5.5 Myr, with $Pr(\tau^2) = 0.67$.

3. Radiative-convective gap

An interesting feature of Figure 1 is the paucity of stars on the PMS isochrone immediately prior to the point where it joins the MS. The gap is clear and wide for the youngest groups, but by the age of h and χ Per has narrowed almost to the point of invisibility. The physical explanation for the gap lies in the change of structure between the fully convective interiors of PMS stars, and the partially radiative ones of MS stars. This drives a change in radius, which happens relatively quickly, leading to a rapid movement to bluer colours in the CMD, as stars move from the PMS to the MS. Hence in Mayne *et al.* (2007) we named this the radiative convective gap.

Clearly this change in the size of the gap with age could be used an age indicator. It has two main advantages over PMS (contraction) ages. First, it uses brighter stars. Second, since one is measuring a distance, rather than position on colour-magnitude space, it is independent of errors in distance or extinction (assuming the latter is uniform). Before it can be used as an age indicator, though, it will need to be calibrated against ages derived from other techniques.

4. Nuclear (not quite turn-off) ages

Before stars turn off the MS, they evolve redwards away from the zero-age MS, driven by their nuclear evolution. This means that in colour-magnitude space the MS, which normally has a positive gradient can, near its high-mass terminus, be vertical or even have a negative gradient (see Figure 2 (right)). Although the effect is subtle, if we can fit it, this should provide an age indicator. The best method for doing this is the τ^2 method we described in Naylor & Jeffries (2006), since this allows for the effects of binarity, gives reliable uncertainties for the parameters, and provides a goodness-of-fit test. Unfortunately the technique as we described it will not work if the isochrone is vertical. Therefore we first outline the improvements which we have made to the method to allow us to address this problem (which will be described in more detail in Naylor in prep.), before moving on to our results.

4.1. *Improvements to the τ^2 technique*

The technique relies on a finely sampled grid (such as the colour scales of Figure 2 (right)), which is a model created from of order a million simulated stars. We refer to this as $\rho(c, m)$ where c and m are the colour and magnitude co-ordinates respectively. For any given datapoint we calculate τ^2 by multiplying this grid on a point-by-point basis with a function representing the datapoint and its uncertainties (typically a two-dimensional Gaussian). We represent this as $U_i(c - c_i, m - m_i)$, where i is the index of a datapoint at co-ordinates (c_i, m_i). If we now sum the resulting points, and repeat this over all datapoints we arrive at the definition of τ^2

$$\tau^2 = -2 \sum_{i=1,N} \ln \int U_i(c - c_i, m - m_i)\rho(c, m)\mathrm{d}c\ \mathrm{d}m. \qquad (4.1)$$

The best-fitting model corresponds to the lowest value of τ^2. For example, a simple fit in distance modulus can be viewed as moving the models up and down in Figure 2 (right) until the "cross correlation" between the datapoints and the model is maximised. In practice the simplest way to find the best fitting model is to calculate τ^2 for a grid of models covering the range of parameters of interest.

There is a question as to how ρ should be normalised. In Naylor & Jeffries (2006) we derived a normalisation such that Equation 4.1 reduced to that for χ^2 for fitting a curve to data with uncertainties in one dimension. Unfortunately, when the isochrone is vertical, this results in an infinity in Equation 17 of that paper making it impractical for post-main-sequence fitting. Instead we now use a normalisation where the integral of ρ between the faintest and brightest datapoints is one. Similarly we demand that the integral of U is one over the entire CMD.

Having found a fit to the data, we must establish whether it is a good fit. We do this by calculating the probability that we would obtain our value of τ^2 from observations, assuming the model was correct. This is $Pr(\tau^2)$, which we showed how to calculate for no free parameters in Naylor & Jeffries (2006). Our suggested correction to allow for free parameters, multiplying the values of τ^2 by $(N - n)/N$ (where N is the number of datapoints and n the number of free parameters), is not invariant under changes in normalisation. A better approximation is to subtract the expectation value of τ^2 before multiplying by $(N - n)/N$, and then add the expectation value on again.

Finally, to find the uncertainties in the parameters we have found a quicker method than that we presented in Naylor & Jeffries (2006). Assuming the minimum value of τ^2 has been found by a grid search, each datapoint in the grid has a probability P associated with it (via the definition of τ^2) of

$$P = e^{-\tau^2/2}. \qquad (4.2)$$

By summing the probability below a given τ^2, and dividing by the probability summed over the entire grid, one can obtain the probability that τ^2 lies below a given value. This allows one to draw a confidence contour in the parameter space, in an identical fashion to that used in χ^2 analysis.

4.2. *Results*

We have fitted UBV photometry for bright stars in NGC6530, NGC2264, σ Ori, λ Ori, NGC2362, Cep OB3b, NGC2547, IC2602 and stars in the vicinity of the Orion Nebula Cluster. For preference we have used data from the 1950s to 1970s of Walker, Johnson and collaborators, since this is a relatively homogeneous group of datasets, and we find our models fit them well. We have used the Geneva-Bessell models described in Mayne & Naylor (2008). We first use a $U - B/B - V$ diagram to determine the extinction. In some cases the extinction is uniform, and we determine its value using τ^2 fitting. In other cases we find the extinction is non-uniform, and we find the extinctions on a star-by-star basis by comparison with the isochrone as described in Mayne & Naylor (2008). This is essentially an updated Q method. We then perform a grid search in both age and distance to derive ages and associated uncertainties. Unsurprisingly our distances are all consistent with those in Mayne & Naylor (2008), but for the groups less than 10 Myr old, we find the ages are a factor 1.5-2.0 larger than those given in Table 1.

Before attaching any significance to this result, we questioned whether it could be due to either our fitting procedure or the models used. We have experimented with models without convective overshoot, which we find give poor fits to the data, and with the Padova models (Girardi *et al.* 2002), which we find give similar answers to the Geneva-Bessell isochrones, though both models have yet to be tested exhaustively. We also find that our age for stars in the vicinity of the ONC (5 Myr) is similar to that found using the same dataset by Meynet *et al.* (1993). For IC2602 Mermilliod (1981) obtains a nuclear age of 35 Myr, which compares favourably with our estimate of about 40 Myr. For NGC 2547 we obtain about 45 Myr, compared with Claria (1982) who obtains 57 Myr. Since our ages are broadly consistent with other turn-off/nuclear ages, we can rule out some systematic effect from our models or fitting procedure. We are, therefore, forced to conclude that this is a genuine discrepancy between PMS ages and nuclear ages.

For the groups older than 10 Myr it is harder to be definitive about any difference. The problem is we need both PMS photometry (to obtain an age on the same scale as Table 1) and good photo-electric photometry to obtain a nuclear age. We can carry out the test for NGC2547 for which we obtain 45 Myr, compared with PMS and Lithium depletion ages of about 38 Myr (Naylor & Jeffries, 2006) suggesting the discrepancy decreases with age. For IC2602 the situation is more ambiguous. The age we obtain (40 Myr) is larger than the PMS age of 25 Myr found by Stauffer *et al.* (1997), but is consistent with the finding of Jeffries *et al.* (2000) that IC2602 is a little older than NGC2547. Thus the question as to whether this discrepancy disappears for older clusters (and if so at what age) must await obtaining further PMS ages for older clusters.

4.3. *Implications*

It appears we have found a genuine difference between the PMS contraction and MS nuclear age scales. It is hard at this stage to decide which scale is correct, but as most modern work relies on the PMS age scale, it is interesting to examine the implications of the nuclear age scale being correct. It would help address two outstanding problems in the area. First, it is well known that that there appears to be a lack of clusters in the age range 5-30 Myr (Jeffries *et al.*, 2007). Changing the ages in the way we suggest would fill that gap, especially if there is a return to the classical age scale for clusters older than 30 Myr. Second there is a problem in the disc-clearing timescale measured from IR observations of young stars (of order 3 Myr, e.g. Briceño *et al.*, 2007) and the time required to form a planet by classical core accretion (perhaps 9 Myr; Pollack *et al.*, 1996). Whilst there are active attempts to solve this problem by reducing the theoretical

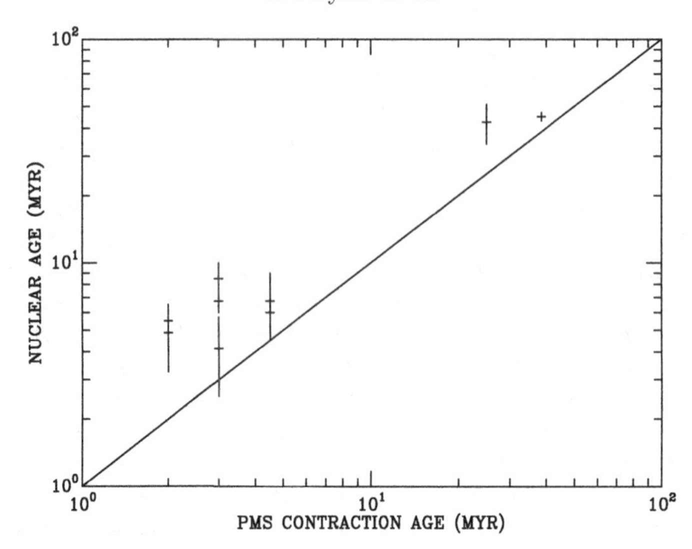

Figure 3. PMS vs nuclear ages for young clusters and associations. The datapoints would lie on the line if they were equal. The error bars show the uncertainties derived from the τ^2 fitting.

timescale (e.g. Dodson-Robinson *et al.*, 2008 and references therein), our work supports a different solution - increasing the observed timescale by a factor two.

References

Bonatto, C., Bica, E., & Girardi, L. 2004, *A&A* 415, 571

Briceño, C., Preibisch, T., Sherry, W. H., Mamajek, E. A., Mathieu, R. D., Walter, F. M., & Zinnecker, H., *in proceedings of Protostars and Planets V* p345

Claria, J. J., 1982, *A&AS* 47, 323

Dodson-Robinson, S. E., Bodenheimer, P., Laughlin, G., Willacy, K., Turner, N. J., & Beichman, C. A., 2008, 2008arXiv0810.0288D

Girardi, L., Bertelli, G., Bressan, A., Chiosi, C., Groenewegen, M. A. T., Marigo, P., Salasnich, B., & Weiss, A. 2002, *A&A* 391, 195

Jeffries, R. D., Totten, E. J., & James, D. J., 2000, *MNRAS* 316, 950

Jeffries, R. D., Oliveira, J. M., Naylor, T., Mayne, N. J., & Littlefair, S. P. 2007, *MNRAS* 376, 580

Jeffries, R.D., Naylor, T., Walter, F. M., Pozzo, M. P., & Devey, C. R., 2008, *MNRAS, accepted*, arXiv:0810.5320

Mayne, N. J. & Naylor, T., 2008, *MNRAS* 386, 261

Mayne, N. J.,Naylor, T., Littlefair, S. P., Saunders, E. S., & Jeffries, R. D., 2007, *MNRAS* 375, 1220

Mermilliod, J. C., 1981, *A&A* 97, 235

Meynet, G., Mermilliod, J.-C., & Maeder, A., 1993, *A&AS* 98, 477

Naylor, T., Totten, E. J., Jeffries, R. D., Pozzo, M., Devey, C. R., & Thompson, S. A. 2002, *MNRAS* 335, 291

Naylor, T. & Jeffries, R. D. 2006, *MNRAS* 373, 1251

Pinsonneault, M. H., Terndrup, D. M., Hanson, R. B., Stauffer, J. R. 2004, *ApJ* 600, 946

Pollack, J. B., Hubickyj, O., Bodenheimer, P., Lissauer, J. J., Podolak, M., & Greenzweig, Y., *Icarus* 124, 62

Siess, L., Dufour, E., & Forestini, M. 2000, *ApJ* 358, 593

Stauffer, J. R., Hartmann, L. W., Prosser, C. F., Randich, S., Balachandran, S., Patten, B. M., Simon, T., & Giampapa, M., 1997, *ApJ* 479, 776

Walker, M. F. 1957, *ApJ* 125, 636

Discussion

E. MAMAJEK: Your ages for high masses are critically dependent on converting the reddening-free Q-parameter and de-reddening the stellar photometry. There are issues with inferring T_{eff} from Q and $(B-V)_0$, i.e., disagreements among various studies. There are also uncertainties in the calibration of T_{eff} for OB stars. Could you discuss how these uncertainties affect the certainty of your conclusions that the pre-MS ages are a bigger problem than the ages of the high mass stars?

T. NAYLOR: So far I have only tested the Padova and Geneva isochrones. They agree which is more than any two PMS isochrones do! If the extinction were a problem I would expect to see a correlation with A_V, which I don't.

F. WALTER: Just to clarify, is this an age offset or a stretching of the PMS time-scale?

T. NAYLOR: At this point it is hard to say whether it's an offset, say 4 Myr or a multiplicative factor which decays to zero by, say, 10 Myr. The statistics are just too poor.

Janusz Kaluzny and Hashima Hasan

Patrick Young

The Ages of Stars
Proceedings IAU Symposium No. 258, 2008
E.E. Mamajek, D.R. Soderblom & R.F.G. Wyse, eds.

doi:10.1017/S1743921309031767

Circumstellar disk evolution: Constraining theories of planet formation

Michael R. Meyer

Steward Observatory, The University of Arizona
933 N. Cherry Avenue, Tucson, AZ 85721 (USA)

Abstract. Observations of circumstellar disks around stars as a function of stellar properties such as mass, metallicity, multiplicity, and age, provide constraints on theories concerning the formation and evolution of planetary systems. Utilizing ground- and space-based data from the far–UV to the millimeter, astronomers can assess the amount, composition, and location of circumstellar gas and dust as a function of time. We review primarily results from the Spitzer Space Telescope, with reference to other ground- and space-based observations. Comparing these results with those from exoplanet search techniques, theoretical models, as well as the inferred history of our solar system, helps us to assess whether planetary systems like our own, and the potential for life that they represent, are common or rare in the Milky Way galaxy.

Keywords. solar system: formation, stars: circumstellar matter, pre–main-sequence, planetary systems: protoplanetary disks, planetary systems: formation

---

1. Introduction

Are there multitudes of planetary systems capable of harboring life, like our own Solar System? Answering this question motivates the research activities of a great number of astronomers, as well as scientists of many disciplines. Yet the answer depends on which aspect of our solar system to which one is comparing the physical properties of other systems. Extrapolation of radial velocity results to 20 AU suggests that planets with mass at least a third that of Jupiter's surround 15–20 % of sun-like stars (Cumming *et al.* 2008). Yet lower mass planets might turn out to be even more common (e.g. Mayor *et al.* 2009). Enormous progress has been made in the past several years on many aspects of circumstellar disk evolution (e.g. Meyer *et al.* 2007), especially those that can be addressed with observations from the Spitzer Space Telescope (Werner *et al.* 2006). In this review we explore answers to three key questions: 1) What is the time available to form gas giant planets? 2) What is the history of planetesimal collisions versus radius? 3) How do answers to the above vary with stellar properties? Because the answers to these questions are subtle, one needs large stellar samples with reliable stellar ages from the youngest pre–main sequence stars to the oldest stars known in the galactic disk. In our attempt to study important evolutionary processes in the formation and evolution of planetary systems, we assemble groups of stars with like properties (such as a narrow range in stellar mass) as a function of age, the main topic of this symposium. We hope that by studying the mean (as well as the dispersion) in those properties of the circumstellar environment as a function of time, we can create a "movie" (or range of plausible trajectories) that helps tell the story of how our solar system might have formed. Once completed for one range in stellar mass, we can attempt to repeat the study for other stellar masses. Examining the differences in circumstellar disk evolution as a function of stellar mass may be our best tool in delineating the most important physical processes in planet formation. It is a lofty goal, and often strong assumptions are required to make progress. We can only hope that most of these assumptions represent hypotheses we can test in the near future.

Observations of circumstellar gas and dust, both its amount and geometrical distribution, can be compared to theoretical timescales for its expected evolution. Keplerian orbits can range from days to millennia. The viscous timescale in the context of an α disk model depends on the orbital radius and can be < 1 Myr within 10 AU for reasonable parameters (Hartmann 1998). Preliminary results suggest that disk chemistry proceeds more slowly than relevant dynamical times indicating that mixing could be important (Bergin *et al.* 2007). The inward migration of solids in the disk results in the loss of planet–building material and remains a serious problem on many scales: a) gas drag on meter–sized bodies can reach 1 AU/century (Weidenschilling, 1977); b) Type I migration of lunar–mass planetary embryos on timescales of 10^5 yrs; and c) Type II migration of forming gas giant planets on timescales proportional to the viscous time (e.g. Ida & Lin, 2008 and references therein). The timescale for orderly growth of bodies through collisions (e.g. Goldreich *et al.* 2004) is proportional to the product of the radius and volume density of typical particles divided by the product of the mass surface density of solids and the orbital frequency. The timescale for radiogenic heating of forming planetesimals is set by the relative abundances of radioactive nuclides as well their half-lives (Sanders & Taylor, 2005). Current models suggest that photoevaporation of gas from illumination by EUV, FUV, and x–ray emission can disperse primordial gas disks around sun-like stars on timescales < 10 Myr (Gorti & Hollenbach 2009; Ercolano *et al.* 2009). Additional physical processes are important in debris disk evolution. In most (if not all) debris disks studied to date (our solar system dust disk being a notable exception), mutual collisions between dust grains reduce particle sizes to the blow–out limit where radiation pressure efficiently removes them from the system (Wyatt 2008). The Lyapunov timescale characterizes divergence of orbital elements in a chaotic system and is related to the timescale for instability in planetary systems (Murison *et al.* 1994). In this context, we note that our own planetary system is thought to be stable on timescales comparable to its present age of 4.56 Gyr (Laskar 1994) while the main sequence lifetime of the Sun is approximately twice this.

In what follows, we describe some of the observational evidence for primordial disk evolution and resulting constraints on theories of planet formation. We follow with a brief discussion of planet formation, the evolution of planetesimals belts, and the dust debris they generate. We conclude with a summary and look forward to exciting developments we can anticipate in the years ahead.

2. Primordial circumstellar disks

It is now well established that most sun–like stars form surrounded by circumstellar disks. These disks are primordial mixtures of gas and dust initially resembling the composition of the interstellar medium from which the star-disk systems form. Near–infrared excess emission traces the hottest inner disk structures within 0.1 AU. While the inner edge of the dust disk is determined by the location at which the dust particles sublimate ($\sim$ 1400 K for typical silicates; Muzerolle *et al.* 2003), the gas disk can extend inward, perhaps terminating near the boundary set by magnetospheric accretion theory (e.g. Shu *et al.* 1994). The pioneering work of Strom *et al.* (1989) suggested that inner disks dissipate on timescales of 10 Myr from observations of K–band excess toward T Tauri stars in the Taurus dark cloud. Haisch *et al.* (2001) surveyed near–IR excess emission using color–color diagrams toward hundreds of stars in several young clusters concluding that the mean inner disk lifetime is approximately 3 Myr. Yet if the spectral type of star is known, and multi–color optical/infrared photometry is available, one can carefully separate the effects of intrinsic stellar colors, reddening due to dust along the line of sight toward the star, and circumstellar disk excess determining with greater precision the magnitude of any excess emission (e.g. Meyer *et al.* 1997). This approach was used by Hillenbrand (2008) and colleagues to construct the evolutionary diagram shown in Figure 1 which we have adapted to include: a) relative formation timescales for solar system objects as determined from measurements of extinct radioactive nuclides from meteorite samples (Scott, 2007; Jacobsen, 2005); and b) the frequency distribution of inner disk lifetimes based on these evolutionary diagrams. While the mean disk lifetime is 3 Myr, there is a large *dispersion* of inner disk lifetimes. Andrews & Williams (2005) have observed the distribution of disk masses for young stars ranges over two orders of magnitude from millimeter wave continuum data. Perhaps the distribution of initial conditions (specific angular momenta in collapsing cloud cores) results in a distribution of disk masses and thus disk lifetimes.

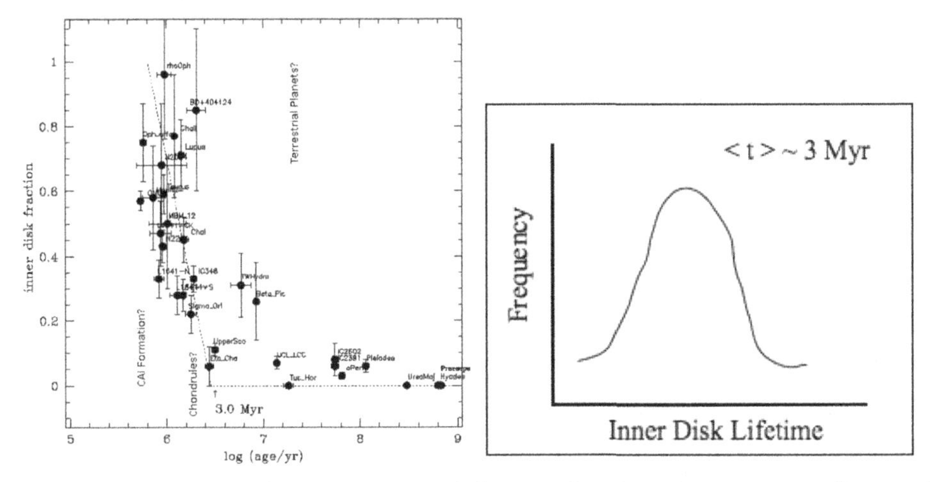

Figure 1. *Left:* Fraction of inner accretion disks as a function of time surrounding sun-like stars (roughly 0.5–2 $M_{\odot}$) as traced by near–IR excess emission observed toward stars with known spectral types. Figure adapted from Hillenbrand (2008) with the approximate timescales for the formation of solar system objects indicated from study of extinct radioactive nuclides. *Right:* Schematic representation of *distribution* of inner disk lifetimes obtained by subtracting adjacent bins in the left panel.

How long does it take for material in the inner disk to transition from optically–thick to optically–thin? Skrutskie *et al.* (1990) provided some preliminary answers based on 10 μm observations of T Tauri stars. They found that most stars lacking 2 μm excess also lacked mid–infrared excess emission. A small handful of objects, the so–called transition objects, exhibited modest mid–IR excess but lacked evidence for hot dust. From the ratio of the number of objects in transition compared to samples of T Tauri stars (few % with respect to T Tauri stars with primordial accretion disks, CTTS, or a smaller fraction of young stars overall) times the typical age of the sample stars (1–3 Myr), they estimated the duration of the transition phase to be very short ($\sim 10^5$ years). Subsequent work has verified these estimates. For example, Silverstone *et al.* (2006) found no examples of inner dust disks in systems that lack signatures of on–going gas accretion from the disk onto the star (cf. Cieza *et al.* 2007). However with surveys of star-forming regions enabled by the Spitzer Space Telescope, dozens of these rare objects can now be identified (e.g. Merin *et al.* 2008).

Much attention is focused on understanding the nature of transition disks (Najita *et al.* 2007; Alexander & Armitage 2007). Some appear to have inner cleared regions due to the presence of previously unseen faint companions at small orbital radii (Ireland & Kraus, 2008). Work continues to understand the detailed distribution of gas and dust in these systems (e.g. Espaillat *et al.* 2007). One troubling feature in all of these disk studies is the general *lack* of correlation between key physical variables and observational properties of the disks (Watson *et al.* 2009). For example, Pascucci *et al.* (2008) found no difference in dust properties between binary and single stars. Some disks can be quite long lived (more than 10 Myr old) and still appear very similar to typically much younger classical T Tauri stars (e.g. PDS 66; Cortes *et al.* 2009). Often morphological ordering of disk spectra or other properties can be quite compelling (e.g. Bouwman *et al.* 2008). But more often than not, ranking by estimates of stellar age appear to disturb the apparent evolutionary sequences. Although errors in determining stellar ages are a likely culprit, it does seem that there are "hidden variables" contributing significantly to primordial disk evolution.

As a complement to studies of the dust, can we probe further the evolution of the gas from which giant planets might form? For young stars in the Taurus dark cloud, there is an excellent correspondence between spectroscopic signatures of gas accretion from the inner disk onto the surface of the star and the presence of near–IR excess emission (Hartigan, Edwards, and Ghandour, 1995). Indeed the evolution of accretion rates appears to mimic to some extent the evolution of hot dust (Hartmann *et al.* 1998; Gatti *et al.* 2008). Spitzer spectra are starting to reveal interesting chemistry in the disks young young stars (Carr & Najita, 2008; Pascucci *et al.* 2009). High resolution near–infrared spectroscopy combined with high spatial resolution can reveal offsets between the emitting regions of various molecules (Pontoppidan *et al.* 2008). Millimeter wave observations are required to trace rotational transitions of cool gas at large radii (e.g. Dutrey *et al.* 2007).

Yet the bulk of the mass in these disks is in molecular hydrogen, which is difficult to detect. Some efforts have been made to trace molecular hydrogen directly by observing the pure rotational lines in the mid–infrared from the ground at high spectral resolution (Bitner *et al.* 2008). The Spitzer Space Telescope also provided a platform to search for gas using the IRS in high resolution mode. Based on equilibrium chemical models of Gorti & Hollenbach, 2004), Pascucci *et al.* (2006) report non–detections from the FEPS survey (Meyer *et al.* 2006) for stars with ages between 3–100 Myr that lack signatures of accretion, but possess optically–thin dust emission. The upper limits place constraints of < 10 Myr on the timescale to form gas giants in these systems. These timescales are of interest for comparison to models of gas giant planet formation through classical core accretion (e.g. Lissauer & Stevenson 2007) as well as gravitational fragmentation which proceeds more quickly (Durisen *et al.* 2007). Future observations with Herschel and SOFIA will be powerful probes of even small amounts of residual emission perhaps placing constraints on the formation of super-earths and ice giants. We note for completeness that circumstellar gas has been detected for some debris disks (e.g. Dent *et al.* 2005), which are the subject of the next section.

3. Planet formation and generation of debris dust

How does remnant disk material evolve when the bulk of the gas capable of forming giant planets has dissipated? In the classical theory, protoplanets initially grow in an orderly way through collisions of comparable sized bodies. As gravitational focusing becomes important, the larger bodies grow fastest in a runaway mode until a modest number of "successes" reach the local isolation mass, essentially consuming all material within several times their gravitational sphere of influence (Hill radius). These oligarchs then perturb each other over time resulting in handfuls of collisions that create a small number of planets (e.g. Nagasawa *et al.* 2007). At 1 AU around the Sun, we imagine the Earth as built through the collision of several Mars-sized objects.† It is thought that similar processes operate (in the presence of gas) to form the cores of giant planets. In the first phase, we can estimate the frequency of collisions by knowing the surface density of solids (σ) as well as the orbital timescale (Ω):

$$\tau \sim (R_{body} \times \rho_{body})/(\sigma \times \Omega)$$

where R and ρ are the radius and volume density of the object built through these collisions in time τ. If we combine estimates of the dependence of disk mass on star mass ($\sigma_{disk} \sim M_*$) and orbital radius ($\sigma_{disk} \sim 1/a$) with the orbital frequency we get $\tau \sim a^{5/2} M_*^{-3/2}$ for a constant ρ. This implies, within a fixed time interval, disks around stars of higher mass form: a) more massive planets at a given orbital distance; and b) planets of a given mass at larger orbital separation. Both are consistent with recent radial velocity results (Johnson *et al.* 2007). Yet it appears that primordial disks evolve more quickly around stars of higher mass (Carpenter *et al.* 2006), complicating the implications of this simple picture for planet formation around stars as a function of their mass.

Observations of debris disks have become powerful tools to study the evolution of planetary systems and their value has increased as we have learned more about the asteroid and Kuiper belts in our own Solar System (Wyatt 2008). As cooler dust orbiting at larger radii emits at longer wavelengths, we can use the wavelength dependence of excess emission as a proxy for its location in the absence of resolved images. With the Spitzer Space Telescope, we can detect $> 10^{-5}$ Earth masses of dust if it is found in micron–sized particles from photometric surveys at 24 μm, efficiently tracing material with temperatures above 100 K (often tracing radii $<$10 AU around sun–like stars). Meyer *et al.* (2008) report the results from FEPS for 24 μm excess emission around an unbiased parent sample of 309 stars. They find that 10–20 % of sun–like stars show evidence for 24 μm excess over an age range from 3–300 Myr with a significant drop in the frequency of 24 μm excess for older stars. As we can only observe the product of the frequency of this phenomena and its duration, the interpretation of these results is ambiguous. They also suggest that the frequency of excess *may* be higher for stars in open clusters compared to field stars of comparable age though the evidence for this is not significant. Note that the ages of the open cluster stars are the "gold standard" for stellar ages in these studies and thus it is difficult to compare results between them and field stars. In contrast, Greaves *et al.* (2009) find a surprising lack of emission at 1.3 mm towards sun–like stars in the Pleiades open cluster despite significant improvement in sensitivity compared to previous observations.

The above mentioned Spitzer results are consistent with other surveys for 24 μm excess around FGK stars as a function of time (e.g. Siegler *et al.* 2007). It is worth remembering that these survey limits are only able to detect dust producing planetesimal belts that are about $\times$1000 brighter than our own inner zodiacal dust generated from collisions in the asteroid belt. Wyatt *et al.* (2007) compare these observations to models of collisional evolution for planetesimal belts finding that some warm debris excesses (e.g. HD 69830) are so large that they cannot be the result of pure collisional evolution and must be transient. Although the models of Kenyon & Bromley (2004) explain the observed behavior in a qualitative way, updates to the input physics continue to improve the agreement

† The last of these collisions resulting in the formation of the Moon (e.g. Canup 2004).

between the models and data (Bromley & Kenyon 2008). Models for several debris disks detected with Spitzer require extended distributions of dust (Hillenbrand *et al.* 2008) consistent with resolved images of some sources (Ardila *et al.* 2004).

What (if any) correlation should we expect between the presence of dust debris and the frequency of gas giant planets detected through radial velocity variations? One might speculate that a disk rich in heavy elements capable of forming gas giant planets (e.g. Fischer & Valenti 2005; Santos *et al.* 2004) might initially have a high surface density of solids and thus produce a lot of dust early–on, outshining a system that lacked enough dust to form giant planets. If however, such a system produces several planets unstable to mutual perturbations, dynamical rearrangement could deplete dust-producing planetesimals in the inner and outer regions as suggested by the Nice model for the early evolution of our Solar system (e.g. Tsiganis *et al.* 2005). Such a system could then become a very weak dust producer at late times (Figure 2). It is in part these complexities that prevent us from making comparisons of Spitzer observations to what is known about our Solar system.

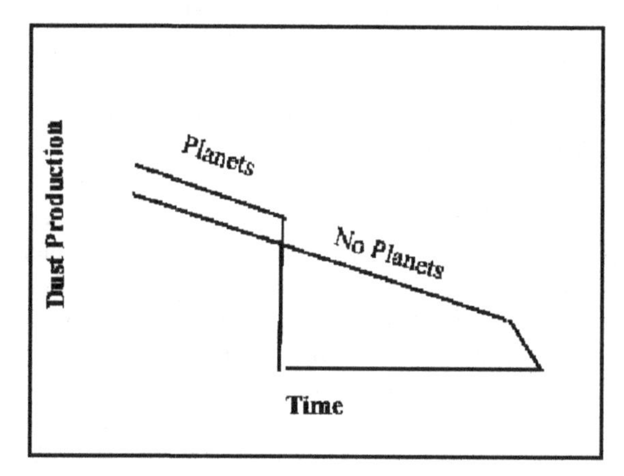

Figure 2. Graphical representation of dust production versus time for the case of a disk with an initially high surface density of solids that forms gas giant planets and another with a lower initial surface density that does not. The planet-forming disk outshines the other disk early on, but then suffers a dynamical event that removes most of the dust-generating planetesimals. The disk that was never able to form giant planets might generate more dust at late times.

Moro–Martin *et al.* (2007a) search for a correlation between the presence or absence of excess with gas giant planets and was unable to confirm any (though the sample size was modest). Apai *et al.* (2008) also searched (without success) for the presence of gas giant planets around stars for which large inner holes in the dust debris had been inferred. There are, however, remarkable examples of systems with evidence for planets that also maintain debris which deserve special study (Moro–Martin *et al.* 2007b; Lovis *et al.* 2006) including the recently announced direct imaging results for Fomalhaut (Kalas *et al.* 2008) and HR 8799 (Marois *et al.* 2008). We can also ask whether debris is correlated with the heavy element abundance of the star as it is famously with the probability of having a gas giant planet. Greaves, Fischer, & Wyatt (2006) find that the metallicity distribution of stars with cold debris detected in the sub–millimeter is consistent with having been drawn from the same distribution as the parent sample of stars. We find a similar result for stars in the FEPS sample (Najita *et al.*, in preparation). Perhaps gas giant planet formation is a *threshold* phenomena with regard to the heavy element content of the primordial circumstellar disk whereas the requirements for ending up with a dust producing debris disk are more forgiving. Bryden *et al.* (2006) estimate that Spitzer only detects the tip of the iceberg in terms of dust debris, and survey results to date are *consistent* with our Solar System IR emission being near the mean for sun–like stars in the Milky Way.

Given the mixed results for the correlation of debris with planets, can we expect more order in the behavior of debris with stellar multiplicity? There is considerable literature on the similarities and differences in disk evolution between single and binary T Tauri stars (Monin *et al.* 2007). With regard to debris around one or both components of a binary system, Trilling *et al.* (2007) report Spitzer results finding a complex mix of behaviors including circumbinary debris disks, circumstellar disks in wide binary systems, as well as the apparition of dust at inferred orbital distances consistent with the separation of the binary companion! At minimum, one can infer that debris dust as detected by Spitzer is not inhibited in multiple systems.

The "last word" from the FEPS program comes from Carpenter *et al.* (2008; 2009) in which the extant database and synthesis of results are described respectively. With the addition of the IRS spectra from 5–33 μm for the sample, debris suspected from the 24 μm photometry can be confirmed with confidence. Even more important, the temperature of the hottest dust detected can be estimated from the broadband spectrophotometry providing estimates of the inner radius of the debris disks in the sample. The detected dust is cooler on average (and thus located at larger radii) than assumed in Meyer *et al.* (2008): the distribution of inner radii for the (often extended) debris disks detected in the survey range from 3–30 AU, with a peak at 10 AU. There is no evidence for evolution of the dust temperature with age, though the magnitude of the observed excess does decline over time (with considerable dispersion at any one age). It appears that some of the 24 μm excess emission observed is the Wien tail of bright, but cool, dust at radii beyond the terrestrial planet zone. Emission evolves from bright to faint (or in some cases hot to cool) within 300 Myr. Yet the picture that emerges overall is one where most debris is cleared out to radii of 10 AU on timescales of 3–10 Myr! Whether this represents complete "mission success" for planet formation between 0.1–10 AU or presents a challenge to current theory remains to be explored. Constraining the outer disk radii will require observations at longer wavelengths utilizing Herschel, SOFIA, or more sensitive sub–millimeter observations (e.g. Roccatagliati *et al.* 2009).

It is extremely tempting to compare these results for FKG stars to those for stars of different temperature & luminosity in order to explore diversity in the formation and evolution of planetary systems as a function of stellar mass. However, such comparisons require caution. For example, in a survey where the sensitivity is limited by photometric precision of detecting an excess in contrast to the flux of the star, the amount of detectable dust around a more luminous star is larger than the detectable amount around one of lower luminosity. Also dust at a given temperature is located at a larger radius around a more luminous star. Any comparison should take this into account in terms of solid angle for the emission as well as the expected orbital period and dust surface density distribution. Overall, one can say that the magnitude of the excesses observed at 24 and 70 μm around A stars are larger, and more common, than those observed around G stars (Su *et al.* 2006), yet the timescale for the duration of the phenomena is shorter around A stars compared to G stars (Currie *et al.* 2007; cf. Trilling *et al.* 2008). Gautier *et al.* (2007) find that debris emission around M dwarfs is even weaker.

4. Summary and future work

So are planetary systems like our own are common or rare among sun-like stars in the Milky Way galaxy? Unfortunately, we cannot yet answer this question, but we can provide some important constraints that are stepping–stones to future progress. *Primordial Disk Evolution:* Disks around lower mass stars are less massive and live longer than their more massive counterparts. The large observed dispersion in evolutionary times could indicate a dispersion in initial conditions. Overall, disk evolution appears to proceed from inside–out as expected. *Change you can believe in:* The duration of the transition time from primordial to debris is $\sim 10^5$ yrs. Planetesimal belts evolve quickly out to 3–30 AU. Evidence for 24 μm excess is largely gone by 300 Myr. There is a hint that such excesses might be more common in open clusters at a given age though more work on this is needed. *Debris Disk Evolution:* Currently detectable extra–solar debris systems are all collision-dominated in their evolution. Debris is brighter and more common around stars of higher mass. Evolutionary paths are diverse but the observed distributions are consistent with our Solar System debris disk being common among stars $>$ 1 Gyr old. The connection between debris and planetary systems is unclear. Yet it may turn out that debris (and perhaps terrestrial planets) are more common than their gas giant counterparts. As we know that nearly all young stars begin their lives surrounded by primordial disks of gas and dust capable of making some sort of planetary system, one wonders: are systems without debris those with dynamically full planetary systems, or those without any planets whatsoever?

Anyone who has taken a turn leading the blind around the proverbial elephant (i.e. modelling spectral energy distributions) knows that a resolved image is worth more than 1024×1024 photometric points on an SED. Constraining model parameters with resolved emission at one or more wavelengths is vital to making progress. In addition to high contrast imaging of disks in scattered light from space and ground–based telescopes equipped with adaptive optics, millimeter wave interferometry will continue to make crucial contributions. The soon to be launched ESA Herschel Space Telescope will also build on the work of Spitzer, providing dozens of newly resolved debris disks in thermal emission. Finally, we look forward to science observations with ALMA and JWST that will expand our understanding of the formation and evolution of planetary systems in ways we can scarcely imagine (provided of course that we first solve the thorny problem of obtaining accurate stellar ages and uncover the nearest, youngest, sun–like stars as prime targets of observation).

Acknowledgement

The author would like to thank many colleagues whose work has contributed to his current understanding regarding the formation and evolution of planetary systems including members of the FEPS/c2D/Glimpse Legacy Science Teams, the MIPS/IRS/IRAC Instrument Teams, and the Spitzer Science Center staff. MRM acknowledges support from the Legacy Science Program through a contract from NASA/JPL as well as the LAPLACE node of the NASA Astrobiology Institute.

References

Alexander, R. D. & Armitage, P. J. 2007, *MNRAS*, 375, 500
Andrews, S. M. & Williams, J. P. 2005, *ApJ*, 631, 1134
Apai, D., *et al.* 2008, *ApJ*, 672, 1196
Ardila, D. R., *et al.* 2004, *ApJ*, 617, L147
Bergin, E., Aikawa, Y., Blake, G., & van Dishoeck, E. 2007, Protostars & Planets V, 751
Bitner, M. A., *et al.* 2008, *ApJ*, 688, 1326
Bouwman, J., *et al.* 2008, *ApJ*, 683, 479
Bryden, G., *et al.* 2006, *ApJ*, 636, 1098

Canup, R. M. 2004, *ARAA*, 42, 441
Carpenter, J. M., Mamajek, E. E., Hillenbrand, L. A., & Meyer, M. R. 2006, *ApJ*, 651, L49
Carpenter, J. M., *et al.* 2008, *ApJS*, 179, 423
Carpenter, J. M., *et al.* 2009, *ApJS*, 181, 197
Carr, J. S. & Najita, J. R. 2008, Science, 319, 1504
Cieza, L., *et al.* 2007, *ApJ*, 667, 308
Cortes, S., *et al.* 2009, *ApJ*, in press (arXiv:0903.3801).
Cumming, A., Butler, R. P., Marcy, G. W., Vogt, S. S., Wright, J. T., & Fischer, D. A. 2008, *PASP*, 120, 531
Currie, T., Kenyon, S. J., Balog, Z., Rieke, G., Bragg, A., & Bromley, B. 2008, *ApJ*, 672, 558
Dent, W. R. F., Greaves, J. S., & Coulson, I. M. 2005, *MNRAS*, 359, 663
Durisen, R. H., Boss, A. P., Mayer, L., Nelson, A. F., Quinn, T., & Rice, W. K. M. 2007, Protostars and Planets V, 607
Dutrey, A., Guilloteau, S. & Ho, P. 2007, Protostars and Planets V, 495
Ercolano, B., Drake, J. J., Raymond, J. C., & Clarke, C. C. 2008, *ApJ*, 688, 398
Espaillat, C., Calvet, N., D'Alessio, P., Hernández, J., Qi, C., Hartmann, L., Furlan, E., & Watson, D. M. 2007, *ApJ*, 670, L135
Fischer, D. A. & Valenti, J. 2005, *ApJ*, 622, 1102
Gatti, T., Natta, A., Randich, S., Testi, L., & Sacco, G. 2008, *A&A*, 481, 423
Gautier, T. N., III, *et al.* 2007, *ApJ*, 667, 527
Goldreich, P., Lithwick, Y., & Sari, R. 2004, *ARAA*, 42, 549
Gorti, U. & Hollenbach, D. 2004, *ApJ*, 613, 424
Gorti, U. & Hollenbach, D. 2009, *ApJ*, 690, 1539
Greaves, J. S., Fischer, D. A., & Wyatt, M. C. 2006, *MNRAS*, 366, 283
Greaves, J. S., Stauffer, J. R., Collier Cameron, A., Meyer, M. R., & Sheehan, C. K. W. 2009, *MNRAS*, 394, L36
Haisch, K. E., Jr., Lada, E. A., & Lada, C. J. 2001, *ApJ*, 553, L153
Hartigan, P., Edwards, S., & Ghandour, L. 1995, *ApJ*, 452, 736
Hartmann, L. 1998, Accretion processes in star formation / Lee Hartmann. Cambridge, UK; New York : Cambridge University Press, 1998. (Cambridge astrophysics series; 32).
Hartmann, L., Calvet, N., Gullbring, E., & D'Alessio, P. 1998, *ApJ*, 495, 385
Hillenbrand, L. A. 2008, Physica Scripta Volume T, 130, 014024
Hillenbrand, L. A., *et al.* 2008, *ApJ*, 677, 630
Ida, S. & Lin, D. N. C. 2008, *ApJ*, 673, 487
Ireland, M. J. & Kraus, A. L. 2008, *ApJ*, 678, L59
Jacobsen, S. B. 2005, Geochimica et Cosmochimica Acta Supplement, 69, 386
Johnson, J. A., Butler, R. P., Marcy, G. W., Fischer, D. A., Vogt, S. S., Wright, J. T., & Peek, K. M. G. 2007, *ApJ*, 670, 833
Kalas, P., *et al.* 2008, Science, 322, 1345
Kenyon, S. J. & Bromley, B. C. 2004, *ApJ*, 602, L133
Kenyon, S. J. & Bromley, B. C. 2008, *ApJS*, 179, 451
Laskar, J. 1994, *A&A*, 287, L9
Lissauer, J. J. & Stevenson, D. J. 2007, Protostars and Planets V, 591
Lovis, C., *et al.* 2006, *Nature*, 441, 305
Marois, C., Macintosh, B., Barman, T., Zuckerman, B., Song, I., Patience, J., Lafrenière, D., & Doyon, R. 2008, Science, 322, 1348
Mayor, M., *et al.* 2009, *A&A*, 493, 639
Merín, B., *et al.* 2008, *ApJS*, 177, 551
Meyer, M. R., Calvet, N., & Hillenbrand, L. A. 1997, *AJ*, 114, 288
Meyer, M. R., *et al.* 2006, *PASP*, 118, 1690
Meyer, M. R., Backman, D. E., Weinberger, A. J., & Wyatt, M. C. 2007, Protostars and Planets V, 573
Meyer, M. R., *et al.* 2008, *ApJ*, 673, L181
Monin, J.-L., Clarke, C. J., Prato, L., & McCabe, C. 2007, Protostars and Planets V, 395
Moro-Martín, A., *et al.* 2007a, *ApJ*, 658, 1312

Moro-Martín, A., *et al.* 2007b, *ApJ*, 668, 1165
Murison, M. A., Lecar, M., & Franklin, F. A. 1994, *AJ*, 108, 2323
Muzerolle, J., Calvet, N., Hartmann, L., & D'Alessio, P. 2003, *ApJ*, 597, L149
Nagasawa, M., Thommes, E. W., Kenyon, S. J., Bromley, B. C., & Lin, D. N. C. 2007, Protostars and Planets V, 639
Najita, J. R., Strom, S. E., & Muzerolle, J. 2007, *MNRAS*, 378, 369
Pascucci, I., *et al.* 2006, *ApJ*, 651, 1177
Pascucci, I., Apai, D., Hardegree-Ullman, E. E., Kim, J. S., Meyer, M. R., & Bouwman, J. 2008a, *ApJ*, 673, 477
Pascucci, I., Apai, D., Luhman, K., Henning, T., Bouwman, J., Meyer, M., Lahuis, F., & Natta, A. 2009, *ApJ*, in press (arXiv:0810.2552)
Pontoppidan, K. M., Blake, G. A., van Dishoeck, E. F., Smette, A., Ireland, M. J., & Brown, J. 2008, *ApJ*, 684, 1323
Roccatagliata, V., Henning, T., Wolf, S., Rodmann, J., Corder, S., Carpenter, J. M., Meyer, M., & Dowell, D. 2009, arXiv:0902.0338
Sanders, I. S. & Taylor, G. J. 2005, Chondrites and the Protoplanetary Disk, 341, 915
Santos, N. C., Israelian, G., & Mayor, M. 2004, *A&A*, 415, 1153
Scott, E. R. D. 2007, Annual Review of Earth and Planetary Sciences, 35, 577
Siegler, N., Muzerolle, J., Young, E. T., Rieke, G. H., Mamajek, E. E., Trilling, D. E., Gorlova, N., & Su, K. Y. L. 2007, *ApJ*, 654, 580
Shu, F., Najita, J., Ostriker, E., Wilkin, F., Ruden, S., & Lizano, S. 1994, *ApJ*, 429, 781
Silverstone, M. D., *et al.* 2006, *ApJ*, 639, 1138
Skrutskie, M. F., Dutkevitch, D., Strom, S. E., Edwards, S., Strom, K. M., & Shure, M. A. 1990, *AJ*, 99, 1187
Strom, K. M., Strom, S. E., Edwards, S., Cabrit, S., & Skrutskie, M. F. 1989, *AJ*, 97, 1451
Su, K. Y. L., *et al.* 2006, *ApJ*, 653, 675
Trilling, D. E., *et al.* 2007, *ApJ*, 658, 1289
Trilling, D. E., *et al.* 2008, *ApJ*, 674, 1086
Tsiganis, K., Gomes, R., Morbidelli, A., & Levison, H. F. 2005, *Nature*, 435, 459
Watson, D. M., *et al.* 2009, *ApJS*, 180, 84
Weidenschilling, S. J. 1977, *MNRAS*, 180, 57
Werner, M., Fazio, G., Rieke, G., Roellig, T. L., & Watson, D. M. 2006, *ARAA*, 44, 269
Wyatt, M. C., Smith, R., Greaves, J. S., Beichman, C. A., Bryden, G., & Lisse, C. M. 2007, *ApJ*, 658, 569
Wyatt, M. C. 2008, *ARAA*, 46, 339

Discussion

M. LIU: Open clusters and young moving groups represent the best "benchmark" systems for understanding evolution, since they are coeval and formed in the same environment. However, even in these very simple samples, debris disk properties are diverse and not easily explained. Do you find this to be a discouraging result in attempting to develop a comprehensive picture of debris disk evolution?

M. MEYER: Evidence from debris disk studies as well observations of exoplanets suggest that the outcomes of the planet formation process are diverse, as are the paths of subsequent evolution. By combining detailed studies of individual objects (those systems that are resolved in scattered light and/or thermal emission) as well as large surveys conducted with ground- and space- based telescopes (Spitzer, Herschel and JWST) we can hopefully discern the overall climates that are conducive to planet formation, in contrast to our inability to predict the prospects of planet formation for any system (the "weather").

R. JEFFRIES: Could the low frequency of debris disks be explained by "episodic" dust production? In other words, could debris disks be more common but only episodically produced?

M. MEYER: It is true that we only observe the product of the frequency of the debris disk phenomenon and the distribution of durations. Indeed, if the warm debris disk epoch lasts a short time, the overall frequency could be much higher than observed (Meyer et al 2008). However observational support for this scenario is lacking (Carpenter *et al.* 2009). Wyatt *et al.* (2007) explore conditions under which an observed debris disk can be unambiguously identified as transient.

Michael Meyer

Antonela Monachesi

The Ages of Stars
Proceedings IAU Symposium No. 258, 2008
E.E. Mamajek, D.R. Soderblom & R.F.G. Wyse, eds.

doi:10.1017/S1743921309031779

Using ages and kinematic traceback: the origin of young local associations

David Fernández, Francesca Figueras and Jordi Torra
Departament d'Astronomia i Meteorologia, IEEC-Universitat de Barcelona
Av. Diagonal 647, E-08028 Barcelona, Spain
email: `david.fernandez@am.ub.es`

Abstract. Over the last decade, several groups of young (mainly low-mass) stars have been discovered in the solar neighbourhood (closer than $\sim$ 100 pc), thanks to cross-correlation between X-ray, optical spectroscopy and kinematic data. These young local associations – including an important fraction whose members are Hipparcos stars – offer insights into the star formation process in low-density environments, shed light on the substellar domain, and could have played an important role in the recent history of the local interstellar medium. Ages estimates for these associations have been derived in the literature by several ways (HR diagram, spectra, Li and Hα widths, expansion motion, etc.). In this work we have studied the kinematic evolution of young local associations and their relation to other young stellar groups and structures in the local interstellar medium, thus casting new light on recent star formation processes in the solar neighbourhood. We compiled the data published in the literature for young local associations, including the astrometric data from the new Hipparcos reduction. Using a realistic Galactic potential we integrated the orbits for these associations and the Sco-Cen complex back in time. Combining these data with the spatial structure of the Local Bubble and the spiral structure of the Galaxy, we propose a recent history of star formation in the solar neighbourhood. We suggest that both the Sco-Cen complex and young local associations originated as a result of the impact of the inner spiral arm shock wave against a giant molecular cloud. The core of the giant molecular cloud formed the Sco-Cen complex, and some small cloudlets in a halo around the giant molecular cloud formed young local associations several million years later. We also propose a supernova in young local associations a few million years ago as the most likely candidate to have reheated the Local Bubble to its present temperature.

Keywords. Galaxy: kinematics and dynamics, Galaxy: solar neighbourhood, Galaxy: open clusters and associations: general, Stars: kinematics, Stars: formation, ISM: individual objects: Local Bubble

1. Introduction

In this work we propose a scenario for the history of the recent star formation (during the last 20–30 Myr) in the nearest solar neighbourhood ($\sim$150 pc), from the study of the spatial and kinematic properties of the members of the so-called young local associations, the Sco-Cen complex and the Local Bubble, the most important structure observed in the local interstellar medium (ISM).

2. The local interstellar medium: the Local Bubble

Locally, within the nearest 100 pc, the ISM is dominated by the Local Bubble (LB). The displacement model (Snowden *et al.* 1998) for this structure assumes that the irregular local HI cavity is filled by an X-ray-emitting plasma, with an emission temperature of $\sim 10^6$ K and a density, n_e, of ~ 0.005 cm^{-3}. Snowden *et al.* (1998) derived an extension for the LB of 40 to 130 pc, it being larger at higher Galactic latitudes and smaller nearer the equator. Lallement *et al.* (2003) obtained the contours of the LB from NaI absorption measurements, tracing them with an estimated precision of $\approx \pm 20$ pc in most directions.

Several models have been presented to explain the origin of the LB. The consensus is reached with a scenario in which about 10–20 supernovae (SNe) formed the local cavity and, after that, a few SNe reheated the LB a few Myr ago, explaining the currently observed temperature of the diffuse soft X-ray background (Breitschwerdt & Cox 2004). Some authors have remarked that there is independent evidence for the occurrence of a close SN (~30 pc) ~5 Myr ago (Knie *et al.* 1999), which could be the best candidate for reheating the LB. Maíz-Apellániz (2001) proposed that the 2 or 3 SNe that reheated the local cavity could have exploded in LCC (one of the OB associations of the Sco-Cen complex), but his results faced some geometrical problems due to the peripheral situation of LCC with respect to the LB.

3. Local associations of young stars

A decade ago, very few PMS stars had been identified less than 100 pc from the Sun. Nearly all the youngest stars ($\leqslant$30 Myr) studied then were located more than 140 pc away in the molecular clouds of Taurus, Chamaeleon, Lupus, Sco-Cen and R CrA. The cross-correlation of the Hipparcos and ROSAT catalogues suddenly changed this; a few stars were identified as very young but closer than 100 pc, where there are no molecular clouds with stellar forming regions (Neuhäuser & Brandner 1998). Two explanations for the existence of these young stars far away from SFR were proposed. Sterzik & Durisen (1995) suggested that the stars were formed in molecular clouds and later ejected as high-velocity stars during the decay of young multiple star systems. Feigelson (1996) suggested that the stars were formed inside small molecular clouds (or *cloudlets*), which later dispersed among the ISM and therefore can no longer be detected.

These young nearby stars were grouped into clusters, associations and moving groups, each with a few dozen members. Different approaches were used in each YLA discovery, but most of them made use of Hipparcos proper motions, X-ray emission, infrared emission and ground-based spectroscopy and photometry. We have compiled all the published YLA data. Astrometric data come from the new Hipparcos reduction (van Leeuwen 2007). Table 1 shows the mean spatial and kinematic properties, ages and number of members for each YLA. The adopted age shown in Table 1 is that assigned for back-tracing the association orbits in the next section. In Fig. 1 the observational errors for the stars of our compilation are shown.

4. Ages and integration of orbits

The integration back in time of the Sco-Cen and YLA orbits allows us to study their origin and possible influence on the local ISM over the last million years. To compute the stellar orbits back in time we used the code developed by Asiain *et al.* (1999) based on the integration of the equations of motion using a realistic model of the Galactic gravitational potential. We decomposed this potential into three components: the general axisymmetric potential $\Phi_{\rm AS}$, the potential due to the spiral structure of the Galaxy $\Phi_{\rm Sp}$, and that due to the central bar $\Phi_{\rm B}$. Details on the method can be found in Fernández *et al.* (2008).

4.1. *Orbits of individual stars: are we able to derive dynamic ages?*

One of the most interesting results that could be derived from the study of the past trajectories of the individual members of the young local associations is the direct

Table 1. Mean spatial coordinates and heliocentric velocity components of the young local associations and the Sco-Cen complex (in the latter case, data from de Zeeuw 1999 (Z99), Madsen *et al.* 2002 (M02) and Sartori *et al.* 2003 (S03)). In brackets, the standard deviation of the sampling distribution. N is the number of known members in each association (N_k with complete kinematic data).

Association		$\overline{\xi'}$ (pc)	$\overline{\eta'}$ (pc)	$\overline{\zeta'}$ (pc)	$\overline{r}$ (pc)	$\overline{U}$ (km s^{-1})	$\overline{V}$ (km s^{-1})	$\overline{W}$ (km s^{-1})	Age (Myr)	N	N_k
TW Hya		$-18_{(12)}$	$-51_{(16)}$	$22_{(7)}$	$60_{(18)}$	$-8.4_{(3.9)}$	$-18.3_{(3.4)}$	$-4.9_{(1.8)}$	8	39	5
Tuc-Hor/GAYA		$-12_{(26)}$	$-26_{(12)}$	$-36_{(13)}$	$53_{(17)}$	$-11.4_{(7.3)}$	$-22.4_{(5.5)}$	$-2.7_{(3.7)}$	20	52	28
β Pic-Cap		$-10_{(26)}$	$-7_{(13)}$	$-14_{(10)}$	$37_{(13)}$	$-10.1_{(3.5)}$	$-15.2_{(3.7)}$	$-10.2_{(2.9)}$	12	33	21
ϵ Cha		$-52_{(3)}$	$-89_{(5)}$	$-27_{(2)}$	$107_{(6)}$	$-10.2_{(0.8)}$	$-18.9_{(0.6)}$	$-10.2_{(2.0)}$	10	16	4
η Cha		-31	-75	-32	88	-10.1	-18.4	-10.0	10	18	1
HD 141569		$-77_{(3)}$	$10_{(8)}$	$64_{(8)}$	$101_{(8)}$	$-5.4_{(1.5)}$	$-15.6_{(2.6)}$	$-4.4_{(0.8)}$	5	5	2
Ext. R CrA		$-81_{(33)}$	$-6_{(6)}$	$-28_{(9)}$	$87_{(33)}$	$-2.2_{(6.1)}$	$-15.8_{(1.4)}$	$-10.9_{(0.8)}$	13	59	2
US	Z99	$-141_{(34)}$	$-22_{(11)}$	$50_{(16)}$	$145_{(2)}$				5-6	120	
	M02[1]	$-138_{(27)}$	$-22_{(10)}$	$49_{(12)}$	$149_{(28)}$	-0.9	-16.9	-5.2		120	
	S03					$-6.7_{(5.9)}$	$-16.0_{(3.5)}$	$-8.0_{(2.7)}$	8-10	155	
UCL	Z99	$-122_{(30)}$	$-69_{(26)}$	$32_{(16)}$	$140_{(2)}$				14-15	221	
	M02[1]	$-121_{(26)}$	$-68_{(21)}$	$32_{(15)}$	$145_{(24)}$	-7.9	-19.0	-5.7		218	
	S03					$-6.8_{(4.6)}$	$-19.3_{(4.7)}$	$-5.7_{(2.5)}$	16-20	262	
LCC	Z99	$-62_{(18)}$	$-102_{(24)}$	$14_{(16)}$	$118_{(2)}$				11-12	180	
	M02[1]	$-61_{(14)}$	$-100_{(15)}$	$14_{(15)}$	$120_{(18)}$	-11.8	-15.0	-6.7		179	
	S03					$-8.2_{(5.1)}$	$-18.6_{(7.3)}$	$-6.4_{(2.6)}$	16-20	192	

[1] M02 derived an internal velocity dispersion of 1.33 km s^{-1} for US, 1.23 km s^{-1} for UCL and 1.13 km s^{-1} for LCC.

determination of their age. This could be achieved if one observes a clear spatial concentration of the members of one specific association in the past.

However, we have not found this concentration in the past. We have computed the trajectories back in time for the stars of the three associations with the largest number of members with complete, high-quality data. In Fig. 2 we show the spatial dispersion of these stars as a function of time in the past, the results are those shown in the figure. As can be seen, during the last 4 million years, the spatial dispersion remains similar to that observed at present, but it grows rapidly to the past. So, it seems that the combination of the observational errors and the intrinsic velocity dispersion prevent us from deriving the ages for the associations. Or maybe there are some problems with the membership assignment of the stars to each association. Then, in our analysis we will only work with the past trajectories of the associations as a whole.

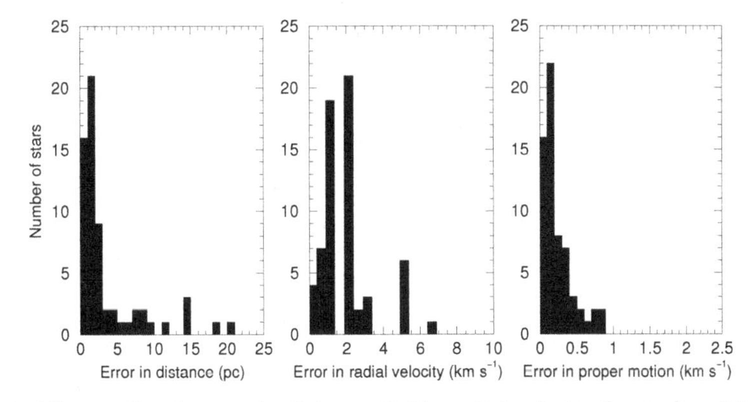

Figure 1. Observational errors in distance (left), radial velocity (center) and tangential velocity (right) for the stars of our compilation of YLA.

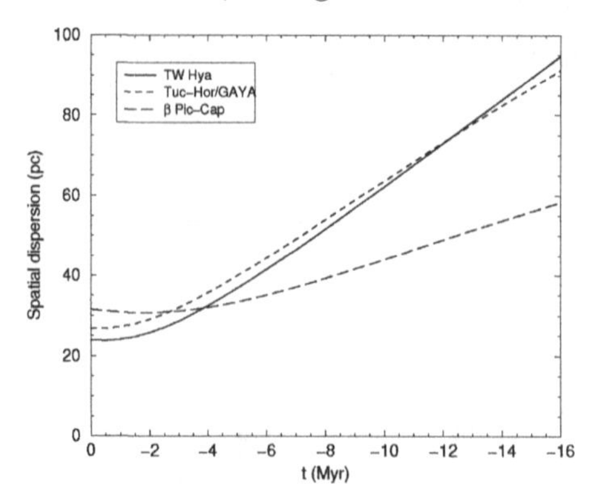

Figure 2. Spatial dispersion of those stars belonging to TW Hya, Tuc-Hor/GAYA and β Pic-Cap during the last Myr.

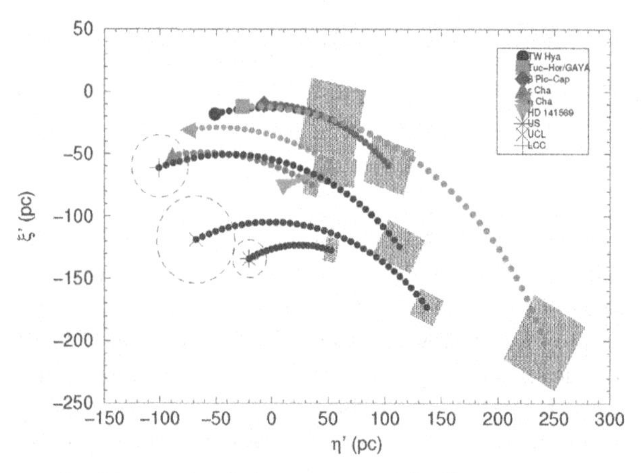

Figure 3. Positions and orbits in the Galactic plane (ξ', η') of YLA and the Sco-Cen complex going back in time to their individual ages. The grey areas show the expected positional errors at birth due to kinematic observational errors. The centre of the (ξ', η') coordinate system is comoving with the LSR.

4.2. *The origin of young local associations*

We have integrated the orbits of the associations using the mean position and velocity for each association (Table 1). The results are presented in Fig. 3. We show the estimated error in the position of the associations at birth (grey area). An error in age shall be read as a displacement of the error areas along the plotted orbits in the figure. The most obvious trend observed is the spatial concentration of all the associations (Sco-Cen complex and YLA) in the first Galactic quadrant in the past. There is a very conspicuous spatial grouping at the time of birth of TW Hya, ϵ Cha, η Cha and HD 141569 (in a sphere 25 pc in radius). At its birth, β Pic-Cap was located about 50 pc from the other YLA. The region where these associations were formed had a size of $50 \times 70 \times 40$ pc (ξ' x η' x ζ'). At present the volume has increased by a factor of ~ 3.5. The errors associated with the mean velocity components of the associations do not have a crucial influence on the previous results: the error areas have typical side lengths of about 10–30 pc.

Mamajek *et al.* (2000) found that extrapolating past motions (assuming linear ballistic trajectories) shows that TW Hya, ϵ Cha, η Cha and the three subgroups of the Sco-Cen complex were closest together about 10–15 Myr ago. They suggest that these three

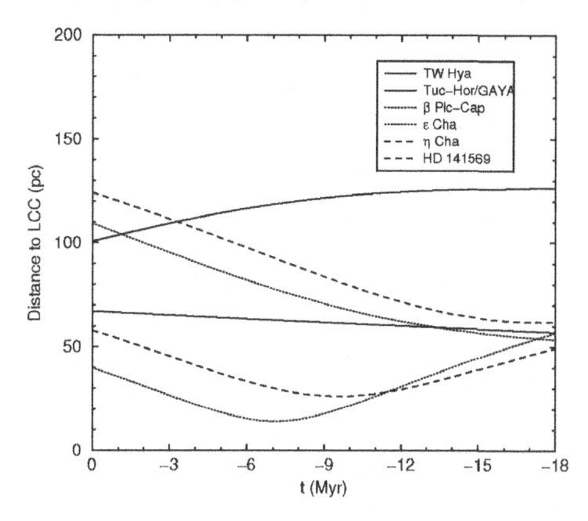

Figure 4. Distances between the centres of YLA and the centre of LCC as a function of time, using the Sco-Cen kinematics in S03.

associations were formed in the progenitor Sco-Cen GMC or in short-lived molecular clouds formed by Sco-Cen superbubbles. Jilinski *et al.* (2005) locate the birthplaces of ϵ Cha and η Cha at the edge of LCC. Ortega *et al.* (2004) suggest that the β Pic moving group was formed near Sco-Cen, probably due to a SN in this complex. Our compendium of all the known members of the whole set of YLA and the integration of their orbits back in time allow us to present here a more detailed analysis of the origin of YLA.

In Fig. 4 the temporal evolution of the distances between each one of the YLA and the centre of LCC (the nearest OB association) is shown. The instant when the distance minima between LCC and YLA occurred is of great interest. In the cases of the η Cha cluster and the ϵ Cha association, minima with distances of 14 and 26 pc, respectively, are obtained for $t \sim -(7\text{–}9)$ Myr. No clear distance minima to LCC are found in the recent past ($t > -20$ Myr) for the other YLA. TW Hya was 62 pc from LCC 8 Myr ago (the estimated age for this association), whereas the HD 141569 system was about 102 pc from LCC 5 Myr ago, continuously decreasing to 62 pc 18 Myr ago. The Tuc-Hor/GAYA association has maintained a distance to LCC of more than 100 pc over the last 20 Myr, but this could be the only YLA considered here not to originate from a SN in LCC, since its estimated age is equal to (or even larger than) that derived for LCC.

We conclude that, at the moment of their birth, YLA were at distances of between 15 and 100 pc from the *centre* of LCC, and even further from the other two Sco-Cen associations. Although observational errors in parallax and velocity components, as well as errors in age estimations, could affect these results, the fact that we work with mean values for distances, velocity components and ages minimises this possibility. Although the present radius of LCC is about 25 pc, it has been continuously expanding since birth, 16-20 Myr ago. Even considering a moderate expansion rate, one should expect an initial radius $\leqslant$20 pc. The distances obtained from Fig. 4, together with the expected reliability of the orbits, lead us to believe that the local associations were not born inside the cloud that formed the Sco-Cen complex, but in small molecular clouds outside it.

One possible scenario for the formation of YLA in these small molecular clouds is the explosion of one or several close SNe, which could have produced compression that triggered star formation. These hypothetical SNe should belong to the Sco-Cen complex. This complex is made up of several thousand stars, more than 300 of which are early-type stars, and around 35 are candidates for Type II SNe. Maíz-Apellániz (2001) estimated the number of past SNe inside the three associations, obtaining 1 SN for US, 13 for UCL and 6 for LCC. The first SN that exploded in each association took place when it was

3–5 Myr old, and the others have been exploding and will continue to explode at a nearly constant rate, for the first ~30 Myr of the complex's life. Even a conservative estimate gives at least 6 SNe in UCL during the last 10–12 Myr, another 6 in LCC during the last 7-9 Myr and at least 1 in US.

The wave front of a SN typically moves at a velocity of a few tens of pc per million years; therefore, a SN explosion in LCC or UCL 9-11 Myr ago may have triggered star formation between 1 and 3 Myr later in small molecular clouds at distances of 15 to 75 pc. These would be the parent clouds of η Cha and ϵ Cha, located at ~20 pc from LCC at their birth. Taking into account that the first SNe in LCC exploded when the associations were 3-5 Myr old (Maíz-Apellániz 2001), this scenario is only possible for a present age of LCC of at least 12 Myr. This is not a problem, since the estimated ages for LCC published in the literature range from 11–12 Myr to 16–20 Myr. If only one SN could explain this star formation outbreak ~8.5–9 Myr ago, this would be the age of η Cha and ϵ Cha, which would have been formed simultaneously. It should be remembered that the estimated ages for η Cha and ϵ Cha are 5–15 and $\leqslant$10 Myr, respectively.

Such a SN could also have triggered the formation of TW Hya, whose estimated age is ~8 Myr. As mentioned above, at that time TW Hya was about 45 pc from the centre of LCC, in perfect agreement with the typical distance at which a SN wave front can trigger star formation in a small molecular cloud. However, it was not necessarily a single SN in LCC or UCL that was the origin of these four YLA. The SN rate in these two associations is ~0.5 Myr^{-1} and, therefore, it is possible that a few SNe in the period $-7 \leqslant t \leqslant -10$ Myr triggered the star formation that resulted in YLA. In any case, from our results we can conclude that these associations definitely did not form inside the associations of Sco-Cen, to be later ejected. They were formed in regions of space far from Sco-Cen, probably in small molecular clouds that were later totally dispersed by the newly born stars and/or by the shock fronts of later SNe in Sco-Cen.

Our results support a star formation scenario for very young stars far away from SFR or molecular clouds, such as that proposed by Feigelson (1996) and not that of Sterzik & Durisen (1995). The latter authors perform numerical simulations to explain the existence of haloes of isolated T Tau stars around SFR. In their simulations a significant number of stars were ejected from these regions at birth with large velocities, allowing trajectories of some tens of pc in a few million years. Meanwhile, Feigelson (1996) proposed another scenario for the formation not only of the haloes of T Tau stars, but also of other completely isolated very young stars that have been discovered. In this model, the isolated T Tau stars form in small, fast-moving, short-lived molecular clouds. The gas remaining after the star formation process is rapidly dispersed by the stellar winds of the new stars. At present the stars are located in regions of space where there is no gas and so, apparently, they have formed far away from any SFR. The case of YLA supports this scenario, since our kinematic study shows that these associations formed far away from the Sco-Cen complex. For the HD 141569 system, a SN in UCL, as opposed to one in LCC, is a more promising candidate to explain its origin. This is because the distance to LCC for the range of ages accepted for this group (2–8 Myr) is between 88 and 116 pc, whereas for UCL it is 72–88 pc.

4.3. *Young local associations and Local Bubble*

If we superimposed on the present LB structure the trajectories back in time for YLA (see Fig. 13 in Fernández *et al.* 2008) we can see that the past orbits of the YLA are closer to the central region of the LB than Sco-Cen associations. To be exact, the trajectories of the centres of the associations TW Hya, Tuc-Hor/GAYA and β Pic-Cap have crossed very near to the geometric centre of the LB in the last ~5 Myr. The uncertainty boxes

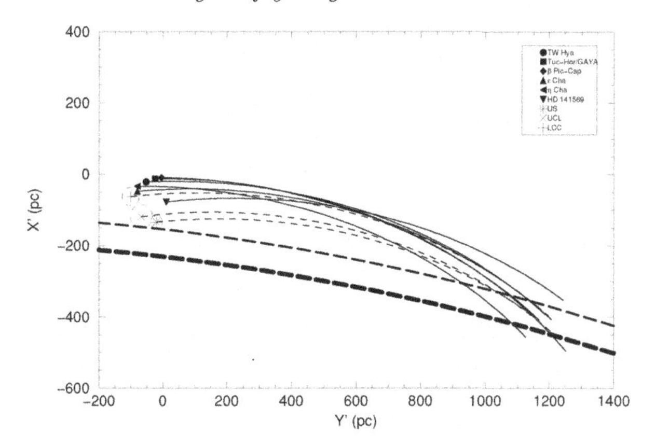

Figure 5. Orbits in the Galactic plane (X'-Y') integrated back in time to $t = -30$ Myr for YLA and the three associations of the Sco-Cen complex. The thick-dashed line shows the position of the minimum of the spiral potential (Fernández et al. 2001). The thin-dashed line is the position of the phase of the spiral structure $\psi = 10°$.

in position on the Galactic plane do not exceed a few tens of pc. On the other hand, errors in age estimates for the YLA result in positional errors along the trajectories back in time. Even considering the large uncertainties in age obtained for some of these YLA, ages are not expected to exceed 20 Myr for any of them (except for Tuc-Hor/GAYA). We can therefore conclude that the YLA have been moving inside the LB for (at least) most of their lifetime, and can question whether the presence of these young stars inside the LB bears any relation to its origin and/or evolution.

All the YLA except TW Hya and β Pic-Cap contain B-type stars (13 stars from a total of 223 members; see Table 6 in Fernández *et al.* 2008). It is not possible to derive the number of stars earlier than B2.5 that were born in the local associations, since we do not know their total mass precisely. However, the fact that at present we observe one SN candidate (a B2IV star), and more than a dozen stars of spectral types between B5 and B9, allows us to affirm that it is possible that one or more of these associations has sheltered a SN in the recent past (the last 10 Myr). As there is direct evidence for an explosion of a SN at a distance of $\sim$30 pc, $\sim$5 Myr ago, several pieces of the same puzzle seem to support the theory of a recent SN in the nearest solar neighbourhood originating from a parent star belonging to a YLA, probably Tuc-Hor/GAYA or the extended R CrA association (which currently show the highest content of B-type stars). This near and recent SN would have been responsible for the reheating of the gas inside the LB needed to achieve the currently observed temperature of the diffuse soft X-ray background.

As we mention above, there is no agreement in the literature on the number of SNe needed to form the local cavity. If only one was enough, the SN we propose would be the most promising candidate, since it would be placed very near the geometric centre of the LB, explaining in a natural way its present spatial structure. If more SNe are needed (as recent works suggest), we could consider other stars in the vicinity of LCC and UCL, as proposed by Fuchs *et al.* (2006).

5. A scenario for the local and recent star formation

If the impact of the spiral arm shock wave was the initial cause of star formation in the Sco-Cen region (see details in Fernández *et al.* 2008), then the history of the nearest solar neighbourhood during the last few tens of Myr would have been as follows. 30 Myr ago the GMC that became the parent of Sco-Cen was in the Galactic plane with coordinates

$(X', Y') \sim (-400, 1200)$ pc (see Fig. 5). The arrival of the potential minimum of the inner spiral arm triggered star formation in the region. At the same time it disturbed the cloud's motion, whose velocity vector became directed in the opposite direction to Galactic rotation and away from Galactic centre. The compression due to the spiral arm did not necessarily trigger star formation in the whole cloud, but perhaps only in the regions with the largest densities. This would be favoured by the smaller relative velocity between the shock wave and the RSR. The regions where star formation began must be those which generated UCL, LCC and, probably, the Tuc-Hor/GAYA association, which were all born at nearly the same time: about 16–20 Myr ago. The stellar winds from the first massive stars began to compress the gas of the neighbouring regions, maybe causing them to fragment into small molecular clouds that moved away from the central region of the parent cloud. About 9 Myr ago, a SN in LCC or UCL triggered star formation in these small molecular clouds, giving birth to the majority of YLA, as we saw in the previous section. The stellar winds of the newly born stars rapidly expelled the remaining gas from these small clouds (the *cloudlets* proposed by Feigelson 1996), completely erasing every trace of them and leading to our observation that there is no gas in these regions at present. YLA may have had a crucial influence on the history of the LB. We suggest that one or two SNe in these associations were responsible for reheating the LB a few million years ago. This hypothesis seems to be reinforced by the evidence of a very near SN about 5 Myr ago (Knie *et al.* 1999). At about the same time, as proposed by Preibisch & Zinnecker (1999), the shock front of a SN in UCL would have triggered star formation in US about 6 Myr ago. Only 1.5 Myr ago, the most massive star in US would have gone SN and its shock front would now be reaching the molecular cloud of ρ Oph, triggering the beginning of the star formation process there.

References

Asiain, R., Figueras, F., & Torra, J. 1999, *A&A*, 350, 434

Breitschwerdt, D. & Cox, D. P. 2004, in: E. J. Alfaro, E. Pérez, & J. Franco (eds.), *How does the Galaxy work? A Galactic Tertulia with Don Cox and Ron Reynolds*, Astrophysics and Space Science Library, vol. 315 (Dordrecht: Kluwer), p. 391

Feigelson, E. D. 1996, *ApJ*, 468, 306

Fernández, D., Figueras, F., & Torra, J. 2001, *A&A*, 372, 833

Fernández, D., Figueras, F., & Torra, J. 2008, *A&A*, 480, 735

Fuchs, B., Breitschwerdt, D., de Avillez, M. A., Dettbarn, C., & Flynn, C. 2006, *MNRAS*, 373, 993

Jilinski, E., Ortega, V. G., & de la Reza, R. 2005, *ApJ* 619, 945

Knie, K., Korschinek, G., Faestermann, T., Wallner, C., Scholten, J., & Hillebrandt, W. 1999, *Phys. Rev. Lett.*, 81, 18

Lallement, R., Welsh, B. Y., Vergely, J. L., Crifo, F., & Sfeir, D. M. 2003, *A&A*, 411, 447

van Leeuwen, F. 2007, *Hipparcos, the New Reduction of the Raw Data*, Astrophysics and Space Science Library, vol. 350 (Dordrecht: Springer)

Madsen, S., Dravins, D., & Lindegren, L. 2002, *A&A*, 381, 446

Mamajek, E. E., Lawson, W. A., & Feigelson, E. D. 2000, *ApJ*, 544, 356

Maíz-Apellániz, J. 2001, *ApJ*, 560, 83

Neuhäuser, R. & Brandner, W. 1998, *A&A*, 330, L29

Ortega, V. G., de la Reza, R., Jilinski, E., & Bazzanella, B. 2004, *ApJ*, 609, 243

Preibisch, T. & Zinnecker, H. 1999, *AJ*, 117, 2381

Sartori, M. J., Lépine, J. R. D., & Dias, W. S. 2003, *A&A*, 404, 913

Snowden, S. L., Egger, R., Finkbeiner, D. P., Freyberg, M. J., & Plucinsky, P. P. 1998, *ApJ*, 493, 715

Sterzik, M. & Durisen, R. 1995, *A&A*, 304, L9

de Zeeuw, P. T., Hoogerwerf, R., Bruijne, J. H. J., Brown, A. G. A., & Blaauw, A. 1999, *AJ*, 117, 354

Discussion

MELBOURNE: What effect do these SNe have on the Earth? Are there tracers on Earth that suggest recent SNe explosions?

FERNÁNDEZ: In a paper published in 1999 by Knie and collaborators, they claimed to have found in a deep ocean crust evidences of the explosion of a SN in the recent past. They proposed that this SN exploded 5 Myr ago at a distance of about 30 pc.

MAMAJEK: Other researchers have calculated dynamical ages based on the time of minimum pass between some groups in the past. We are all here because we want improved ages for stars. How useful do you think these dynamical ages are?

FERNÁNDEZ: We have found that deriving dynamic ages is not a reliable aging methodology, al least considering the present membership assignment for each one of the associations. Other authors have derived dynamic ages, but using what they called "core stars"; that is, selecting those stars with the minimum dispersion in their velocity components. I strongly believe that the present membership assignment must be revised to be able to derive reliable dynamic ages for each one of these young local associations.

NAYLOR: You have a diagram which shows the motion of the local associations after they passed through the spiral arm. Where does the Sun move on this diagram? How long have we been near the local associations?

FERNÁNDEZ: We have not computed the orbit of the Sun in this figure, so I do not know exactly how long the Sun has been near the young local associations.

ROBERTO: Are the counts of low-mass stars in local associations compatible with the number of SNe?

FERNÁNDEZ: The problem with the counts of low-mass stars is that they are not complete and it is difficult to say if these counts are compatible with the number of SNe we propose (one or two). However, the counts of early-type stars, which are complete, seems to support the hypothesis that a SN exploded in the recent past in the young local associations.

John Stauffer and Marc Pinsonneault

Jeff Valenti, LOC Chair

The Ages of Stars
Proceedings IAU Symposium No. 258, 2008
E.E. Mamajek, D.R. Soderblom & R.F.G. Wyse, eds.

doi:10.1017/S1743921309031780

On the use of lithium to derive the ages of stars like our Sun

Sofia Randich

INAF-Osservatorio Astrofisico di Arcetri
Largo E. Fermi, 5 I-50125, Firenze, Italy
email: randich@arcetri.astro.it

Abstract. Along with chromospheric emission, lithium abundances are widely used to infer the ages of solar-like. We re-assess the validity and limits of this approach, based on new high quality Li measurements in seven open clusters observed with Giraffe on the ESO VLT.

Keywords. stars: abundances, stars: interiors, Galaxy: open clusters and associations: general

1. Introduction

Stars like our Sun† deplete lithium (Li) during their permanence on the main sequence, due to a mixing process that brings surface material down in the stellar interior where the temperature is high enough for Li burning to occur. Since the base of the convective zone of the Sun (and similar stars) does not reach the temperature needed for Li reactions, different extra-mixing processes have been proposed in the literature. Whereas no consensus has been reached on the nature of the extra-mixing mechanism, this can be very efficient, as indicated by the solar Li abundance. The Sun has in fact depleted a factor of $\sim$ 100 Li with respect to the initial meteoritic abundance. Also, the solar Li abundance is much lower than that of similar stars in young open clusters, such as the Pleiades.

Based on the assumption that the Sun and its Li content are representative of stars with similar age and temperature, and that Li abundance decreases as stars get older, the obvious conclusion was drawn that Li could be used to age date solar-type stars. Several papers in the 80's indeed focused on the determination of qualitative or even quantitative Li-age relationships. Among these works, we mention Herbig (1965), Skumanich (1972), Duncan (1981), Soderblom (1983), Boesgaard & Tripicco (1987). Note that all these studies were based on field stars and on a few young clusters such as the Pleiades (100 Myr), with a huge age gap. Noticeably, during the same years several examples of old (as indicated by the chromospheric emission levels and/or rotational velocities), but Li-rich stars were reported (Duncan 1981; Spite & Spite 1982; Pallavicini *et al.* 1987). Pallavicini *et al.* indeed concluded that *a high Li abundance is a necessary, but not sufficient condition for a star to be young.*

Whereas results on field stars might be affected by their uncertain age determination, similar findings were also reported for open clusters, whose age is more securely constrained. In particular, different studies (Spite *et al.* 1987; García López *et al.* 1988; Pasquini *et al.* 1997; Jones *et al.* 1999) pointed out that the Li distribution of solar-type

† In this paper we indicate as 'solar like' stars with temperature, but not necessarily mass, close to the solar value.

Table 1. The sample clusters. We list ages, metallicities, and number of members for which we derived Li abundances.

Cluster	age (Gyr)	[Fe/H]	N_{stars}
NGC 3960	0.7	0.02±0.04	36
NGC 2477	1.0	0.07±0.04	73
NGC 2506	2.2	−0.20 ± 0.02	71
NGC 6253	3.0	0.36±0.07	54
Melotte 66	4.0	−0.33 ± 0.03	53
Be 32	6.0	−0.29 ± 0.04	57
Cr 261	8.0	0.13±0.05	135

members of the solar age, solar metallicity cluster M 67 appears to be bimodal; along with stars showing an amount of Li depletion comparable to the solar one, several otherwise similar cluster members are present with a factor of about 10 higher Li. More recently, Randich *et al.* (2003) measured Li in 11 solar-like members of the 6–8 Gyr old NGC 188, finding that all the stars are 10–20 times more Li-rich than the Sun.

Both results are in agreement with those for field stars and evidence the existence of a population of old, but not necessarily Li-poor stars; in turn, this casts doubts both on the assumption that the solar Li is typical for a star of that age, and on the use of Li as an age tracer. In this context, we report here the results of a Li survey among a sample of open clusters well sampling the age-metallicity plane, aimed to put tighter constraints on the evolution of Li abundance with age and its possible dependence on metallicity,

2. The FLAMES survey

We have used the multiplex facility FLAMES on ESO VLT to perform a spectroscopic survey of a large sample of Galactic open clusters, addressing a variety of scientific goals (Randich *et al.* 2005; Pallavicini *et al.* 2006). In particular, we have used the Giraffe spectrograph to obtain high resolution spectra ($R \sim 20,000$) of unevolved cluster candidates; our primary goals were membership determination and Li measurements among confirmed members. A total of 11 clusters were observed; nine of these were close enough to allow us to observe solar-type stars. In Table 1 we list the seven sample clusters whose Li analysis has been completed. Analysis for the remaining two clusters (NGC 2324 and To 2) is in progress. As the table shows, the clusters span the age interval between ∼0.7 and 8 Gyr and the metallicity range [Fe/H]= −0.38 – +0.35. After excluding radial velocity non members, each cluster sample typically consists of ∼40–140 stars.

The analysis was carried out as described in Sestito & Randich (2005) and Randich *et al.* (2009). Briefly, effective temperatures (T_{eff}) were derived employing the calibration T_{eff} vs. B–V of Soderblom *et al.* (1993a) for solar metallicity stars or that of Alonso *et al.* (1996) for stars with metallicity different from solar.

At our resolution, the Li I 670.8 nm feature is blended with the Fe I 670.74 nm line. The contribution of the latter to the Li feature was estimated using the analytical expression of Soderblom *et al.* (1993b) in the case of solar metallicity clusters; for clusters with over-/under-solar metallicity we instead measured the equivalent width of the Fe line on a grid of synthetic spectra with the appropriate metallicity and different temperatures. Finally, Li abundances (log n(Li)) were computed from deblended equivalent widths and using curves of growths of Soderblom *et al.* (1993b).

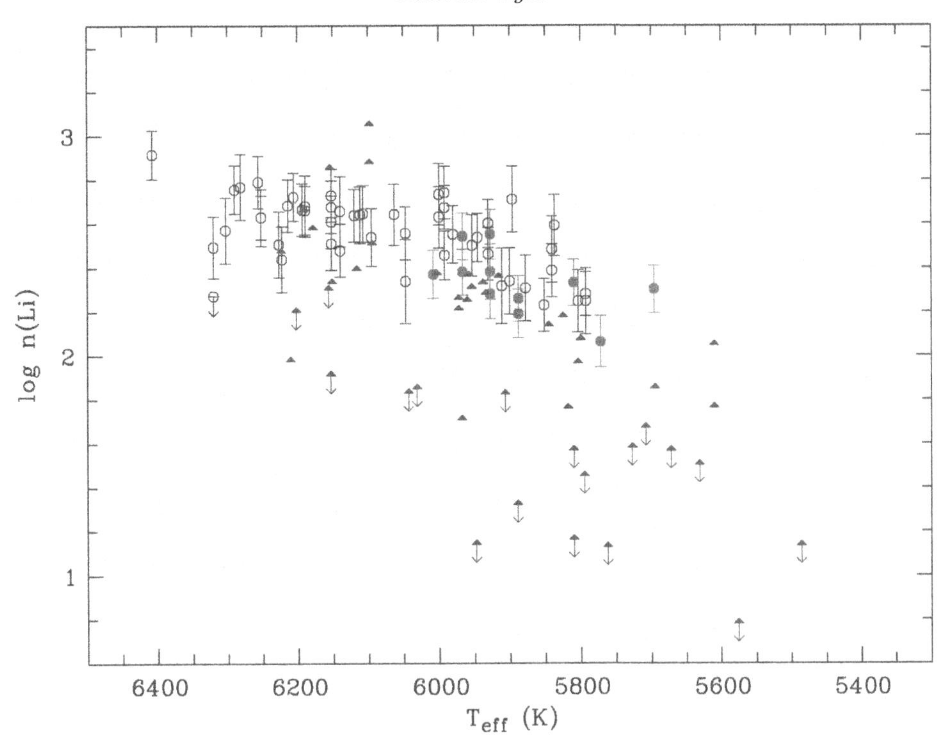

Figure 1. Li abundances (in the usual scale log n(Li)=log n(Li)/n(H)+12) as a function of effective temperature for Be 32 (open circles-present sample), NGC 188 (filled circles), and M 67 (filled triangles). Li abundances for NGC 188 and M 67 have been retrieved from Sestito & Randich (2005) and had been derived in the same fashion as for the sample clusters.

3. Li vs. temperature

In Figs. 1, 2, 3 we compare the log n(Li) vs. T_{eff} distributions of three of the sample clusters (Be 32 –Fig. 1; Mel 66 –Fig. 2; Cr 261 –Fig. 3) with data from the literature for M 67 and NGC 188. All the five clusters are as old as the Sun or older.

The comparison of the three figures clearly shows that both the upper envelope of the Li vs. T_{eff} distribution and the maximum Li abundance is similar in the five clusters; however, each of the three clusters Be 32, Mel 66, and Cr 261 behaves in a different way. The distribution of Be 32 is similar to that of NGC 188: at variance with M 67, it is characterized by no dispersion and all stars have a Li abundance more than a factor of 10 larger than the Sun. Vice versa, most members of Mel 66 are heavily Li depleted and only five stars (out of 53) have log n(Li)$>$ 2. Finally, Cr 261 exhibits an intermediate behaviour, more similar to the pattern of M 67: a fraction of stars have a Li content similar to the Sun, while another fraction shows a much higher Li. As to the other four sample clusters, three show almost no dispersion and all their members are Li-rich, while a scatter is observed in NGC 6253.

In conclusion, our results confirm on solid grounds and based on large number statistics that old stars are not necessarily Li poor as the Sun is. Old open cluster members show a variety of Li patterns and there is not a 'standard'. In a few clusters Li-rich and Li-poor stars are both present, while in others only the Li rich ones are present. Based on the

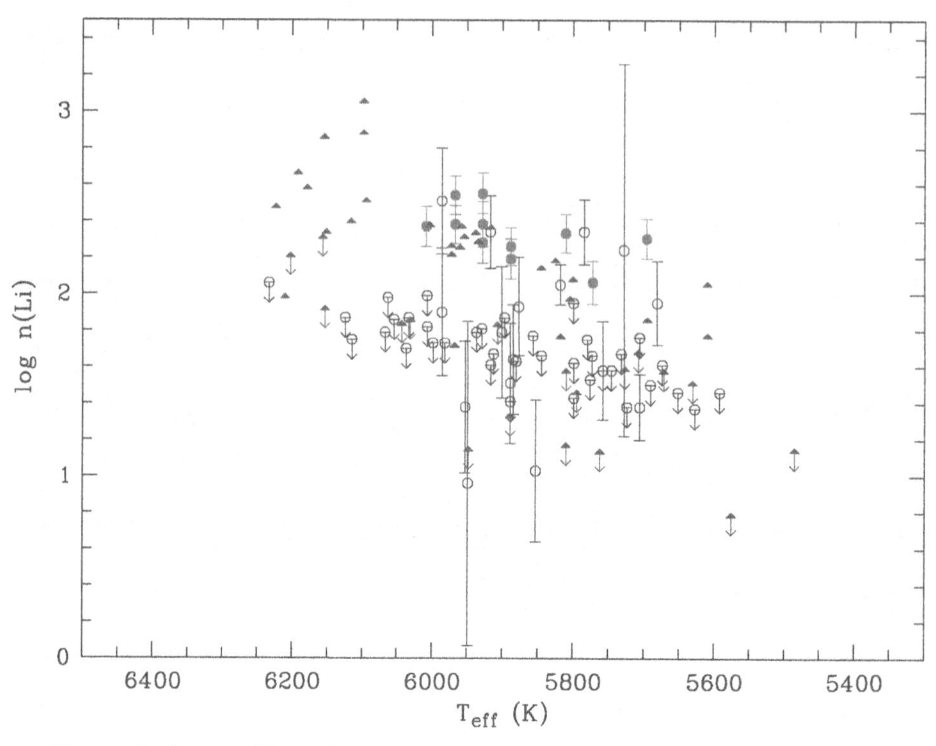

Figure 2. Same as Fig. 1, but NGC 188 and M 67 are compared to Melotte 66.

whole sample, we find that the Li patterns do not seem to depend on obvious cluster properties, such as, for example, metallicity.

4. Evolution of Li with age

In Fig. 4 we summarize the results on the evolution of Li with age, plotting the average Li abundance as a function of age; for each cluster we considered stars in the temperature interval 5750 K $\leqslant T_{\rm eff} \leqslant$ 6050 K. Open symbols denote clusters from the literature re-analyzed by Sestito & Randich (2005), while filled symbols indicate the present sample. Symbols at the same age do not indicate different clusters, but the average of the upper and lower envelopes of the Li vs. $T_{\rm eff}$ distribution of clusters characterized by a dispersion. The Sun is also plotted in the figure. Fig. 4 shows that solar-type stars undergo very little depletion (less than a factor of 2) up to 100 Myr. Then, they suffer continuous depletion up to an age of about 1 Gyr; the decay of Li abundance is well fitted with an exponential law log n(Li)$\propto t^{-\alpha}$ with $\alpha = -0.68 \pm 0.09$. After 1 Gyr depletion becomes bimodal: part of the stars do not undergo any additional depletion and Li abundances converge towards a plateau value†. Another fraction of the stars, including the Sun, instead continue depleting Li at a very fast rate ($\alpha = -1.63 \pm 0.08$).

The reason why otherwise similar stars deplete Li at different rates after 1 Gyr is so far not understood. However, this empirical evidence allows us to draw definitive conclusions on the use of Li to derive the age of solar-type stars. Namely, Li is a good

† This plateau value is surprisingly similar to the Spite plateau of Pop. II stars.

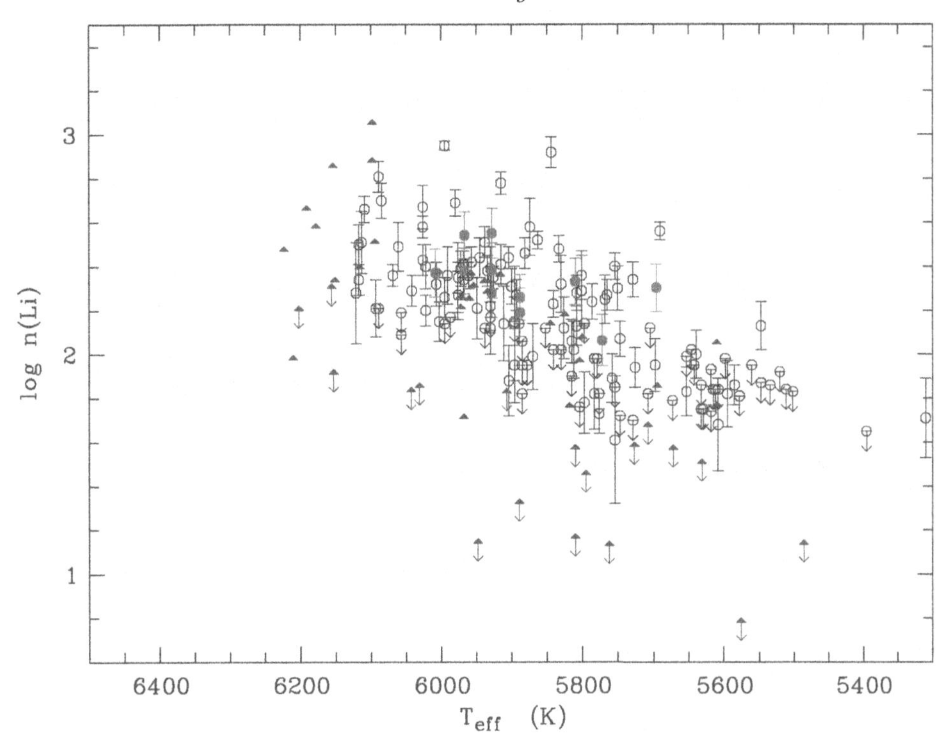

Figure 3. Same as Figs. 1 and 2, but NGC 188 and M 67 are compared to Collinder 261.

age tracer for these stars up to about 1 Gyr. After that, a 'low' solar-like Li abundance is indicative of an old age, plus, possibly, a peculiar evolution. Vice versa, a 'high' Li content ($\sim$ 10 times the solar value) only allows deriving a lower limit to the star age. A Li abundance log n(Li) $\sim$ 2.4 does not allow discerning whether a star is 1 or 8 Gyr old. These conclusions should be kept in mind when using Li to derive the ages of field stars, in particular of exo-planet host stars.

References

Alonso, A., Arribas, S., & Martinez-Roger, C. 1996, *A&A*, 313, 873
Boesgaard, A. M. & Tripicco, M. J. 1987, *ApJ*, 313, 389
Duncan, D. K. 1981, *ApJ*, 248, 651
García López, R. J., Rebolo, R., & Beckmann, J. E. 1988, *PASP*, 100, 1489
Herbig, G. H. 1965, *ApJ*, 141, 588
Jones, B. F., Fisher, D., & Soderblom, D. R. 1999, *AJ*, 117, 330
Pallavicini, R., Cerruti-Sola, M., & Duncan, D. K. 1987, *ApJ*, 174, 116
Pallavicini, R., *et al.* 2006, *In Chemical Abundances and Mixing in the Milky Way and its Satellites*, ESO Astrophysics Symposia, S. Randich & L. Pasquini (eds), Springer-Verlag, p. 181
Pasquini, L., Randich, S., & Pallavicini, R. 1997, *A&A*, 325, 535
Randich, S., Sestito, P., & Pallavicini, R. 2003, *A&A*, 399, 133
Randich, S., *et al.* 2005, *ESO Messenger*, 121, 18
Randich, S., Pace, G., Pastori, L., & Bragaglia, A. 2009, *A&A*, in press
Sestito, P. & Randich, S. 2005, *A&A*, 442, 615

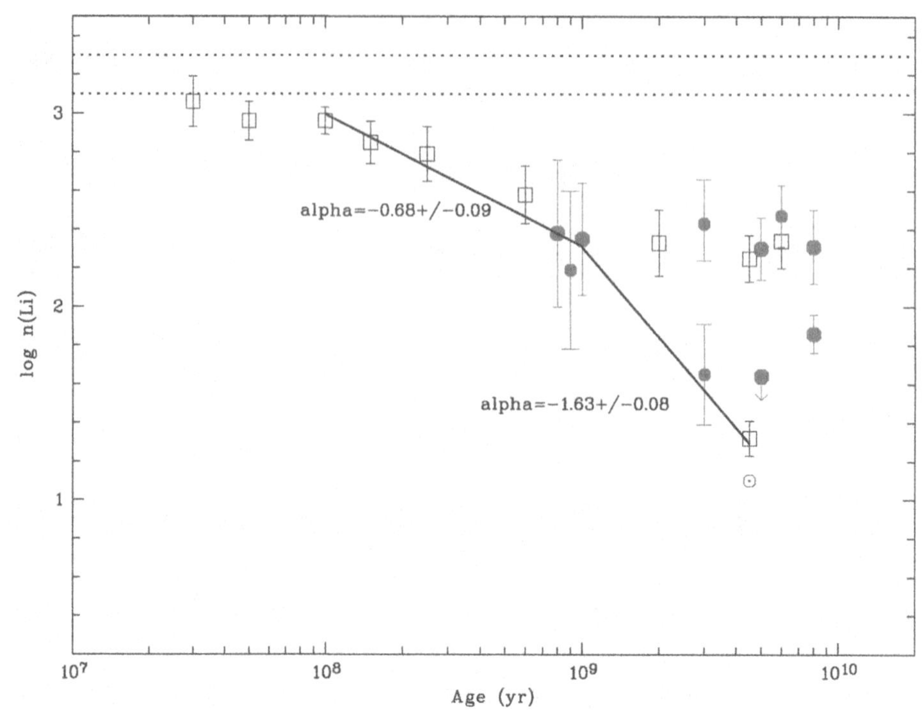

Figure 4. Average Li abundance as a function of age for solar-type stars (5750 K $\leqslant$ T_{eff} $\leqslant$ 6050 K). Open symbols indicate clusters from Sestito & Randich (2005), while filled circles denote the present sample. Error bars indicate the 1σ deviation from the average. Symbols at the same age indicate the average of the upper and lower envelopes of the Li vs. T_{eff} distribution of clusters characterized by a dispersion. The two dotted lines limit the range of initial Li abundances for Pop. I stars. The best fit exponential decays (log n(Li)$\propto t^{-\alpha}$) between 100 Myr and 1 Gyr, and between 1 and 4.5 Gyr are also shown.

Skumanich, A. 1972, *ApJ*, 171, 565
Soderblom, D. R 1983, *ApJS*, 171, 565
Soderblom, D. R, Stauffer, J. R., Hudon, J. D., & Jones, B. F. 1993a, *ApJS*, 85, 113
Soderblom, D. R., Jones, B. F., Balachandran, S. C. *et al.* 1993b, *AJ*, 106, 1059
Spite, M. & Spite, F. 1982, *A&A*, 115, 351
Spite, F., Spite, M., Peterson, R. C., & Chaffee, F. H., Jr. 1987, *A&A* 171, L8

Discussion

E. Mamajek: This is a question for Sofia and the previous speaker (Deliyannis). Why is the Sun so weird (Li poor)? Follow up: Are exoplanet host stars also anomalously Li poor?

S. Randich: 1) Actually we do not know – several people have suggested that it could be because of the fact that the Sun hosts a planetary system. 2) Yes, on average they are.

R. Mathieu: Following on Marc Pinsonneault's thought regarding the effect of planets on disks and thus lithium enrichment, might binary companions play the same role? Of order one-third of stars in NGC 188 and M67 are spectroscopic binaries, and presumably undetected wider binaries are also present.

S. Randich: Yes, I agree.

L. Hillenbrand: I also wanted to ask about the bimodality in $N(\mathrm{Li})$ vs. age. Are there additional clusters in the 1–10 Gyr age range that might someday be placed on this figure to test bimodality vs. scatter between the two limits defined by 1) the flat relation and 2) the solar track?

S. Randich: Yes, there are a few clusters within reach. But I think that the observational pattern is now rather well defined. We should try understanding what could be the reason of bimodality, before aiming for more observing time.

C. Deliyannis: What is your favorite Galactic Li production mechanism?

S. Randich: AGB stars with a superwind.

M. Pinsonneault: You've demonstrated that Li depletion is highly variable, making Li a poor age diagnostic for field stars. There is also a theoretical framework in which Li depletion is expected for stars which truncated their accretion disks early.

S. Randich: OK. Thanks for mentioning this.

N. Panagia: Judging from your last plot I would conclude that the "normal" Li depletion is slowly evolving with time (less than a factor of 10 in 10 Gyr) whereas "exceptional" stars evolve faster (due to rotation?).

S. Randich: Yes, the majority of stars shows slowly evolving Li and converge towards a plateau at old ages.

H. Richer: Have people looked for planets or binary companions to the lithium-depleted stars?

S. Randich: We do have a program (just started) to do that. The PI is Luca Pasquini.

Sofia Randich

Christine Chen

The Ages of Stars
Proceedings IAU Symposium No. 258, 2008
E.E. Mamajek, D.R. Soderblom & R.F.G. Wyse, eds.

doi:10.1017/S1743921309031792

Observational problems in determining the ages of open clusters

Elizabeth J. Jeffery

Department of Astronomy, The University of Texas at Austin,
Austin, TX 78712, USA
email: ejeffery@astro.as.utexas.edu

Abstract. Open clusters have long been objects of interest in astronomy. As a good approximation of essentially pure stellar populations, they have proved very useful for studies in a wide range of astrophysically interesting questions, including stellar evolution and atmospheres, the chemical and dynamical evolution of our Galaxy, and the structure of our Galaxy. Of fundamental importance to our understanding of open clusters is accurate determinations of cluster ages. Currently there are two main techniques for independently determining the ages of stellar populations: main sequence evolution theory (via cluster isochrones) and white dwarf cooling theory. We will provide an overview of these two methods, the current level of agreement between them, as well as a look to the current state of increasing precision in the determination of each. Particularly I will discuss the comprehensive data set collection that is being done by the WIYN Open Cluster Study, as well as a new Bayesian statistical technique that has been developed by our group and its applications in improving and determining white dwarf ages of open clusters. I will review the so-called bright white dwarf technique, a new way of measuring cluster ages with just the bright white dwarfs. I will discuss the first application of the Bayesian technique to the Hyades, also demonstrating the first successful application of the bright white dwarf technique. These results bring the white dwarf age of the Hyades into agreement with the main sequence turn off age for the first time.

Keywords. open clusters and associations: general, white dwarfs

---

1. Introduction

Age measurements are a fundamental problem in astronomy and essential for a number of astrophysically interesting problems. From the fundamental questions of the formation of the Universe to the creation of planets, knowing and understanding ages of astronomical objects is important. Answering the questions of *when* (for example, "*When* did the Universe form?" or "*When* did the halo/bulge/disk form in relation to each other?" or "*When* did the various elements form?" or "*When* do planets form?") is a crucial step in understanding the questions of *how* these phenomena occur.

Open star clusters have long been objects of interest in astronomy. They are a good approximation of essentially pure stellar populations and have proved useful for studies in a wide range of astrophysically interesting questions. They are useful in understanding stellar evolution and atmospheres, the chemical and dynamical evolution of our Galaxy, as well as its structure, and are essential to distance scale studies. In order to gain the most information from open clusters, accurate ages are essential.

Currently there are two main techniques for independently determining the ages of stellar populations: main sequence (MS) evolutionary theory (via cluster isochrones; e.g., Chaboyer *et al.* 1996) and white dwarf (WD) cooling theory (e.g., Winget *et al.* 1987). Open clusters provide the ideal environment for the calibration of these two important clocks, as well as providing the unique opportunity to test theory against theory and therefore increase our understanding of both.

Each of these two techniques is accompanied by both theoretical and observational obstacles to determining ages. In this paper I will focus mainly on the observational challenges in determining accurate cluster ages, as well as some of the current efforts to minimize these issues as best as can be currently done.

This paper is organized as follows: in Section 2 I will review the techniques for measuring MSTO and WD ages, and examine some of the major observational challenges in determining these ages accurately for open clusters. In Section 3 I will discuss a few solutions to alleviate some of the observational challenges discussed. These include improving cluster observations by collecting comprehensive data sets, such as is being done by the WIYN Open Cluster Study (WOCS) (Section 3.1); improving techniques for objectively measuring cluster ages, specifically discussing a new Bayesian statistical technique (Section 3.2); and finally, I will discuss a new technique to measure cluster WD ages, based on using the bright cluster WDs alone (Section 3.3). I will review the first application of the Bayesian technique and the bright WD technique to real data (the Hyades) by DeGennaro *et al.* (2009) and discuss those results in Section 4.

2. Open cluster ages: Techniques and observational challenges

As mentioned in Section 1, there are two main techniques used to measure open cluster ages: main sequence turn off (MSTO) ages and white dwarf (WD) cooling ages. I discuss each of these below.

2.1. *Main sequence turn off ages*

MSTO techniques are among the most mature methods for determining the ages of open clusters. A cluster's age is determined by fitting an isochrone to the MSTO on the color-magnitude diagram (CMD) (see, for example, Figure 1, taken from Sarajedini *et al.* 1999). Theoretical problems still exist in such models (as can be seen in Figure 1, for example, imperfect fits to the red giant branch (RGB), as well as different isochrones giving a best fit to the MSTO in different colors).

Some of the observational challenges in determining MSTO age include field stars and unresolved binaries, especially in the TO region. This gives a certain amount of width to MSTO (see Figure 1), and makes fitting isochrones by eye more difficult. Additionally, there are errors in the actual observations themselves, as well as errors in cluster distance, metallicity, and reddening (and, in some clusters, differential reddening; e.g., NGC 2477 (Hartnick *et al.* 1972). Because the position of a star on the CMD is affected by not only its color and magnitude, but also its distance, metallicity, and reddening, the effects of these properties can be confused for one another and can be difficult to disentangle, based on the CMD alone. A large uncertainty in one may cause an incorrect determination in another, and together may affect the best isochrone fit, and therefore the accuracy of the MSTO age.

2.2. *White dwarf cooling ages*

If an open cluster is sufficiently old, some of its members will be WDs. There is a simple relationship between a WD's luminosity and its cooling time (i.e., age) (Mestel, 1952). Cluster WDs show a low luminosity terminus whose location is determined by the age (Claver, 1995; von Hippel *et al.* 1995). I illustrate this in Figure 2; the figure shows a simulated CMD for a cluster of 3 Gyr. Over plotted are theoretical cooling tracks for WDs of several masses (ranging here from 0.5 $M_{\odot}$ to $0.9 M_{\odot}$). WDs follow simple cooling

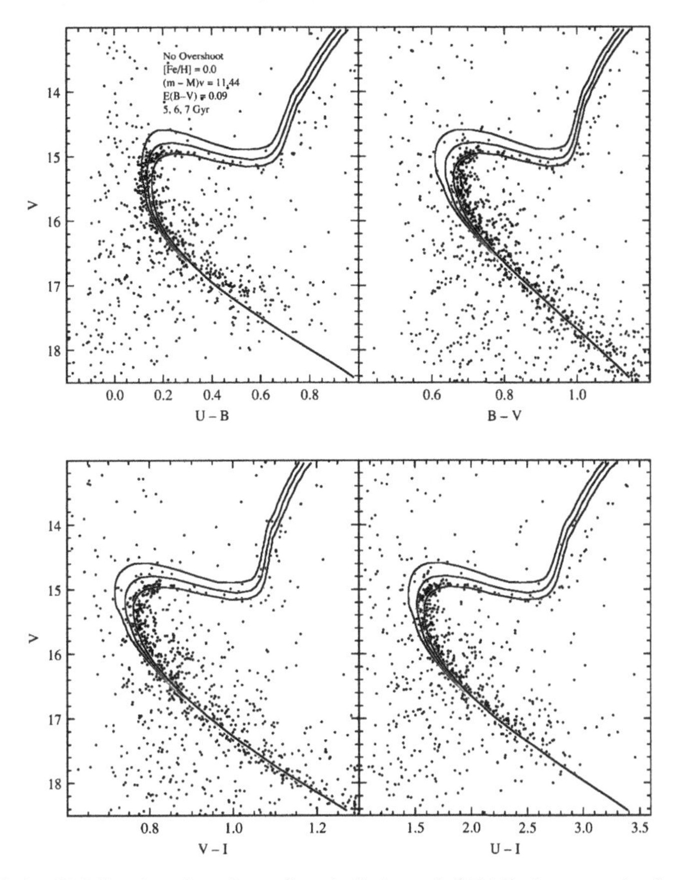

Figure 1. CMD for NGC 188, taken from Sarajedini *et al.* (1999), in several colors with theoretical isochrones overlaid. Note that no single isochrone gives a perfect fit in all colors, illustrating the difficulties that still exist in determining MSTO ages.

laws and therefore the rate at which a WD cools slows with time. This causes a "pile up" for WDs of different masses near the terminus of the WD sequence, causing the observed "hook" shape.

Like MSTO ages, the determination of cluster WD ages is also affected by observational errors, and errors in cluster distance and reddening. (They are less affected by errors in cluster metallicity.) Additional problems also include contamination from field stars (especially from field WDs) in the CMD and from background galaxies with similar colors. Cluster WDs are also intrinsically faint and therefore require large (or space-based) telescopes to observe. In addition to these observational challenges, theoretical uncertainties still exist in the models for the coolest WDs, namely issues related to crystallization, phase separation (Isern *et al.* 2000; Metcalfe *et al.* 2004), and collision-induced absorption (CIA) in the coolest WD atmospheres (Frommhold 1993; Kilic *et al.* 2006).

2.3. *Effect of observational errors on age*

As mentioned above, errors in cluster distance, metallicity, and reddening can affect the derived MSTO and WD age. For example, assuming no error in other parameters (observations, metallicity, and reddening) an error in the cluster distance modulus of 0.1 magnitudes translates to an error of 7% in the measured MSTO age. Similarly, a

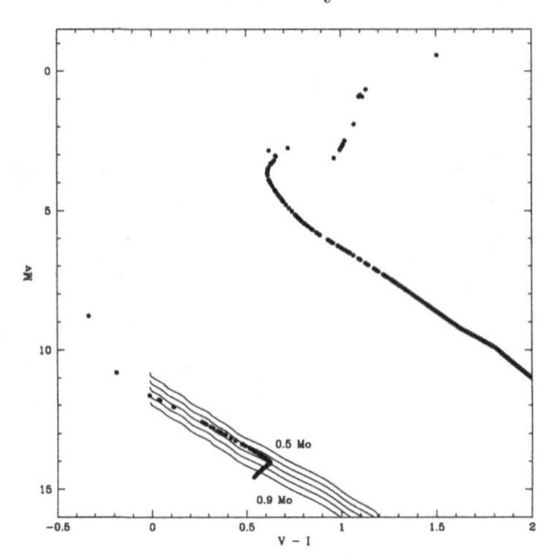

Figure 2. A simulated 3 Gyr open cluster. The figure illustrates where the WD cooling sequence falls on the CMD. The over plotted lines are cooling tracks for WDs of several masses. The hook shape of the WD terminus is explained by the "pile up" of WDs at the terminus.

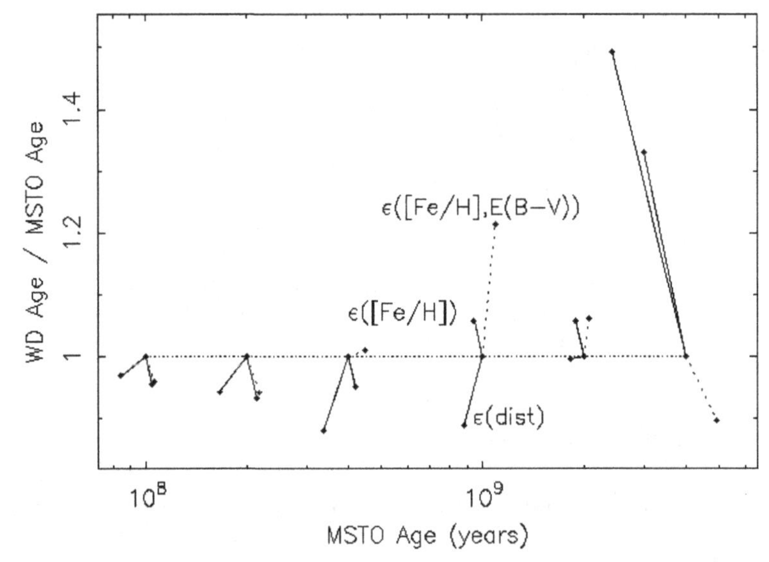

Figure 3. Effect of three types of errors on the MSTO and WD ages of open clusters, taken from von Hippel (2005). Thick lines show a distance overestimation of 0.2 magnitudes, the thin lines an overestimation of metallicity by 0.2 dex, and the dotted lines an overestimation of metallicity by 0.2 dex then compensated by a decrease in reddening enough to keep the MS at the same color.

distance modulus error of 0.2 and 0.5 magnitudes means an age error of 17% and 45%, respectively.

This point was explored further by von Hippel (2005); Figure 3 is taken from that study. He illustrated how the WD and MSTO age are affected by errors in distance (thick solid line) of 0.2 magnitudes, metallicity (thin solid line) of 0.2 dex, and an error in metallicity of 0.2 dex then compensated for by decreasing the reddening to keep the MS at the same color (dotted line).

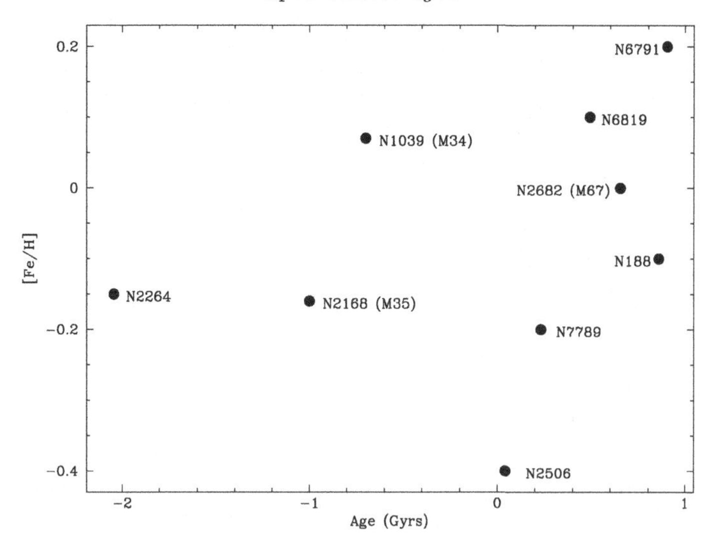

Figure 4. WOCS key clusters in age and metallicity space.

2.4. *Current agreement and calibration*

A calibration is needed to put these two chronometers on the same absolute scale. Open clusters provide the ideal environment for this calibration. This can be done by measuring the WD age and the MSTO age in a number of open clusters. Additionally, the WDs provide an independent check on MS models and the MS models check the WD models. This allows us to better understand the physics of each. Recent studies show good agreement up to 4 Gyr for open clusters (von Hippel, 2005).

3. Challenges and solutions

A few of the solutions to the observational challenges I have discussed in the previous section include collecting comprehensive data sets, improved statistical interpretations of CMDs, and a new technique to determine cluster WD ages. In the following sections, I examine each of these points.

3.1. *WOCS: A comprehensive study of open clusters*

Tools for the observational study of open clusters continue to improve and increase observational efficiency with (for example) mosaic CCDs and multi-object spectrographs. Photometric, spectroscopic, and astrometric measurements can be made with higher precision than before. The WIYN Open Cluster Study (WOCS) is a collaboration formed with the goal to create a comprehensive and definitive photometric, spectroscopic, and astrometric database for several key clusters (see Figure 4). Such a comprehensive and complete data set allows for the investigation of critical astrophysical problems through the study of open clusters.

Efforts are being done by WOCS on nearly all fronts of data acquisition. These efforts include: absolute photometry to obtain carefully calibrated CMDs (e.g., Sarajedini *et al.* 1999); monitoring of relative photometry to find cluster variable stars (e.g., Kafka & Honeycutt 2003); spectroscopic observations to determine metallicities and elemental abundances, including lithium (e.g., Steinhauer & Deliyannis 2004); radial velocity studies as a means of finding binary stars and studying cluster dynamics (e.g., Geller *et al.* 2008); and astrometry to measure proper motions, separating cluster members from contaminating field stars (e.g., Platais *et al.* 2003). When used together, the information gleaned from these multiple studies provide valuable insights to many astrophysical problems through the study of open clusters.

3.2. *Measuring ages using Bayesian statistics*

Despite many high quality datasets having been collected on open clusters, age precision of better than 10–20% is still generally out of reach. The greatest gains that can currently be made in age precision will require improved modeling techniques (see also work done by Tosi *et al.* 1991, Tosi *et al.* 2007; Hernandez & Valls-Gabaud 2008).

A new technique has been developed to determine cluster ages using Bayesian statistics (see von Hippel *et al.* 2006; Jeffery *et al.* 2007; and DeGennaro *et al.* 2009, and in the poster at this conference by DeGennaro *et al.*). Briefly, the Bayesian technique derives a posterior probability distribution for a cluster's age, metallicity, distance, and line-of-sight absorption by objectively incorporating our prior knowledge of stellar evolution, star cluster properties, and data quality estimates. It incorporates a Miller & Scalo (1979) initial mass function, MS and giant branch stellar evolution time scales of Girardi *et al.* (2000), the initial-final mass relation (IFMR) from Weidemann (2000), WD cooling timescales of Wood (1992), and WD atmosphere colors from Bergeron *et al.* (1995). Two additional sets of MS models have also been added – Yale-Yonsei (Yi *et al.* 2001), and a finer grid of models from the Dartmouth Stellar Evolution Database (DSED, Dotter *et al.* 2008), as well as updated versions of Bergeron WD atmosphere models.

3.3. *The bright white dwarf technique*

Jeffery *et al.* (2007) demonstrated the theoretical feasibility of determining WD ages of clusters using the brighter WDs alone. Briefly, this technique relies on the subtle differences in slope and position of the WD cooling sequence relative to the MS for clusters of different ages (as illustrated in Figure 5).

For an individual cluster WD, its total age (i.e., the age of the cluster) is its WD cooling time plus the lifetime of its MS counterpart. The cooler WDs are more massive and hence evolved off the MS more quickly; i.e., they have spent a larger fraction of their total lifetime as WDs. In these cases, their WD cooling time is good measure of the total cluster age. I have plotted the relationship between the ratio of a WD's cooling time and the total cluster age as a function of absolute magnitude in Figure 6.

The bright WD technique assumes that the IFMR is universal and single-valued, which is the general consensus among researchers (Weidemann 2000). Because of this dependence on the IFMR, this technique is a relative age indicator and requires calibration. As will be discussed in the next section, recent work on the Hyades is the first step in such a calibration.

4. The white dwarf age of the Hyades

The Hyades is one of the most well-studied open clusters in the sky. Perryman *et al.* (1998) report a MSTO age for this cluster of 625 $\pm$ 50 Myr and a distance to the center of the cluster (based on trigonometric parallaxes from Hipparcos) of $m - M = 3.33 \pm 0.01$. High resolution spectroscopy has been used to determine the metallicity to high accuracy, [Fe/H] = +0.103 $\pm$ 0.008 (Taylor & Joner 2005, based on their re-analysis of Paulson *et al.* 2003). In order to apply the Bayesian technique to real data for the first time, this cluster was the logical choice.

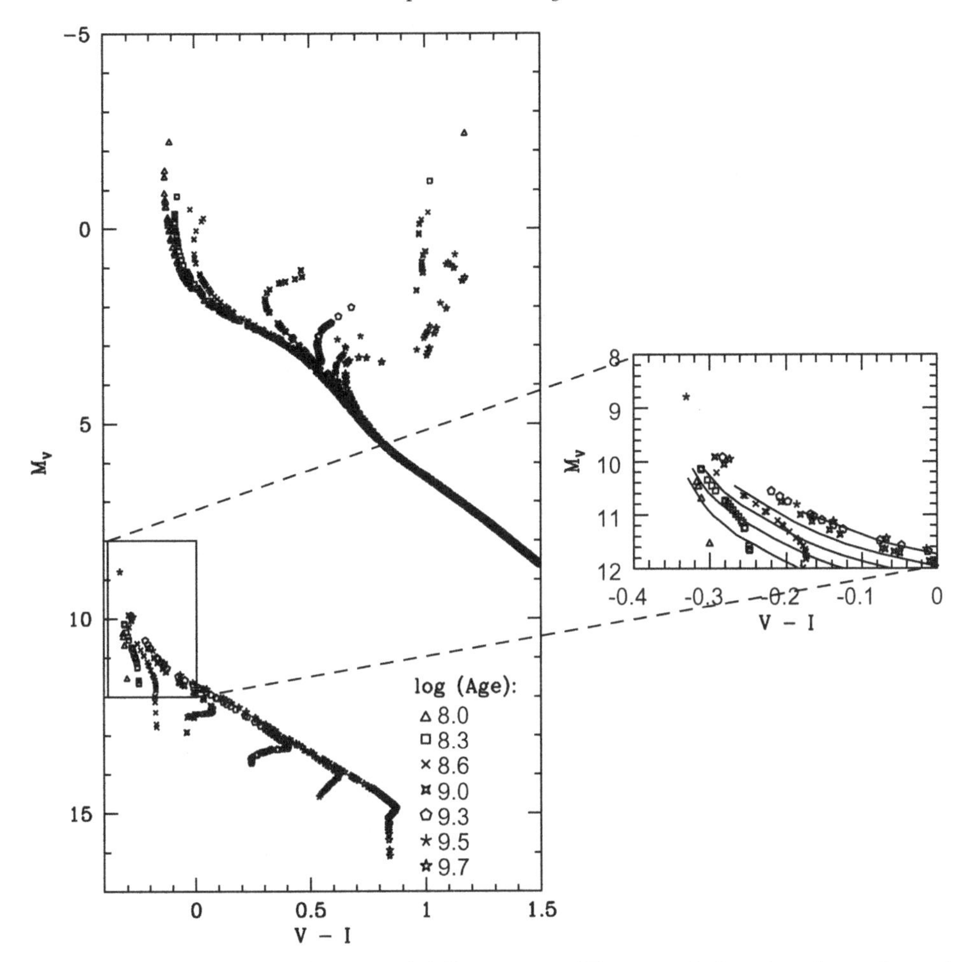

Figure 5. Simulated clusters for several different ages. The expanded region shows the regime of the brighter WDs, clearly showing the subtle differences in the slopes and positions of the WD cooling sequences relative to the MS for clusters of different ages. This makes it possible to extract age information without observing the faintest WDs.

Previous studies to determine the WD age of the Hyades cluster have produced a result (300 Myr; Weidemann 1992) that is about half the measured MSTO age (625 Myr; Perryman *et al.* 1998). Weidemann (1992) suggested that this discrepancy is due to the dynamical evaporation of stars from this cluster; the coolest WDs are no longer present. This is illustrated in Figure 7. This is the CMD for the WD region of the Hyades with a WD isochrone for the MSTO age of the cluster over plotted, demonstrating the lack of cool WDs that are expected to populate the bottom of the WD cooling sequence. In the absence of any data on these missing faint WDs, traditional techniques to determine WD ages can provide at best a lower limit to the WD age. (In reference to this figure and the results, however, it should be noted that when doing a best fit of the data, the Bayesian technique fits the data (UBV photometry was used for this study) in magnitude-magnitude space, not color-magnitude space, as is plotted in Figure 7.)

As summarized earlier, in Jeffery *et al.* (2007) showed the possibility of determining cluster WD ages from just the bright WDs, when the coolest WDs are not observed.

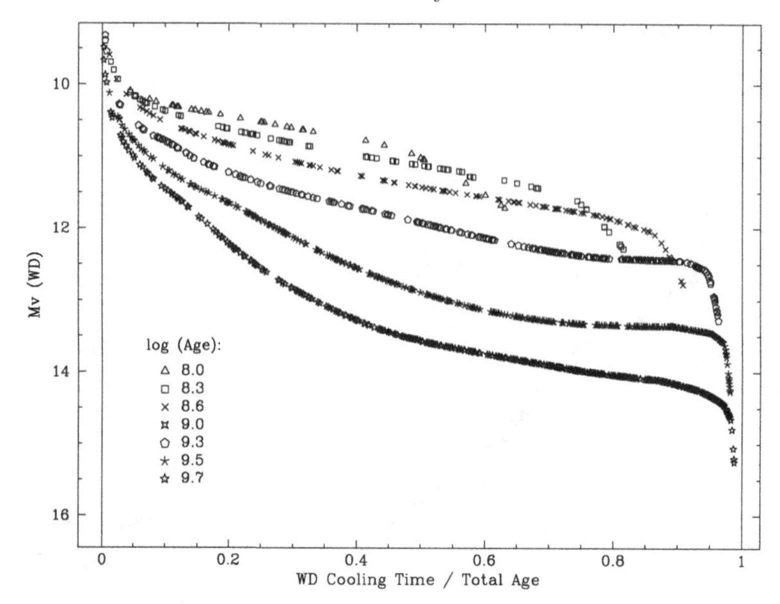

Figure 6. The relationship between the ratio of an individual WD's cooling time and the total cluster age and the absolute magnitude of that WD.

Because the coolest WDs are missing from the Hyades, the bright WD technique is required to measure the true WD age (rather than a lower limit, as was done previously).

DeGennaro *et al.* (2009) applied the Bayesian technique to the Hyades, the first application of the technique to real data, and has proven the technique successful. This is also the first empirical evidence that the bright WD technique yields reasonable and precise ages for real data, as well as providing an important step in calibrating the technique. (See Figure 8.)

5. Conclusions

Age is a quantity of importance to all areas of astrophysics. The study of open clusters, as nearly pure stellar populations, continues to be an important area of study in stellar astronomy. They provide the environment for answering many questions of astrophysical import and measuring accurate ages for open clusters is essential to fully leveraging this potential.

MSTO and WD cooling ages are the two main techniques used to determine ages of these stellar populations. The CMD is the primary tool for these age determinations. Observational challenges for each method include observational errors, as well as errors in cluster parameters (that is, distance, metallicity, and reddening). Challenges to fitting MSTO isochrones come as a result of field star contamination and unresolved cluster binaries. For WDs, there is also the issue of contamination from field stars (especially field WDs in the WD region) and background galaxies. Also, cluster WDs are faint, thus requiring large or space-based telescopes to observe them.

These problems can be alleviated in part by comprehensive data sets, such as are being assembled by WOCS. WOCS is creating a comprehensive and definitive photometric, spectroscopic, and astrometric database for fundamental clusters, covering a large range

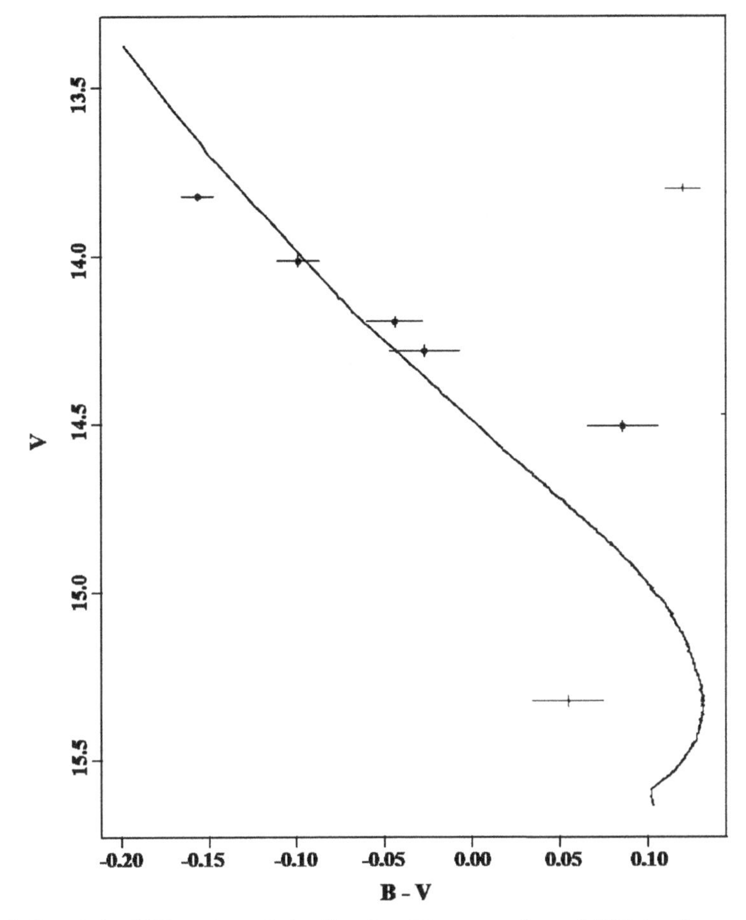

Figure 7. CMD of the WD region of the Hyades. The over plotted WD isochrone is for an age equal to the MSTO age. This demonstrates the lack of cool WDs that are expected to populate the bottom of the WD cooling sequence. However, note that the Bayesian technique does a best fit to the data based on magnitude-magnitude plots, not CMDs. In this analysis, UBV colors were used.

of ages and metallicities. Many critical astrophysical problems are being addressed as a result of the data being gathered.

We are continuing an effort to calibrate WD and MSTO ages using open clusters. In order to improve accuracy of the measured ages, our group has developed a technique using Bayesian statistics. This technique is proving successful in determining cluster WD ages to higher precision than before.

Additionally, here I have reviewed the so-called bright WD technique, a new technique to determine cluster WD ages using just the bright WDs (i.e., not observing the faintest/coolest cluster WDs). DeGennaro *et al.* (2009) have demonstrated the technique by applying it on the Hyades as a test case, whose cool WDs are missing due to the dynamical evaporation of the cluster. These results bring the WD age of the Hyades in agreement with the MSTO age for the first time, as well as provide a first step in calibrating the bright WD technique. (For further discussion of this result, see the poster by DeGennaro *et al.* at this conference.)

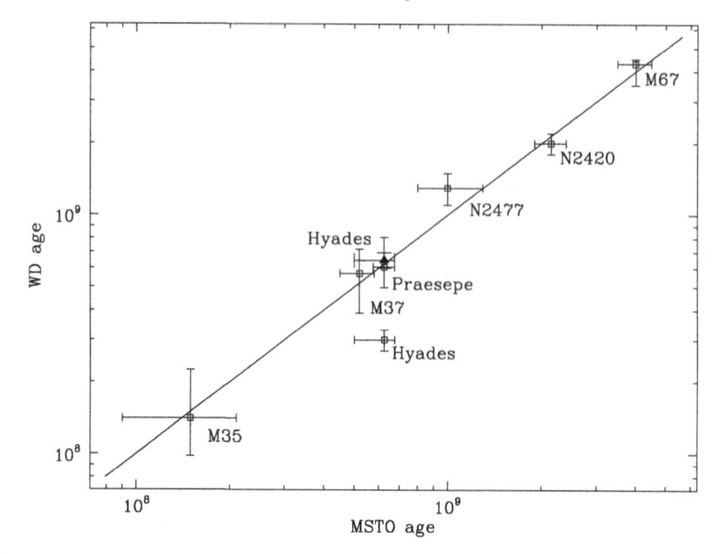

Figure 8. WD versus MSTO age for seven clusters, adapted and updated from von Hippel (2005). The age derived from the WDs in the Hyades by DeGennaro *et al.* (2009) using the bright WD technique brings the WD age of the Hyades into agreement with the MSTO age for the first time (solid triangle). The solid line shows the one-to-one correspondence between the WD and MSTO ages, and the gray point shows the most reliable WD age of the Hyades prior to this work (Weidemann 1992).

Acknowledgments

This material is based upon work supported by the National Aeronautics and Space Administration under Grant No. NAG5-13070 issued through the Office of Space Science.

References

Bergeron, P., Wesemael, F., & Beauchamp, A. 1995, *pasp*, 107, 1047

Chaboyer, B., Demarque, P., & Sarajedini, A. 1996, *ApJ*, 459, 558

Claver, C. F. 1995, PhD. Thesis, The University of Texas at Austin

DeGennaro, S., von Hippel, T., Jefferys, W. H., Stein, N., van Dyk, D., & Jeffery, E. J., *ApJ*, in press

Dotter, A., Chaboyer, B., Jevremovic, D., Kostov, V., Baron, E., & Ferguson, J. W. 2008, *ApJS*, in press, arXiv:0804.4473

Frommhold, L. 1993, Collision-Induced Absorption in Gases (Cambridge: Cambridge Univ. Press)

Geller, A. M., Mathieu, R. D., Harris, H. C., & McClure, R. D. 2008, *AJ*, 135, 2264

Girardi, L., Bressan, A., Bertelli, G., & Chiosi, C. 2000, *A&AS*, 141, 371

Hartnick, F. D. A., Hesser, J. E., & McClure, R. D. 1972, *ApJ*, 174, 554

Hernandez, X. & Valls-Gabaud, D. 2008, *MNRAS*, 383, 1603

Isern, J., Garca-Berro, E., Hernanz, M., & Chabrier, G. 2000, *ApJ*, 528, 397

Jeffery, E. J., von Hippel, T., Jefferys, W. H., Winget, D. E., Stein, N., & DeGennaro, S. 2007, *ApJ*, 658, 391

Kafka, S. & Honeycutt, R. K. 2003, *AJ*, 126, 276

Kilic, M., von Hippel, T., Mullally, F., Reach, W. T., Kuchner, M. J., Winget, D. E., & Burrows, A. 2006, *ApJ*, 642, 1051

Mestel, L. 1952, *MNRAS*, 112, 583

Metcalfe, T. S., Montgomery, M. H., & Kanaan, A. 2004, *ApJ*, 605, L133
Miller, G. E. & Scalo, J. M. 1979, *ApJS*, 41, 513
Paulson, D. B., Sneden, C., & Cochran, W. D. 2003, *AJ*, 125, 3185
Perryman, M. A. C. *et al.* 1998, *A&A*, 331, 81
Platais, I., Kozhurina-Platais, V., Mathieu, R. D., Girard, T. M., & van Altena, W. F. 2003, *AJ*, 126, 2922
Sarajedini, A., von Hippel, T., Kozhurina-Platais, V., & Demarque, P. 1999, *AJ*, 118, 2894
Steinhauer, A. & Deliyannis, C. P. 2004, *ApJ*, 614, 65
Taylor, B. J. & Joner, M. D. 2005, *ApJS*, 159, 100
Tosi, M., Greggio, L., Marconi, G., & Focardi, P. 1991 *AJ*, 102, 951
Tosi, M., Bragaglia, A., & Cignoni, M. 2007, *MNRAS*, 378, 730
von Hippel, T., Gilmore, G., & Jones, D. H. P. 1995, *MNRAS*, 273, L39
von Hippel, T. 2005, *ApJ*, 622, 565
von Hippel, T., Jefferys, W. H., Scott, J., Stein, N., Winget, D. E., DeGennaro, S., Dam, A., & Jeffery, E. J. 2006, *ApJ*, 645, 1436
Winget, D. E., Hansen, C. J., Liebert, J., van Horn, H. M., Fontaine, G., Nather, R. E., Kepler, S. O., & Lamb, D. Q. 1987, *ApJ*, 315, L77
Weidemann, V., Jordan, S., Iben, I. J., & Casertano, S. 1992, *AJ*, 104, 1876
Weidemann, V. 2000, *A&A*, 363, 647
Wood, M. A. 1992, *ApJ*, 386, 539
Yi, S., Demarque, P., Kim, Y.-C., Lee, Y.-W., Ree, C. H., Lejeune, T., & Barnes, S. 2001, *ApJS*, 136, 417

Discussion

M. Tosi: Could you explain how the applications of the Bayesian method has allowed to reconcile the Hyades age taking evaporation into account? What is exactly that brings the new age up on the curve fitting all the other clusters?

E. Jeffrey: The Bayesian technique measures the age of the Hyades by fitting the white dwarfs that are available – that is, the bright (hot) white dwarfs. We've shown that it is theoretically possible to do this, and the case of the Hyades shows that this technique works on real data. The age is determined by doing a best fit to the white dwarf sequence (in UBV), which exploits the slope of the white dwarf sequence and its relative distance from the main sequence.

R. Jeffries: How can the age estimates or errors in age estimates from the WD cooling tracks (for the Hyades in particular) be believed when the "best fit" model in V vs. $(B - V)$ is clearly a poor fit?

E. Jeffery: The plot shown here (V vs. $(B - V)$) is just for illustrative purposes, since most of us are used to thinking in the color-magnitude plane. However, there are a couple of things to note: The first is that BV data is not all that was used to fit and measure this age; we also had U data. Second is that the Bayesian algorithm fits the data in the magnitude-magnitude plane (that is, U vs. B, B vs. V, etc.), not the color-magnitude plane. Taking all three filters into account, the age we fit is the best fit. (The poster by DeGennaro, as well as his recently submitted paper explaining this result, expounds on this further.) Also, the Hyades has a problem that most open clusters do not, that is, the depth of the cluster (about 10% of the distance) can affect individual stars, making them appear brighter or fainter than we'd expect (therefore affecting their position slightly in the color- magnitude plane, making some appear slightly off the isochrone in one particular color, as seen here).

A. WEST: How much of your age uncertainty is due to uncertainty in the WD progenitor lifetimes?

E. JEFFERY: Uncertainty in cluster ages from the cooling times of the coolest white dwarfs depends little on their progenitor lifetimes. This is because the coolest white dwarfs came from the massive stars whose progenitor lifetimes are small compared to their cooling times. Therefore we can take their cooling time to be the cluster age. For the hot (bright) white dwarfs, the cooling age of an individual star is a smaller fraction of its total age (i.e., the cluster age). However, because we are fitting multiple white dwarfs along the white dwarf sequence (rather than an individual white dwarf), we don't need to worry about progenitor lifetimes. Although, that said, the bright white dwarf technique assumes a universal and single-valued initial-final mass relation, and requires calibration. Our Hyades results presented here are the first step in that calibration.

H. RICHER: Have you tried your technique on the bright end of the white dwarf cooling sequence in the two globular clusters with lots of white dwarfs?

E. JEFFERY: We have not yet, but we hope to in the future. However, the technique hasn't been theoretically tested at ages that great. It receives leverage from the mass spread in the upper white dwarf sequence and there is not much of a mass spread in clusters of globular cluster age. But we are interested in testing it.

G. DE MARCHI: You said that the coolest WDs in the Hyades are no longer there because they evaporated. But dynamical evaporation should start from the bottom of the main sequence, so say 0.3 $M_{\odot}$ stars should have long ago evaporated from the Hyades well before WDs, but they are still there?

E. JEFFERY: That's an interesting dynamical question. I'm not sure why it is that way.

I. KING: You cited the slope of the brighter part of the WD sequence as one method of age dating, and statistical fitting of theoretical isochrones as another. But you showed a slide in which the stars had a quite different slope from that of the isochrone. How do you explain this discrepancy?

E. JEFFERY: Just to clarify, we fit isochrones to the slope of the cooling sequence. (That is, they are not different methods; the different method comes from fitting the entire cooling sequence vs. fitting just the bright portion.) As for why the isochrone appears to be a poor fit to the data in the V vs. $B-V$ color-magnitude diagram, I refer my answer to Rob Jeffries's question.

The Ages of Stars
Proceedings IAU Symposium No. 258, 2008
E.E. Mamajek, D.R. Soderblom & R.F.G. Wyse, eds.

doi:10.1017/S1743921309031809

The Bologna Open Cluster Chemical Evolution project: a large, homogeneous sample of Galactic open clusters

Angela Bragaglia

INAF-Osservatorio Astronomico di Bologna
via Ranzani 1, I-40127 Bologna, Italy
email: angela.bragaglia@oabo.inaf.it

Abstract. The Bologna Open Cluster Chemical Evolution (BOCCE) project is a photometric and spectroscopic survey of open clusters, to be used as tracers of the Galactic disk properties and evolution. The clusters parameters (age, distance, reddening, metallicity, and detailed abundances) are derived in a precise and homogeneous way. This will contribute to a solid, reliable description of the disk: the clusters parameters will be used, for instance, to determine the metallicity distribution in the Galactic disk and how it has evolved with time. We have concentrated on old open clusters and we have presently in our hands data for about 40 open clusters; we have fully analyzed the photometric data for about one half of them and the spectra for one quarter of them.

Keywords. techniques: photometric, techniques: spectroscopic, stars: abundances, Hertzsprung-Russell diagram, Galaxy: abundances, Galaxy: disk, open clusters and associations: general

1. Introduction

The distribution of metals with Galactocentric distance is a signature of how the disk formed (inside-out? with infalls, outflows, mergers, etc ?). To derive it we need to know, for a given population, metallicity, distance, and age. The metallicity distribution can be traced for instance by young stars and Cepheids, but they can only be used to describe the present-day situation. Planetary nebulae go further in the past; however, there is still no consensus on the radial gradient based on them, or on the time evolution, see e.g., Maciel *et al.* (2003) and Stanghellini *et al.* (2006) for opposite ideas.

Open clusters (OCs) are very useful tracers of the properties of the Galactic disk (e.g., Friel 1995) because i) they are seen over the entire disk, ii) their ages and distances are generally measurable with better precision than those of isolated field stars up to large distances, and iii) they cover the entire range of metallicities and ages of the disk. Studying OCs we may understand whether and how the metallicity distribution changed with time.

OCs have been used in the past to define the metallicity distribution, but also in this case results were not conclusive. Most researchers found a negative radial gradient (e.g., Friel *et al.* 2002, based on homogeneous, low resolution spectroscopy), but an alternative picture has been presented: Twarog, Ashman & Anthony-Twarog (1997) invoked two flat distributions at about solar and sub-solar metallicity, with the discontinuity near a Galactocentric distance (R_{GC}) of 10 kpc. More recently, the observations of OCs at large R_{GC}'s, up to more than 20 kpc, seem to indicate a negative gradient in the inner region and a flattening in the outer disk, with the transition at 10-14 kpc: see Yong, Carney & Teixera de Almeida (2005), Carraro *et al.* (2007), Sestito *et al.* (2008). Figure 1 shows

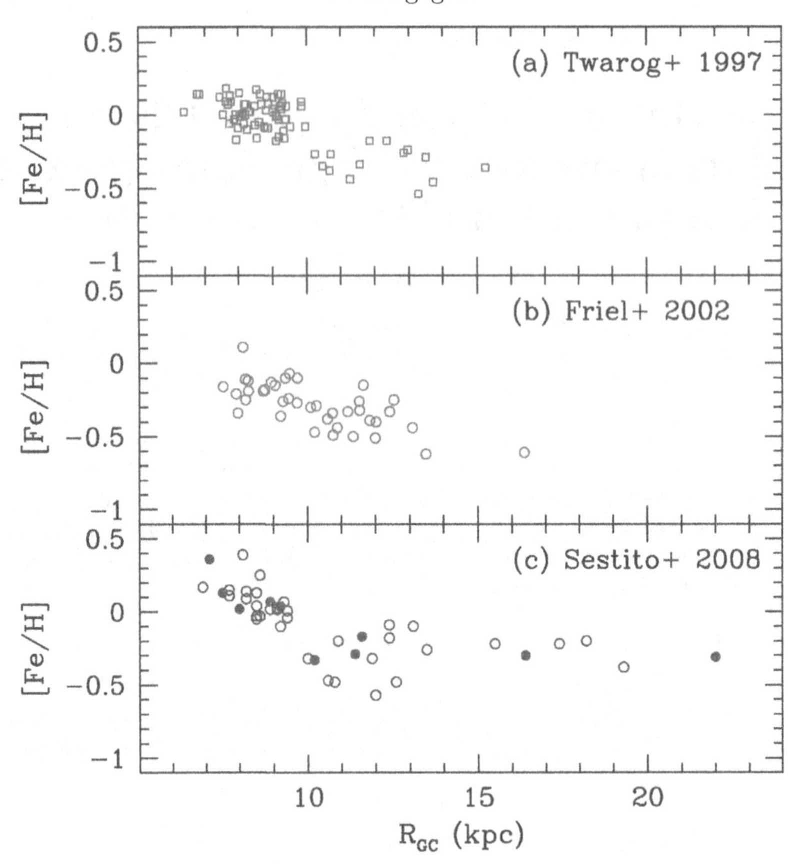

Figure 1. Radial metallicity gradient measured from OCs. (a) Results based (mostly) on DDO photometry from Twarog, Ashman & Anthony-Twarog (1997), that the authors interpret as a step distribution, with OC's inside R_{GC} = 10 kpc with solar [Fe/H] and OCs farther away with [Fe/H]$\simeq -0.3$. (b) Results based on low resolution spectroscopy from Friel *et al.* (2002), where a single radial gradient seems evident. (c) Results based on high resolution spectroscopy, adapted from Sestito *et al.* (2008). The open circles represent literature values (see the paper for all references), while the filled dots indicate the homogeneous analysis of 10 OCs observed with FLAMES, see Sestito *et al.* (2006), Sestito, Randich & Bragaglia (2007), Bragaglia *et al.* (2008), Sestito *et al.* (2008) for details. This large (but still largely inhomogeneous) dataset seems to indicate a rather steep gradient for the inner part of the disk, followed by a flattening in the outer part.

these three possibilities; the lower panel, including only results based on high resolution spectroscopy, is adapted from Sestito *et al.* (2008) and includes more than 40 OCs, one quarter of them with metallicities derived in an homogeneous fashion by the same group.

2. The BOCCE survey

Even in the last case described above we are dealing with a rather inhomogeneous sample; systematic effects can be present and maybe blur the picture. To avoid this, we have started a survey of OCs to derive in the most precise and homogeneous way their main parameters: age, distance, reddening, metallicity (and detailed abundances). We

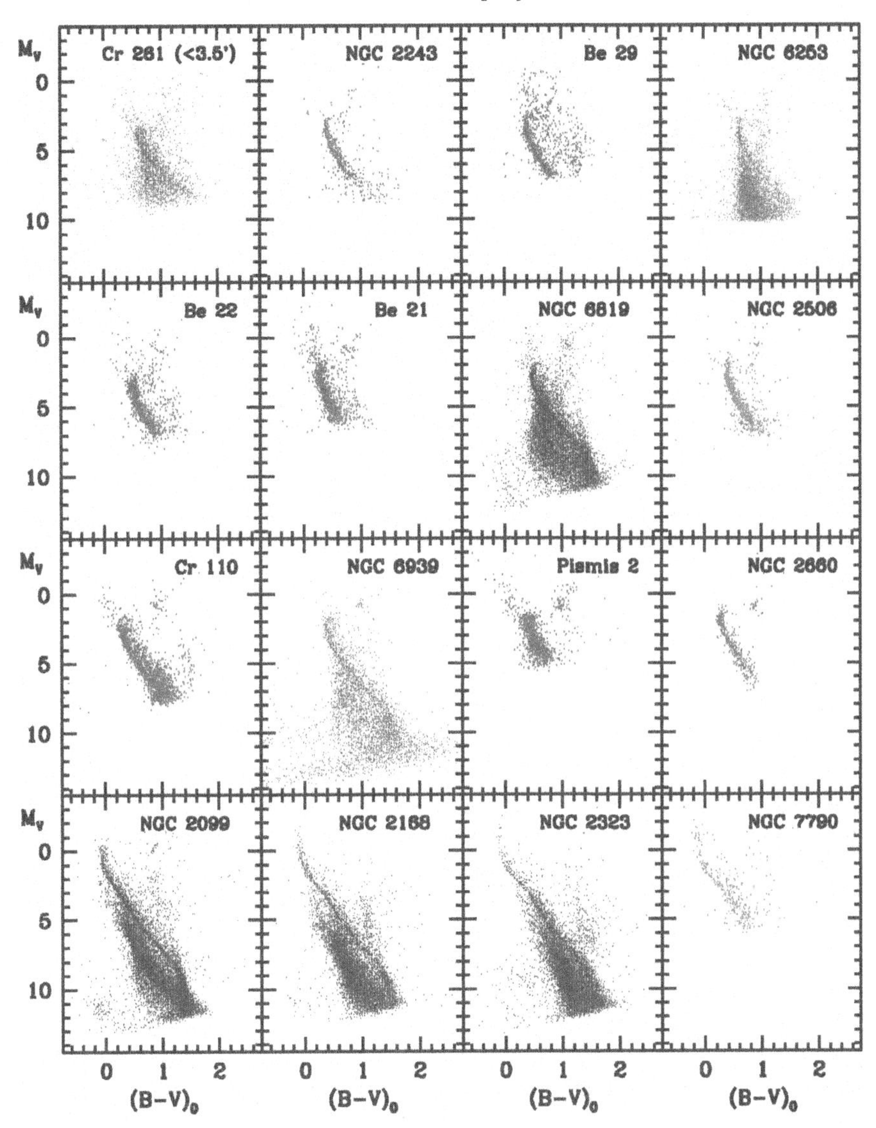

Figure 2. CMDs of the 16 BOCCE OCs presented in Bragaglia & Tosi (2006), shown in decreasing age order. The different colors indicate the source of data: red for ESO telescopes, magenta for the TNG, blue for the CFHT, grey for mixed, green for Loiano.

named our survey the "Bologna Open Clusters Chemical Evolution" project (BOCCE in short), since one of our main interests is the study of the chemical evolution of the disk. We are building a sample large enough (about 40 OCs) to be representative of the whole cluster population (in age, metallicity and position in the Galaxy). We employ:

• Photometry and the Synthetic Colour-Magnitude diagrams (CMD) technique to derive at the same time age, distance, reddening and a first indication of metallicity; for a review of the method and results see Bragaglia & Tosi (2006). This is the more advanced part of our survey; the CMDs for the first 20 OCs analyzed are shown in

Fig. 2. To build the synthetic CMDs we take into account the photometric errors, the completeness factors, the presence of binaries and the contamination due to field stars. To maintain the highest homogeneity we always use the same three sets of evolutionary tracks†, without overshooting and with two treatments of it. With our method we are able to quantify the systematics between results based on different stellar models, and we can also get an homogeneous ranking of the OC population by choosing results based on (any) one of the models.

• Medium resolution spectroscopy to derive radial velocities, hence membership, for stars in crucial evolutionary phases, like the main sequence Turn-Off or the Red Giant Branch; for an application, see e.g. D'Orazi *et al.* (2006). This is a secondary part of our project.

• High resolution spectroscopy to measure chemical abundances, using both equivalent widths and spectrum synthesis; for a presentation of the method, see Bragaglia *et al.* (2001), Carretta *et al.* (2004). We obtain spectra of a few stars per cluster, usually on the Red Clump, and possibly already known to be cluster members. We try to maintain also in this case the most homogeneous procedure, using the same line lists, gf's, model atmospheres, solar reference values, way to measure equivalent widths, synthesis, and kind of stars. A smaller part of the spectroscopic work has been completed (see Tab. 1 in Bragaglia *et al.* 2009 for an update). However, recent and planned observations have doubled the set of OCs, and we plan to use also archive data and homogenize results obtained within a parallel program (see Randich *et al.* 2005 and the already mentioned Sestito *et al.* 2008).

We have observed clusters with ages from 0.1 to 9 Gyr, R_{GC} from about 7 to 21 kpc, and metallicity from less than half solar to more than double solar. Fig. 3 (left panels) shows more CMDs for clusters recently analyzed or on which we are presently working. Since we want to study the history of the disk we concentrated on old OCs: we have collected data for a fair fraction of the about 120 OCs with age larger than 1 Gyr found in the Dias *et al.* (2002) catalogue, see Fig. 3 (right panels).

We have a few very interesting objects in our sample, like Be 17 and NGC 6791‡, which are the oldest OCs known, with an age near 9 Gyr. We also analyzed Be 20 and Be 29, with R_{GC} of about 16 and 21 kpc, or Cr 261 and NGC 6253, which lie inside the solar circle; they are very important to define the metallicity gradient. Note also that two of these clusters, NGC 6791 and NGC 6253, are the most metal-rich OCs presently known (with [Fe/H]$\geqslant$+0.4 dex, Carretta, Bragaglia & Gratton 2007).

Our effort to build a large, significant sample of open clusters with ages, distances, metallicities and detailed abundances measured with homogeneous methods is well under way and has already produced interesting results. While we have not yet succeeded in having all the clusters properties measured on the same scale (to guarantee that features are not created or lost because of systematic differences between analyses), we

† Since *homogeneity* is a key requirement of the project, we choose not to use other models or newer versions of our basic set; we are more interested in the ranking than in the absolute values. If really better tracks appear on the market we can switch to them, but in this case we have to uniformly apply them to the whole sample.

‡ We are presently working on NGC 6791, using CFHT data, see Kalirai *et al.* (2007), that we combine with literature information on other photometric bands and on membership (both from radial velocities and proper motions, the latter thanks to private communication kindly provided by K. Cudworth). Given the extreme metallicity of this cluster, we will have to use new tracks, computed on purpose using the same -or very similar- codes as the other ones.

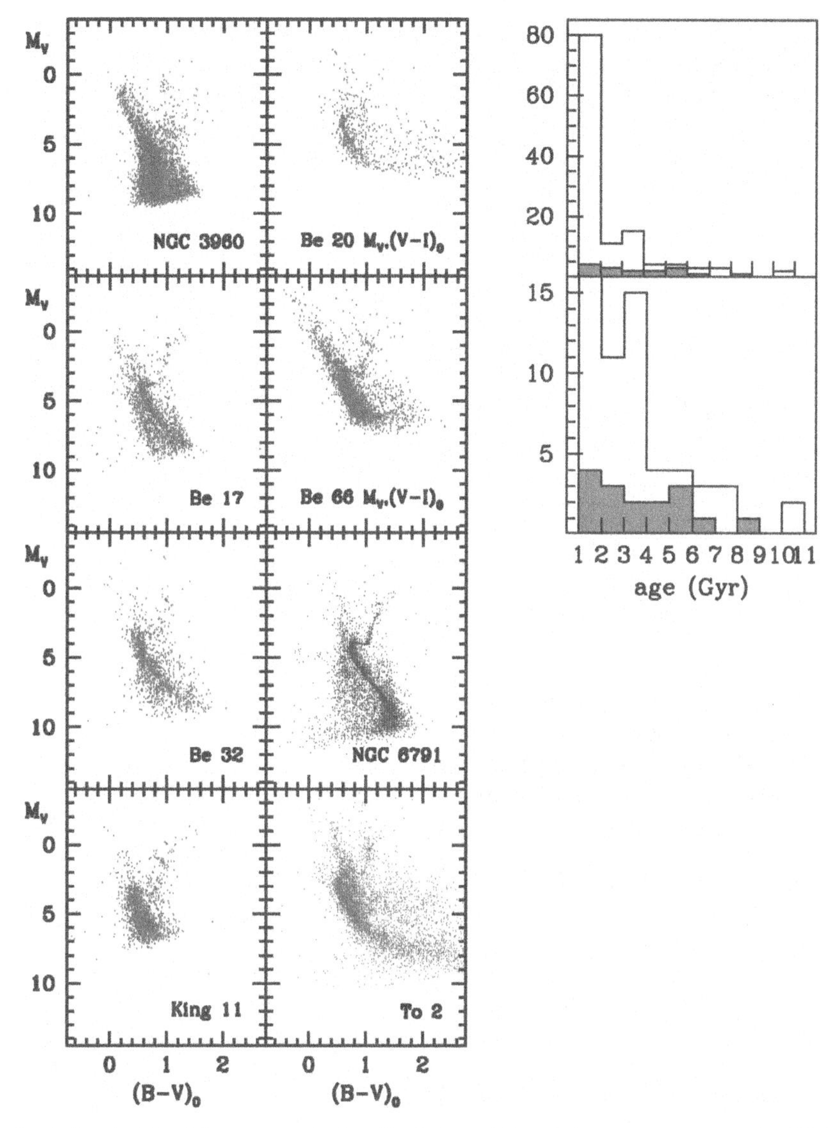

Figure 3. Left: CMD of four more BOCCE OCs published in refereed journals (NGC 3960, Bragaglia *et al.* 2006a; Be 17, Bragaglia et al 2006b; Be 32 and King 11, Tosi *et al.* 2007), of two clusters presented at a conference (Be 20, Be 66, Andreuzzi *et al.* 2008), and of two OCs on which we are presently working (NGC 6791, To 2). Right: Age distribution of the about 120 OCs with ages larger than 1 Gyr in the Dias *et al.* (2002) catalogue (open histogram) and of the 16 OCs in the BOCCE sample already analyzed and older than 1 Gyr (filled histogram); the lower panel shows an enlargement.

are struggling towards this goal and hope to be able to present a completely homogeneous picture of our sample soon.

The work described is a collaboration of many researchers: Monica Tosi and Eugenio Carretta (INAF – Bologna Observatory), Raffaele Gratton (INAF – Padova Observatory), Gloria Andreuzzi and Luca Di Fabrizio (INAF – Fundación Galilei), Michele Cignoni (Bologna University), Jason Kalirai (STScI), and Gianni Marconi (ESO Chile).

We thank Paola Sestito and Sofia Randich for useful discussions; we make use here of results obtained in collaboration with them. The WEBDA (created by J.-C. Mermilliod and presently developed by E. Paunzen at http://www.univie.ac.at/webda/) was and will be of invaluable help. Generous allocation of observing time at Italian telescopes (TNG and Loiano), at the CFHT, and at ESO telescopes (La Silla and Paranal) is acknowledged.

References

Andreuzzi, G., Bragaglia, A., & Tosi, M. 2008, *MemSAI*, 79, 657

Bragaglia, A., *et al.* 2001, *AJ*, 121, 327

Bragaglia, A. & Tosi, M. 2006, *AJ*, 131, 1544

Bragaglia, A., Tosi, M., Carretta, E., Gratton, R. G., Marconi, G., & Pompei, E. 2006a, *MNRAS*, 366, 1493

Bragaglia, A., Tosi, M., Andreuzzi, G., & Marconi, G. 2006b, *MNRAS*, 368, 1971

Bragaglia, A., Sestito, P., Villanova, S., Carretta, E., Randich, S., & Tosi, M. 2008, *A&A*, 480, 79

Bragaglia, A., Carretta, E., Gratton. R., & Tosi, M. 2009, in: J. Andersen, J. Bland-Hawthorn & B. Nordström (eds.), *The Galaxy Disk in Cosmological Context*, in press

Carraro, G., Geisler, D., Villanova, S., Frinchaboy, P. M., & Majewski, S. R. 2007, *A&A*, 476, 217

Carretta, E., Bragaglia, A., Gratton, R. G., & Tosi, M. 2004, *A&A*, 422, 951

Carretta, E., Bragaglia, A., & Gratton, R. G. 2007, *A&A*, 473, 129

D'Orazi, V., Bragaglia, A., Tosi, M., Di Fabrizio, L., & Held, E. V. 2006, *MNRAS*, 368, 471

Dias, W. S., Alessi, B. S., Moitinho, A., & Lépine, J. R. D. 2002, *A&A*, 389, 871

Friel, E. D. 1995, *ARAA*, 33, 381

Friel, E. D., Janes, K. A., Tavarez, M., Scott, J., Katsanis, R., Lotz, J., Hong, L., & Miller, N. 2002, *AJ*, 124, 2693

Kalirai, J. S., Bergeron, P., Hansen, B. M. S., Kelson, D. D., Reitzel, D. B., Rich, R. M., & Richer, H. B. 2007, *ApJ*, 671, 748

Maciel, W. J., Costa, R. D. D., & Uchida, M. M. M. 2003, *A&A*, 397, 667

Randich, S., *et al.* 2005, *ESO Messenger*, 121, 18

Sestito, P., Bragaglia, A., Randich, S., Carretta, E., Prisinzano, L., & Tosi, M. 2006, *A&A*, 458, 121

Sestito, P., Randich, S., Bragaglia, A. 2007, *A&A*, 465, 185

Sestito, P., Bragaglia, A., Randich, S., Pallavicini, R., Andrievski, S. M., & Korotin, S. A. 2008, *A&A*, 488, 943

Stanghellini, L., Guerrero, M. A., Cunha, C., Manchado, A., & Villaver, E. 2006, *ApJ*, 651, 898

Tosi, M., Bragaglia, A., & Cignoni, M. 2007, *MNRAS*, 378, 730

Twarog, B. A., Ashman, K. M., & Anthony-Twarog, B. J. 1997, *AJ*, 114, 2556

Yong, D., Carney, B. W., & Teixera de Almeida, M. L. 2005, *AJ*, 130, 597

Discussion

B. Weaver: Your metallicity errors seem very small. How do you include uncertainties in atomic constants and atmosphere models?

A. Bragaglia: The internal errors are small. These uncertainties make the systematic part. But as long as we compare clusters analyzed in the same way, the internal errors are the most important. This is why we try to obtain all parameters for the whole sample homogeneously.

B. NORDSTRÖM: The metallicity gradient that goes up towards the Galactic center seems to be flat from about 10 kpc out. Could you please comment on that? Will yo have more data to see if this is a real effect?

A. BRAGAGLIA: The flattening is seen in other works on open clusters, like those of Yong *et al.* 2005 (AJ, 130, 597), or Carraro *et al.* 2007 (A&A, 476, 217). The exact position of the flattening is still uncertain, at about 10-12 kpc. We are obtaining new data near that R_{GC} to better define it.

Roelef de Jong

Pierre Demarque and Andrea Dupree

The Ages of Stars
Proceedings IAU Symposium No. 258, 2008
E.E. Mamajek, D.R. Soderblom & R.F.G. Wyse, eds.

doi:10.1017/S1743921309031810

Eclipsing binary stars as tests of stellar evolutionary models and stellar ages

Keivan G. Stassun[1], Leslie Hebb[2], Mercedes López-Morales[3], and Andrej Prša[4]

[1]Physics & Astronomy Dept., Vanderbilt University, VU Station B 1807, Nashville, TN 37235
email: keivan.stassun@vanderbilt.edu
[2]University of St. Andrews, [3]Carnegie Institution of Washington, [4]Villanova University

Abstract. Eclipsing binary stars provide highly accurate measurements of the fundamental physical properties of stars. They therefore serve as stringent tests of the predictions of evolutionary models upon which most stellar age determinations are based. Models generally perform very well in predicting coeval ages for eclipsing binaries with main-sequence components more massive than $\approx$1.2 $M_\odot$; relative ages are good to $\sim$5% or better in this mass regime. Low-mass main-sequence stars ($M < 0.8$ $M_\odot$) reveal large discrepancies in the model predicted ages, primarily due to magnetic activity in the observed stars that appears to inhibit convection and likely causes the radii to be 10–20% larger than predicted. In mass-radius diagrams these stars thus appear 50–90% older or younger than they really are. Aside from these activity-related effects, low-mass pre–main-sequence stars at ages $\sim$1 Myr can also show non-coevality of $\sim$30% due to star formation effects, however these effects are largely erased after $\sim$10 Myr.

Keywords. binaries: eclipsing, stars: activity, atmospheres, evolution, formation, fundamental parameters, low-mass, brown dwarfs

1. Introduction

Eclipsing binary stars are one of nature's best laboratories for determining the fundamental physical properties of stars and thus for testing the predictions of theoretical models. Detached, double-lined eclipsing binaries (hereafter EBs) yield direct and accurate measures of the masses, radii, surface gravities, temperatures, and luminosities of the two stars. These are measured directly via combined analysis of multi-band light curves and radial velocity measurements (Wilson & Devinney 1971; Prša & Zwitter 2005).

Knowledge of the distance to the EB is not required, and thus the physical properties of the stars can be measured with exquisite accuracy. As an example, Morales *et al.* (2008) have measured the component masses and radii of the low-mass EB CM Dra (Fig. 1) with an accuracy better than 0.5%, perhaps the most accurate measurements ever made for a low-mass EB. Similar accuracy has been achieved for the high mass β Aur (Southworth *et al.* 2007). Indeed, at this level of precision, a non-negligible contributor to the error budget is the uncertainty on Newton's gravitational constant (Torres & Ribas 2002).

Such high-quality measurements allow the predictions of theoretical models to be rigorously tested. For a main-sequence star of a given mass and metallicity, the radius is a monotonic function of age. Thus the models should assign the same age to the components of an EB (i.e., they should lie on a single model isochrone in, e.g., the M–R plane), assuming that the components formed from the same material at the same time. The apparent difference in age, $\Delta\tau$, of the two components is thus a direct measure of the error in the age calibration of the models. As we now discuss, the accuracy of the age

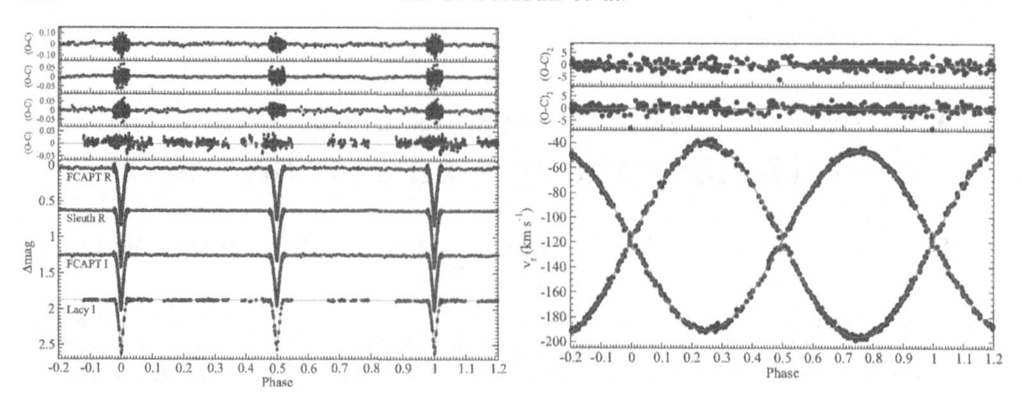

Figure 1. Simultaneous analysis of multi-band light curves (left) and radial velocities (right) of CM Dra by Morales *et al.* (2008). The resulting masses and radii of the stellar components are determined with an accuracy better than 0.5%.

calibration is principally a function of stellar mass, varying from ∼5% for $M > 1.2$ M$_\odot$, to ∼10% for $M \approx 1$ M$_\odot$, to 50–90% for $M < 0.8$ M$_\odot$ (see Table 1).

2. Accuracy of the stellar-age calibration as a function of stellar mass

2.1. *Massive stars ($M > 1.2$ M$_\odot$)*

In general, theoretical models perform best in predicting coeval ages in main-sequence EBs with M > 1.2 M$_\odot$. For example, Young & Arnett (2005) have performed a comprehensive re-analysis of the 20 EBs with $22 < $ M/M$_\odot < 1.2$ that were included in the seminal review of Andersen (1991). Their TYCHO models incorporate updated abundances and, most importantly, improved treatment of interior-mixing physics such as core convective-overshooting. They find $\Delta\tau < 5\%$ for the typical case and $\Delta\tau < 10\%$ for all of the EBs in their sample.

This excellent performance of the models includes a few EBs near the terminal-age main sequence (TAMS), where the stars are evolving very rapidly toward the red-giant phase and for which any discrepancies in the models are amplified. For example, Fig. 2 shows the case of ξ Phe, a particularly challenging EB with a 2.6 M$_\odot$ secondary and a 3.9 M$_\odot$ primary that is leaving the main sequence. An *ad hoc* decrease in the metallicity

Table 1. Accuracy of stellar-age calibrations from eclipsing binaries.

Regime	Accuracy $\Delta\tau^\dagger$	Limiting Physics and/or Data	Exemplar(s)	Refs.
M > 1.2 M$_\odot$	∼5%	modeling of core overshooting, mixing	ξ Phe	1
∼1 M$_\odot$	∼10%	modeling of convection, activity	CV Boo	2
< 0.8 M$_\odot$	50–90%	activity calibration, abundance measurements	V1174 Ori YY Gem V818 Tau V1061 Cyg	3 4 5 6
PMS stars, $\tau < 10$ Myr††	∼50%	star-formation effects (e.g. accretion history)	Par 1802	7

Notes:
†Apparent age difference of presumably coeval stellar components. ††Convolved with mass-dependent effects.
[1]Young & Arnett (2005), [2]Torres *et al.* (2008), [3]Stassun *et al.* (2004), [4,5]Torres & Ribas (2002), [6]Torres *et al.* (2006), [7]Stassun *et al.* (2008)

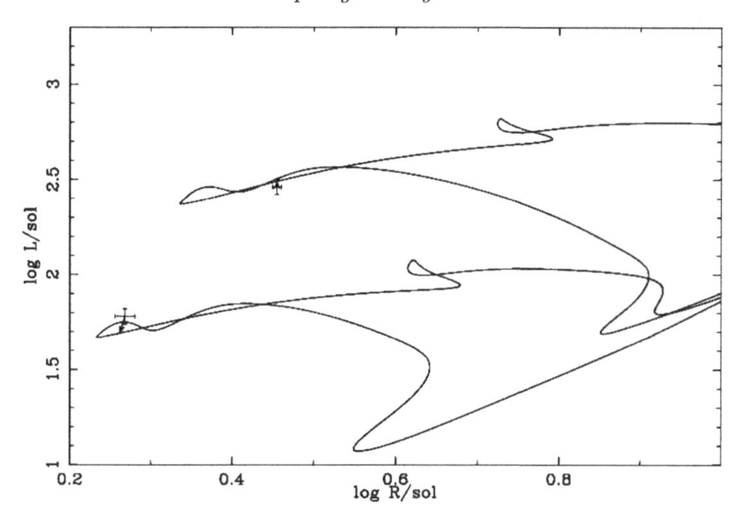

Figure 2. Theoretical evolutionary model fits to the components of the EB ξ Phe by Young & Arnett (2005). This massive, slightly evolved EB is a challenging case, yet even without *ad hoc* adjustments to the fitting parameters achieves $\Delta\tau \approx 3\%$.

of the secondary is required to improve the fit, but even without such an adjustment the fit to both components is marginally acceptable ($\chi^2_\nu \approx 4$) and has $\Delta\tau \approx 3\%$.

2.2. *Solar-mass stars ($M \approx 1$ $M_\odot$)*

At masses of approximately 1 $M_\odot$, the theoretical models begin to show larger systematic discrepancies in the predicted ages. A good example is CV Boo, an old main-sequence EB with component masses of 1.03 and 0.97 $M_\odot$ (Torres *et al.* 2008). The primary shows evidence for having entered the H shell-burning stage, for which the predicted model age of 8 Gyr is in good agreement (Fig. 3). However, the secondary appears to be 25% older due to its radius being ~10% larger than predicted by the 8 Gyr isochrone. The oversized radius of the secondary is likely due to its magnetic activity (see §3).

Of course, the Sun is the only star for which an absolute age can be determined directly (e.g., chemical dating of meteorites). While the Sun's physical properties can be matched to better than 1% at the solar age by models that incorporate all of the observational constraints (including, e.g., helioseismology), the Sun's age cannot be predicted to better than ~7% if given only its observed mass, radius, temperature, and metallicity (Young & Arnett 2005). This is likely to be the best absolute accuracy that can be achieved with current models applied to EBs with a similar set of observational constraints.

2.3. *Low-mass stars ($M < 0.8 M_\odot$)*

The past several years have seen rapid progress in the number of low-mass EBs that have been discovered and their components analyzed. A consistent finding among these studies is that the observed stellar radii are 10–20% larger than predicted by the models. For example, López-Morales (2007) and Ribas *et al.* (2008) have compiled the literature data for low-mass EBs with $0.2 < M/M_\odot < 0.8$. They find that in virtually all cases the theoretical main-sequence predicts radii smaller than those observed (Fig. 4). These oversized radii make the stars appear 50–90% older or younger than expected (depending on whether post– or pre–main-sequence models are used; see also Fig. 3).

Importantly, there are now several low-mass EBs for which there exist independent age constraints (e.g., YY Gem, V818 Tau, V1061 Cyg), and in these systems the same age discrepancies are verified (Fig. 5). A few EBs in young open clusters have also been

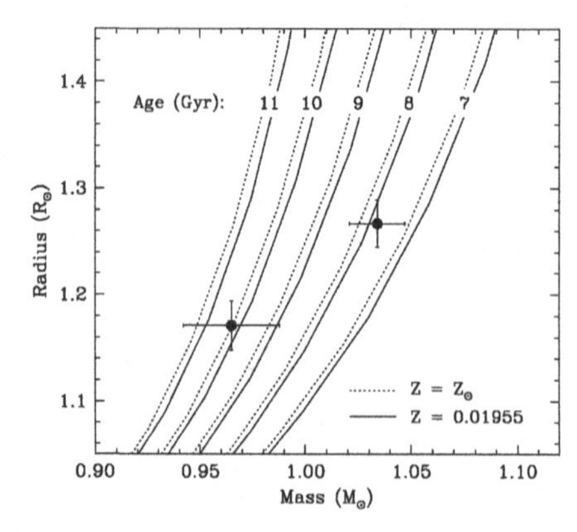

Figure 3. Theoretical evolutionary model fits to the components of the EB CV Boo by Torres *et al.* (2008). The active secondary of this solar-mass system is 10% larger than predicted by the 8 Gyr model isochrone, leading to a large age discrepancy, $\Delta\tau \approx 25\%$, between the components.

found (e.g. Hebb *et al.* 2006; Southworth *et al.* 2004), again verifying these trends. More EBs such as these with independent age determinations are very much needed.

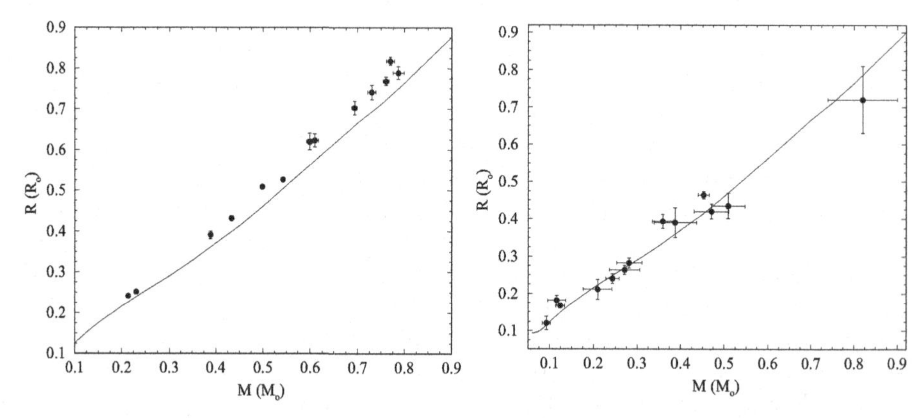

Figure 4. *Left:* Compilation of low-mass EB measurements, showing that the observed radii of these active stars are systematically larger, by 10–20%, than predicted. The solid line is a 1 Gyr isochrone from the models of Baraffe *et al.* (1998). *Right:* Same, but for single-lined EBs, which are effectively single stars from the standpoint of tidal effects which may induce activity. These inactive stars' radii agree much better with predictions. Note that the masses of single-lined EBs are model dependent and hence less accurate. Adapted from Ribas *et al.* (2008).

3. The effect of activity in low-mass stars

There is now very good evidence that the unexpectedly large radii of low-mass EBs is related directly or indirectly to magnetic activity on these stars. Several of the authors who published the original analyses of low-mass EBs had noted that the stars showing larger-than-predicted radii also show evidence for activity, in the form of Hα emission, X-rays, spot-modulated light curves, and other tracers.

More recently, López-Morales (2007) has demonstrated the relationship explicitly (Fig. 6). This is very good news, not only because it points clearly to an underlying cause for the observed oversized radii, but also because the tight correlation with X-ray luminosity suggests that this effect can be calibrated and the ages corrected. Indeed, single-lined EBs—which can be regarded as effectively single stars and which are thus

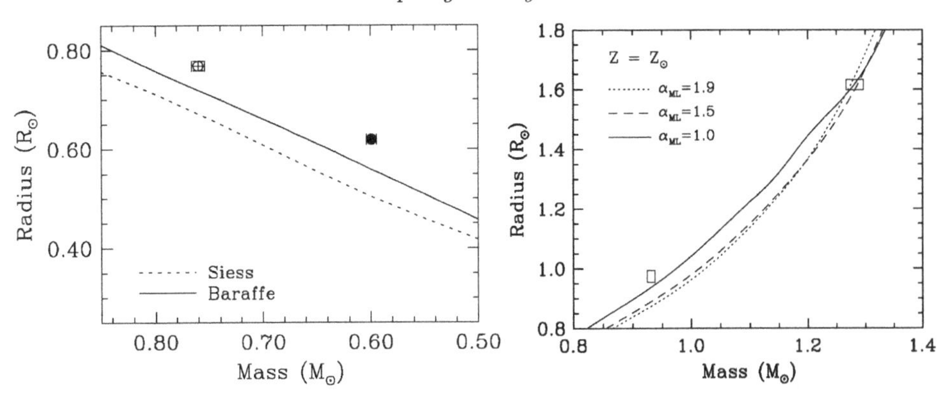

Figure 5. *Left:* Oversized radii are confirmed for the active, low-mass components of the EBs YY Gem (filled) and V818 Tau (open), for which independent age estimates have been made from their membership in young comoving groups. Adapted from Torres & Ribas (2002). *Right:* The components of V1061 Cyg are compared with isochrones from Baraffe *et al.* (1998) with different values of the convective mixing length, α. The oversized radius of the low-mass secondary requires suppressed convection (small value of α). Adapted from Torres *et al.* (2006).

less likely to have magnetic activity driven through interactions with a companion—do not show systematically oversized radii (Fig. 4, right).

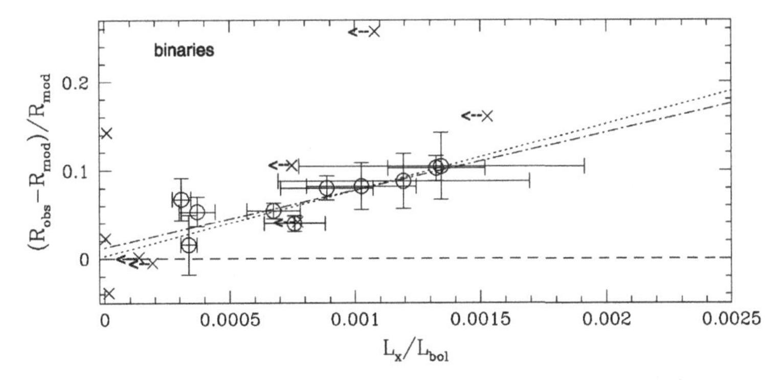

Figure 6. Correlation between X-ray luminosity and fractional discrepancy between measured and predicted radii for low-mass eclipsing binaries. Adapted from López-Morales (2007).

In addition, recent modeling that incorporates the effects of magnetically suppressed convection in low-mass stars due to magnetically active surfaces is now able to fit the observed oversized radii of active EBs extremely well (Fig. 7). In addition, these models simultaneously can explain the systematically low values of the effective temperatures of these stars.

In these new models, strong magnetic fields cause a suppression of convection near the surface. Heat flow to the surface is inhibited (by analogy to dark sunspots on the Sun), resulting in a decrease in the star's effective temperature. However, the star's overall luminosity is roughly fixed by internal processes far removed from the surface boundary condition, and thus the star's radius adjusts to a larger size in order to radiate the flux.

It should be stressed that at present these models use parametrizations of surface spots and of suppressed convection, in place of a full physical treatment of convection and surface fields. Even so, several additional lines of evidence corroborate this general picture. First, the observed properties of young, low-mass EBs are in general best fit by model isochrones with low convective efficiency, $\alpha \sim 1$ (e.g. Mathieu *et al.* 2007).

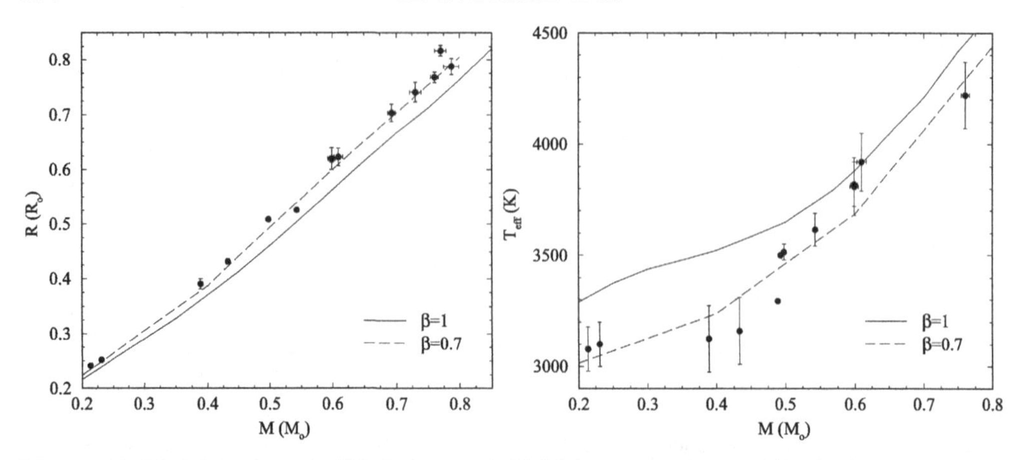

Figure 7. Model isochrones (Chabrier *et al.* 2007) incorporating activity in low-mass stars. *Left:* Oversized radii of active, low-mass EBs are well fit by a 1 Gyr isochrone from Baraffe *et al.* (1998) adopting a spot-covering fraction of 0.3 (i.e. 70% of the stellar surface is free of spots). *Right:* Same, but for effective temperature. Adapted from Ribas *et al.* (2008).

Second, the observed surface lithium abundances of young, low-mass EBs clearly indicate weak convective mixing (e.g. Stassun *et al.* 2004). Third, analyses of low-mass EBs have found that indeed the luminosities of the stars are in good agreement with the models even when the radii and temperatures are very discrepant. For example, in the brown-dwarf EB 2M0535–05 (Stassun *et al.* 2006), the brown dwarfs display ~10% oversized radii, and the temperature of the very active primary (Reiners *et al.* 2007) has been so severely suppressed that it is in fact cooler than the lower-mass secondary. However, the luminosities remain in good agreement with model predictions for brown dwarfs at an age of ~1 Myr (Stassun *et al.* 2007).

Finally, these findings have implications for low-mass stars more generally. First, because activity has the effect of decreasing the effective temperature but leaving the luminosity relatively unaffected, we can expect to see these stars scattered horizontally in the H-R diagram. Second, these effects will need to be taken into account when deriving ages from other means, such as age-activity relations and surface lithium abundances.

4. Star-formation effects at very young ages

Testing the accuracy of stellar evolutionary models via the $\Delta\tau$ test, as we have done above, assumes that EBs represent *coeval* systems of two stars that formed from the same material at the same time. Indeed, in many cases, this assumption of the coevality of EB components has been used to calibrate the various input parameters of the evolutionary tracks. For example, Young & Arnett (2005) have adjusted model parameters such as core overshooting, and have determined secondary stellar properties such as metallicities, on the basis of requiring that the evolutionary tracks yield the same model ages for the two stars of an EB. Similarly, Luhman (1999) has adjusted the temperature scale of young, low-mass stars on the basis of requiring that pre–main-sequence evolutionary tracks yield coeval ages for the components of pre–main-sequence binaries.

We now have evidence that, in at least some cases, the components of very young binaries may not in fact be strictly coeval. In particular, Par 1802 is a recently discovered EB in the Orion Nebula, with a mean age of ~1 Myr, whose components are identical in mass to within 2% ($M_1 = M_2 = 0.41\ M_\odot$; Stassun *et al.* 2008). Having the same mass,

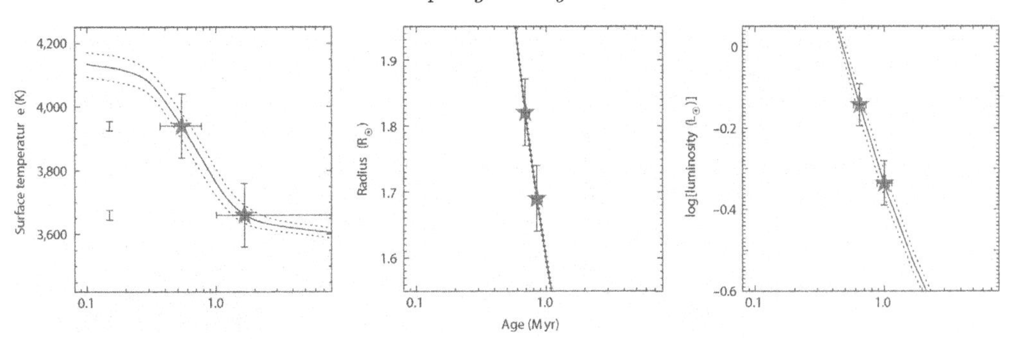

Figure 8. Comparison of physical properties of Par 1802 with theoretical predictions. In each panel, the solid line shows the predicted evolution of a 0.41 $M_\odot$ star from the models of D'Antona & Mazzitelli (1997). Dotted lines show the result of changing the stellar mass by the 0.015 $M_\odot$ uncertainty in the measured masses. Vertical error bars on the points represent the combination of measurement and systematic uncertainties. Horizontal error bars represent the range of ages for which the theoretical models are consistent with the measurements within the uncertainties (including systematic uncertainties). Note that the uncertainties in the temperatures, radii and luminosities are not independent between the two stars, because they are connected by precisely determined ratios; thus, for example, the primary star cannot be forced cooler while simultaneously forcing the secondary warmer. The nominal age of the Orion nebula cluster in which this EB is found is ~1 Myr. Adapted from Stassun *et al.* (2008).

these 'identical twin' stars are predicted by the models to have identical temperatures, radii, and luminosities. However, the components of Par 1802 are found to have different temperatures ($\Delta T \approx$ 300K, or about 10%), radii that differ by 5%, and luminosities that differ by a factor of ~2 (Fig. 8). These surprising dissimilarities between the two stars can be interpreted as a difference in age of $\Delta\tau$~30%. It has been speculated that this age difference likely reflects differences in the star-formation history of the two stars, differences that may be specific to binary-star formation (Simon *et al.* (2009)).

Unfortunately, if such non-coevality turns out to be a common feature of young binaries, then Par 1802 suggests that using very young EBs to calibrate the evolutionary model ages may be limited to ~30% accuracy. Fortunately, these effects are largely erased after ~10 Myr. For example, Stempels *et al.* (2007) find that the components of ASAS J052821+0338.5, a solar-mass EB with an age of 12 Myr, are coeval to $\Delta\tau \sim$ 10%.

5. The future of eclipsing binaries with large surveys

The central importance of EBs for stellar-age determinations implies an ongoing need for precise and accurate EB data. As sky surveys are gaining on both precision and diversity, and since more and more medium size observatories are being refurbished into fully robotic telescopes, there is a "fire-hose" of photometric and spectroscopic data coming our way. Methods to reduce and analyze the data thus cannot rely on manual labor any longer; rather, automatic approaches must be devised to face the challenge of sheer data quantity. Pioneering efforts of automating the analysis of survey data by several groups, most notably Wyithe & Wilson (2001, 2002), Wyrzykowski *et al.* (2003), Devor (2005), and Tamuz *et al.* (2006). These are reviewed in Prša & Zwitter (2007).

A recent stab at automation is implemented within the Eclipsing Binaries via Artificial Intelligence project (EBAI; Prša *et al.* 2008). A back-propagating neural network is applied as a non-linear regression tool that maps EB light curves onto a subset of parameter space that is sensitive to photometric data. Its performance has been thoroughly tested on detached EB light curves (Fig. 9) and applied successfully to OGLE data. In a matter of seconds, the network is able to provide principal parameters of tens of thousands of EBs. The results that come from such an engine may be readily used to select those EBs that are most interesting for the studies of stellar formation and evolution. Given the number of surveys, we are talking thousands of interesting EBs! Since our understanding relies on these systems, such a disproportionate jump in data quantity will surely provide further insights and enhance the statistical significance of our results.

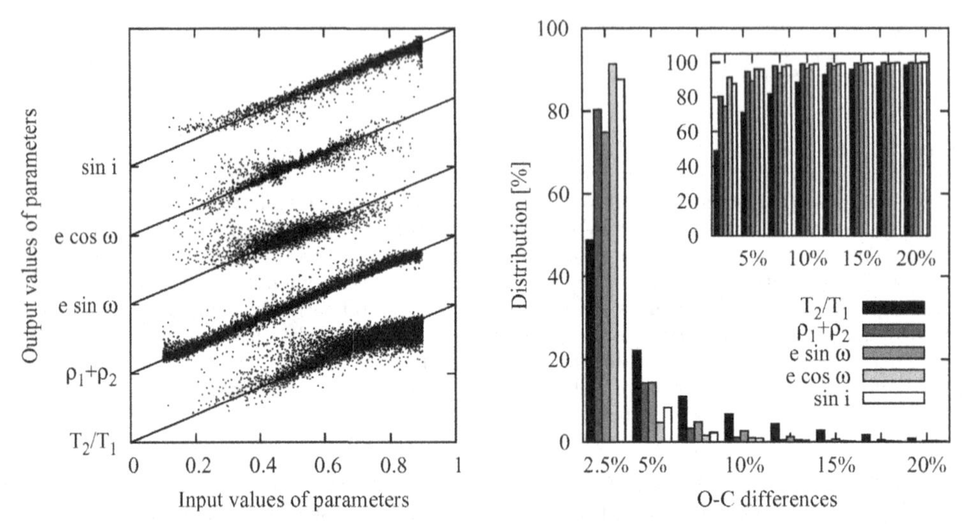

Figure 9. Neural-network recognition performance on 10,000 detached EB light curves. *Left:* comparison between input and output values of parameters. *Right:* distribution of differences (main panel) and their cumulative distribution (inset). Adapted from Prša *et al.* (2008).

References

Devor, J. 2005, *ApJ*, 628, 411

Hebb, L., *et al.* 2006, *AJ*, 131, 555

López-Morales, M. 2007, *ApJ*, 660, 732L

Luhman, K. L. 1999, *Apj*, 525, 466

Mathieu, R. D., *et al.* 2007, in *Protostars & Planets V*, 411

Morales, J. C., *et al.* 2008, *in press*

Prša, A. & Zwitter, T. 2005, *ApJ*, 628, 426

Prša, A. & Zwitter, T. 2007, in *IAU Symp. 240*, Binary Stars as Critical Tools & Tests in Contemporary Astrophysics, eds W. I. Hartkopf, E. F. Guinan & P. Harmanec, p217

Prša, A., *et al.* 2008, *ApJ*, 687, 542

Ribas, I., *et al.* 2008, *MmSAI*, 79, 562

Simon, M., *et al.* 2009, *submitted*

Southworth, J., Bruntt, H. & Buzasi, D. L. 2007, *A&A*, 467, 1215

Southworth, J., *et al.* 2004, *MNRAS*, 355, 986

Stassun, K. G., *et al.* 2008, *Nature*, 453, 1079

Stassun, K. G., Mathieu, R. D. & Valenti, J. 2006, *Nature*, 440, 311

Stassun, K. G., Mathieu, R. D. & Valenti, J. 2007, *ApJ*, 664, 1154

Stassun, K. G., *et al.* 2004, *ApJS*, 127, 3537

Stempels, H. C., *et al.* 2007, *A&A*, 481, 747

Tamuz, O., *et al.* 2006, *MNRAS*, 367, 1521

Torres, G. & Ribas, I. 2002, *ApJ*, 567, 1140

Torres, G., *et al.* 2006, *ApJ*, 640, 1018

Torres, G., *et al.* 2008, *AJ*, 136, 2158

Wilson, R. E. & Devinney, E. J. 1971, *ApJ*, 166, 605
Wyithe, J. S. B. & Wilson, R. E. 2001, *ApJ*, 559, 260
Wyithe, J. S. B. & Wilson, R. E. 2002, *ApJ*, 571, 293
Wyrzykowski, L., *et al.* 2003, *Acta Astronomica*, 53, 1
Young, P. A. & Arnett, D. 2005, *ApJ*, 618, 908

Discussion

E. JENSEN: If the effects of stellar activity can be corrected for, what other effects are then the next most important for obtaining agreement between models and data?

K. STASSUN: The next most important effect is metallicity. There are relatively few EBs for which a good, independent, metallicity constraint is available. In part this is because of the complexity of disentangling multiple-component spectra, but such techniques are improving. Another issue is that the surface abundances of many EBs may be complicated by rapid rotation and activity. In this respect, EBs that are associated with a cluster are ideal, as metallicities can be determined from the other stars.

M. PINSONNEAULT: Spot filling-factors can produce blocking at about the expected level, but one expects a smaller impact for fully convective stars. Is this observed?

K. STASSUN: The recent models of Chabrier *et al.* (2007) indeed use large spot covering-fractions, but critically it is also important to invoke less efficient convection ($\alpha \sim 1$). Probably these effects are inter-related through the strong surface fields that drive activity. The observations do not support a strong dependence of the fractional radius discrepancy on mass; fully-convective stars (all the way down to the brown-dwarf eclipsing binary) show oversized radii of ~10%. On the other hand, the degree of radius discrepancy does correlate very well with activity level.

L. HILLENBRAND: You mentioned evidence for a 50% age spread in pre-MS binaries younger than 10 Myr. Can you give a "35 second" version of the evidence for this claim?

K. STASSUN: Thank you for the excellent question. (I will give you your $20 for asking this question later.) The evidence is that in Par 1802 (a ~1 Myr-old EB in Orion) the component stars have masses that are identical to about 2%, yet their temperatures differ by 10%, the radii differ by 5%, and the luminosities differ by 50%. These surprising differences can be explained if one component is a few times 10^5 yr older than its companion (see Stassun *et al.* 2008).

Ivan King

Kurtis Williams

The Ages of Stars
Proceedings IAU Symposium No. 258, 2008
E.E. Mamajek, D.R. Soderblom & R.F.G. Wyse, eds.

doi:10.1017/S1743921309031822

Globular cluster ages from main sequence fitting and detached, eclipsing binaries: The case of 47 Tuc

Aaron Dotter[1], **Janusz Kaluzny**[2], **and Ian B. Thompson**[3]

[1]Dept. of Physics & Astronomy, University of Victoria
Victoria BC V8P 5C2 Canada
email: dotter@uvic.ca

[2]Copernicus Astronomical Center
Bartycka 18, 00-716 Warsaw, Poland
email: jka@camk.edu.pl
[3]Carnegie Observatories
813 Santa Barbara St., Pasadena, CA 91101-1292
email: ian@ociw.edu

Abstract. Age constraints are most often placed on globular clusters by comparing their CMDs with theoretical isochrones. The recent discoveries of detached, eclipsing binaries in such systems by the Cluster AgeS Experiment (CASE) provide new insights into their ages and, at the same time, provide much-needed tests of stellar evolution models. We describe efforts to model the properties of the detached, eclipsing binary V69 in 47 Tuc and compare age constraints derived from stellar evolution models of V69A and B with ages obtained from fitting isochrones to the cluster CMD. We determine whether or not, under reasonable assumptions of distance, reddening, and metallicity, it is possible to simultaneously constrain the age and He content of 47 Tuc.

Keywords. binaries: eclipsing, globular clusters: individual: (47 Tuc), stars: evolution

1. Background and previous results

47 Tuc is among the closest and most carefully studied globular clusters in the Galaxy. Table 1 lists the best available estimates of [Fe/H], [α/Fe], distance, and reddening along with a representative sample of ages from recent studies.

2. Stellar evolution models

The stellar evolution models utilized in this contribution were computed using the Dartmouth Stellar Evolution Program (DSEP; Dotter *et al.* 2007, 2008). The models include the effects of partially inhibited microscopic diffusion of He and metals (Chaboyer *et al.* 2001). The models were converted to the observational plane using PHOENIX synthetic spectra. Stellar evolution tracks and isochrones were computed for [Fe/H] = −0.8, −0.75, and −0.7; [α/Fe] = 0, +0.2, and +0.4; and Y = 0.24, 0.255, 0.27, 0.285, and 0.3. The color transformations include the effects of α-enhancement but all assume Y$\sim$0.25.

3. Age constraints from the CMD

The analysis performed in this section uses the color-magnitude diagram (CMD) of 47 Tuc from the ACS Survey of Galactic Globular Clusters (Sarajedini *et al.* 2007; Anderson

Table 1. 47 Tuc: Basic Parameters and Previous Age Results

Parameter	**Value**	**Source**
[Fe/H]	–0.76±0.01±0.04	Koch & McWilliam (2008)
	–0.75±0.01±0.04	McWilliam & Bernstein (2008)
[α/Fe]	~0.4	Koch & McWilliam (2008)
	~0.3	McWilliam & Bernstein (2008)
DM_V	13.35±0.08	Thompson *et al.* (2009)
E(B–V)	0.0320±0.0004	Schlegel *et al.* (1998)
Age (Gyr)	10–13	Salaris *et al.* (2007)
	~11.3	Gratton *et al.* (2003)
	11.0±1.4	Percival *et al.* (2002)
	10.7±1.0	Salaris & Weiss (2002)
	11.5±0.8	VandenBerg (2000)
	12.5±1.5	Liu & Chaboyer (2000)

et al. 2008). The data have been culled (by removing stars with large photometric errors, see §7 of Anderson *et al.* 2008) in order to more clearly delineate the main sequence, but the final CMD still contains more than 50,000 stars.

In order to simplify the analysis, the apparent distance modulus and reddening were fixed to those values listed in Table 1. Hence the only uncertainties in the age derived from isochrone fitting are due to the inherent scatter in the data and the allowed range of metallicities.

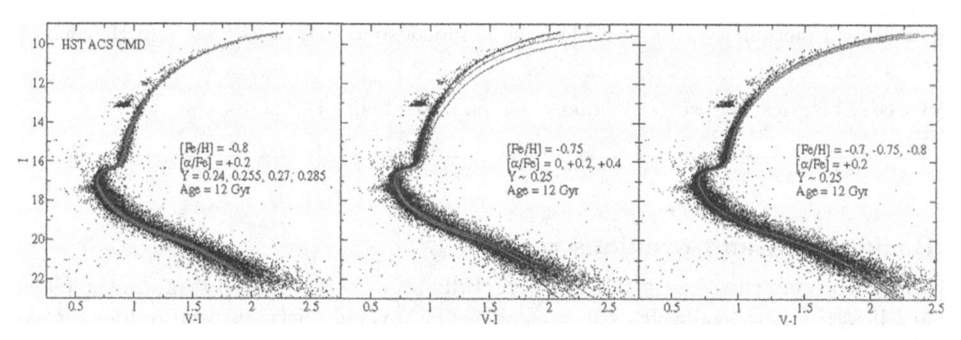

Figure 1. In each of three panels we show isochrones with one varying parameter compared to the 47 Tuc CMD.

The basic results from fitting isochrones to the CMD are as follows: (i) the main sequence and red giant branch impose little or no constraint on Y (see Figure 1, left panel); (ii) the models favor a level of [α/Fe] that is lower than the spectroscopic value (see Figure 1, center panel) and indeed appear to rule out [α/Fe] = +0.4; and (iii) the range of possible [Fe/H] values listed in Table 1 all give acceptable fits to the CMD, given the preferred value of [α/Fe] (see Figure 1, right panel).

Figure 2 shows how age and [Fe/H] (assuming [α/Fe] = +0.2) are correlated. The solid line is the best-fit value as a function of [Fe/H] and the dashed lines represent the error bars that arise only from the finite width of the subgiant branch. A value of Y = 0.255 was adopted for the plot, but any Y value within the assumed range would give a similar result. The isochrone fits to the CMD yield an age of 11.5 ± 0.75 Gyr where the uncertainty due to the fitting procedure is ~0.3 Gyr and the rest is due to the ~0.05 dex uncertainty in [Fe/H].

Table 2. Parameters of V69 from Thompson *et al.* (2009)

Parameter	Primary	Secondary
$M/M_{\odot}$	0.8762±0.0048	0.8588±0.0060
$R/R_{\odot}$	1.3148±0.0051	1.1616±0.0062
$L/L_{\odot}$	1.94±0.21	1.53±0.17

4. Age constraints from the binary V69

A careful analysis by Thompson *et al.* (2009) of spectroscopic radial-velocity measurements and photometric light curves provides high precision estimates of the fundamental parameters of V69, see Table 2. While the masses and radii are measured to better than 1%, the luminosities are measured to no better than ~10% at present (though additional near-infrared light curves will substantially improve this number).

In order to measure the age of 47 Tuc using V69, individual stellar evolution tracks were computed for six different masses: three for each star, encompassing the central value and quoted uncertainties in Table 2. These models were calculated for the same grid as the isochrones described in section 2. The mass-radius (M-R) and mass-luminosity (M-L) relations in the tracks were used to determine the range of age for which each track lay within the error bars for the corresponding parameter (R or L) of each component. We make two important general comments: (i) the ages derived from each relation are in excellent agreement between the two stars and (ii) the agreement between the two methods in a given star are quite sensitive to Y. For either star, the maximum Y value for which the M-R and M-L results overlap increases with [Fe/H].

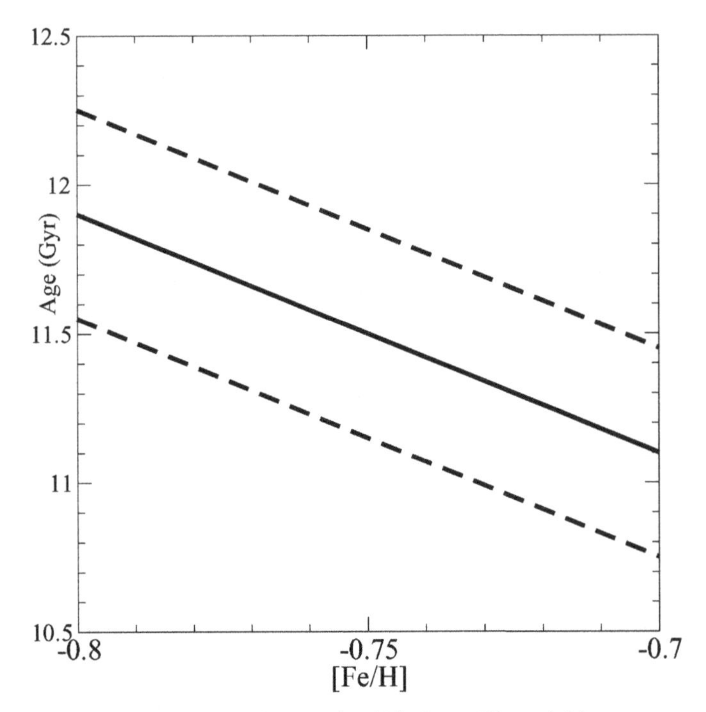

Figure 2. Isochrone fitting results in the age–[Fe/H] plane. The solid line represents the best fits while the dashed lines indicate the uncertainties due only to the width of the subgiant branch.

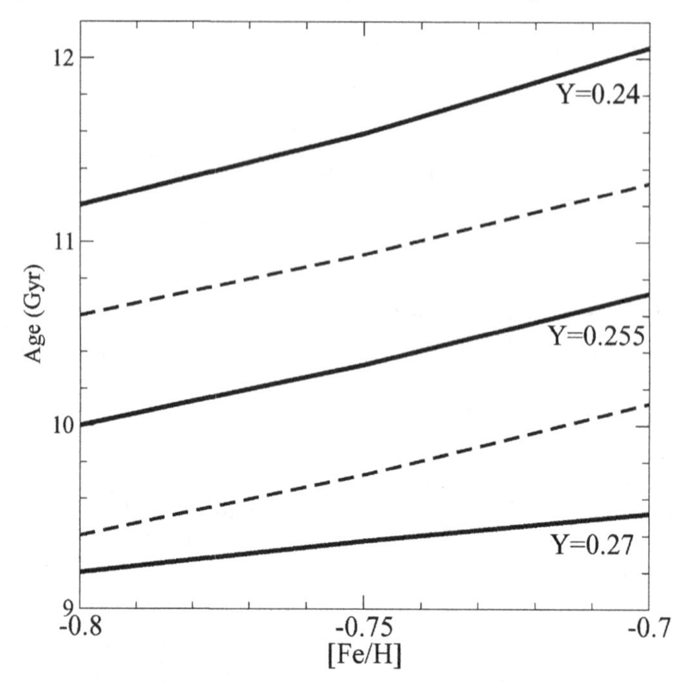

Figure 3. V69 results in the age–[Fe/H] plane. The solid lines correspond to results from the labeled Y values. The dashed lines about the Y = 0.255 line indicate the errors for that particular case which are typical of all cases.

The results of the V69 age analysis are presented in the Figure 3. The solid lines indicate how the measured age varies with [Fe/H], for the Y value listed on the figure. The dashed lines above and below the line for Y = 0.255 indicate the combined uncertainties in M, R, and L, with ~0.5 Gyr due to the luminosity uncertainty alone. Figure 3 clearly shows that the age-[Fe/H] relation is complementary to that of the CMD method (Figure 2). Of equal importance is the fact that since the masses of the stars are known, there is a significant sensitivity to the He abundance that is lacking in the CMD method.

5. Combined constraints

It was demonstrated in §3 that the CMD age-[Fe/H] relation (assuming that both the distance and reddening are well-constrained) is linear for $-0.8 <$ [Fe/H] < -0.7 and that age decreases as [Fe/H] increases. Conversely, the result from §4 showed that while the age-[Fe/H] relation derived from V69 was also linear over the range of [Fe/H] considered, the age increased as [Fe/H] increased. Combining the two methods is therefore quite powerful because the slopes have opposite signs. The region where the two overlap has a finite size and gives preferred values for age, [Fe/H], and Y simultaneously, though ultimately the size of the overlap region is dominated by the uncertainties in both methods.

Figure 4 presents the results given in Figures 2 and 3 together so that the situation described in the preceding paragraph can be clearly seen. The figure indicates that the binary and CMD methods agree best if the He abundance is low (Y ~ 0.24) while a value of Y ~ 0.25 would have been more preferable given the constraints from WMAP and Big Bang nucleosynthesis, see Spergel et al. (2003).

If the bias towards $[\alpha/\text{Fe}] = +0.2$ derived from isochrone fits to the CMD in section 2 is relaxed and the spectroscopic value (+0.4) is adopted instead, then the combined

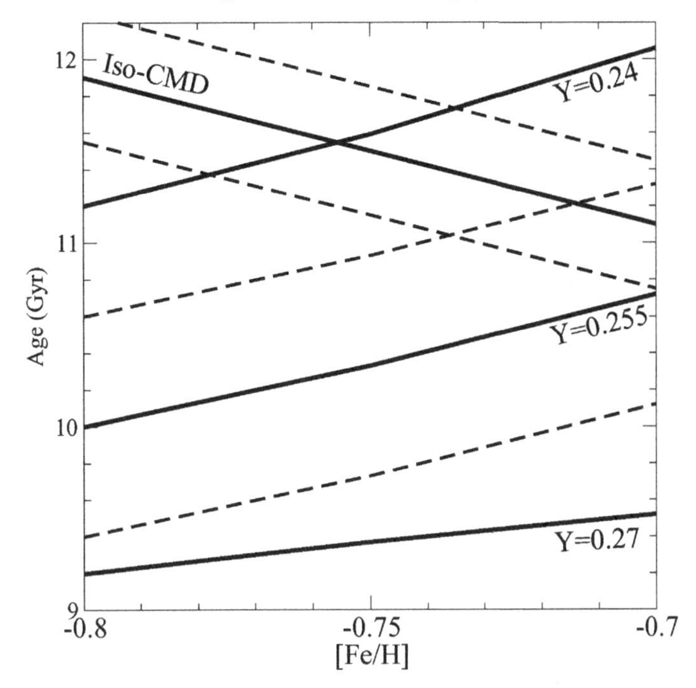

Figure 4. Combined isochrone-CMD and V69 results in the age–[Fe/H] plane.

results favor Y $\sim$ 0.25. However, it is difficult to reconcile the larger [α/Fe] ratio given the precise reddening.

Further improvements to the distance and luminosity of V69 from near-infrared light curves will significantly reduce the uncertainties and provide a much more stringent test of the stellar evolution models' ability to simultaneously fit the components of V69 and yield an age consistent with the turnoff age. As such, this ongoing project represents the most exacting test of stellar evolution models in a globular cluster to date and is certain to provide further insights into the ages of such systems.

Acknowledgements

Support for AD is provided by CITA and an NSERC grant to Don VandenBerg. Research of JK is supported by the Foundation for Polish Science through the grant MISTRZ. IBT is supported by NSF grant AST-0507325.

References

Anderson, J. *et al.* 2008, *AJ*, 135, 2114
Chaboyer, B. *et al.* 2001, *ApJ*, 562, 521
Dotter, A. *et al.* 2007, *AJ*, 134, 376
Dotter, A. *et al.* 2008, *ApJS*, 178, 89
Gratton, R. G. *et al.* 2003, *A&A*, 408, 529
Koch, A. & McWilliam, A. 2008, *AJ*, 136, 518
Liu, W. M. & Chaboyer, B. 2000, *ApJ*, 544, 818
McWilliam, A. & Bernstein, R. A. 2008, *ApJ*, 684, 362
Percival, S. M. *et al.* 2002, *ApJ*, 573, 174
Salaris, M. *et al.* 2007, *A&A*, 476, 243
Salaris, M. & Weiss, A. 2002, *A&A*, 388, 492
Sarajedini, A. *et al.* 2007, *AJ*, 133, 1658

Schlegel, D. J. *et al.* 1998, *ApJ*, 500, 525
Spergel, D. N. *et al.* 2003, *ApJS*, 148, 175
Thompson, I. B. *et al.* 2009, *AJ*, submitted
VandenBerg, D. A. 2000, *ApJS*, 129, 315

Discussion

M. TOSI: I thought that the HB luminosity and its difference from the main sequence turnoff luminosity could put tighter constraints on both the age and the He content. Could you please explain why you prefer not to use it?

A. DOTTER: Two reasons: first, time limitations prohibited a more detailed analysis; second, I was particularly interested in age-sensitive regions of the CMD. The point is well taken, however, and a more careful analysis will certainly include the HB.

J. KALIRAI: What fraction of 47 Tuc's literature age range simply results from the adoption of different groups theoretical models? Is the intrinsic error-bar from your study meaningful, given differences between different models? Fitting the same CMD with other models may give an absolute age that differs from your results by more than the error bar?

A. DOTTER: The agreement among different groups' models is actually quite good in the case of the binary V69, with the one caveat that different groups adopt a different primordial Y as well as $\Delta Y/\Delta Z$. When the Y-dependence is accounted for, the age agreement is quite good.

Aaron Dotter

The Ages of Stars
Proceedings IAU Symposium No. 258, 2008
E.E. Mamajek, D.R. Soderblom & R.F.G. Wyse, eds.

doi:10.1017/S1743921309031834

Models for Pop I stars: implications for age determinations

Georges Meynet[1], Patrick Eggenberger[2], Nami Mowlavi[1] and André Maeder[1]

[1]Geneva University, Geneva Observatory
CH-1290 Versoix, Switzerland
email: georges.meynet@unige.ch

[2]Institut d'Astrophysique et de Géophysique de l'Université de Liège, Allée du 6 Août, 17
B-4000 Liège, Belgium eggenberger@astro.ulg.ac.be

Abstract. Starting from a few topical astrophysical questions which require the knowledge of the age of Pop I stars, we discuss the needed precision on the age in order to make progresses in these areas of research. Then we review the effects of various inputs of the stellar models on the age determination and try to identify those affecting the most the lifetimes of stars.

Keywords. convection, diffusion; Stars: rotation, mass loss, magnetic fields

---

1. Importance of reliable age-calibrations for Pop I stars

Pop I stars cover a wide range of ages, from millions to billions of years, and their age determination is useful for studying the evolution of processes with very different timescales, going from the evolution of planetary systems to the analysis of powerful starbursts in remote galaxies. A few examples are given below:

• How does star formation propagate around young star-forming regions? What are the timescales for star formation to trigger star formation (see for instance the very young associations observed in the vicinity of ηCar, discussed in the review by Smith & Brooks 2008)? In order to study the time sequence between different associations, relative ages with an accuracy better than about 10–20% would be quite useful, and that means an absolute accuracy of a few 100,000 years on ages of a few Myr.

• What are the ages of young, powerful starbursts in remote galaxies? In distant galaxies, where individual stars cannot be resolved, emission-line ratios in the spectrum of the integrated light can be used to determine the age of starbursts (see the review by Leitherer 2005). The typical ages are of a few Myr and the needed precision is of the same order as the one indicated above for the age determination of resolved associations. Let us however stress that here, in addition to the uncertainties pertaining the stellar models, those due to the duration of the burst of star formation, to the stellar Initial Mass Function and to the possibility of superpositions of many starbursts makes the exercise still more difficult.

• What is the upper mass limit of the progenitors of White Dwarfs (WD)? To answer this question one needs to establish the initial-final mass relationship of WD. This can be done by determining the age of the open clusters where WD are observed. This age is then used, together with the cooling age of the WD, to estimate the mass of the WD progenitor using stellar models. In this process age determination enters in three ways: first through the isochrone fitting of the cluster, second through the cooling age of the WD and finally through an age-mass relation (Weidemann 2000; see also the talks by Kalirai and Richer in this volume). To determine the upper mass limit for the progenitor of WD, accurate determination of the age of clusters with a mass at the turn off around 8 $M_{\odot}$ are needed, which means ages of the order of a few tens of Myr. In this age range, an uncertainty of 20% on the age translates into an uncertainty of about 1 $M_{\odot}$ on the mass (thus an uncertainty of ~10% on the mass).

• What is the lifetime of a very-hot Jupiter? These giant planets orbit their host stars with orbital periods below 3 days. Given their proximity to their host stars, these planets should undergo some degree of evaporation due to the heating by stellar UV photons. Some models predict a catastrophic destiny for these planets, being completely evaporated in a relatively short timescale. If true, very-hot Jupiter could then only be observed around the youngest stars. Is this the case? Melo *et al.* (2006) find that none of the stars studied in their paper seem to be younger than 0.5 Gyr. Only lower limits for most of the ages of these stars are obtained. To make progresses in this area, restricted ranges of ages for the planet host stars should be obtained.

• What was the evolution of the chemical gradients in the Milky Way? Nowadays the Milky Way presents a gradient d lg(O/H)/dR (where R is the Galactocentric radius) of between −0.07 dex/kpc and −0.04 (see e.g. Daflon & Cunha 2004). Was this gradient steeper or shallower in the past? From a theoretical point of view the answer is uncertain. In order to answer such a question, one needs objects for which the metallicity and the age can be measured. Open clusters and planetary nebulae are the objects that can be used as probes of the largest part of the history of the Galaxy (from a time when its age was about 6 Gyr until today), while simultaneously spanning a wide range of Galactocentric radii. In order to improve our present knowledge of the evolution of this gradient, a precision better than one half Gyr should be obtained on the ages (precision of about 10% needed).

• What is the form of the age-metallicity relation for the thin disc stars of the Galaxy? For instance Nordström *et al.* (2004) have found a small change of the mean metallicity of the thin disk since its formation and a very substantial scatter in metallicity at all ages. Again an accuracy better than about 10% on the age would allow us to sharpen this picture.

While age determinations may reach an internal precision sometimes better than 10%, systematic effects may prevent a similar level of accuracy being reached in absolute ages. In the following, we shall focus on the systematic effects on age estimates due to uncertainties pertaining to various physical effects accounted for in stellar models.

2. Comparison of different models for near-solar metallicity

In table 1 below, we list some grids of non-rotating stellar models, covering the case of Pop I stars (the list is not exhaustive and refers to only one paper per group!). To make a first comparison, we can plot, for different models, the relation between the bolometric magnitude at the end of the Main-Sequence phase and the corresponding age. This is done in the left panel of Fig. 1. Overall there is good agreement between the plotted models. The slope of the relation [$M_{\rm bol} \propto \sim 15/4\,\lg(\rm age)$] agrees well with analytical estimates based on the mass-luminosity relation ($L \propto (\mu^4 M^3)/\kappa$, where μ is the mean molecular weight, M the total mass and κ, the mean opacity inside the star) and the mass-age relation ($\tau_{\rm MS} \propto (qXMfc^2)/(\mu^4 M^3/\kappa)$, where q is the mass fraction of the total mass of the star where nuclear reactions occur, X the mass fraction of hydrogen, c the velocity of light, f the fraction of the initial mass of hydrogen transformed into energy when H-burning occurs, f is equal to 0.007). Thus an error of 0.1 magnitude implies an error on the age of less than 3%.

Table 1. Some grids of non-rotating stellar models for near solar metallicity.

Reference	Masses	Y	Z	α_{ov}
Bono *et al.* 2000[1]	3 - 15	0.27	0.02	0.0
Claret 2004	0.8 - 125	0.28	0.02	0.2
Demarque *et al.* 2004[1]	4 - 52	0.279	0.02	(6)
Dominguez *et al.* 1999	1.2 - 9	0.28	0.02	0.0
Dotter *et al.* 2007[1]	0.1 - 1.8	0.274	0.0189	0.2 see (3)
Girardi *et al.* 2000[1]	0.15 - 7	0.273	0.019	(5)
Pietrinferni *et al.* 2004[1]	0.5 - 10	0.273	0.0198	0.2 see (3) & (4)
Schaller *et al.* 1992[1]	0.8 -120	0.30	0.02	0.2
VandenBerg *et al.* 2006[1]	0.5 - 2.4	0.2715	0.0188	see(2)
Ventura *et al.* 1998	0.6 - 15	0.274	0.017	FST(7)

[1] Models for other compositions are available in that paper.
[2] Roxburgh criterion calibrated using binaries and clusters.
[3] Value adopted in models with well developed convective cores. Lower values are used for small convective cores see Dotter *et al.* (2007)
[4] Models without overshooting with the same initial composition are also available.
[5] The overshooting is accounted for using the formalism of Bressan *et al.* (1981).
[6] The overshooting is accounted for using the formalism of Demarque *et al.* (2004).
[7] Convection is treated according to the Full Spectrum Turbulence model.

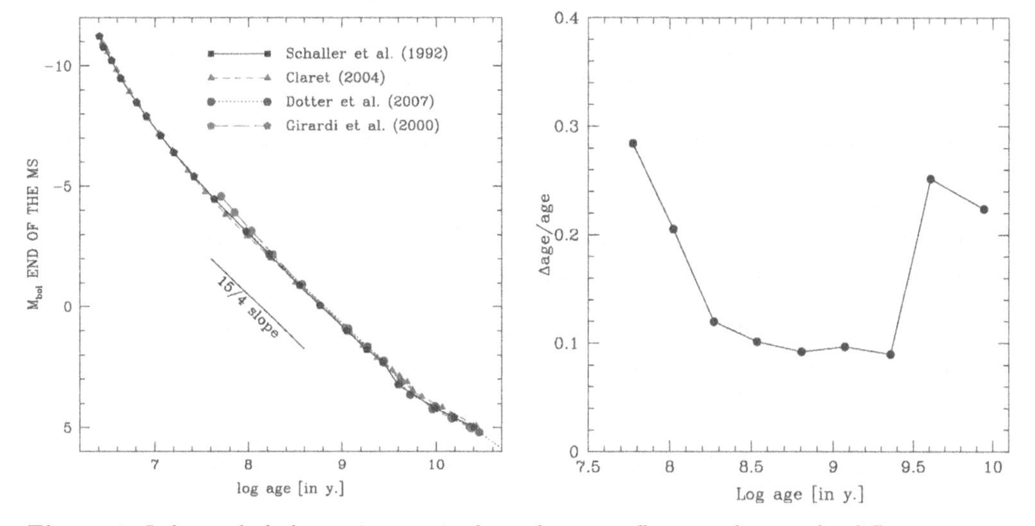

Figure 1. *Left panels*: bolometric magnitude at the turn-off versus the age, for different non-rotating stellar models with similar physical ingredients. *Right panel*: dispersion of the ages due to the use of the different stellar models listed on the left panel.

In the right panel of Fig. 1, we indicate the difference between the minimum age and the maximum age given by these models for a given magnitude, normalised to the mean age, the mean age being (maximum – minimum age)/2. We see that the age dispersion is of the order of 10% for ages between 200 Myr and 2 Gyr. The dispersion increases up to values between 25% and 30% for greater and smaller ages. Most of the differences come from different ways of treating the way the overshooting parameter evolves as a function of the initial mass (see the respective references describing the models).

3. Effects of a change in the initial abundances

In non-rotating models, the metallicity affects the evolution of stars mainly through its impact on the radiative opacities, the equation of state and the nuclear reaction rates.

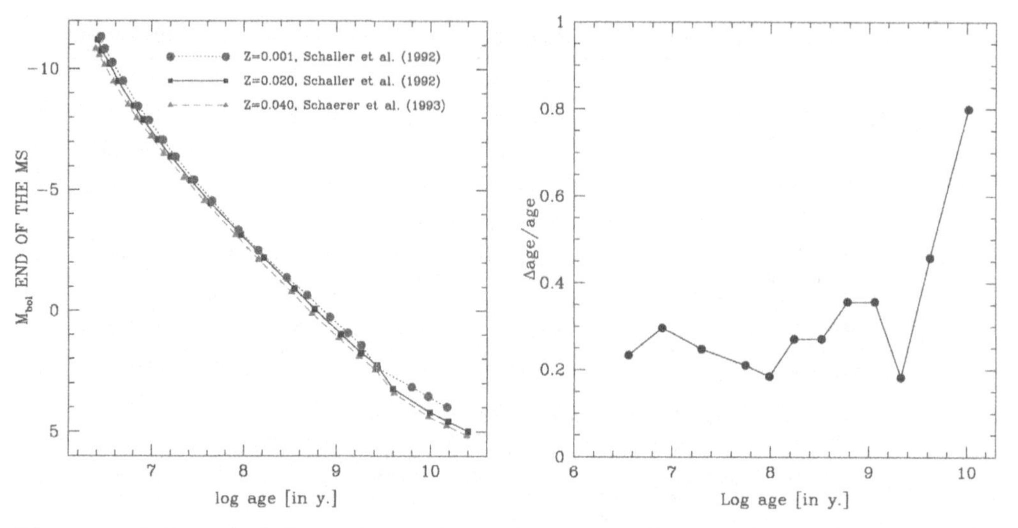

Figure 2. *Left panels*: bolometric magnitude at the turn off versus the age, for non-rotating stellar models at different metallicities. *Right panel*: dispersion of the ages due to a change of metallicity (see left panel for the list of models used).

For stars with initial mass greater than about 30 $M_\odot$, mass loss becomes an important ingredient already during the MS phase and the effects of metallicity on the mass-loss rates have to be taken into account (see the effect on age determination below). These effects of metallicity on stellar models are discussed in detail in Mowlavi *et al.* (1998). Metallicity also affects the transport mechanisms induced by rotation (Maeder & Meynet 2001). Typically a star of given initial mass, starting its evolution with a given initial rotation velocity, will be more efficiently mixed by rotation at low metallicity than at high metallicity.

The most important effect of a change of metallicity occurs through the effect on the opacity (at least for a large range of metallicities and ages). In general, the opacities increase with increasing metallicity. This is the case for opacity due to bound-free and free-free transitions. Using the mass-luminosity relation seen above, one immediately deduces that the increase of the opacity produces a decrease of the luminosity of a given initial mass model. In contrast, for massive stars free-electron scattering is the main opacity source. This opacity depends only on X [$\kappa_e \simeq 0.20(1+X)$], which is approximately constant for $Z \leqslant 0.01$ and decreases with increasing Z at $Z \geqslant 0.01$. Thus at high Z the luminosity of a massive star increases with metallicity.

In Fig. 2 (left panel) the relation between the bolometric magnitude at the turn off and age are shown, for various metallicities (see references on the figure). When the metallicity increases, a given magnitude is achieved by a higher initial stellar mass, since, for the Z range considered here, the luminosity of the tracks for a given initial mass are decreased. Thus smaller ages are obtained at higher Z for a given value of the turn-off magnitude. In Fig. 2 (right panel) we have plotted the age dispersion which would result from using tracks corresponding to different initial metallicities in the range between Z = 0.001 and Z = 0.040 (i.e. for [Fe/H] between about −1.3 and +0.3). We see that the dispersion is higher than that resulting from different stellar grids with similar metallicities. It amounts to about 30% for ages below about 3 Gyr. It increases a lot for greater values of the ages, reflecting the increased sensitivity of the luminosity on Z in the low-mass range. This demonstrates that precise determination of the age needs relatively precise determination of the metallicity.

Dotter *et al.* (2007) have recently studied the impact of individual changes in the abundances of some elements. They computed models for stars with masses between 0.5 $M_\odot$ and 3 $M_\odot$ enhancing the abundance of one element, keeping X, Y and Z constant. Of course enhancing one element, at constant Z, must be done at the expense of all other elements. This work shows that the elements which have the most important effects are oxygen and iron. Varying their abundances by a factor of two produces changes in the MS lifetime, for a star of given initial mass, of 15% (decrease for O and increase for Fe, see their Fig. 13). Let us note that the mass fraction of helium also has a big impact on the structure and the evolution of stars. For instance, models by Claret (1997) or Bono *et al.* (2000) have been computed with different He mass fractions. We refer the reader to these works for more details on that question (see also the paper by Decressin *et al.* in the present volume).

4. Treatment of convection

The treatment of convection, and more generally of all turbulent processes, remains one of the biggest difficulties in stellar modeling. Depending on the criterion chosen for the set up of the convective instabilities (Schwarzschild or Ledoux criterion), on the efficiency of semiconvection, and on the amplitude of the overshooting effects, the quantity of fuel, the luminosity, and therefore the MS lifetime, can vary a lot. For instance, a moderate overshoot (0.2 H_P) at the border of the Schwarzschild convective core associates an age at turn off luminosities equal to 2 [log $L/L_\odot$] (turn off mass around 3 $M_\odot$), 1 (1.6 $M_\odot$) and 0.5 (1.3 $M_\odot$) which are respectively 40%, a factor 2.5 and a factor 3.2 greater than the lifetimes obtained from models without overshoot. One sees that compared to the age dispersion arising from different grids of models (but with similar physical ingredients) or from grids at various metallicities, the age dispersion due to overshoot is much greater, especially in the low-mass range.

It has to be noted that the luminosity at the turn off is particularly sensitive to the amount of overshoot. This property leads many authors to use the width of the observed MS to calibrate the overshoot (see for instance Maeder & Meynet 1989). Note that the temperature of the MS termination is much less sensitive to overshoot. This is due to the fact that overshooting increases the age but also pushes the MS termination to the red, in such a way that the log $T_{\rm eff}$ at the MS termination versus age relation is nearly unchanged (see Figs 17 & 18 in Maeder & Meynet 1989). However, if the overshoot does not change the age assigned to a given turn-off effective temperature, it does affect the initial mass associated to it!

Asteroseismology will likely help to resolve the question of the size of the convective core in massive stars. Some first results are presented in Aerts (2008): in five B-type stars, the size of the convective core has been deduced from asteroseismology, and an overshoot parameter between 0.10 and 0.44 H_P has been found.

In low-mass stars, non-adiabatic convection occurs in the outer layers. The extent of the convective zone is governed by the choice of the mixing length parameter α, the value of it being fixed by calibrating solar models and/or from the position of the red giant branch in the HR diagram. Of course (as for the overshooting parameter) α is not a fundamental constant of nature and it may vary with mass, metallicity, and even the evolutionary phase. Thus it would be advantageous to have a means of constraining this quantity in other objects. Asteroseismology may be the tool to do that. Eggenberger *et al.* (2008) show that the observed parameters of 70 Oph A can be reproduced by two sets of very different models: one with an initial helium value of 0.266, a mixing length parameter equal to 1.7998 (obtained from a solar model) and an age of 6.2 Gyr, another with an initial helium value of 0.240, a mixing length parameter equal to 2.2497 (previous value multiplied by 1.25) and an age of 10.5 Gyr. Thus we see that the uncertainties on α may have a large impact on the age determination. While these two models show the same mean large separations, the mean small separation of the model with the higher initial helium abundance is significantly larger (3μHz) than the one of the model with the lower initial helium abundance. Thus having a precise observed value of the mean small separation will allow us to obtain an independent determination of the age, of the mixing-length parameter and of the helium abundance.

5. Rotation

Rotation affects all the outputs of the stellar models and in particular changes the age derived for a star of given initial mass (Heger & Langer 2000; Meynet & Maeder 2000). By inducing internal mixing, rotation modifies both the total quantity of fuel available and the luminosity. The amplitude of the changes depends on the nature of the instabilities (induced by rotation) which are considered in the model. The two most important instabilities are shear instabilities and meridional currents. They both transport chemical species and angular momentum. They are much less efficient in transporting energy since their timescale is in general longer than the thermal diffusion timescale (see Zahn 1992).

Before discussing the implications of rotation on the age determinations, let us say a few words about the observations which can be used to constrain rotational mixing. A rotating star is predicted to show some nitrogen surface enrichment already during the main sequence. The amplitude of the nitrogen enrichment at the surface depends on the initial mass (increases with the mass), the age (increases with the age) and the initial rotational velocity. This is correct as long as we consider stars with a given initial composition (rotational mixing is more efficient at low Z) and whose evolution is not affected by a close binary companion. Thus we see that the nitrogen surface abundance is a function of at least three parameters: mass, age and velocity. To see a relation between N-enrichment and velocity, it is necessary to use stars with different rotational velocities but having similar masses and ages.

When data samples limited in mass and ages are used, a very nice correlation is found between the surface N-enrichment and $v \sin i$ (see Figs. 3 and 4 in Maeder *et al.* 2008), supporting a N-enrichment dependence on rotational velocities. Stars beyond the end of the MS phase do not follow such a relation, since their velocities converge to low values (see Fig. 12 by Meynet & Maeder 2000). A fraction, which we estimate to be $\sim$20 % of the stars, may deviate from the relation as a result of binary evolution, either by tidal mixing or mass transfer.

Rotation modifies the size of the convective core (see Fig. 3 left panel). First the size of the convective core is decreased when rotation is accounted for. Indeed, due to the action of the centrifugal acceleration, the star behaves as one with a lower gravity, or as a star of lower initial mass. Then, when evolution proceeds, rotational diffusion will supply the core with hydrogen and the radiative envelope with helium. The result will be an increase in both the luminosity and the size of the convective core. The MS lifetimes are increased by rotation. During the MS phase, towards the turn-off, the tracks for a given mass in the HR diagram become more luminous and extend further to the cool part. In that respect, the effects of rotation are somewhat similar to that of overshooting (but of course the effects of rotation cannot be modeled by adding overshooting; for instance, as indicated above, rotation modifies the surface abundances, while overshooting does not). A given width of the MS can be obtained by a combination of rotation velocity and overshooting parameter (see Talon *et al.* 1997). For example, in the temperature range

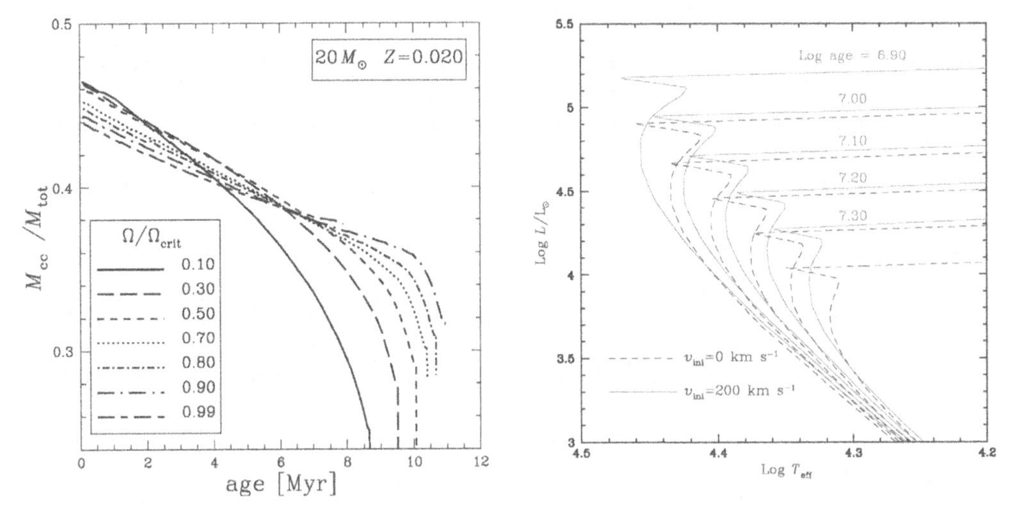

Figure 3. *Left panels*: Evolution of the size of the convective core during the MS, for the various values of rotational rate, in a 20 M$_\odot$ star, at standard metallicity. Figure taken from Ekström *et al.* (2008). *Right panel*: Isochrones computed from stellar evolutionary tracks for solar metallicity. The dashed and continuous lines correspond to the case of non–rotating and rotating stellar models respectively. In this last case, models have an initial velocity $v_{\rm ini}$ of 200 km s^{-1}. The logarithms of the ages (in years) label the isochrones computed from the models with rotation. Figure taken from Meynet & Maeder (2000)

between $\log T_{\rm eff} = 4.1$ and 4.3 (i.e. in the mass range between about 5 and 20 M$_\odot$), a similar MS width can be obtained either by non-rotating models with an overshooting parameter value of 0.2, or by rotating models with an average rotational velocity on the MS around 100 km s^{-1} and an overshooting parameter value of 0.1 (estimates based on the models discussed in Ekström *et al.* 2008). Rotation and overshooting thus concur to make the MS band wider. Calibrating the overshooting parameter in non-rotating stellar models, by adjusting the MS band-width thus overestimates it. How can we disentangle the two processes? Comparisons of the core size, obtained by asteroseismology, in stars of similar mass, metallicity and age but having different rotational velocity will give precious constraints. Interestingly, the asteroseismic signature of overshooting, producing a sharp chemical gradient at the border of the convective core, is different from the signature of the smoother gradient produced by rotational mixing (see Montalban *et al.* 2008). Another approach would be to calibrate the overshooting parameter using stars massive enough to have a convective core, but with a velocity distribution biased toward slow rotators. Typically, early type F-stars would match these conditions (masses between 1.3 and 1.5 M$_\odot$), although recent rotational velocity determinations (see Royer *et al.* 2007) show that the mean velocity of these stars is not as low as previously thought: according to Royer *et al.* (2007), the mean velocity of these stars is in the range of 150 km/s, instead of 80–100 km/s. Thus the effects of rotation are probably not completely absent (although much less developed than in massive stars). Another difficulty with this mass range comes from the fact that probably the overshooting parameter varies with mass, as has been found by many authors using non-rotating models (see e.g. Girardi *et al.* 2000; Demarque *et al.* 2004). Thus the problem is not so easily resolved and its solution should probably await more stringent constraints coming from asteroseismology (see the talk by Vauclair in the present volume).

To give an idea of the effects of rotation on the age determination of young clusters, Fig. 3, right panel, compares isochrones obtained from non-rotating and rotating models.

The rotating models have a rotational velocity of ~150 km/s, on average, during the MS phase. We see that the age determined from rotating models are ~25% higher than those obtained from non-rotating ones (Meynet & Maeder 2000).

When the rotation is high enough, significant changes in the shape of the star are expected. Typically, for a 20 $M_\odot$ star, when $\Omega/\Omega_c > 0.7$ (average velocity on the MS superior to 280 km/s), the equatorial radius is more than 10% larger than the polar radius. The von Zeipel theorem (1924) then implies that the radiative flux in the polar region is higher than in the equatorial region. This can be translated into a change of the effective temperature with colatitude. A consequence is that fast rotators will be characterized by different values of the luminosity and of the effective temperature, depending on the viewing angle. For a fast rotator (initial velocities on the ZAMS such that $\Omega/\Omega_{\rm crit} \sim 0.90$), the dispersion in age due to the dispersion in viewing angle can amount to about 40% (using the luminosity at the turn-off for determining the age, higher when the star is seen pole-on).

In a cluster there will be a distribution of rotational velocities and of the directions of the rotational axis. Thus we expect that the above effects will induce some dispersion in the position of the stars in the HR diagram around the turn off. Such effects should probably be accounted for when determining the age of clusters containing significant proportions of Be-stars (i.e. B-type stars with emission lines, where the emission is due to the presence of an expanding equatorial disk, whose origin is commonly attributed to the very fast rotation of the star). This is the case, for instance, in NGC 330, a cluster in the SMC (see e.g. Keller *et al.* 2000) in which 40% of the MS stars which are two magnitudes below the turn off are Be stars. Note that in the bin half-a-magnitude wide just below the turn-off, 80% of the stars are Be stars!

6. Microscopic diffusion

Microscopic diffusion appears as soon as a fluid is out of thermodynamic equilibrium. Any gradient of temperature, density, external forces like gravity or radiative forces, will induce a diffusion velocity. In most cases, the diffusion velocities are much weaker than other velocities in the medium, therefore microscopic diffusion can only have a significant effect in very stable media.

In non-rotating solar-mass models, microscopic diffusion tends to reduce the MS lifetime by about 10% (based on a model computed by P. Eggenberger for the present paper). The lifetime is reduced mainly because helium diffuses in the H-burning region, and hydrogen is pushed out from the core by mass conservation. Diffusion thus decreases the amount of available fuel for the star and thus its lifetime. When a small level of rotation is taken into account (initial velocity of about 50 km s^{-1}), the lifetime increases, almost back to the same value as the model without diffusion and without rotation. This is due to the fact that the two effects nearly compensate, diffusion decreasing the quantity of fuel in the core and rotation increasing it. Eggenberger *et al.* (2008) find that accounting or not for the effect of diffusion changes the age estimate of the binary system 70 Oph AB (masses of about 0.9 and 0.7 $M_\odot$) by about 1 Gyr, a relative change of about 14–16%.

7. Mass loss

The effects of mass loss during the MS phase begins to become significant for stars more massive than about 30 $M_\odot$, at solar metallicity (that means ages less than about 5 Myr). Mass loss makes a star of a given mass follow an evolution in the HR diagram similar to that of a star of a lower initial mass. This tends thus to reduce the luminosity and

Table 2. Tentative estimates of the dispersion of ages obtained from models of different authors or computed with different physical ingredients.

Cause	Range of ages	Relative error	Remarks
Origin of stellar models	all	10-30%	for similar input physics
metallicity	6.5 - 9.5	20-35%	for $-1.3 < [Fe/H] < 0.3$
	9.5 - 10.0	35-80%	for $-1.3 < [Fe/H] < 0.3$
		15%	changing O or Fe by a factor 2
overshooting	6.5 - 9.5	40-320%	for $0 < d_{ov}/H_P \leqslant 0.20$
mixing-length	>9.5	70%	based on the study of 70 Oph A
Rotation	<9.5	25%	for normal rotational velocities
	<9.5	40%	mostly due to viewing angle (fast rotators only)
Microscopic diffusion	>9.5	15%	based on the study of 70 Oph AB
Mass loss	<6.5	25%	difference between 1 X and 2 X dM/dt

to increase the MS lifetime. It may also make stars that are initially very massive enter into the WR phase early, even during the MS phase. The MS termination then occurs at younger ages and at lower luminosities than in the case without mass loss. Mass loss significantly affects the relationship between the magnitude at the main sequence turn-off and age, only for ages less than about 3.3 million years. High mass-loss rates reduces the age corresponding to a given value of the magnitude at the main sequence turn-off. For example, for M_{bol} at the turn-off equal to -10.8, the difference between the ages derived using models with either normal or enhanced mass-loss rates amounts to 0.73 Myr, going from 3.55 Myr (normal mass-loss, Schaller *et al.* 1992) to 2.62 Myr (enhanced mass-loss, Meynet *et al.* 1994).

Depending on the intensity of mass loss, the evolution with time after a starburst episode of the fraction by number of WR to O-type stars varies. Thus an age determination based on the derivation of this fraction, estimated from emission-line ratios, is affected by uncertainties in the mass-loss rates. For instance, a young starburst showing a WR/O number ratio of 0.20, at $Z = 0.02$, may have an age either of 4.25 or of 3.25 Myr, depending on whether standard or enhanced mass-loss rates are used, respectively (Meynet 1995). In the case of starbursts, other factors also have an impact on the age determination, such as the timescale for the star formation episode (in the example above, we supposed an instantaneous episode), or the slope of the IMF.

8. Conclusions

From Table 2 above we can estimate the accuracy which would be needed for various physical ingredients of the models, in order to reach, let's say, a 10% accuracy in the age. We see that a precision better than 0.12 dex in [Fe/H] is needed to assure a 10% precision on the age, over the whole range of ages, and that individual abundances of oxygen and of iron should be known with an accuracy better than a factor of 2. The size of the convective core should be known with a very high precision, especially in the lower mass range. As explained above, this question is intimately related to the effects of rotation. The value of l/H_P should be known with an accuracy better that 5-10%, and the mass-loss rate with a precision better than about 20%.

As usual, improvements will be brought by progress both in observations and in our physical understanding of turbulence under stellar conditions. In the observational context, asteroseismology will play a key role. This technique allows us to probe the stellar interiors and provides, in addition to the classical constraints such as gravity and effective temperature, new observables, reducing significantly the possible range of values for the free parameters. From the point of view of theory, multi-dimensional computations of hydrodynamic processes like convection will provide more thoughtful recipes to take account of the effects of turbulence in stellar interiors.

The list of effects in Table 2 is not exhaustive: what are the consequences of magnetic fields, of internal gravity waves, of tidal mixing in close binaries, or of accretion during the pre-MS phase of massive stars? All these effects still need to be studied.

References

Aerts, C. 2008, in Massive Stars as Cosmic Engines, Proceedings of the International Astronomical Union, IAU Symposium, Volume 250, p. 237

Bono, G., Caputo, F., Cassisi, S., Marconi, M., Piersanti, L., & Tornambe, A. 2000, *ApJ*, 543, 955

Claret, A. 1997, *A&AS*, 125, 439

Claret, A. 2004, *A&A*, 424, 919

Daflon, S. & Cunha, K. 2004, *ApJ*, 617, 1115

Demarque, P., Woo, J.-H., Kim, Y.-C., & Yi, S. K. 2004, *ApJS*, 155, 667

Dominguez, I., Chieffi, A., Limongi, M., & Straniero, O. 1999, *ApJ*, 524, 226

Dotter, A., Chaboyer, B., Jevremovic', D., Baron, E., Ferguson, J. W., Sarajedini, A., & Anderson, J. 2007a, *AJ*, 134, 376

Dotter, A., Chaboyer, B., Ferguson, J. W., Lee, H.-c., Worthey, G., Jevremovic', D., & Baron, E. 2007b, *ApJ*, 666, 403

Eggenberger, P., Miglio, A., Carrier, F., Fernandes, J., & Santos, N. C. 2008, *A&A*, 482, 631

Ekström, S., Meynet, G., Maeder, A., & Barblan, F. 2008a, *A&A*, 478, 467

Girardi, L., Bressan, A., Bertelli, G., & Chiosi, C. 2000, *A&AS*, 141, 371

Heger, A. & Langer, N. 2000, *ApJ*, 544, 1016

Hunter, I., Brott, I., Lennon, D.J. *et al.* 2008, *ApJ*, 676, L29

Keller, S. C., Bessell, M. S., & Da Costa, G. S. 2000, *AJ*, 119, 1748

Leitherer, C. 2005, in The Evolution of Starbursts: The 331st Wilhelm and Else Heraeus Seminar. AIP Conference Proceedings, Volume 783, p. 280

Maeder, A. & Meynet, G. 1989, *A&A*, 210, 155

Maeder, A. & Meynet, G. 2001, *A&A*, 373, 555

Maeder, A., Meynet, G., Ekstrom, S., & Georgy, C. 2008, Comm. in Asteroseismology, Contribution to the Proceedings of the 38th LIAC, HELAS-ESTA, BAG, in press (arXiv:0810.0657)

Melo, C., Santos, N. C., Pont, F., Guillot, T., Israelian, G., Mayor, M., Queloz, D., & Udry, S. 2006, *A&A*, 460, 251

Meynet, G. 1995, *A&A*, 298, 767

Meynet, G. & Maeder, A. 2000, *A&A*, 361, 101

Meynet, G., Maeder, A., Schaller, G., Schaerer, D., & Charbonnel, C. 1994, *A&AS*, 103, 97

Montalban, J., Miglio, A., Eggenberger, P., & Noels, A. 2008, *Astronomische Nachrichten*, 329, 535

Mowlavi, N., Meynet, G., Maeder, A., Schaerer, D., & Charbonnel, C. 1998, *A&A*, 335, 573

Nordström, B., Mayor, M., Andersen, J., Holmberg, J., Pont, F., Jorgensen, B. R., Olsen, E. H., Udry, S., & Mowlavi, N. 2004, *A&A*, 418, 989

Pietrinferni, A., Cassisi, S., Salaris, M., & Castelli, F. 2004, *ApJ*, 612, 168

Royer, F., Zorec, J., & Gomez, A. E. 2007, *A&A*, 463, 671

Schaerer, D., Charbonnel, C., Meynet, G., Maeder, A., & Schaller, G. 1993, *A&AS*, 102, 339

Schaller, G., Schaerer, D., Meynet, G., & Maeder, A. 1992, *A&AS*, 96, 269

Smith, N. & Brooks, K.J. 2008, Handbook of Star Forming Regions, Vol. II, Bo Reipurth, ed, in press (arXiv:0809.5081)

Talon, S., Zahn, J.-P., Maeder, A., & Meynet, G. 1997, *A&A*, 322, 209

VandenBerg, D. A., Bergbusch, P. A., & Dowler, P. D. 2006, *ApJS*, 162, 375

Ventura, P., Zeppieri, A., Mazzitelli, I., & D'Antona, F. 1998, *A&A*, 334, 953

von Zeipel, H. 1924, *MNRAS*, 84, 665
Weidemann, V. 2000, *A&A*, 363, 647
Zahn, J.-P. 1992, *A&A*, 265, 115

Discussion

K. Covey: Do you track the angular-momentum evolution of the star over its lifetime in your models? If so, how do you characterize the initial (or final) angular-momentum content of each model?

G. Meynet: Yes, the models self-consistently follow the evolution of the angular momentum in the interior. On the ZAMS, the star is supposed to have solid-body rotation, then under the action of convection, contraction/expansion, shear turbulence and meridional circulation, angular momentum evolves in each mass shell. Angular momentum may be lost at the surface as a result of stellar winds. It is to be noted that the efficiency of the angular-momentum transport depends on the angular velocity and the gradient of the angular velocity. Thus the rotation at a given time step is obtained by an iterative procedure. (This is what is meant by self-consistently above.)

C. Deliyannis: When you indicate that rotation increases the ages of massive stars by 25%, do you include the effects of rotation on the convective core, and do you also include mixing (e.g., due to circulation) bringing in more hydrogen to the core?

G. Meynet: In the models I presented, the increase in the size of the convective core in the last part of the MS phase results from rotational mixing in the radiative zone. Rotational mixing provides fuel to the core by making H diffuse from the radiative envelope into the convective core. Rotational mixing also makes the He produced in the core diffuse into the radiative envelope, lowering its opacity.

S. Vauclair: This is a comment about seismic constraints on core overshooting. With a graduate student we have studied in detail a solar-type star for which we have good seismic data, (μ Arae), with the aim of constraining the size of the convective core, including overshooting. This is possible because the star is at the end of the main sequence, so that the core is He-rich, with a clear signature in the oscillation frequencies. We find that overshooting is very small or absent (less than $\alpha_{OS} = 0.02$; our paper is in preparation).

G. Meynet: This is a indeed a great result if it is possible from asteroseismology to make progress on the question of the size of the convective core (whose size may result from various processes, rotation and overshooting being two of them). In the transition mass range between stars with no convective core ($M < 1.1\ M_\odot$) and stars with a well-developed convective core ($M > 1.3\ M_\odot$) – these limits depend on metallicity – probably overshooting increases progressively, being nearly non-existent around a very small core and then reaching a moderate value above $M > 1.3\ M_\odot$. Therefore in the star you analyzed, if its mass is sufficiently small, I would not be surprised that very small overshooting is found.

A. Dotter: Why is the post-main sequence evolution of $\sim$10 $M_\odot$ solar-Z models (especially the extent of the blue loop) so strongly influenced by rotation?

G. Meynet: Rotation modifies the sizes of the cores (increasing them). The presence of blue loops and their extension depend sensitively on the size of the cores (the bigger they are, the shorter for instance are the blue loops; see the discussion in Maeder & Meynet 2001, A&A, 373, 555). Therefore the blue loop for a star of 9 $M_\odot$ at $Z = 0.020$, computed with rotation is less extended than the blue loop for the same model without rotation. In the case of 12 $M_\odot$, the blue loop is even suppressed in the rotating models (see DHR in Meynet & Maeder 2000, A&A, 361, 101).

Georges Meynet

The Ages of Stars
Proceedings IAU Symposium No. 258, 2008
E.E. Mamajek, D.R. Soderblom & R.F.G. Wyse, eds.

doi:10.1017/S1743921309031846

A new method to estimate the ages of globular clusters: the case of NGC 3201

A. Calamida[1], G. Bono[2,3], P. B. Stetson[4], M. Dall'Ora[5], M. Monelli[6], C.E. Corsi[3], P. G. Prada Moroni[7], S. Degl'Innocenti[7], A. Dotter[8], C. Brasseur[8], P. Amico[1], E. Marchetti[1], R. Buonanno[2], A. Di Cecco[2], S. D'Odorico[1], I. Ferraro[3], G. Iannicola[3], M. Nonino[7], M. Romaniello[1], N. Sanna[2], D. A. Vandenberg[8], M. Zoccali[9], and A. Walker[10]

[1]ESO, Karl-Schwarzschild-Str. 2, D-85748 Garching bei München, Germany
email: acalamid@eso.org

[2]Universita' di Roma Tor Vergata, Via della Ricerca Scientifica 1, 00133 Rome, Italy

[3]INAF-Osservatorio Astronomico di Roma, Via Frascati 33, 00040, Monte Porzio Catone, Italy

[4]DAO, HIA-NRC, 5071 W. Saanich Road, Victoria, BC V9E 2E7, Canada

[5]INAF-Osservatorio Astronomico di Capodimonte, Via Moiariello 16, 80131 Napoli, Italy

[6]IAC - Instituto de Astrofisica de Canarias, Calle Via Lactea, E38200 La Laguna, Tenerife, Spain

[7]INAF-Osservatorio Astronomico di Trieste, Via G.B. Tiepolo 11, 40131 Trieste, Italy

[8]UVIC, Victoria, BC V8W 3P6, Canada

[9]Universidad de Concepcion, Departamento de Fisica, Casilla 106-C, Concepcion, Chile

[10]Cerro Totolo Inter-American Observatory, Chile

Abstract. We devised a new method to estimate globular cluster absolute ages by adopting the knee of the bending of the lower main-sequence (MS) in the Near-Infrared (NIR) $J, J - K_s$ color-magnitude diagram. The color difference between this feature and the Turn-Off point is strongly correlated to the cluster age. This method is marginally affected by distance and reddening uncertainties, and by the possible occurrence of differential reddening. Furthermore, the knee location does not depend on the cluster age and it is a robust theoretical prediction. We adopted accurate J, K_s-band photometry collected with both MAD/VLT and SOFI/NTT for the Galactic globular cluster NGC 3201 to identify the location of the knee at $J \sim 19.90 \pm 0.03$ and $J - K_s \sim 0.76 \pm 0.02$ mag. The comparison with different sets of cluster isochrones, transformed adopting different Color–Temperature–Relations (CTRs), shows that the models are slightly redder than the observations for $J > 19$ mag. This difference could be due to the presence of a calibration drift or to a problem of the CTRs in this magnitude range.

Keywords. globular clusters: general, globular clusters: individual (NGC 3201)

---

1. Introduction

The absolute age of Galactic Globular Clusters (GGCs) is a crossroad of several astrophysical problems (Vandenberg *et al.* 1996; Chaboyer 1998; Castellani 1999). This parameter provides: *i)* a lower limit to the age of the Universe (Buonanno *et al.* 1998; Stetson *et al.* 1999; De Angeli *et al.* 2005); *ii)* robust constraint on the input physics adopted in stellar evolutionary models (Castellani & Degl'Innocenti 1999; Vandenberg *et al.* 2002, 2008; Dotter *et al.* 2007), and *iii)* the chronology for the assembly of the halo, the bulge and the disk of the Milky Way (Rosenberg *et al.* 1999, Zoccali *et al.*

2003). However, estimates of GC ages are still hampered by uncertainties affecting both the distance modulus and the reddening (Renzini 1991; Bono *et al.* 2008), together with uncertainties on the chemical composition (Gratton *et al.* 2004), on the metallicity scale (Rutledge *et al.* 1997; Kraft & Ivans 2003), and on the photometric zero-points (Stetson 2005).

The GGC NGC 3201 presents some interesting features. It is located at relatively short distance ($\mu = 13.32 \pm 0.06, E(B-V) = 0.30 \pm 0.03$, Piersimoni *et al.* 2002; $\mu = 13.36 \pm 0.06, E(B-V) = 0.25 \pm 0.02$, Layden & Sarajedini 2003; Mazur *et al.* 2003), and has accurate estimate of both iron ($[Fe/H] = -1.54 \pm 0.10$ dex, Kraft & Ivans 2003; Covey *et al.* 2003) and α-elements ($[\alpha/Fe] \sim 0.2 - 0.4$, Pritzl *et al.* 2005). Moreover, NGC 3201 presents a low central density, a large tidal radius ($\log \rho_V = 2.69 L_\odot pc^{-3}$, $r_t \sim 28$ arcmin, Harris 2003), and a highly-retrograde orbit (van den Bergh 1993), thus suggesting that it might not be a typical member of the Galactic halo. It has been associated to different Galactic stellar streams (Grillmair 2006; Belokurov *et al.* 2007), but these associations were ruled out by Casetti-Dinescu *et al.* (2007), on the basis of its distance on the Galactic plane. The main drawbacks of NGC 3201 are that it is affected by field contamination, presents a relatively high reddening ($E(B-V) = 0.25 - 0.30$), and is also affected by differential reddening. Owing to the quoted problems we still lack an accurate estimate of the absolute age of NGC 3201.

2. Observations and data reduction

The B, V, I-band data considered in this investigation come from the database of original and archival observations which have been collected, reduced and calibrated by P. B. Stetson (see his proceeding in this book). For NGC 3201 we rely on a catalog with 112,238 stars having at least two measurements in each of the three optical bands. The accuracy of the zero-points in the quoted bands is of the order or better than 0.01 mag.

Near-Infrared (NIR) data (J, K_s) were collected in two observing runs with different pointings of the NIR camera SOFI (Field of View, FoV $\sim 5 \times 5$ arcmin^2; pixel scale = 0.29 arcsec/pixel) available at the New Technology Telescope (NTT; ESO, La Silla). The total exposure time per band are $\sim$10 (J) and $\sim$40 (K_s) minutes, and the images cover an area of $\approx$20 $\times$ 18 arcmin across the cluster center (a more detailed discussion concerning this data set will be given in a forthcoming paper).

These data were supplemented with deep NIR data (J, K_s) collected with the Multi-Conjugate Adaptive Optics Demonstrator (MAD) available at the Very Large Telescope (VLT; ESO, Paranal). MAD is a prototype instrument performing wide FoV real-time correction for atmospheric turbulence (Marchetti *et al.* 2006, Gilmozzi & Spyromilio 2007). MAD is equipped with an infrared imaging camera, CAMCAO, based on a 2048$\times$2048 pixels Hawaii2 infrared detector with a pixel scale of 0.028 arcsec/pixel for a total FoV of 1 squared arcminute. During the first on-sky demonstration run of MAD, four 1 $\times$ 1 arcminutes fields were observed in the region located in the S-W corner of NGC 3201. For wavefront sensing, five guide stars with visual magnitude ranging from 11.8 and 12.9, equally distributed on a circle of 2 arcmin diameter concentric to the field were used, and the Multi-Conjugate Adaptive Optics (MCAO) loop was closed at a correction frequency of 400 Hz. For each pointing we collected three J-band and five K_s-band images of 240 sec ($DIT = 10, NDIT = 24$). The seeing during the observations changed from 0.6 to 0.8 arcsec (J-band) and from 0.8 to 1.3 arcsec (K_s-band). The full-Width Half Maximum (FWHM) measured on the images ranges from 0.07 to 0.10 arcsec. Details concerning the pre-reduction strategy adopted will be discussed in a forthcoming paper (Dall'Ora *et al.* 2009, in preparation).

Photometry on individual images was performed with DAOPHOT *IV*/ALLSTAR, followed by simultaneous photometry over 358 NIR images with ALLFRAME (Stetson 1994). The instrumental magnitudes were transformed into the 2MASS photometric system adopting a large sample of local standards. We ended-up with a catalog of ~29,000 stars with at least one measurement in each of the two NIR bands.

3. Results and discussion

In order to validate the cluster isochrones we first performed two independent fits using optical bands. The top panel of Fig. 1 shows the comparison, in the $V, B - I$ color-magnitude diagram (CMD), between selected stars and cluster isochrones at fixed chemical compositions and two different ages (Dotter *et al.* 2008). Theory was transformed into the observational plane using either the Phoenix atmosphere models (red and purple lines, Brott & Hauschildt 2005) or with the semi-empirical relations by Vandenberg & Clem (2003, hereinafter VC03, green and violet lines). Data plotted in this figure show, within the errors, that this cluster has an absolute age of 12 ± 1 Gyr. The plausibility of the adopted distance modulus and reddening is supported by the good agreement between predicted and Zero-Age-Horizontal-Branch (ZAHB) and HB stars.

As a further test for the theoretical validation we performed a similar comparison, but using different sets of cluster isochrones (Victoria-Regina: Vandenberg *et al.* 2006; Dartmouth: Dotter *et al.* 2008; Pisa: Cariulo *et al.* 2004) constructed by adopting different physical assumptions, but transformed into the observational plane using the same Color–Temperature–Relations (CTRs, Phoenix models). The NIR CMDs present some advantages when compared to optical CMDs: 1) they are less affected by uncertain and differential reddening; 2) stellar isochrones plotted in Fig. 2 show that the MS in the low-mass regime presents a well-defined bending. This feature is mainly caused by Collisional Induced Absorption (CIA) of molecular hydrogen at NIR wavelengths (Saumon *et al.* 1994). According to theory, the color and the shape of the bending depend on the metal content (see Fig. 2). However, the magnitude and the color of the bending are, at fixed chemical composition, minimally affected by cluster age. This feature offers the unique opportunity to anchor cluster isochrones, and in turn to estimate the absolute age as a color difference between the knee of the bending and the cluster Turn-Off (TO, Fig. 2). The bending is a robust prediction, since the stellar structures, when moving from this region toward lower stellar masses, are minimally affected by uncertainties in the treatment of the convective transport. The convective motions are indeed adiabatic (Saumon & Marley 2008). We note that a similar feature has already been detected in two other GGCs (ω Cen, Pulone *et al.* 1998; M4, Pulone *et al.* 1999) and in the Galactic bulge (Zoccali *et al.* 2000), using deep NIR (J, H) data collected with NICMOS on the Hubble space Telescope.

According to this evidence we took advantage of the deep NIR CMD (Fig. 3) based on images collected with both SOFI and MAD. Data plotted in this figure display a well defined bending along the MS for $J > 19.5$ mag. To properly define the bending, the stellar distribution along the MS was smoothed according to a Gaussian Kernel with standard deviations equal to the photometric error in magnitude and in color. We defined the knee of the bend as the reddest point and we found that it is located at $J \sim 19.90 \pm 0.03$ and $J - K_s \sim 0.76 \pm 0.02$ mag (see the plus sign in Fig. 3). We overplotted the same isochrones, for a fixed chemical composition and ages of 11, 13 Gyr, that we have validated on the optical CMDs, and we adopted the same distance modulus but a different reddening, $E(B - V) = 0.19$. The ~30% decrease of the adopted

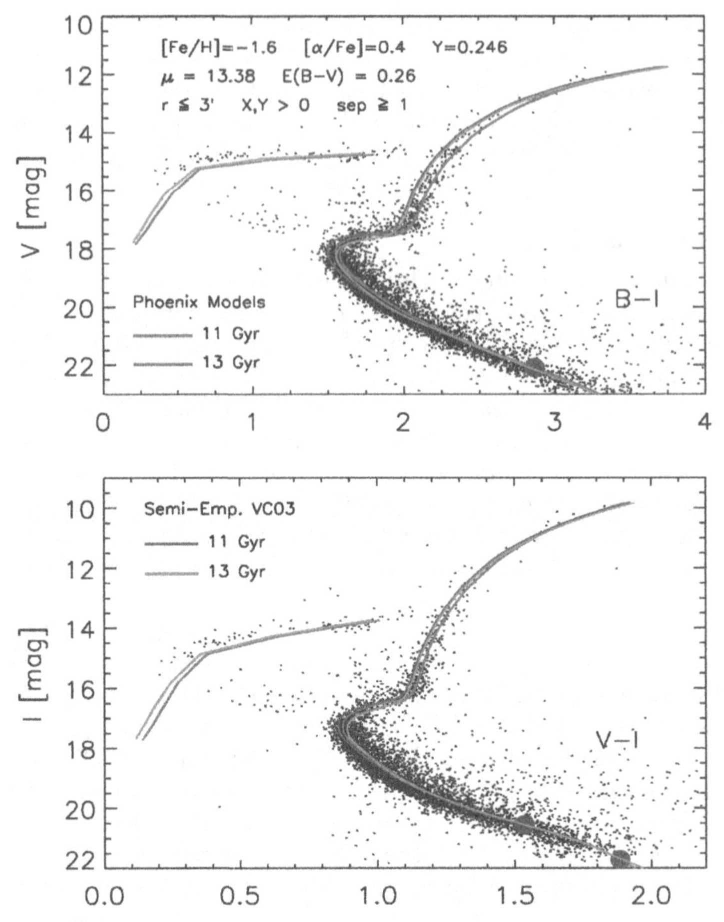

Figure 1. – Top: $V, B-I$ CMD of NGC 3201 based on data collected with ground-based and space telescopes. Stars plotted in this CMD were selected adopting different criteria (separation, position). The red and the purple lines show two cluster isochrones from the Dartmouth database at fixed chemical composition and different ages. These isochrones were transformed into the observational plane using the Phoenix atmosphere models, while the violet and the green isochrones were transformed using the semi-empirical transformation by VC03. The ZA-HBs plotted in this CMD have been transformed using the same CTRs. The adopted true distance modulus and cluster reddening are labeled. – Bottom: Same as top, but for the $I, V-I$ CMD. The two large blue circles mark the position of two stellar structures with $M = 0.50$ and $0.37 M_{\odot}$, respectively.

reddening in order to fit the NIR CMD could be due to a problem in the reddening law (Cardelli *et al.* 1989; Fitzpatrick 1999). Moreover, Fig. 3 shows that current isochrones are slightly redder than observations for $J > 19$ mag. This could be caused by a drift in the photometric calibration or by a problem with the NIR CTRs in the lower MS regime. However, we need to improve both the accuracy of the photometry and the CTRs in this faint magnitude range.

References

Belokurov, V., *et al.* 2007, *ApJ*, 658, 337
Bono, G., *et al.* 2008, *ApJ*, 686, L87

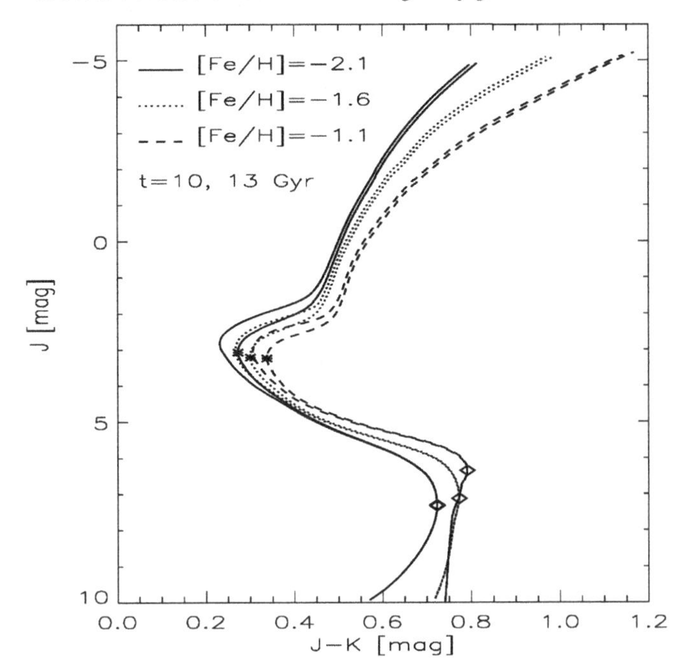

Figure 2. Stellar isochrones from the Dartmouth database with different ages and chemical compositions plotted in the $J, J-K$ CMD. Predictions were transformed into the 2MASS photometric system using Phoenix atmosphere models. The lower MS shows a well defined bending affected by the chemical composition, but independent of cluster age. The asterisks mark the TO position for the isochrones with $t = 13Gyr$, while the diamonds in the lower MS marks the knee of the bending.

Brott, I. & Hauschildt, P. H. 2005, *ESASP*, 576, 565

Buonanno, R., *et al.* 1998, *ApJ*, 501, L33

Cardelli, J. A., Clayton, & G. C., Mathis, J. S. 1989, *ApJ*, 345, 245

Cariulo, P., Degl'Innocenti S., & Castellani, V. 2004, *A&A* 421, 1121

Casetti-Dinescu, D. I., *et al.* 2007, *AJ*, 134, 103

Castellani, V. 1999, in *Globular clusters*, 10th Canary Islands Winter School of Astrophysics, 109

Castellani, V. & Degl'Innocenti, S. 1999, *A&A*, 344, 97

Covey, K. R., *et al.* 2003, *PASP*, 115, 819

Chaboyer, B. 1998, *PhR*, 307, 23

De Angeli, *et al.* 2005, *AJ*, 130, 116

Dotter, A., *et al.* 2007, *AJ*, 134, 376

Dotter, A., *et al.* 2008, *ApJS*, 178, 89

Fitzpatrick, E. L. 1999, *PASP*, 111, 63

Gilmozzi, R. & Spyromilio, J. 2007, Msngr, 127, 11

Gratton, R., Sneden, C., & Carretta, E. 2004, *ARA&A*, 42, 385

Grillmair, C. J. 2006, *ApJ*, 651, 29

Harris, W. E. 2003, Catalog of Parameters for Milky Way Globular Clusters: The Database Hamilton: McMaster Univ., *http://physun.physics.mcmaster.ca/ harris/mwgc.dat*

Kraft, R. & Ivans, I. 2003, *PASP*, 115, 143

Layden, A. C. & Sarajedini, A. 2003, *AJ*, 125, 208

Mazur, B., Krzemiski, W., & Thompson, I. B. 2003, *MNRAS*, 340, 1205

Marchetti, E., *et al.* 2006, *SPIE*, 6272, 21

Piersimoni, A., Bono, G., & Ripepi, V. 2002, *AJ*, 124, 1528

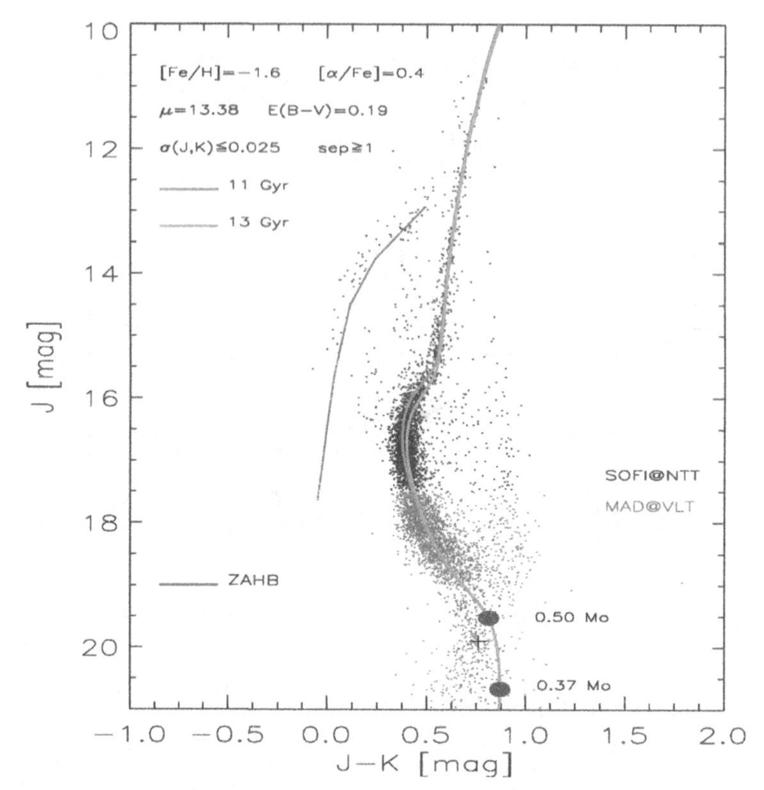

Figure 3. NIR $J, J-K$ CMD of NGC 3201. Black and red dots display stars observed with SOFI/NTT and with MAD/VLT. The green and the yellow lines show two cluster isochrones from the Dartmouth database at fixed chemical composition. The purple line shows the ZAHB for the same chemical composition. The adopted distance modulus and reddening are labeled. The two large blue circles mark the position of two stellar structures with $M = 0.50$ and $0.37 M_{\odot}$, respectively.

Pulone, L. *et al.* 1998, *ApJ*, 492, 41

Pulone, L., De Marchi, G., & Paresce, F. 1999, *A&A*, 342, 440

Pritzl, B. J., Venn, K. A., & Irwin, M. 2005, *AJ*, 130, 2140

Renzini, A. 1991, otci.conf., 131

Rosenberg, A., *et al.* 1999, *AJ*, 118, 230

Rutledge, *et al.* 1997, *PASP*, 109, 907

Saumon, D., *et al.* 1994, *ApJ*, 424, 333

Saumon, D. & Marley, M. S. 2008, *ApJ*, 689, 1327

Stetson, P. B. 1994, *PASP*, 106, 250

Stetson, P. B., *et al.* 1999, *AJ*, 117, 247

Stetson, P. B. 2005, *PASP*, 117, 563, *http://www1.cadc-ccda.hia-iha.nrc-cnrc.gc.ca/ community/STETSON/standards*

Vandenberg, D., Stetson, P. B., & Bolte, M. 1996, *ARA&A*, 34, 461

Vandenberg, D., *et al.* 2002, *ApJ*, 571, 487

Vandenberg, D. & Clem, J. L. 2003, *AJ*, 126, 778

Vandenberg, D., *et al.* 2008, *ApJ*, 675, 746

van den Bergh, S. 1993, *AJ*, 105, 971

Zoccali, M., *et al.* 2000, *ApJ*, 538, 289

Zoccali, M., *et al.* 2003, *A&A*, 399, 931

Discussion

A. Sarajedini: Why do you refer to your method as an absolute age indicator? You rely on color differences which depend sensitively on the model colors. Getting an absolute age from these models is especially uncertain.

A. Calamida: We can anchor the isochrones adopting two different reference points, in the near-infrared color-magnitude diagram: the turn-off point that depends on the cluster age and the "knee" of the lower main sequence bending that does not depend on cluster age (see Fig. 2 of the text). Moreover, the bending is a robust theoretical prediction since stellar structure in this region is minimally affected by uncertainties in the treatment of convective transport (see text for details). More importantly, this method is not affected by the uncertainties in the estimate of the cluster distance modulus and by the possible occurrence of differential reddening. We also adopted different sets of cluster isochrones computed in Pisa (Cariulo *et al.* 2004) and in Victoria (VandenBerg *et al.* 2006) and different color-temperature relations (Phoenix and the semi-empirical ones of VandenBerg & Clem 2003) for our near-IR CMD fits. They are in agreement within the uncertainties. This method is much more robust that the classical isochrone fits that rely only on the color and the magnitude of the turn-off point, which is affected by the uncertainties in the cluster distance, reddening, and differential reddening.

Annalisa Calamida

Fred Walter

Gail Schaefer

The Ages of Stars
Proceedings IAU Symposium No. 258, 2008
E.E. Mamajek, D.R. Soderblom & R.F.G. Wyse, eds.

doi:10.1017/S1743921309031858

Homogeneous photometry of globular clusters—a progress report

Peter B. Stetson

Dominion Astrophysical Observatory, Herzberg Institute of Astrophysics, National Research Council, 5071 West Saanich Road, Victoria BC V9E 2E7
email: Peter.Stetson@nrc-cnrc.gc.ca

Abstract. Classical broad-band photometry can provide direct comparisons of star clusters both with each other and with theoretical models of stellar evolution. The confidence with which conclusions can be drawn is often limited by the accuracy of the measurements. The present work is part of a long-term attempt to improve photometric calibrations.

Keywords. Techniques: photometric, standards, globular clusters: general

---

1. Introduction

It has long been a goal of mine to eliminate photometric uncertainty as a major source of confusion in the analysis of Galactic star clusters and nearby galaxies. If we can reach the point where the uncertainty in the calibration of a particular photometric study is negligible compared to, say, the uncertainty in the foreground reddening of the target, I will consider that the job has been done.

For more than a third of a century, our species' first line of defense against photometric error has been the work of Arlo Landolt. His 1973 paper laid out a network of standards in the Johnson UBV system that were in a magnitude range ($7 \lesssim V \lesssim 14$) suitable for use with photomultipliers on the smallest to the largest research telescopes in use at the time. In 1983 and 1992 papers, he added measurements in the R and I photometric bandpasses of Cousins (1976), which was based on a similar system established by Kron, White, and Gascoigne (1953).

If one makes the arbitrary assumption that, to be reliable, a photometric standard star must have at least five observations in each filter and should have a standard error of the mean magnitude no larger than 0.02 mag in each index, then the three Landolt papers, combined, have about 275 stars meeting these criteria in at least the B, V, and I filters. These stars span a broad range of color: $-0.7 < B{-}I < +6.0$ (approximately equivalent to $-0.3 < B{-}V < +2.3$, or $-0.4 < V{-}I < +4.0$).

Nearly all of Arlo's standards are quite close to the celestial equator. This is valuable, because they are accessible to observers in both hemispheres, making it possible to relate observations over the entire sky to a single, common photometric system. However, this also means that the standard stars never pass truly overhead for most ground-based observatories, and from any given site they can be observed only over a restricted range of azimuth. As a result, long slews are often required to move from a target field to a standard field and back; this tends to discourage frequent observations of standard stars throughout a high-quality night, especially on larger telescopes. In addition, Landolt's best-observed fields—located at three-hour intervals—contain ∼17 stars each, spread over an area 20 or 30 arcminutes on a side. Thus, there are relatively few opportunities to get more than a few standards onto a typical CCD at a time, and it turns out that there are hardly any opportunities to observe, simultaneously, multiple stars of highly

Table 1. Individuals contributing CCD data for defining secondary photometric standards.

Abi Saha	Elena Pancino	Mike Bolte
Alfred Rosenberg	Howard Bond	Nancy Silbermann
Alistair Walker	Judy Cohen	Nick Suntzeff
Andy Layden	Luigi Bedin	Noelia Noël
Bart Pritzl	Manuela Zoccali	Peter Bergbusch
Carme Gallart	Márcio Catelan	Randy Zingle
Don Hamilton	Matteo Monelli	

Table 2. Archives utilized during the course of this work.

DAO
ESO
CFHT
Isaac Newton Group
Subaru
Telescopio Nazionale Galileo

dissimilar colors. Finally, not many of these stars are faint enough to be used effectively with modern 8–10 m telescopes, or your typical diffraction-limited 2.4 m telescope in space.

2. The project

When my colleagues and I began observing star clusters with CCDs on 0.9 m–4 m telescopes in the early 1980's, we made a point of observing those few asterisms where two or more of Arlo's equatorial standard stars could be placed on the chip at a given time. In many cases we had enough observations that other stars falling in the same images could be turned into *secondary* standards. Clearly, since photometric indices could only be assigned to these stars by reference to Landolt's primary standards, such secondary standards would not contribute to the absolute calibration of our observations to Arlo's photometric system. However, they could be useful for expanding the basis of comparison whereby data from different nights, different observing runs and, most notably, different telescopes could be placed on a common system. One example was the attempt by Stetson and Harris (1988) to define new secondary standards in some Landolt fields as well as in their target star clusters. In particular, they used the Kitt Peak 0.9 m telescope to define new faint secondary standards that could augment the comparatively few primary standards faint enough to be observed with the 4 m telescope. I now know that this attempt was not entirely successful: too great a faith placed in stars observed too few times produced an internal photometric system that was able to drift roughly 0.02 mag away from the true Landolt system.

Since then, the body of data that I am using to define secondary standards has grown by about three orders of magnitude. At first, the images mostly came from observing runs personally carried out by my collaborators and me. Then friends, colleagues and well-wishers (see Table 1 for a partial list) began contributing their data to the cause. More recently, I have been mining the international data archives that have become increasingly valuable over the years (Table 2).

As of now, I have acquired and reduced 1,195 observation sets, where an "observation set" may be loosely defined as a corpus of data obtained from one CCD on one photometric night *or* on one or more consecutive, usable, but not strictly photometric nights with the same instrumental setup. (Data from poorer-quality nights can be used to improve the relative photometry for multiple stars recorded in the same images, but not to intercompare photometric indices among stars in different images.) Note that I treat data from the individual CCDs of a mosaic camera as coming from *independent* photometers: *i.e.*, the CFH12k mosaic produced twelve data sets per night of observing, and the ESO 2.2m+WFI produces eight.

The current data sets contain 9,288,741 individual measurements of 99,054 distinct stars, virtually all of which have been individually chosen by hand and eye from deep, stacked images. Among these nearly 10^5 stars, 48,768 have at least five observations *and* standard error of the mean magnitude $\sigma < 0.02$ mag in B, V, and I, and show no evidence of intrinsic variability as large as 0.05 mag (r.m.s.). Fig. 1 is a color-magnitude diagram (CMD) for these stars; the larger filled squares represent the 275 Landolt standards that meet these same selection criteria. The new secondary standards reach about seven magnitudes fainter than Arlo's primary photometric standards. One may also note that Arlo has done a good job of including in his sample standards that are as blue as the bluest stars known. However, there are in the Solar Neighborhood at least a few stars significantly redder than the reddest stars that he observed; my own results for these stars obviously depend upon extrapolation (or, rather, the average of many extrapolations) of Arlo's photometric system.

3. How standard are these stars?

The obvious question is, of course, "How can one claim to define a homogeneous photometric system from heterogeneous data?" The *sine qua non* for defining a homogeneous photometric system has always been to restrict oneself to a single detector and a single

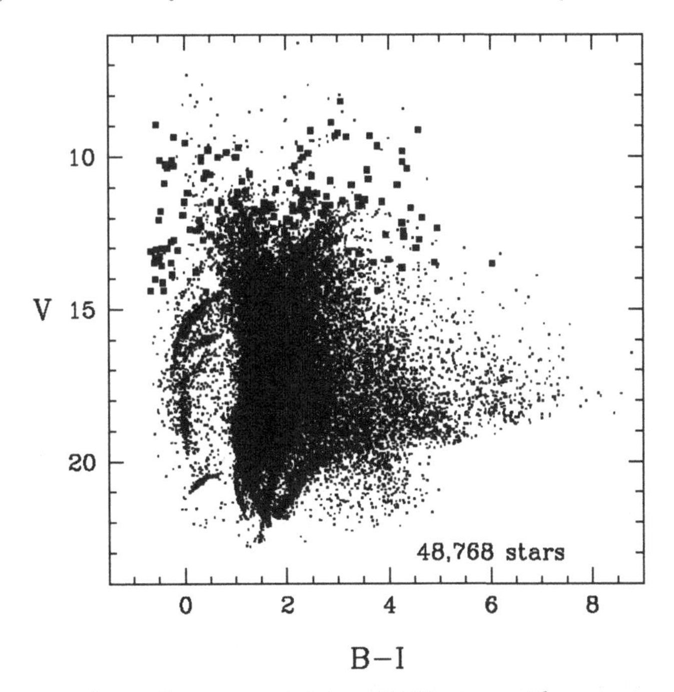

Figure 1. Color-magnitude diagram containing 48,768 stars with measurements considered good enough to serve as secondary photometric standards: at least five observations on fully photometric occasions in *each* of B, V, and I, standard errors of the mean magnitude no larger than 0.02 mag in each filter, and no evidence of intrinsic variability exceeding 0.05 mag r.m.s. Larger filled squares represent 275 Landolt stars meeting the same acceptance criteria.

set of filters, or—at the very least—to detectors and filter sets that have been carefully designed to be as similar as humanly possible. However, when you do not control your own telescope, you usually do not have this luxury; you request your observing time, and if the time is granted and scheduled, you show up at the telescope and use the equipment that has been provided. This equipment is not always a perfect match to that with which the photometric system was defined, or—indeed—to the equipment that was available for your previous observing runs.

A filter combines with the detector to define each photometric bandpass: the sensitivity of the system to incoming photons as a function of their wavelength. In some cases, these bandpasses can also be significantly altered by the passage of the light through Earth's atmosphere and the telescope optics. It requires only a few minutes of surfing the internet or perusing the relevant literature to learn that it is quite difficult to reproduce in detail a given sensitivity curve with different photodetectors and pieces of colored glass. Since a magnitude measurement is the product of a stellar spectral-energy distribution (SED) multiplied by a sensitivity curve and integrated over wavelength, it is clear that slightly different bandpasses can produce different magnitudes for the same star; two similar but not identical stars may produce the *same* magnitude measurement for one photometer, and *different* magnitudes for another.

However, stellar SEDs are not completely arbitrary functions of wavelength; rather, they form a rather well-defined, nearly one-parameter family of curves. The single dominant parameter determining the form of a star's SED, of course, is its effective temperature. Smaller perturbations to the SED corresponding to a particular temperature are produced by, for instance, the star's effective surface gravity, chemical composition, and rotation speed. Foreground reddening also alters the perceived SED is a fashion that is very similar, but not quite identical, to a reduction in the star's effective temperature.

Since the family of stellar SEDs is comparatively well behaved, it is generally found that for most practical purposes one can model empirically the differences between a particular filter-detector combination and the corresponding standard photometric bandpass. You use direct observations of standard stars to determing the fitting parameters a_i in transformation equations of the form

$$v' \equiv v(\text{observed}) - k_V \cdot X = V + a_0 + a_1 \cdot (\text{COLOR}) + a_2 \cdot (\text{COLOR})^2 + \cdots ,$$

where I use lower-case text to denote an observed magnitude in the particular *instrumental* photometric system defined by the equipment that one is actually using, and upper-case text represents photometric indices in the *standard* system that one is attempting to reproduce. "COLOR" represents a color index—in the standard photometric system—defined at wavelengths near the photometric bandpass in question: for the V bandpass, one might use the B–V color, the V–I color, or whatever is most convenient. The "..." might represent additional high-order terms involving colors, airmass, azimuth, time of night, or anything that might be affecting the atmospheric and instrumental throughput in a systematic way. Thus, one is using observations of stars of known photometric properties to produce an empirical model representing the difference between the ideal and the actual photometric bandpass, in the sense of a Taylor-series expansion in variables describing the morphology of the star's SED. (However, note that it is potentially quite misleading to apply such empirical transformations derived from normal stars to objects having distinctly nonstellar spectral-energy distributions, such as supernovae or quasars.)

I assert that if one assiduously attempts to determine the transformations that correct observations from any one instrumental system to the equivalent indices that *would have been* obtained with Arlo Landolt's equipment, and that if one does this for many different instrumental setups and then averages the results so obtained, then one eventually approaches, asymptotically, a well defined and robust average photometric system.

Fig. 2 shows the absolute differences, on a star-by-star basis, between Arlo's published magnitudes and my own for stars in common. In the left panel, the absolute V-magnitude differences are plotted against the number of photometric measurements in my data for stars that Arlo measured ten or more times; the right panel shows absolute V differences against the number of Arlo's observations, for stars that have at least ten photometric measurements in my data set. Since the x-axes are linear in the square root of the number of observations, if the magnitude differences were due solely to random measuring errors one would expect a wedge-shaped distribution of points declining to the right. This is not seen. In fact, if one ignores the points for stars with fewer than three or four observations in one data set or the other, there is essentially no change in the distribution of photometric errors with increasing number of observations. This implies that random measuring errors are *not* the dominant cause of the perceived magnitude differences. Rather, they must be due to the (small) range of SEDs that are capable of producing the same observed magnitude when integrated over the bandpass of a particular filter/detector combination, but produce different results for a different approximation of the same bandpass. After making a minor correction for that part of the dispersion that *can* be attributed to observational error (readout noise, Poisson photon statistics, unmodeled extinction variations, ...), I find that this irreducible scatter amounts to ~0.012 mag in V, ~0.014 mag in R, and ~0.016 mag in B and I. This represents the irreducible difference between Arlo's *particular* photometric system and the average of many independent attempts to reproduce his system, caused by the variety of stellar SEDs that are actually out there. Presumably, the irreducible differences between Arlo's system and any *one* attempt to reproduce it (*i.e.*, the results of any one given observing run, empirically transformed to

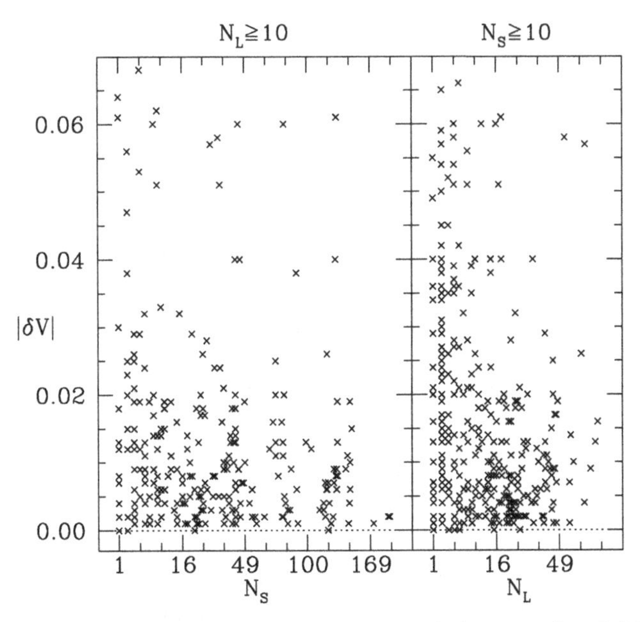

Figure 2. (*Left*) Absolute difference in V-band magnitude between Landolt's published photometry and mine for stars in common; stars observed at least 10 times by Arlo are plotted against the number of photometric observations in my database. (*Right*) The same, except that here stars that I have observed at least ten times are plotted against the number of Landolt observations.

Table 3. Properties of four globular clusters

	$R_\odot$ (Harris)	M_V (Harris)	[Fe/H] (Harris)	$[Fe/H]_{ZW}$ (Rutledge)	$[Fe/H]_{CG}$ (Rutledge)	E_{B-V} (Harris)	E_{B-V} (Schlegel+)
NGC 288	8.8	−6.7	−1.24	−1.40	−1.14	0.03	0.013
NGC 362	8.5	−8.4	−1.16	−1.33	−1.09	0.05	0.032
NGC 1851	12.1	−8.3	−1.22	−1.23	−1.03	0.02	0.037
M5 = NGC 5904	7.5	−8.8	−1.27	−1.38	−1.12	0.03	0.037

Table 4. Observations of four globular clusters

	288	362	1851	5904
observing runs	11	11	13	36
CCD images	386	340	485	2665
B measurements (max)	54	33	78	79
V measurements (max)	84	45	83	136
I measurements (max)	58	36	59	132
median seeing ($''$)	1.1	1.4	1.0	1.1

Arlo's system through observations of his standards) may be expected to be a factor of order $\sqrt{2}$ larger.

4. Example of use

The people that I work with and I have been most interested in producing CMDs for star clusters in the Milky Way Galaxy, and for nearby galaxies that can be resolved into individual stars. Moreover, in attempting to bolster this secondary photometric system I have combed the available data archives for the most popular fields—those having the greatest number of images from the various observatories—and it turns out that most of these are star clusters and nearby galaxies as well. This results in a data set that allows a more critical comparison of the fiducial sequences of different star clusters than has been possible in past.

For instance, consider the four globular clusters NGC 288, NGC 362, NGC 1851, and NGC 5904 (= Messier 5). According to, for instance, the Harris (1996) compilation catalog, these clusters are all comparatively luminous and minimally reddened, and have indistinguishable chemical abundances (Table 3). Here I list, for each of the four clusters, the heliocentric distance, absolute magnitude, and metallicity from Harris's compilation. Since his metallicities are taken from heterogeneous sources, I supplement them with metallicities from a particular single source, namely the Rutledge *et al.* (1997) catalog of values derived from the infrared calcium triplet in giant stars; these latter have been expressed both on the so-called "Zinn-West" scale and the so-called "Carretta-Gratton" scale—which are rather different in this abundance range—but they rest upon the same observational data. It is evident that the range of metallicities among these clusters is small compared to the uncertainty of the estimates. The last two columns of Table 3 give foreground reddening values for each of the clusters, first from the Harris catalog, and then from the all-sky reddening map of Schlegel *et al.* (1998).

Table 4 summarizes the observations available for these clusters in the current body of data. The first two lines give the number of independent observing runs (generally meaning independent instrumental setups, detectors, filters) and number of individual CCD images for each target. Since the various CCDs did not observe exactly the same part of the sky (in particular, with a mosaic camera containing N chips, no given star can fall within more than $1/N$ of the available images) the number of measurements per star can be much less than the total number of images. Accordingly, the next three lines give the *maximum* number of photometric measurements available for any given star in the B, V, and I bandpasses. Finally, the last line gives the median seeing among the images available for each target.

Even given the large body of data available for each of these targets, the quality of the photometry can differ significantly from one star to another. Obviously, the signal-to-noise ratio will decrease for stars of increasing apparent magnitude. Furthermore, crowding and hence photometric reliability will both become systematically worse as the center of the cluster is approached. Finally, the outermost parts of the field will be covered by comparatively few images. Therefore, it is worthwhile to pay attention to how the sample of stars is selected from among all those for which photometry is available.

First, for each cluster I selected out stars in two magnitude ranges, one around the level of the horizontal branch (HB) and one around the level of the main-sequence turnoff (TO). Having determined the center position of each cluster, I then plotted the color uncertainty σ_{B-I} as a function of radius for stars in each of these two magnitude bins. From this I identified a range of clusterocentric radius where the error distribution was independent of radius for both samples. This typically turned out to be something like 2–6 arcmin. The inner limit of each annulus was defined by increasing photometric scatter due to crowding, and the outer limit was defined by increasing scatter due to a smaller number of available observations. Field contamination was not a serious issue for any of these clusters. Second, I divided the stars within the chosen annulus into V-magnitude bins 0.02 mag high; within each such bin I sorted the stars in order of increasing σ_{B-I}. Then I plotted *only* those stars with the smallest values of σ_{B-I} in each magnitude bin. This permits the tracing-out of the principal sequences of each cluster based on only those stars likely to have the best photometry. Fig. 3 shows such CMDs for our four target clusters, based upon the best three stars in each 0.02-mag bin of luminosity.

As has long been known, these clusters—despite their indistinguishable metal abundances—display four distinctly different HB morphologies: NGC 288 has an HB almost

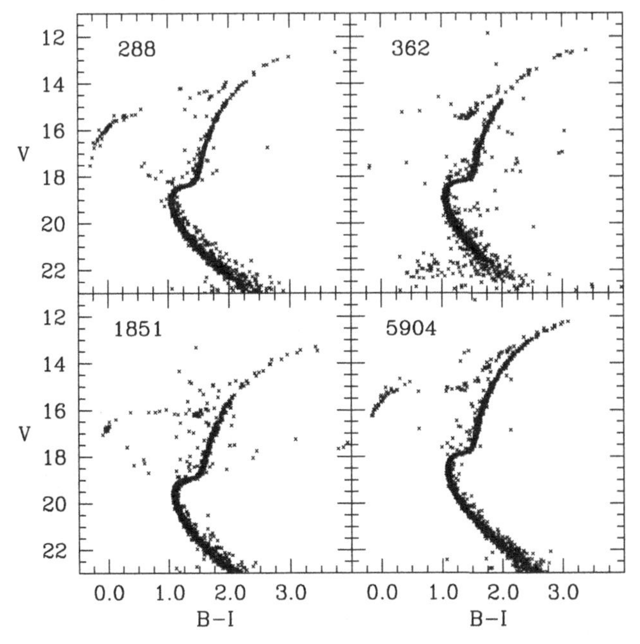

Figure 3. Color-magnitude diagrams for the four globular clusters discussed in the text. The band of faint blue stars in the NGC 362 panel belong to the Small Magellanic Cloud, which lies behind the cluster.

Table 5. Main-sequence TOs and HBs in the four globular clusters

	M_V (TO)	$(B\text{–}I)_{\rm TO}$	$(B\text{–}I)_{\circ,\rm TO}$	V_{HB}	ΔV(TO-HB)
NGC 288	19.049	1.055	1.024	15.58	3.47
NGC 362	18.856	1.056	0.980	15.46	3.40
NGC 1851	19.563	1.089	1.001	16.18	3.38
NGC 5904	18.537	1.089	1.001	15.13	3.41

entirely to the blue of the instability strip; NGC 362 has an HB almost entirely to the red of the instability strip; NGC 1851 has a compact red HB clump and a tight, blue clump with a sparse sprinkling of stars between, including a small number of RR Lyraes (~30 in the whole cluster, only a fraction of these are in the annulus considered here); and NGC 5904 has a continuous HB from the blue to the red, with many RR Lyraes ($\gtrsim$ 130; scaled to the luminosity of NGC 1851, this would correspond to ~85 RR Lyraes, or nearly 3× that cluster's specific frequency). These four clusters therefore illustrate the classical "second-parameter problem" in globular-cluster CMD morphology.

Many authors identify the second parameter with age (the "first parameter," of course, being metal abundance). For instance, Alfred Rosenberg, in his PhD dissertation at the University of La Laguna and in Rosenberg *et al.* (1999) concluded that NGC 288 and NGC 5904 were about the same age, and that NGC 362 and NGC 1851 were also about the same age but both about 2.5 Gyr younger than the other two clusters.

From a visual sliding shift of their HBs to that of NGC 5904 (after correction for the Schlegel *et al.* reddening estimates) I find that NGC 288 is 0.45 mag more distant in apparent modulus; NGC 362 is 0.35 mag more distant; and NGC 1851 is 1.05 mag more distant. I estimate that the uncertainty of each of these modulus differences is of order 0.05 mag, and since the luminosity of the HB is theoretically predicted to be independent

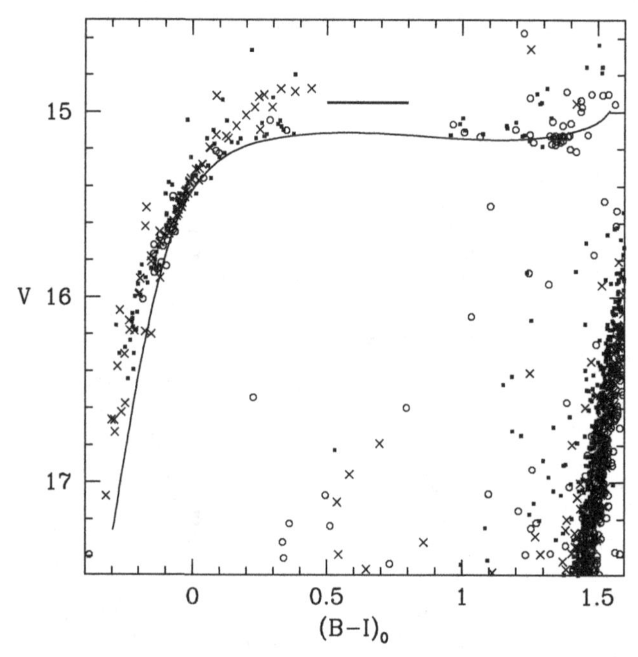

Figure 4. Expanded view of the dereddened HBs of NGC 5904 (small solid squares), NGC 1851 (hollow circles), and NGC 288 (×'s). The Victoria-Regina zero-age HB for [Fe/H] = −1.31, shifted down by 14.45 mag, is shown as a continuous curve. The solid horizontal line represents the zero-age HB level attributed to NGC 288 by Rosenberg *et al.* (1999).

of small age differences, and only weakly sensitive to the minor abundance differences seen among these four clusters, these distance ratios should be quite secure. The case of NGC 288 should be noted in particular. Fig. 4 shows the fit of NGC 288's dereddened HB (×'s) to those of NGC 5904 (small solid squares) and NGC 1851 (hollow circles). With the Schlegel *et al.* reddening values, the blue end of the blue HBs match well with the modulus differences just given. However, as the instability strip is approached from the blue side, the HB stars in NGC 288 tend systematically brighter than those in the other two clusters. The probable explanation is that these stars have evolved away from the zero-age HB, having started from a point rather bluer and fainter. The solid horizontal bar in the figure represents the zero-age HB level attributed to this cluster by Rosenberg *et al.* Evidently not realizing the evolved nature of the reddest HB stars in this cluster, they have overestimated the luminosity of its zero-age HB; this accounts at least in part for the greater age they have assigned to this cluster.

By the way, with the clusters dereddened and matched at the HB, NGC 5904 has an upper giant branch distinctly bluer than that of NGC 1851; the giant branches of NGC 288 and NGC 362 coincide and lie between those of the other two clusters. According to canonical understanding, then, NGC 5904 is the most metal-poor of these four clusters, NGC 1851 is the most metal rich, and NGC 288 and NGC 362 have nearly equal and intermediate abundances based upon this photometric criterion.

By robust weighted fits of parabolas to *all* stars within ± 0.3 mag of the main-sequence TO in the selected annular zones (*i.e.*, not just the stars with the smallest color uncertainties), I obtained the TO magnitudes and colors given in Table 5; again the colors have been dereddened according to the Schlegel reddening estimates, with $E_{B-I} = 2.38\, E_{B-V}$, appropriate for an I photometric bandpass near 800 nm. I estimate the center of the HB in NGC 5904 to lie at $V = 15.13$.† This, combined with the differential moduli given above and the observed apparent magnitudes at the main-sequence TO yields the vertical HB-TO magnitude differences given in the last column of the table.

Curiously, the clusters with the intermediate giant-branch colors, NGC 288 and 362, have the most extreme TO colors, while the most metal-rich cluster and the most metal-poor (by the photometric criterion) have intermediate TO colors. From the Victoria-Regina isochrones (Fig. 5), we would infer that NGC 362 is 1 Gyr younger than NGC 288 from the absolute magnitude of the TO (which, since we have registered the clusters' HBs, is the equivalent of the differential vertical method) or 2 Gyr younger according to the dereddened TO color. NGC 1851 and NGC 5904 appear to be about the same age as NGC 362 by the vertical method, or NGC 5904 may be intermediate between NGC 362 and NGC 1851 on the one hand and NGC 288 on the other, via the TO color. However, allowing 0.05 mag uncertainty in the vertical registration of the CMDs, and uncertainty at a level of 0.01 mag in E_{B-V} ($\sim$0.024 mag in E_{B-I}) in the Schlegel *et al.* reddenings, the error bars are such that age differences of 0 Gyr or 4 Gyr are also allowed at a 1.5σ confidence level—without even allowing for *any* photometric calibration uncertainty.

Ages from the absolute TO color are subject to the assumption that the reddening values are correct. When I attempt to repeat the analysis with the purely differential horizontal method (the color difference between the main-sequence TO and a selected fiducial point on the subgiant branch: Fig. 5, bottom), nonsense results. By this measure,

† The observed vertical shift between the HB in NGC 5904 and the Victoria-Regina (VandenBerg *et al.* 2006) zero-age HB for [Fe/H] = −1.31 is 14.45 mag, with an uncertainty ∼0.02 mag; the model HB has a local luminosity maximum of M_V (predicted) = +0.66 at $(B–I)_\circ = 0.57$ and a local minimum of +0.70 at $(B–I)_\circ = 1.14$; I accordingly adopt $M_V = 0.68$ as the predicted value for the "center" of the theoretical HB. With the 14.45 mag observation-minus-theory difference, this yields 15.13 as the apparent magnitude at the center of NGC 5904's HB.

NGC 362 and NGC 1851 have ages roughly consistent with those from the vertical method and the absolute TO color, but NGC 288 and NGC 5904 have ages that are off the scale of the Victoria-Regina isochrones: at least 4 Gyr older than the other two. The ages of the clusters could be reconciled if the second (or third) parameter is some detailed abundance ratio, such as [CNO/Fe], [Mg/Fe], [Na/Fe], or even—dare I say it—Y.

References

Cousins, A. W. 1996, *MNRAS*, 81, 25
Harris, W. E. 1996, *AJ*, 112, 1487 (on-line catalogue, revision 2/2003)
Kron, G. E., White, H. S., & Gascoigne, S. C. B. 1953, *ApJ*, 118, 502
Landolt, A. U. 1973, *AJ*, 78, 959
Landolt, A. U. 1983, *AJ*, 88, 439
Landolt, A. U. 1992, *AJ*, 104, 340
Rosenberg, A., Saviane, I., Piotto, G., & Aparicio, A., 1999, *AJ*, 118, 2306
Rutledge, G. A., Hesser, J. E., & Stetson, P. B. 1997, *PASP*, 109, 907
Schlegel, D. J., Finkbeiner, D. P., & Davis, M. 1998, *ApJ*, 500, 525
Stetson, P. B. & Harris, W. E. 1988, *AJ*, 96, 909
VandenBerg, D. A., Bergbusch, P. A., & Dowler, P. D. 2006, *ApJS*, 162 375

Discussion

R. MATHIEU: Was your decision not to include a U band philosophical or operational? Follow up: Given that U-like wavelengths are astrophysically valuable, how should the community proceed?

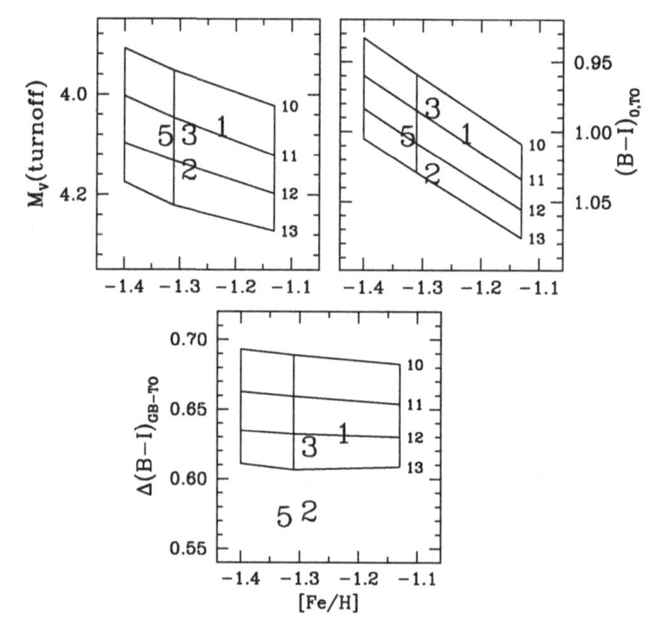

Figure 5. Age diagnostics as a function of metallicity: (*top left*) absolute visual magnitude of the TO; (*top right*) absolute B–I color of the TO; (*bottom*) color difference between the TO and a selected point on the subgiant branch. Numbers designate the different clusters: 1 = NGC 1851, 2 = NGC 288, 3 = NGC 362, 5 = M5 = NGC 5904. I have notionally assigned a lower metal abundance for NGC 5904, a higher abundance for NGC 1851, and intermediate and equal abundances for NGC 288 and NGC 362 as suggested by their giant-branch colors. Solid curves represent predictions of the Victoria-Regina isochrones for the ages shown, in Gyr. The exact placement of these grids is not necessarily correct, due to uncertain physics in the theoretical models, but the relative spacings and slopes should be valid.

P. STETSON: Both. The Johnson U filter is badly behaved. The long-wavelength side is filter-defined but the short- wavelength side is atmosphere-defined. So, the same filter and detector at different observatories will have a different bandpass. Second, CCDs have historically been insensitive at U, so people have avoided it. Since most of my data come from other observers, or archives, I have few U-band observations available to me. (Follow-up) First, CCDs should become sensitive at short wavelengths. Second, observers should agree on a U bandpass and use it consistently. I suggest Thuan-Gunn u or Sloan u.

G. PIOTTO: First of all I want to thank you for your superb talk, which showed us a very rare example of what good photometry means, as well as the limitations intrinsic to good photometry. I have a comment on the inconsistencies you found in the ages you had for your target clusters. I think they are the consequence of the complex chemical composition, peculiar to each cluster, and that we don't know sufficient details, yet.

P. STETSON: I agree with you completely. We are at a conference on stellar ages, and details of chemical abundance ratios, I think, currently represent a limit to our ability to measure globular-cluster ages in units of Earth-orbits around the Sun.

B. WEAVER: How sure are you that you have distinguished yet-unknown CCD photometry problems vs. cosmic scatter? How do you do that?

P. STETSON: To the extent that the known problems are constant throughout the course of a night, and can be parameterized in terms of the intrinsic color and altitude of the star, these effects are removed empirically by the nightly transformation equations. Any remaining effects contribute to the observational scatter, and should decline when averaged over many observing runs. The cosmic scatter is what does not decrease when the number of observing runs increases.

C. CORBALLY: Are there any conclusions you can share with us on the merits of different filter sets?

P. STETSON: No.

Peter Stetson

The Ages of Stars
Proceedings IAU Symposium No. 258, 2008
E.E. Mamajek, D.R. Soderblom & R.F.G. Wyse, eds.

doi:10.1017/S174392130903186X

The ages of stars: The horizontal branch

M. Catelan†‡

Pontificia Universidad Católica de Chile, Departamento de Astronomía y Astrofísica,
Av. Vicuña Mackenna 4860, 782-0436 Macul, Santiago, Chile
email: mcatelan@astro.puc.cl

Abstract. Horizontal branch (HB) stars play a particularly important role in the "age debate," since they are at the very center of the long-standing "second parameter" problem. In this review, I discuss some recent progress in our understanding of the nature and origin of HB stars.

Keywords. stars: abundances, evolution, Hertzsprung-Russell diagram, horizontal branch, mass loss, variables: other, Galaxy: formation, globular clusters: general, globular clusters: individual (M3, M13, NGC 1851), galaxies: dwarf

---

1. Introduction

Horizontal branch (HB) stars have long played a central role in the age debate. These low-mass stars, which burn helium in their core and hydrogen in a shell, are the immediate progeny of the luminous, vigorously mass-losing red giant branch (RGB) stars. Most importantly in the present context, their temperatures depend strongly on their total masses. More specifically, the lower the mass of an HB star, the bluer it becomes, by the time it reaches the zero-age HB (ZAHB). Therefore, the HB morphology in globular clusters (GC's) is naturally expected to become bluer with age.

It has long been known that the *first parameter* controlling HB morphology is actually metallicity, with more metal-rich GC's presenting redder HB's than their more metal-poor counterparts. Still, Sandage & Wallerstein (1960) first realized, based mainly on the early observations of the GC's M3 (NGC 5272), M13 (NGC 6205), and M22 (NGC 6656) by Sandage (1953), Arp & Johnson (1955), and Arp & Melbourne (1959), that GC's with a *given* metallicity might also present widely different HB types, due to the action of an unknown "second parameter." We quote from their study:

> "... the character of the horizontal branch is spoiled by the two clusters M13 and M22. (...) M13 appear[s] to be metal-rich, whereas the character of the horizontal branch simulates that of the very weak-lined group (M15, M92, NGC 5897). (...) M13 is younger than M2 or M5 (...) Consequently, in addition to chemical composition, the second parameter of age may be affecting the correlations."

(Note that the sense of the correlation between age and HB morphology suggested by Sandage & Wallerstein is the *opposite* of what modern studies indicate to be necessary to account for the second-parameter phenomenon.)

It soon became clear that age was not the only possible second-parameter candidate. By the early 1970's, the list of candidates had increased sharply, and already included, in addition to age, the helium abundance and the abundances of the CNO elements (Rood 1973). While the "age as the second parameter" scenario was to gain an important boost with the work by Searle & Zinn (1978), who noted that HB morphology tends to

† John Simon Guggenheim Memorial Foundation Fellow.
‡ On sabbatical leave at Catholic University of America, Dept. of Physics, Washington, DC.

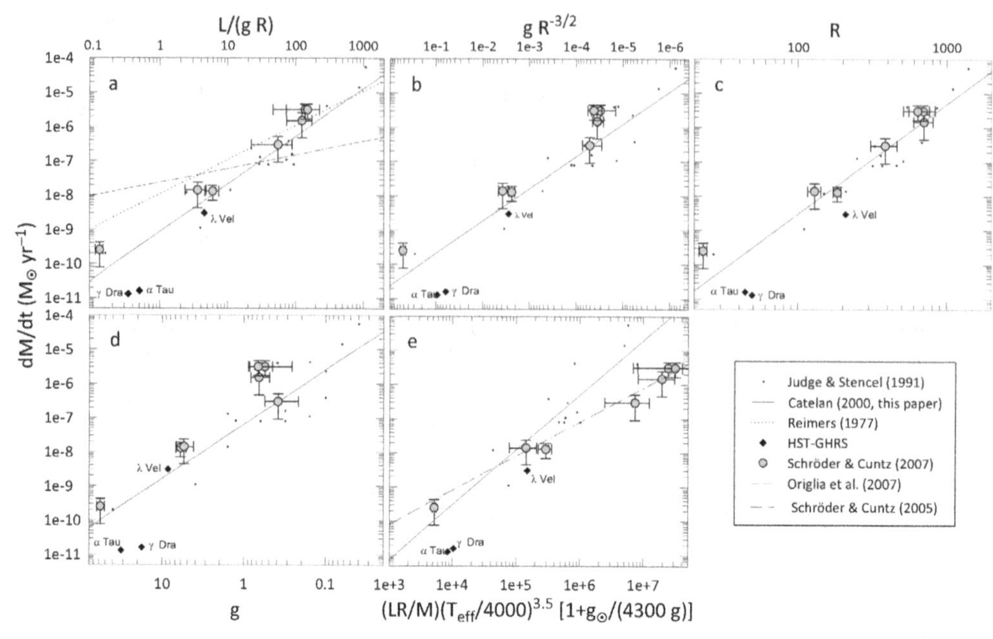

Figure 1. Comparison between different mass-loss recipes and the empirical data. The small crosses represent data from Judge & Stencel (1991), selected according to Catelan (2000). Filled diamonds correspond to the HST-GHRS results by Robinson *et al.* (1998) and Mullan *et al.* (1998). Gray symbols with error bars represent data from the recent compilation by Schröder & Cuntz (2007). In all panels, the *solid lines* show the fits derived by Catelan (2000) from the Judge & Stencel data (using the different combinations of physical parameters indicated in the x-axis of each plot as the independent variable), except for panel e, where the fit is presented here for the first time. In panel a, the *dotted line* represents the predicted mass loss rates according to the Reimers (1977) formula, whereas the *dashed line* indicates the predicted mass-loss rates according to Origlia *et al.* (2007). In panel e, the *dash-dotted line* indicates the mass-loss rates predicted by the Schröder & Cuntz (2005) formula.

become redder with increasing Galactocentric distance – which was interpreted as an age effect, with more distant clusters being younger on average, and having possibly been accreted from "protogalactic fragments" of external origin over the Galaxy's lifetime – many other second parameter candidates have also surfaced over the years. These include, among others, cluster concentration, total mass, and ellipticity; stellar rotation; magnetic fields; planetary systems; and mass loss on the RGB (see Catelan 2008, for extensive references).

While it seems clear now that age does play an important role, it has also become evident that it is not the only parameter involved. Indeed, the presence of bimodal HB's in such GC's as NGC 2808 have long pointed to the need for other second parameters in addition to age (e.g., Rood *et al.* 1993). Recent, deep CMD studies have revealed that some of the most massive globulars, NGC 2808 included, present a surprisingly complex history of star formation, with the presence of extreme levels of helium enhancement among at least some of their stars (e.g., Norris 2004; D'Antona *et al.* 2005; Piotto *et al.* 2005, 2007). As noted by these authors, such He enhancement would provide a natural explanation for the presence of hot HB stars in these clusters. High helium abundances also appear to provide a natural explanation for some of the peculiarities observed in the CMDs and RR Lyrae properties in the globular clusters NGC 6388 and NGC 6441 (e.g., Catelan *et al.* 2006; Caloi & D'Antona 2007, and references therein).

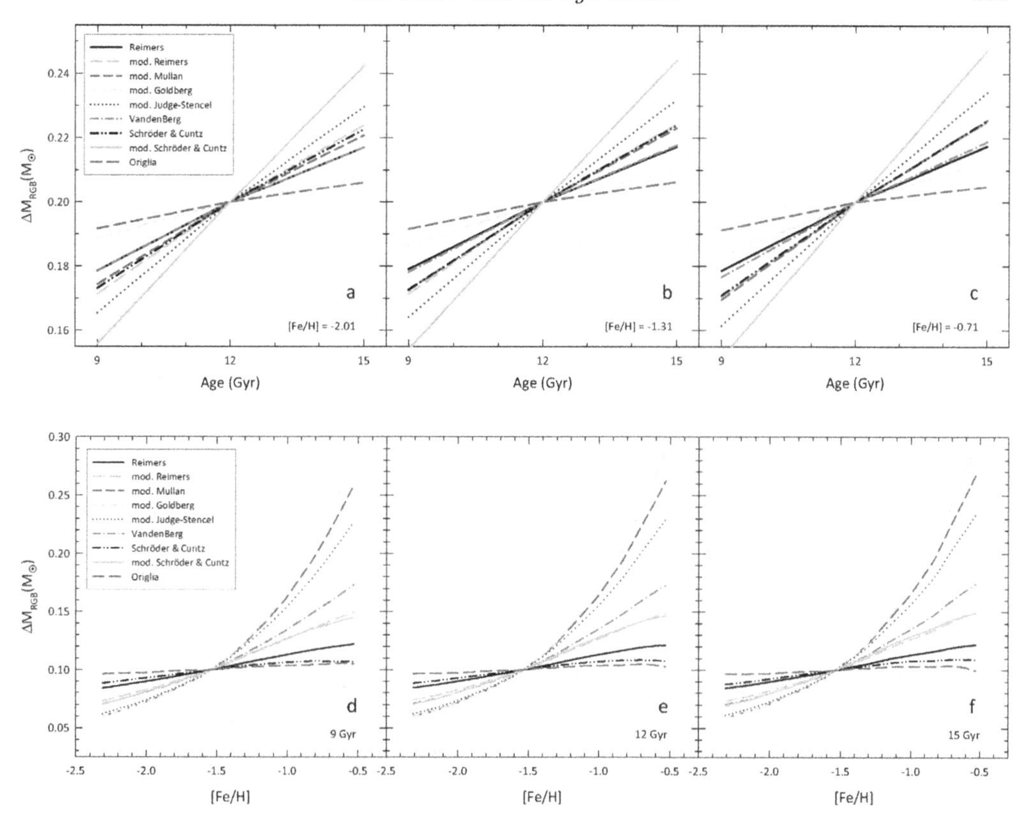

Figure 2. Dependence of the integrated RGB mass loss upon age for fixed metallicity (panel *a*: [Fe/H] = −2.01; panel *b*: [Fe/H] = −1.31; panel *c*: [Fe/H] = −0.71) and upon metallicity for a fixed age (panel *d*: 9 Gyr; panel *e*: 12 Gyr; panel *f*: 15 Gyr), for the different mass loss recipes indicated. In panels *a* through *c*, the total mass loss has been normalized to a value of $0.20\,M_\odot$ at 12 Gyr; in panels *d* through *f*, in turn, the integrated mass loss has been normalized to a value of $0.10\,M_\odot$ at [Fe/H] = −1.54.

In the next few sections, we will address some empirical constraints that may be posed on some of these second parameter candidates.

2. Mass loss in red giants

In order to reliably predict the temperature of an HB star of a given composition and age, we must know how much mass it loses on the RGB. Unfortunately, our knowledge of RGB mass loss remains rather limited. Most studies adopt the Reimers (1977) mass-loss formula to predict the integrated mass loss along the RGB. However, recent evidence indicates that the Reimers formula is not a reliable description of RGB mass loss (e.g., Catelan 2000, 2008; Schröder & Cuntz 2005, 2007). In addition, there are several alternative mass-loss formulations which may better describe the available data. We illustrate this point by comparing, in Figure 1, some empirical mass-loss rates with the predicted rates from several of these alternative mass-loss formulae (see Catelan 2000, for extensive references). While the Reimers formula is clearly inconsistent with the data, the empirical data cannot conclusively distinguish among these alternative formulations.†

† Note that, while the Origlia *et al.* (2007) mass-loss formula is inconsistent with the plotted data, this should not be taken as evidence against its validity, since this formula has been

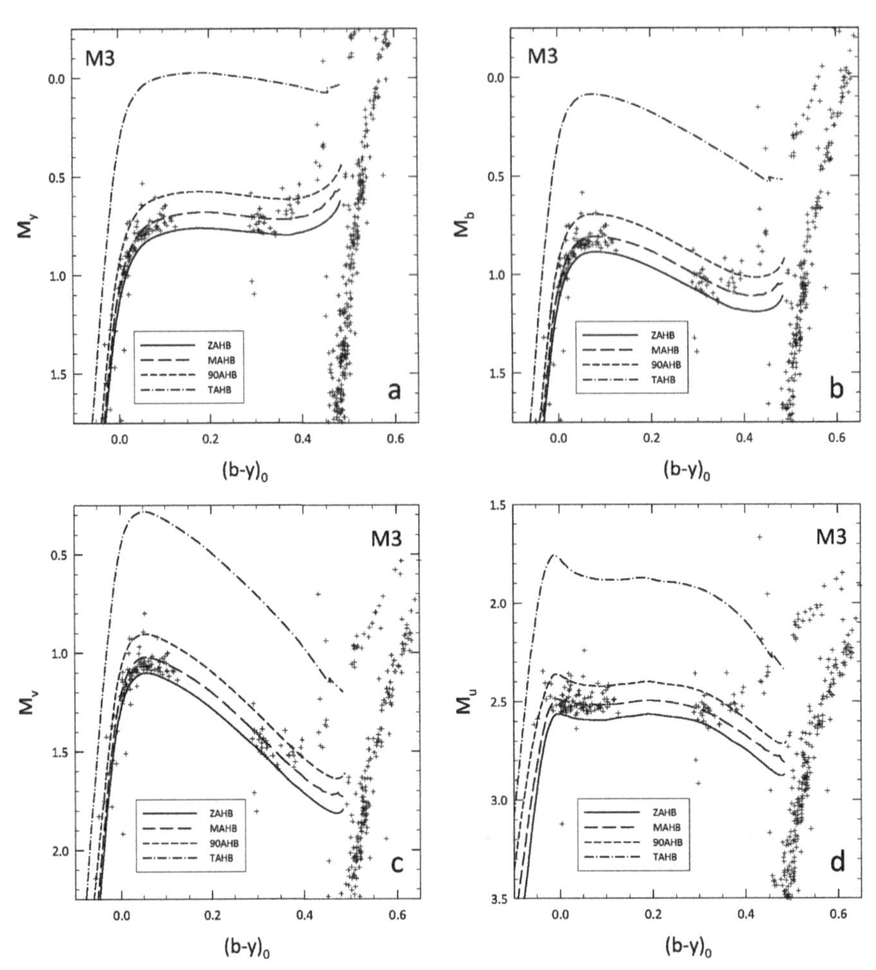

Figure 3. Comparison between Strömgren (1963) photometry for M3, as derived by Grundahl *et al.* (1998, 1999), with the predictions of theoretical models for $Y_{\rm MS} = 0.23$, $Z = 0.002$. The observed data were shifted vertically so as to lead to a good match with the theoretical ZAHB at the red HB.

The serious problem which this uncertainty in the RGB mass loss poses for our understanding of HB morphology and its dependence on age and metallicity is apparent from Figure 2, where the integrated RGB mass loss is plotted as a function of the age for fixed metallicity (panels *a* through *c*) and as a function of metallicity for fixed age (panels *d* through *f*). More specifically, we know that only a very mild ΔM – [Fe/H] dependence can account for the observed relation between HB type and [Fe/H] without resorting to a significant age-metallicity relation (see, e.g., Fig. 1a in Lee *et al.* 1994). In this sense, we find that the Origlia *et al.* (2007) and Schröder & Cuntz (2005) mass-loss formulae lead to the weakest ΔM – [Fe/H] dependence. All other formulae that we have tested lead to steeper dependencies between ΔM and [Fe/H] than the Reimers (1977) relation, thus implying steeper dependencies between age and metallicity as well. The precise dependence between ΔM and metallicity is also important in terms of explaining

suggested to apply exclusively to low-metallicity stars. Still, some caveats regarding the Origlia *et al.* study have recently been raised (see Boyer *et al.* 2008).

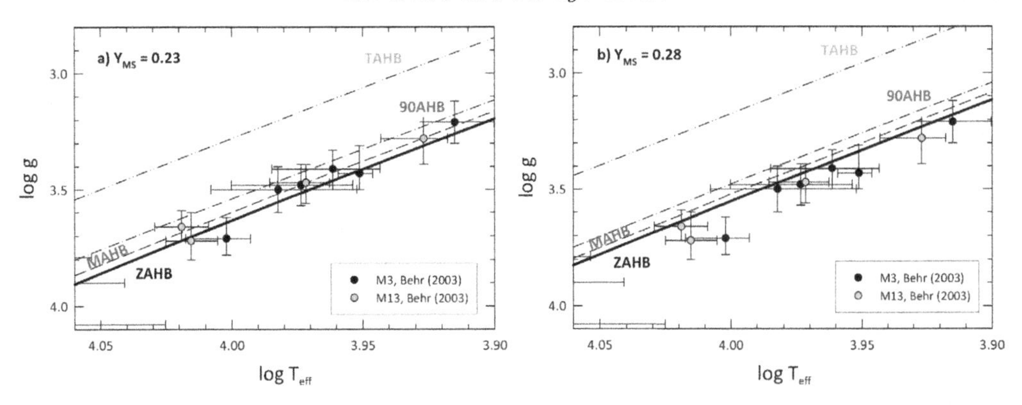

Figure 4. Comparison between spectroscopically derived gravities for blue HB stars in M3 and M13 (from Behr 2003) with the same theoretical models as before, but for two different $Y_{\rm MS}$ values: 23% (panel *a*) and 28% (panel *b*).

the ultraviolet-upturn phenomenon of elliptical galaxies and spiral bulges (see Catelan 2007, for a recent review).

Similarly, a stronger dependence between ΔM and age at fixed [Fe/H] makes it easier to account for a given second-parameter pair in terms solely of an age difference. According to Figure 2, the equation that is most successful in this regard is a modified version of the Schröder & Cuntz (2005) formula, in which the adopted power-law exponents are obtained by a least-squares fit to the Judge & Stencel (1991) data, selected as in Catelan (2000) (see Fig. 1*e*). While not the steepest, the original Schröder & Cuntz formula provides a stronger dependence between ΔM and age than does the Reimers (1977) formula, which should reduce the required age difference between second-parameter pairs. By contrast, the Origlia *et al.* (2007) equation shows the weakest dependence, with a remarkably constant integrated ΔM value over a wide range in ages.

3. Helium enrichment in globular clusters

As previously noted (§1), high levels of helium enrichment have been detected among some of the most massive Galactic GC's. Very recently, it has been suggested that such helium enhancements are in fact not the exception, but indeed the rule, among Galactic GC's (D'Antona & Caloi 2008). Here we provide a first test of this scenario, in the case of the GC's M3 and NGC 1851.

3.1. *The case of M3*

D'Antona & Caloi (2008) and Caloi & D'Antona (2008) claim that the blue HB component in M3 owes its origin to a moderate level of He enhancement in the cluster, between 2% and 6%. Is this supported by the available data?

To answer this question, we compare, in Figure 3, canonical theoretical predictions from Catelan *et al.* (1998) and Sweigart & Catelan (1998), for a helium abundance of $Y_{\rm MS} = 0.23$ and a metallicity $Z = 0.002$, with high-precision photometry in the Strömgren (1963) system, from Grundahl *et al.* (1998, 1999). The empirical data were corrected for reddening following Harris (1996). In these plots, the lines represent different fiducial loci, as follows: ZAHB, middle-age HB (MAHB, or average locus occupied by the HB stars), 90%-age HB (90AHB, or locus below which one should expect to find about 90% of all HB stars), and terminal-age HB (TAHB, or He exhaustion locus). Except for a discrepancy between the predicted and observed numbers of highly evolved stars (both on the blue

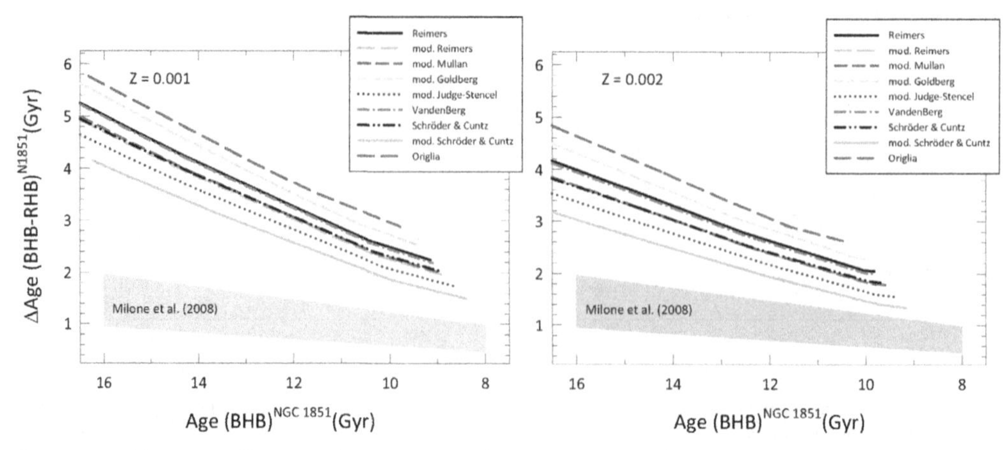

Figure 5. The age difference that is required to account for the difference in HB morphology between the blue and red components of NGC 1851's HB (*lines*, based on the different mass-loss formulae discussed in §2) is compared with the age difference that is estimated from the observed split on the SGB (*gray band*).

and red HB sides of the distribution) that was previously noted by Valcarce & Catelan (2008), one finds remarkable agreement between the model predictions for a constant Y and the observations. Such an agreement is also confirmed by the spectroscopic data from Behr (2003), as can be seen from Figure 4, where we limit the comparison to temperatures lower than 11,500 K due to the well-known complications brought about by the "Grundahl jump" phenomenon (Grundahl *et al.* 1999). Interestingly, this plot also appears to support a similar helium abundance between M3's blue HB stars and the redder blue HB stars in M13.

3.2. *The case of NGC 1851*

Milone *et al.* (2008) have recently discovered that the subgiant branch (SGB) of NGC 1851 is actually split into two separate components, which may be linked to the cluster's well-known bimodal HB morphology. The most straightforward explanation for this split would be a difference in age by 1.0 ± 0.4 Gyr. However, as shown in Figure 5, this is inconsistent with the age difference that would be required to fully account for the separation between the blue and red HB components of the cluster, irrespective of the mass-loss formula (§2) used. A difference in metallicity between the two components is also ruled out by recent spectroscopic data (Yong & Grundahl 2008). Here we apply the same CMD test as in the previous section to constrain the possibility of a difference in Y between the two components as being responsible for the well-known bimodal nature of the cluster's HB.

The result is shown in Figure 6. While the quality of the data is not as high as in the case of M3, one is still able to derive some general conclusions. First, the same theoretical ZAHB does appear to provide a reasonable description of the lower boundary of the data, both for the red and blue HB components – which suggests that at least some of the stars on the blue HB have the same Y as do the red HB stars. Second, there is a predominance of overluminous stars on the blue HB, at colors around $(b-y)_0 \approx 0.05 - 0.15$. While these might in principle be interpreted in terms of a moderate level of helium enrichment, perhaps of the order of 3% – 4% on average (see Fig. 7), the more straightforward explanation is that these stars actually represent the well-evolved progeny of the blue ZAHB stars that are found at higher temperatures. If so, this would again suggest that

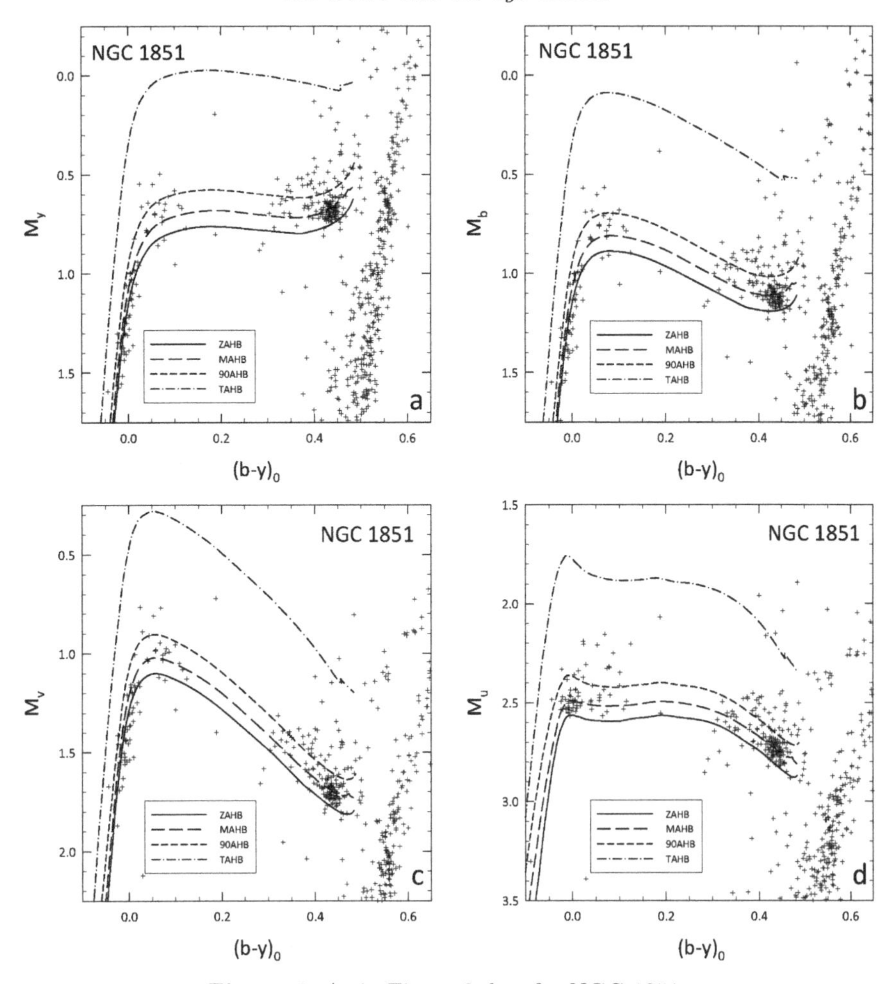

Figure 6. As in Figure 3, but for NGC 1851.

the late stages of HB evolution are somehow significantly underestimated by present-day HB tracks, similar to what was previously found elsewhere (Catelan *et al.* 2001; Valcarce & Catelan 2008, and references therein).

Clearly, more work is needed before we are able to conclusively establish the nature of NGC 1851's bimodal HB and SGB (see also Cassisi *et al.* 2008; Salaris *et al.* 2008).

4. The Oosterhoff dichotomy and the formation of the Milky Way

Irrespective of our ability to properly model HB stars, we can use RR Lyrae stars to derive entirely empirical constraints on the process of formation of the Milky Way. In the Searle & Zinn (1978) scenario, much like in modern ΛCDM cosmology, one expects galaxies such as our own to have formed by the accretion of "protogalactic fragments" that may have resembled the early counterparts of the Milky Way's present-day dwarf satellite galaxies. Useful constraints on recent accretion events may be posed by the presence of younger stellar populations in several of these galaxies (Unavane *et al.* 1996). Still, in order to probe what happened *very early on*, we must look at the ancient components – and RR Lyrae stars are especially useful in that regard (e.g., Catelan 2008).

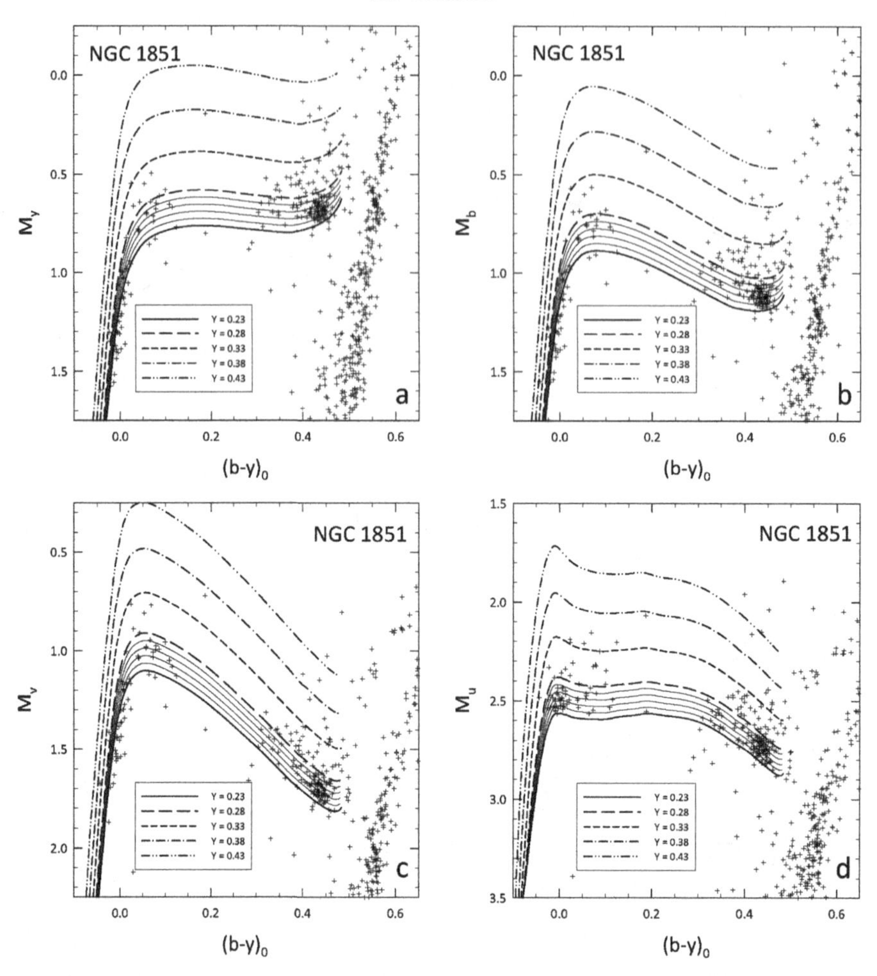

Figure 7. As in Figure 6, but showing ZAHB's for the several different indicated $Y_{\rm MS}$ values. Interpolated ZAHB's are also plotted between the 23% and 28% loci, at intervals of 1%.

Are the ancient populations in the Milky Way's dwarf satellites, as traced by their RR Lyrae pulsators, consistent with the Galactic spheroid having been built therefrom? The answer is provided in Figure 8, where we compare the average properties of the fundamental-mode (ab-type) RR Lyrae stars in Galactic (*left panel*) vs. nearby extragalactic (*right panel*) GC's and field populations. While the Galactic distribution clearly presents the so-called *Oosterhoff dichotomy*, with a tendency for systems to clump around the "Oosterhoff I" (OoI) and "Oosterhoff II" (OoII) regions (see also Miceli *et al.* 2008, for the case of halo field stars), the opposite happens in the case of nearby extragalactic systems, which tend to be preferentially *Oosterhoff-intermediate*. This strongly suggests that the oldest components of the Galaxy cannot have been formed by accretion of even the early counterparts of its present-day dwarf galaxy satellites.

As indicated in Figure 8, at least one of the newly discovered SDSS dwarf galaxies (e.g., Belokurov *et al.* 2006, 2007), CVn I, is Oosterhoff intermediate (Kuehn *et al.* 2008), whereas the Bootes dwarf is OoII (Dall'Ora *et al.* 2006; Siegel 2006). Unfortunately, some of the low-mass SDSS galaxies seem to harbor a mere one or two RR Lyrae stars, which makes it more difficult to assign a conclusive Oosterhoff status to them. Indeed, due to

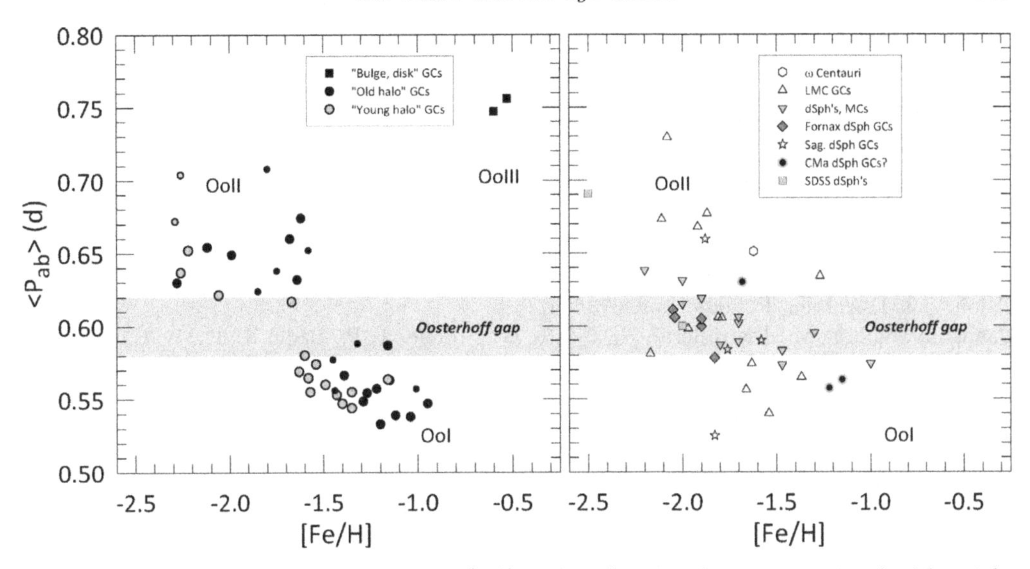

Figure 8. Distribution of Galactic GC's (*left*) and stellar populations associated with neighboring dwarf galaxies (*right*) in the average ab-type RR Lyrae period $\langle P_{\rm ab} \rangle$ vs. [Fe/H] plane. Galactic GC's are classified into "bulge/disk," "young halo," and "old halo" subsystems following Mackey & van den Bergh (2005). See Catelan (2008) for further details.

statistical fluctuations, and since the HB lifetime is of order 100 Myr, it is not entirely clear whether the same Oosterhoff types would necessarily be inferred for these galaxies if they were observed, say, a few hundred Myr in the future (which is very little, in terms of Galactic history), when these HB stars will have long left the HB phase, to be replaced by an entirely new generation of HB stars. Still, the present-day properties for several of the RR Lyrae stars that are found in these very low-mass galaxies do appear to be consistent with an OoII status (e.g., Greco *et al.* 2008).

5. Conclusions

• HB stars play a central role in the age debate. Still, before we are able to predict how (ZA)HB temperature changes with age, we must properly describe RGB mass loss.

• In studies of HB morphology, it is not sufficient anymore to analyze solely the "horizontal" HB morphology, meaning the temperature and/or color distribution of HB stars: *"vertical"* HB morphology, or the distribution of HB stars in luminosity *at fixed temperature (or color)*, provides us with unique information to help us constrain theoretical scenarios for the origin and evolution of these stars. Indeed, the available data appears to strongly constrain, if not conclusively rule out, the possibility of significant He enhancements among M3's blue HB stars, while at the same time suggesting that canonical HB models underestimate the duration of the late stages of HB evolution.

• Irrespective of our ability to model them, HB stars – and RR Lyrae in particular – represent invaluable tools to probe into the Milky Way's early formation history.

Acknowledgments

I would like to warmly thank Gisella Clementini, Frank Grundahl, Bob Rood, Horace Smith, Allen Sweigart, and Aldo Valcarce for useful discussions, comments and suggestions. Support for this work is provided by Proyecto Basal PFB-06/2007, by FONDAP Centro de Astrofísica 15010003, by Proyecto FONDECYT #1071002, and by a John Simon Guggenheim Memorial Foundation Fellowship.

References

Arp, H. C. & Johnson, H. L. 1955, *ApJ*, 122, 171

Arp, H. C. & Melbourne, W. G. 1959, *AJ*, 64, 28

Behr, B. B. 2003, *ApJS*, 149, 67

Belokurov, V., *et al.* 2006, *ApJ*, 647, L111

Belokurov, V., *et al.* 2007, *ApJ*, 654, 897

Boyer, M. L., McDonald, I., van Loon, J. Th., Woodward, C. E., Gehrz, R. D., Evans, A., & Dupree, A. K. 2008, *AJ*, 135, 1395

Caloi, V. & D'Antona, F. 2007, *A&A*, 463, 949

Caloi, V. & D'Antona, F. 2008, *ApJ*, 673, 847

Cassisi, S., Salaris, M., Pietrinferni, A., Piotto, G., Milone, A. P., Bedin, L. R., & Anderson, J. 2008, *ApJ*, 672, L115

Catelan, M. 2000, *ApJ*, 531, 826

Catelan, M. 2007, in: New Quests in Stellar Astrophysics II: The Ultraviolet Properties of Evolved Stellar Populations, in press (astro-ph/0708.2445)

Catelan, M. 2008, to appear in *Ap&SS*(astro-ph/0507464)

Catelan, M., Bellazzini, M., Landsman, W. B., Ferraro, F. R., Fusi Pecci, F., & Galleti, S. 2001, *AJ*, 122, 3171

Catelan, M., Borissova, J., Sweigart, A. V., & Spassova, N. 1998, *ApJ*, 494, 265

Catelan, M., Stetson, P. B., Pritzl, B. J., Smith, H. A., Kinemuchi, K., Layden, A. C., Sweigart, A. V., & Rich, R. M. 2006, *ApJ*, 651, L133

Dall'Ora, M., *et al.* 2006, *ApJ*, 653, L109

D'Antona, F., Bellazzini, M., Caloi, V., Fusi Pecci, F., Galleti, S., & Rood, R. T. 2005, *ApJ*, 631, 868

D'Antona, F. & Caloi, V. 2008, *MNRAS*, 390, 693

Greco, C., Dall'Ora, M., Clementini, G., *et al.* 2008, *ApJ*, 675, L73

Grundahl, F., Catelan, M., Landsman, W. B., Stetson, P. B., & Andersen, M. I. 1999, *ApJ*, 524, 242

Grundahl, F., VandenBerg, D. A., & Andersen, M. I. 1998, *ApJ*, 500, L179

Harris, W. E. 1996, *AJ*, 112, 1487

Judge, P. G. & Stencel, R. E. 1991, *ApJ*, 371, 357

Kuehn, C., Kinemuchi, K., Ripepi, V., *et al.* 2008, *ApJ*, 674, L81

Lee, Y.-W., Demarque, P., & Zinn, R. 1994, *ApJ*, 423, 248

Mackey, A. D. & van den Bergh, S. 2005, *MNRAS*, 360, 631

Miceli, A., Rest, A., Stubbs, C. W., *et al.* 2008, *ApJ*, 678, 865

Milone, A. P., Bedin, L. R., Piotto, G., & Anderson, J. 2008, *A&A*, in press (astro-ph/0810.2558)

Mullan, D. J., Carpenter, K. G., & Robinson, R. D. 1998, *ApJ*, 495, 927

Muñoz, R. R., Carlin, J. L., Frinchaboy, P. M., Nidever, D. L., Majewski, S. R., & Patterson, R. J. 2006, *ApJ*, 650, L51

Norris, J. E. 2004, *ApJ*, 612, L25

Origlia, L, Rood, R. T., Fabbri, S., Ferraro, F. R., Fusi Pecci, F., & Rich, R. M. 2007, *ApJ*, 667, L85

Piotto, G., Bedin, L. R., Anderson, J., *et al.* 2007, *ApJ*, 661, L53

Piotto, G., Villanova, S., Bedin, L. R., *et al.* 2005, *ApJ*, 621, 777

Reimers, D. 1977, *A&A*, 57, 395

Robinson, R. D., Carpenter, K. G., & Brown, A. 1998, *ApJ*, 503, 396

Rood, R. T. 1973, *ApJ*, 184, 815

Rood, R. T., Crocker, D. A., Fusi Pecci, F., & Buonanno, R. 1993, in: G. H. Smith & J. P. Brodie (eds.), Proc. ASP Conf. Ser., Vol. 48, The Globular Cluster-Galaxy Connection, (San Francisco: ASP), p. 218

Salaris, M., Cassisi, S., & Pietrinferni, A. 2008, *ApJ*, 678, L25

Sandage, A. R. 1953, *AJ*, 58, 61

Sandage, A. & Wallerstein, G. 1960, *ApJ*, 131, 598

Schröder, K.-P. & Cuntz, M. 2005, *ApJ*, 630, L73

Schröder, K.-P. & Cuntz, M. 2007, *A&A*, 465, 593

Searle, L. & Zinn, R. 1978, *ApJ*, 225, 357
Siegel, M. H. 2006, *ApJ*, 649, L83
Strömgren, B. 1963, *QJRAS*, 4, 8
Sweigart, A. V. & Catelan, M. 1998, *ApJ*, 501, L63
Unavane, M., Wyse, R. F. G., & Gilmore, G. 1996, *MNRAS*, 278, 727
Valcarce, A. A. R. & Catelan, M. 2008, *A&A*, 487, 185
Yong, D. & Grundahl, F. 2008, *ApJ*, 672, L29

Discussion

A. DOTTER: I would like to suggest that the information required to determine mass loss is not provided in canonical stellar models (R, L, M, $\log g$), and so we do not know, for instance, if it has a dependence on pulsation.

M. CATELAN: I agree: any simple analytical fit, such as the ones I have discussed in my talk, must then reflect averages over sufficiently long timespans, covering several pulsation periods. If so, such formulae may still provide useful descriptions of the integrated mass-loss amounts in the course of their evolution.

C. GALLART: How young do you think an RR Lyrae star can actually be?

M. CATELAN: It has been traditionally thought that the youngest object harboring RR Lyrae stars is NGC 121 in the SMC, with an age around 10 Gyr (e.g., Glatt *et al.* 2008, AJ, 135, 1106, and references therein). RR Lyrae stars (much) younger than this must be exceedingly rare, although it is possible, in principle, for younger stars to become RR Lyrae stars, provided they meet the associated mass-loss requirements. Even the Sun could in principle become an RR Lyrae star, but again it is extremely unlikely that it will lose sufficient mass for this to happen.

A. DUPREE: Is there any connection between rotation or binarity on the HB that could affect the distribution of stars on the horizontal branch?

M. CATELAN: As a matter of fact, the pattern of rotation is also dramatically affected by the Grundahl jump phenomenon, with some stars with temperatures below 11,500 K showing considerable rotation, whereas all stars with higher temperatures basically show no rotation. To me the most likely explanation for this is the one put forward by Allen Sweigart, whereby a stellar wind that is triggered by the onset of radiative levitation at $T > 11,500$ K carries away angular momentum and spins down those stars. As to binarity, we have recently shown (Moni Bidin *et al.* 2008, A&A, 480, L1) that the fraction of close binaries among the EHB stars in NGC 6752 at least is remarkably low, unlike what happens among field sdB stars. Note also that Davis *et al.* (2008, AJ, 135, 2155) have recently presented some intriguing evidence that the initial binary fraction in globulars may be very low as well.

G. PIOTTO: I basically agree with your conclusions about the He enhancement. Still, we know that all clusters that have been looked at show a Na-O anticorrelation and that it is present down to the turn-off, implying that the entire star is formed with material created through hot hydrogen burning. Don't you think this material should also be enhanced?

M. CATELAN: As you may recall, I was in fact one of the first to call attention to the fact that the globulars with the strongest abundance anomalies, as traced by the presence of super- oxygen-poor stars, tended to have the bluest HBs (see Catelan & de Freitas Pacheco 1995, A&A, 297, 345). So perhaps these clusters with the strongest abundance anomalies do indeed harbor populations with strongly enhanced helium, but it is also conceivable that globulars with lesser anomalies have greatly reduced levels of helium enhancement, if any. In addition, it is possible that the abundance anomalies in the RGB atmospheres may lead to enhanced mass-loss rates, and that this may help drive bluer HBs.

R. WYSE: A comment and a question. In Unavane *et al.* we did state explicitly that we were constraining the assembly of the stellar halo after the formation of intermediate-age stars in dwarf spheroidals after redshift $\sim$1. My question is, what are the prospects for gaining sufficient physical understanding to decide among the various mass-loss laws you showed?

M. CATELAN: In regard to your comment, I fully agree. As to your question, unfortunately progress has been quite slow, though the recent studies by Schröder & Cuntz and by Origlia *et al.* that I mentioned in my talk should give one some reason to be hopeful that there will be some breakthroughs in the not-too-distant future. Perhaps Andrea Dupree could add some comments in this respect.

A. SARAJEDINI: First, I have a comment. We should stop using NGC 1851 as a "bridge" between NGC 288 and NGC 362 because NGC 1851 is obviously a complicated stellar system and cannot be considered analogous to a typical globular cluster. I suggest we use NGC 1261 instead. My question is: What observations do we need to constrain mass loss in the age and abundance regimes of the Galactic GCs?

M. CATELAN: I agree that NGC 1261 could be a possible alternative to NGC 1851, as far as the bridge method is concerned, although the latter cluster can still be used, provided due account is taken of possible systematic effects and an increased error budget. As to your question, I defer it to the true expert in the audience, Andrea Dupree. But you may also want to check the recent paper by Origlia *et al.* that I mentioned in my talk for a recent effort to derive a mass-loss formula that is specifically aimed at describing low-metallicity red giants in GCs.

M. TOSI: Concerning the Oosterhoff dichotomy and the newly discovered SDSS satellites, I understand that the lowest-mass ones do avoid the Oosterhoff gap. Do you think these types of satellites could be Galaxy building blocks?

M. CATELAN: It is not clear to me whether an Oosterhoff type can be conclusively assigned to a system with just one or two of RR Lyrae stars, so I am afraid I cannot provide you with a conclusive answer to your question.

P. STETSON: I just wanted to add quickly that it has been known for decades that the binary fraction among RR Lyraes, both in clusters and in the field, is effectively zero. You would see them in the timing of the light curves.

M. CATELAN: Indeed, such systems are exceedingly rare, although a number of RR Lyrae stars in binary systems were discovered by the OGLE team in the LMC (Soszynski *et al.* 2003, AcA, 53, 93).

The Ages of Stars
Proceedings IAU Symposium No. 258, 2008
E.E. Mamajek, D.R. Soderblom & R.F.G. Wyse, eds.

doi:10.1017/S1743921309031871

Relative and absolute ages of Galactic globular clusters

Ata Sarajedini

Department of Astronomy, University of Florida, Gainesville, FL USA
email: ata@astro.ufl.edu

Abstract. We present a review of the latest work concerned with the relative and absolute ages of the Galactic globular clusters (GCs). Relative age-dating techniques generally divide into two types - those that measure a magnitude difference between two features in the color-magnitude diagram (i.e. Vertical Methods) and those that rely on color differences in the color-magnitude diagram (i.e. Horizontal Methods). Both types of diagnostics have been successfully applied and generally reach the same conclusions. Galactic GCs exhibit a mean age range of ~3 Gyr, smaller (or nonexistent) for metal-poor clusters and larger (as much as 6 Gyr) for metal-rich ones. Generally speaking, the inner-halo GCs are older and more uniform in age as compared with those outside of the solar circle. Furthermore, the tendency of GCs with predominantly red horizontal branches (HBs) located in the outer halo to be preferentially younger than those with bluer HBs closer to the Galactic center suggests that age is the second parameter which, in addition to metal abundance, controls the HB morphology. In particular, we present additional compelling evidence supporting this assertion using a detailed examination of new photometry for the classic second-parameter cluster pair NGC 288 and NGC 362. Moving on to the absolute ages, we note that the absolute ages of the most metal-poor Galactic GCs sets a lower limit on the age of the Universe. The preferred age indicator for absolute ages is the luminosity of the main-sequence turnoff because most theoretical models agree on the onset of hydrogen exhaustion in the cores of low-mass stars. Based on the technique of main-sequence fitting to field subdwarfs with *Hipparcos* parallaxes, we find an age of $11.6^{+1.4}_{-1.1}$ Gyr for four metal-poor GCs with deep color-magnitude diagrams on a consistent photometric scale; this age is consistent with the results of a number of previous investigations.

Keywords. Hertzsprung-Russell diagram, stars: Population II, Galaxy: formation, globular clusters: general, globular clusters: individual (NGC 288, NGC 362, NGC 5466, NGC 6341 (M92), NGC 5053, NGC 5024 (M53)), Galaxy: halo,

---

1. Introduction

Authors often state that the Galactic globular clusters (GCs) represent the 'fossil record' of the formation process that created the Milky Way halo. One could argue that this analogy is flawed because, while fossils represent plants and animals that are now dead, the GCs are still alive and continuing to evolve. However, it is important to realize that the Galactic GCs (and the metal-poor field stars) are the only signatures we have remaining of the early epochs of star formation in the Milky Way. Certainly, in this sense, the 'fossil record' analogy is a valid one to employ.

What can we learn by studying the ages of the Galactic GCs? The relative ages and their behavior with chemical composition, Galactocentric distance, and horizontal branch (HB) morphology provide insight into the formation chronology of the Milky Way's halo. The absolute ages of the oldest globular clusters sets a lower limit on the age of the Universe. This therefore provides another constraint for the process of galaxy formation in the early Universe and implications for the broader questions of cosmology.

2. Relative ages

Techniques to estimate relative ages fall into two broad categories - those that rely on the measurement of magnitude differences in the color-magnitude diagram (CMD), also known as Vertical Methods (e.g. ΔV Method), and those that use color differences, similarly known as Horizontal Methods (e.g. Δ(B–V) Method). The Vertical method, which dates back to the work of Sandage (1981) and Iben & Renzini (1984), typically involves the measurement of the horizontal branch (HB) level, which can include the mean magnitude of the RR Lyrae variables, combined with the magnitude of the main sequence turnoff (MSTO). While this method is largely sound within a theoretical framework (see Sec. 3), problems occur in the identification of the MSTO's precise location given its vertical morphology. This inherent flaw can lead to an uncertainty of ~0.1 mag in the level of the MSTO which translates to an uncertainty of ~1 Gyr in the age of a cluster. In contrast, the horizontal method (Sarajedini & Demarque 1990; VandenBerg *et al.* 1990), which relies on the color difference between the MSTO and the red giant branch (RGB) is relatively straightforward to measure but suffers from theoretical uncertainties that arise from our limited knowledge of the model T_{eff} scale and the color-T_{eff} relation for stellar isochrones.

There are a number of other relative-age methods. Among these, the most common are Isochrone Fitting (VandenBerg 2000) and Main Sequence Fitting (Marin-Franch *et al.* 2009), both of which rely on estimating the distance to a cluster and then using the absolute magnitude of the MSTO to obtain an age. This process is more commonly used to infer absolute ages (Sec. 3) but it has seen limited application to relative ages as well. Other techniques that have seen limited use include one originally proposed by Paczynski (1984) and later advocated by Jimenez & Padoan (1996). This method involves matching theoretical luminosity functions (LF) to observed ones in order to infer an age. Jimenez & Padoan (1996) claimed to be able to measure ages to a precision of ~0.2 Gyr. Another method, pioneered by Bergbusch & VandenBerg (1997), but that has not seen wide acceptance, is based on the construction of color functions in the magnitude range of the MSTO, subgiant branch (SGB), and lower RGB. The shape of the color function, which is the color histogram from the turnoff to the lower RGB, is sensitive to age and is used by comparisons with theoretical color-functions of known age and composition. One additional method that deserves comment is represented by the work of Sarajedini & Mighell (1996). They proposed the use of the magnitude difference between the unevolved main sequence and the subgiant branch. This method has multiple advantages such as ease of measurement and insensitivity to the metal abundance of the clusters being considered. However, it does require precise photometry that reaches ~2 magnitudes below the MSTO and a knowledge of the relative reddenings of the clusters to better than $\pm$ 0.02 mag in E(B–V). Furthermore, it should be noted that all of these relative age methods can be combined in order to minimize their uncertainties and maximize their age sensitivity. Meissner & Weiss (2006) provide a detailed discussion of this topic.

Since the pioneering work of VandenBerg *et al.* (1990), it has become increasingly clear that the metal-poor globular clusters (i.e. [Fe/H]$\lesssim$–1.5) have the same age to within ~1 Gyr. Furthermore, as metallicity increases, so does the age dispersion of the clusters. This trend has been confirmed by the results of Chaboyer *et al.* (1996), Rosenberg *et al.* (1999), Salaris & Weiss (2002), De Angeli *et al.* (2005), and Marin-Franch *et al.* (2009). This last study is based on imaging with the Advanced Camera for Surveys onboard the Hubble Space Telescope (HST) as part of the Galactic Globular Cluster Treasury project (GO-10775, hereafter referred to as GGCTP, see Sarajedini *et al.* 2007 for more details). The work of Marin-Franch *et al.* (2009) reinforces the results from many of these same papers with regard to the relation between age and Galactocentric distance (R_{GC}). In particular, it seems that old clusters dominate near the Galaxy's center with the age dispersion increasing for those in the middle and outer halo. This behavior favors the 'inside-out' formation scenario of the Milky Way halo (Kepner 1999, and references therein). In addition, a relation between age and metal abundance, with younger clusters being more metal-rich, is present for GCs that appear to have originated in disrupted dwarf satellites of the Galaxy. For example, Terzan 7, Terzan 8, Arp 2, M54, Palomar 12, and NGC 4147 trace their origins to the Sagittarius dwarf spheroidal galaxy and exhibit an age-metallicity relation.

It is very important to note that the vast majority of papers that have examined the global properties of globular cluster ages have *excluded* the outer halo clusters - those beyond ~40 kpc from the Galactic center. These clusters include NGC 7006, Palomar 15, Pyxis, Palomar 3, Palomar 4, Palomar 14, Eridanus, and AM-1. Why are these clusters important in this regard? As noted by several authors, most notably Searle & Zinn (1978) and Lee *et al.* (1994), the so-called 'second parameter effect' is dependent on Galactocentric distance, being most extreme for clusters outside of R_{GC} ~40 kpc. This effect was first identified by Sandage & Wallerstein (1960) and describes the tendency of globular cluster HB morphologies at constant metallicity to become redder at larger values of R_{GC}. This therefore suggests that another parameter in addition to metal abundance is required to describe the HB and that the prominence of this parameter grows with R_{GC}. Therefore, as pointed out by Sarajedini *et al.* (1997), whatever this second parameter is, it must vary in a reasonable fashion on a global spatial scale, being negligible at small R_{GC} and overwhelming at large values of R_{GC}. Given this requirement, the leading second-parameter candidate ever since the seminal paper of Searle & Zinn (1978) has been cluster age. In this scenario, the mean age of the Galactic globular clusters should become younger as R_{GC} increases and the age dispersion should increase. As a result, any survey of Galactic globular cluster ages is incomplete unless the ages of the outer-halo clusters, which preferentially exhibit red HBs, are included.

Concerning the outer-halo clusters with red HBs, reliable relative ages have been published for Palomar 3, Palomar 4, Palomar 14, Eridanus, AM-1, and Pyxis. Stetson *et al.* (1999) used deep HST imaging with the Wide Field Planetary Camera 2 (WFPC2) to construct CMDs for Palomar 3, Palomar 4, and Eridanus. They compared these clusters to ground-based photometry for M5 and M3 and concluded that the outer-halo clusters are some 1.5 to 2 Gyr younger than the comparison clusters. A similar conclusion was reached by Dotter *et al.* (2008) in examining the ages of AM-1 and Palomar 14 as compared with M3 (see also Sarajedini 1997). Sarajedini & Geisler (1996) compared their CMD of Pyxis with that of NGC 362 and found them to be co-eval. Since it has long been established that NGC 362 is 1.5 to 2 Gyr younger than other GCs at its metallicity, it follows that the same is true for Pyxis.

This leads us to our next topic, which is concerned with the ages of the classic second-parameter pair NGC 362 and NGC 288. The former has a completely red HB while the latter has one that is predominantly blue and yet their metallicities are essentially identical to within 0.05 dex (De Angeli *et al.* 2005, and references therein). There is a substantial body of literature supporting the idea that NGC 288 is 2 to 3 Gyr older than NGC 362, precisely as implied by their HB morphologies if age were the second parameter which, in addition to metal abundance, influences the HB morphology (Bolte 1989; VandenBerg *et al.* 1990; Sarajedini & Demarque 1990; Green & Norris 1990; Rosenberg *et al.* 1999). However, a minority of authors (VandenBerg 2000; Stetson 2009) have recently pointed out that this conclusion is no longer on a firm footing.

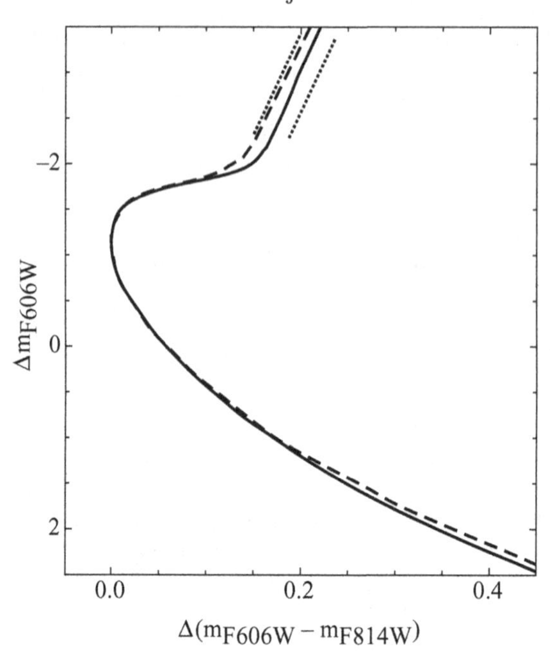

Figure 1. A comparison of the fiducial sequences of NGC 288 (dashed) and NGC 362 (solid line) as derived by Marin-Franch *et al.* (2009). They have been shifted using the prescription of VandenBerg *et al.* (1990) wherein the colors of the turnoffs and the magnitudes +0.05 mag redder than the turnoff are matched. The dotted lines represent age differences of ± 2 Gyr as determined from the isochrones of Dotter *et al.* (2007).

With the advent of high-quality, deep photometry for a large set of clusters obtained as part of the GGCTP and the growing set of well calibrated GC photometry being made available by Stetson (2009), we can revisit the question of the age difference between NGC 288 and NGC 362 with renewed hope that a more definitive answer will emerge. First, we will employ the cluster fiducial sequences of the MS / SGB / lower RGB constructed by Marin-Franch *et al.* (2009) from the GGCTP data. Figure 1 shows these sequences for NGC 288 and NGC 362 registered using the prescription advocated by VanderBerg *et al.* (1990). In this method, which is a version of the Horizontal Method described above, the two fiducials are matched in color at the MSTO and in magnitude at a point +0.05 mag redward of the MSTO. Once shifted and matched in this way, older clusters will have RGBs that are bluer, and lower MSs that are redder, than those of younger clusters. This is precisely what is observed in Fig. 1. The dotted lines in Fig. 1 indicate the locations of RGBs for clusters that are 2 Gyr older (blueward) and younger (redward) than the comparison cluster. Therefore, the comparison shown in Fig. 1 suggests that NGC 288 is ∼2 Gyr older than NGC 362. It should be emphasized that this result does not depend on distance or reddening, and that our value for the age difference agrees with the majority of previous investigators.

A second approach to the age difference between NGC 288 and 362 is illustrated in Fig. 2. We begin in the lower panel where we have adopted the reddenings for these two clusters from the maps of Schlegel *et al.* (1998). Note that NGC 288, being at $b = +89^o$, is very close to the north Galactic pole, making its reddening the smallest of any of the known Galactic GCs. After shifting the cluster fiducials in color to correct for reddening, we then offset them in magnitude in order to match their lower main sequences, under the constraint that their distance moduli differ by 0.03 mag, with NGC 288 having the

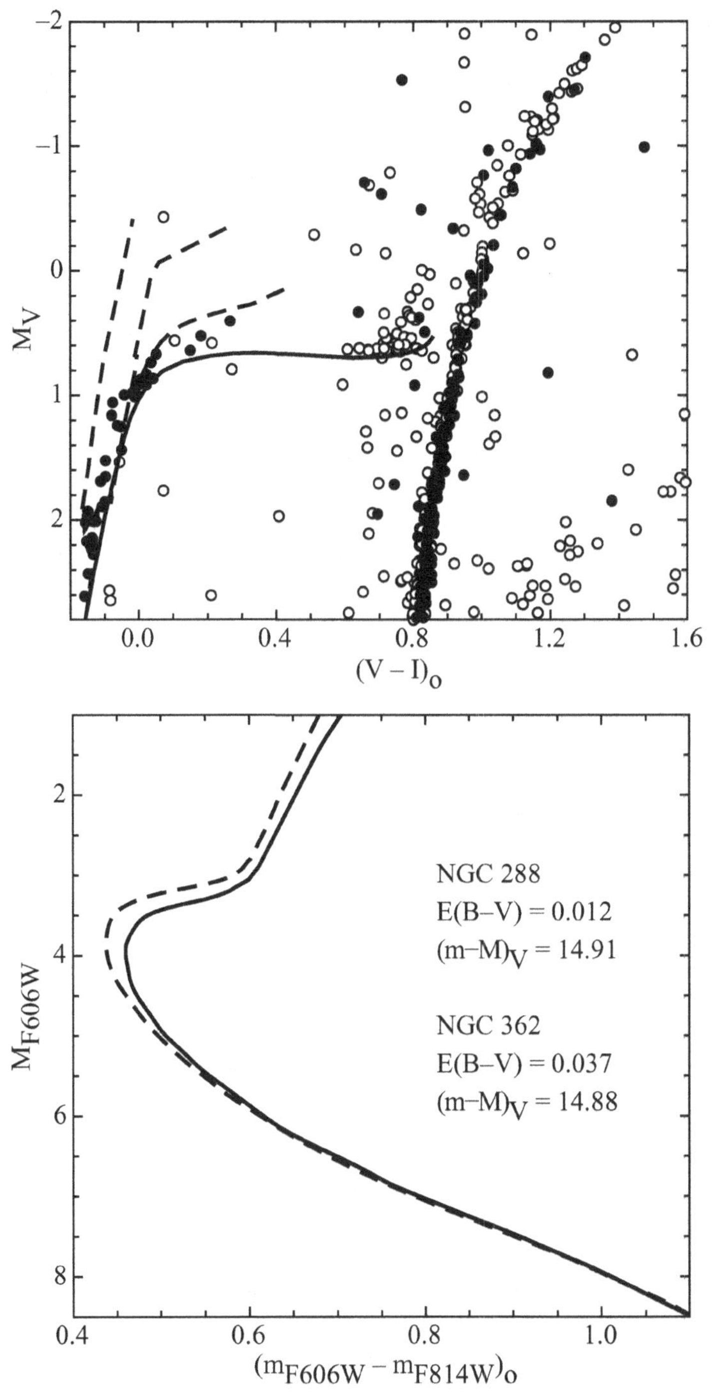

Figure 2. The lower panel shows the fiducial sequences of NGC 288 (solid) and NGC 362 (dashed) as derived by Marin-Franch *et al.* (2009) shifted as described in the text using the reddenings and distance moduli shown in the figure. The upper panel displays the high quality photometry for these two clusters (NGC 288: filled circles, NGC 362: open circles) from Stetson (2009) shifted using the same offsets as in the lower panel. The solid line is the theoretical zero age horizontal branch (ZAHB) from VandenBerg (2000) while the dashed lines are representative HB tracks from Dotter *et al.* (2007) showing the evolutionary paths of low mass HB stars from the ZAHB.

Table 1. Stellar Evolution Models

Name	Nickname	Reference
Padova		Bertelli *et al.* (2008)
Dartmouth	'DSEP'	Dotter *et al.* (2007)
Teramo	'BaSTI'	Cordier *et al.* (2007)
Victoria-Regina	'VandenBerg'	VandenBerg *et al.* (2006)
Yale-Yonsei	'Y^2'	Yi *et al.* (2004)
Geneva		Lejeune & Schaerer (2001)

larger apparent modulus. This constraint is based on the results of Carretta *et al.* (2000) who derived distances for both of these clusters by fitting their fiducials to field subdwarfs with *Hipparcos* parallaxes. Once the fiducials have been matched using this procedure, we see in Fig. 2 that the lower main sequences line up beautifully as they should if the metallicities of the clusters are identical; furthermore, the relative locations of the MSTO / SGB / lower RGB regions suggest, once again, that NGC 288 is older than NGC 362.

To achieve consistency, we must also ensure that the same color and magnitude offsets that yield the comparison seen in the lower panel of Fig. 2 also provide a reasonable match of the brighter sequences in the CMD such as the HB and upper RGB. This comparison is displayed in the upper panel of Fig. 2 wherein the filled circles are NGC 288 and the open circles are NGC 362, both coming from the high-quality photometry of Stetson (2009). The solid line in Fig. 2 is a zero-age horizontal-branch (ZAHB) model from VandenBerg (2000) for [Fe/H] = –1.3 and $[\alpha/\mathrm{Fe}] = +0.3$, showing that the red HB of NGC 362 and the predominantly blue HB of NGC 288 are consistently located relative to each other. Furthermore, the dashed lines are post-ZAHB evolutionary tracks from Dotter *et al.* (2007) shifted to match the VandenBerg (2000) ZAHB locus. These tracks suggest that the HB stars at $M_V \sim +0.6$ and $(V-I)_o \sim 0.2$ are above the ZAHB because they have evolved from the bluer portions of the HB.

It is very important to realize that the morphological differences between NGC 288 and NGC 362 in the region of the MSTO, as illustrated in Figs. 1 and 2, *cannot* be the result of differences in CNO or α-element abundances. For this to be the case, NGC 288 must have a higher CNO or α-element abundance than NGC 362. This would make the morphology of the NGC 288 HB redder than the RR Lyrae instability strip, which is inconsistent with the purely blue HB of NGC 288.

In the past, comparisons such as the one shown in Fig. 2 used an additional cluster, NGC 1851, as a 'bridge' to help clarify the interpretation of such a diagram (e.g. Stetson *et al.* 1996). It was argued that since NGC 1851 has a bi-modal HB with stars on both the blue and red side of the instability strip, it would help identify the expected locations of the blue HB of NGC 288 and the red HB of NGC 362 relative to each other. However, the recent discovery by Milone *et al.* (2008) that NGC 1851 exhibits two subgiant branches throws significant doubt on its utility in comparisons of NGC 288 and NGC 362. The origin of the two SGBs is unclear but it certainly underscores the fact that NGC 1851 is not a 'simple' stellar populaton and therefore disqualifies it from being used as such.

3. Absolute ages

There are three ingredients required in the determination of GC absolute ages. First, investigators have historically used the magnitude of the MSTO as an absolute age indicator because most theoretical models agree on the onset of hydrogen exhaustion in the cores of low-mass stars. High-quality precision photometry, well-calibrated to a standard system, is required to achieve the most reliable determination of the MSTO

Table 2. Metal-Poor Clusters

Cluster	$[Fe/H]_{CG}$	$E(B-V)_{SFD}$
NGC 5466	−2.20	0.017
NGC 6341 (M92)	−2.16	0.022
NGC 5053	−1.98	0.017
NGC 5024 (M53)	−1.86	0.021

magnitude. Next, the apparent magnitude of the MSTO must be converted to an absolute magnitude and this, of course, requires knowledge of the cluster distance. Lastly, the absolute magnitude of the MSTO $[M_V(TO)]$ is converted to an age using theoretical stellar-evolutionary models. A range of models are available from several different groups, and these are listed in Table 1, wherein a representative reference is also given. The procedure outlined above has been followed by a number of investigators with very similar results. The first paper to apply the 'new and improved' *Hipparcos* parallaxes of field subdwarfs to the problem of the absolute ages of GCs was that of Reid (1997). He used 15 of the nearest subdwarfs and fit the fiducial sequences of 7 Galactic GCs to the main sequence defined by these stars. Surprisingly, Reid (1997) concluded that the GCs are closer and therefore younger than previously thought; in particular, he claimed that the age of the most metal-poor GCs is between 11 and 13 Gyr. Chaboyer *et al.* (1998) also present an analysis of the absolute ages of the oldest Galactic GCs but with a new and important wrinkle. They constructed Monte Carlo simulations in which they vary the theoretical inputs into the models over a reasonable range for each parameter. Each set of inputs then results in a specific isochrone, of which there are over 10,000 realizations. They applied these simulations to the ΔV values of 17 metal-poor GCs in order to derive a distribution of ages for these clusters, which yields a mean age of 11.5 ± 1.3 Gyr.

This result was echoed by the work of Carretta *et al.* (2000); however, they discovered that their mean age for the oldest Galactic GCs and the associated distance scale resulted in a distance modulus for the Large Magellanic Cloud that is too large by ~0.1 mag. After adjusting their distance scale to an adopted value of $(m - M)_o = 18.54 \pm 0.07$ for the LMC, they find a mean age of 12.9 ± 1.5 Gyr for the oldest Galactic GCs.

We have performed our own analysis of the absolute ages of the metal-poor GCs, using fiducial sequences from the GGCTP. Table 2 lists the four GCs we have selected, based on their similarly low metallicities (all on the scale of Carretta & Gratton (1997)) and low reddenings, all taken from Schlegel *et al.* (1998). Among these, NGC 5024 (M53) has been selected as a reference cluster because its metal abundance overlaps the range of values for field subdwarfs with well-measured parallaxes from *Hipparcos.* For the other three clusters, we have shifted their $(m_{F606W} - m_{F814W})$ VegaMAG fiducials in the color direction to account for reddening and for the metallicity difference with M53 and then in magnitude to match the unevolved main sequence of M53. The sensitivity of the fiducial color to metal abundance has been established using the isochrones of Dotter *et al.* (2007). For each cluster, this process yields a distance modulus relative to M53. The four independent measures of the MSTO mag are shifted to the distance scale of M53 and averaged together.

The last step involves measuring the distance of M53. We proceeded by first selecting from Table 2 of Carretta *et al.* (2000) the subdwarfs with *Hipparcos* parallaxes and with absolute magnitudes in the range $5 < M_V < 8$. We avoided all suspected binaries, yielding a total of 21 subdwarfs. We compared the *Hipparcos* V–I colors to the B–V values from Carretta *et al.* (2000) and found a tight relation, with three stars appearing as outliers in this plot - HIP 46120, 57939, and 79537. The V–I colors of these stars were offset by −0.07 mag to bring them onto the relation defined by the remaining 18 stars. We then

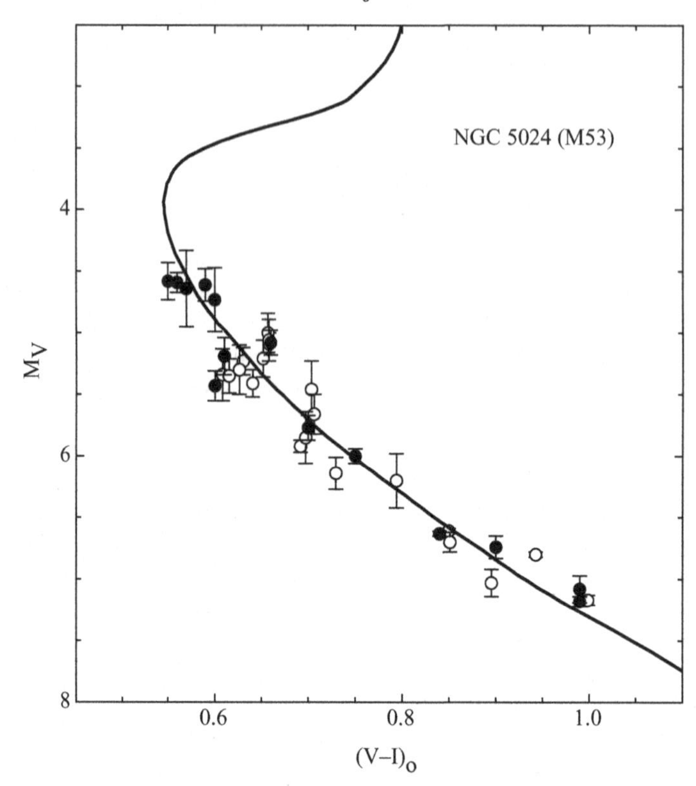

Figure 3. The solid line represents the fiducial sequence of NGC 5024 (M53) from the work of Marin-Franch *et al.* (2009). This has been fitted to the open circles which are field subdwarfs with *Hipparcos* parallaxes as detailed in the text. For comparison, the filled circles are the subdwarfs from the work of Sandquist *et al.* (1999) scaled to a metal abundance of [Fe/H] = –1.91.

compared the V–I colors from *Hipparcos* to those from Sandquist *et al.* (1999). For the seven stars in common between them, we find a constant offset of 0.10 ± 0.01 mag in the sense that the *Hipparcos* V–I colors are too blue. In this way, we have generated a set of 21 single-star subdwarfs with well-determined V–I colors, along with absolute magnitudes, reddenings, and metallicities from Table 2 of Carretta *et al.* (2000). The colors of these stars are adjusted to the metal abundance of M53 ([Fe/H] = –1.86) as shown in Figure 3 using an equation derived from the isochrones of Girardi *et al.* (2000). As a consistency check, the filled circles in Figure 3 are the subdwarfs used by Sanquist *et al.* (1999) at a metal abundance of [Fe/H] = –1.91. The penultimate step involves shifting the M53 V–I fiducial in color, to account for the reddening, and then in magnitude, to fit the subdwarfs in order to determine the distance modulus. We find $(m - M)_V = 16.37 \pm 0.05$ for M53. Applying this to the relative distances of the other three clusters yields a mean MSTO magnitude of $\langle M_V(TO) \rangle = 3.967 \pm 0.064$. The final step then requires us to use this MSTO magnitude along with the metallicity of M53 and equation (3) of Chaboyer *et al.* (1998), which includes the results of their isochrone Monte Carlo simulations, to calculate the mean age of these four metal-poor clusters. This procedure yields an age of $11.2^{+1.4}_{-1.1}$ Gyr. The Chaboyer *et al.* (1998) models include the full effects of Helium diffusion, which spectroscopic evidence suggests is an overestimate of reality (Korn *et al.* 2007). Chaboyer *et al.* (2001) note that ages from isochrones that include full diffusion are about 4% lower than isochrones with diffusion inhibited in the surface layers. As a result, our absolute age for the four GCs in Table 2 turns out to be $11.6^{+1.4}_{-1.1}$ Gyr.

4. Acknowledgments

I would like to thank my collaborators on the HST/ACS Globular Cluster Treasury project (GO-10775) from which many of the results in this contribution originated. Don VandenBerg kindly provided the ZAHB model shown in Fig. 2. I am also grateful to Aaron Dotter and Karen Kinemuchi for a careful reading of this manuscript. Some of this research was funded through Program number GO-10775 provided by NASA through a grant from the Space Telescope Science Institute, which is operated by the Association of Universities for Research in Astronomy, Incorporated, under NASA contract NAS5-26555.

References

Bergbusch, P. A. & VandenBerg, D. A. 1997, *AJ*, 114, 2604
Bolte, M. J. 1989, *AJ*, 97, 1688
Carretta, E. & Gratton, R. G. 1997, *A&AS*, 121, 95
Carretta, E., Gratton, R. G., Clementini, G., & Fusi Pecci, F. 2000, *ApJ*, 533, 215
Chaboyer, B., Demarque, P., & Sarajedini, A. 1996, *ApJ*, 459, 558
Chaboyer, B., Demarque, P., Kernan, P. J., & Krauss, L. M. 1998, *ApJ*, 494, 96
Chaboyer, B., Fenton, W. H., Nelan, J. E., Patnaude, D. J., & Simon, F. E. 2001, *ApJ*, 562, 521
De Angeli, F., *et al.* 2005, *AJ*, 130, 116
Dotter, A., Chaboyer, B., Jevremović, D., Baron, E., Ferguson, J. W., Sarajedini, A., & Anderson, J. 2007, *AJ*, 134, 376
Dotter, A., Sarajedini, A., & Yang, S. C. 2008, *AJ*, 136, 1407
Girardi, L., Bressan, A., Bertelli, G., & Chiosi, C. 2000, *A&AS*, 141, 371
Green, E. M. & Norris, J. E. 1990, *ApJ*, 353, L17
Iben, I. & Renzini, A. 1984, *Phys. Rep.*, 105, 329
Jimenez, R. & Padoan, P. 1996, *ApJ*, 463, L17
Kepner, J. V. 1999 *ApJ*, 117, 2063
Korn, A. J., Grundahl, F., Richard, O., Mashonkina, L., Barklem, P. S., Collet, R., Gustafsson, B., & Piskunov, N. 2007, *ApJ*, 671, 402
Lee, Y. -W., Demarque, P., & Zinn, R. J. 1994, *ApJ*, 423, 248
Marin-Franch, A. *et al.* 2009, *AJ*, submitted
Meissner, F. & Weiss, A. 2006, *A&A*, 456, 1085
Milone, A. *et al.* 2008, *ApJ*, 673, 241
Paczynski, B. 1984, *ApJ*, 284, 670
Reid, I. N. 1997, *AJ*, 114, 161
Rosenberg, A., Saviane, I., Piotto, G., & Aparicio, A 1999 *AJ*, 118, 2306
Salaris, M. & Weiss, A. 2002 *A&A*, 388, 492
Sandage, A. 1981, *ApJS*, 46, 41
Sandage, A. & Wallerstein, G. 1960, *ApJ*, 131, 598
Sarajedini, A. 1997, *AJ*, 113, 682
Sarajedini, A. *et al.* 2007, *AJ*, 133, 1658
Sarajedini, A., Chaboyer, B., & Demarque, P. 1997, *PASP*, 109, 1321
Sarajedini, A. & Demarque, P. 1990, *ApJ*, 365, 219
Sarajedini, A. & Geisler, D. 1996 *AJ*, 112, 2013
Sarajedini, A. & Mighell, K. J. 1996, unpublished
Schlegel, D. J., Finkbeiner, D. P., & Davis, M. 1998, *ApJ*, 500, 525
Searle, L. & Zinn, R. J. 1978, *ApJ*, 225, 357
Stetson, P. B. 2009, This Volume
Stetson, P. B., Vandenberg, D. A., Bolte, M. 1996, *PASP*, 108, 560
Stetson, P. B., *et al.* 1999, *AJ*, 117, 247
VandenBerg, D. A. 2000, *ApJS*, 129, 315
VandenBerg, D. A., Bolte, M., & Stetson, P. B. 1990, *AJ*, 100, 445

Discussion

C. GALLART: You have shown just in passing a result that I think is amazing, which is the fact that you find a flat age-metallicity relation for the majority of clusters and that the "outliers" of younger age seem to be related to disrupted dwarfs. This would point to a basically instantaneous formation of the Galactic halo, and then a mysterious age-metallicity relation for the accreted clusters. How do you interpret the latter?

A. SARAJEDINI: The ages of the clusters in our sample and how they vary with metallicity and Galactocentric distance suggest two modes of cluster formation. First, a group that formed via "rapid hierarchical collapse" which resulted in a small age range and no age-metallicity relation. Second, a group that formed via a slower hierarchical process that resulted in a range of ages and an age-metallicity relation. The latter appear to be associated with disrupted dwarf satellites of the Milky Way. It is not immediately clear how to interpret the age-metallicity relation other than to say that we need to measure ages in other ways to confirm this.

J. KALUZNY: I would like to point out that yet another method for determination of globular cluster ages was proposed by Pacynski. It is based on eclipsing detached binaries as age and distance indicators. With this method, masses and luminosities of turn-off stars can be determined directly. A practical application of this idea is presented in a poster by Thompson *et al.*

A. SARAJEDINI: Thank you for pointing this out. I have not covered this method in my talk because I decided to discuss methods that have been applied to many clusters, in a survey approach, so as to derive the global properties of the Milky Way clusters as they pertain to their ages.

J. MELBOURNE: I was wondering about the correction of the distance scale based on the LMC. Did you use this correction in your analysis and should we be using the LMC for this?

A. SARAJEDINI: No, I have not used this correction to the LMC in my absolute-age analysis. I think we should insist on consistency between nearby distance indicators and more distant ones. At the moment, there is enough uncertainty in the LMC distance that I would not use it to anchor the subdwarf distance scale.

M. CATELAN: I just wanted to mention that NGC 5286 is another cluster that has been associated with the Canis Major dwarf spheroidal. We've recently studied its CMD and determined an age for the cluster (see paper by Zorotovic *et al.* 2008, arXiv:0810.0682), finding it to be about 2 Gyr older than M3. This seems to deviate from the trend that you've shown for other Canis Major globulars. My question: Do you think Canis Major is indeed a dwarf spheroidal?

A. SARAJEDINI: I don't think I'm in a position to answer your question with any certainty. Your age result on NGC 5286 seems to be verified by our HST Treasury data (GO-10775). We need to revisit the question of the relative cluster ages using other techniques.

J. CHRISTENSEN-DALSGAARD: The mixing-length parameter is of course only a simplistic way to parameterize our ignorance about convection. However, we are getting to the point where hydrodynamical simulations can determine the properties of corrective envelopes, for a variety of stellar parameters.

A. SARAJEDINI: This is very good news, and I look forward to more significant advance in this area.

Ata Sarajedini

Jackie Faherty

Sandy Leggett

The Ages of Stars
Proceedings IAU Symposium No. 258, 2008
E.E. Mamajek. D.R. Soderblom & R.F.G. Wyse, eds.

doi:10.1017/S1743921309031883

Observations of multiple populations in star clusters

Giampaolo Piotto

Dipartimento di Astronomia, Università di Padova,
Vicolo dell'Osservatorio, 3, I-35122, Padova, Italy
email: giampaolo.piotto@unipd.it

Abstract. An increasing number of photometric observations of multiple stellar populations in Galactic globular clusters is seriously challenging the paradigm of GCs hosting single, simple stellar populations. These multiple populations manifest themselves in a split of different evolutionary sequences as observed in the cluster color-magnitude diagrams. Multiple stellar populations have been identified in Galactic and Magellanic Cloud clusters. In this paper we will summarize the observational scenario.

Keywords. globular clusters, stars: horizontal-branch, stellar populations, techniques: photometric, astrometry

---

1. Introduction

Globular star clusters (GC) have occupied a prominent role in our understanding of the structure and evolution of (low mass) stars. At the basis of the use of GCs as templates for stellar models was the assumption that their stars can be idealized as "simple stellar populations" (SSP), i.e. as an assembly of coeval, initially chemically homogeneous, single stars. Thanks to this idea, GCs, and star clusters in general, have been used for decades to test and calibrate synthetic models of stellar populations, a critical tool for studying galaxies at low, as well as at high, redshift.

Color-magnitude diagrams (CMD), like the magnificent CMD of NGC 6397 by Richer *et al.* (2008, see also Anderson *et al.* 2008), fully support the paradigma of GCs hosting simple stellar populations. However, there is a growing body of observational facts which challenge this traditional view. Since the eighties, we have known that GCs show a peculiar pattern in their chemical abundances (see Gratton *et al.* 2004 for a recent review). While they are generally homogeneous insofar Fe-peak elements are considered, they often exhibit large anticorrelations between the abundances of C and N, Na and O, Mg and Al. These anticorrelations are attributed to the presence at the stellar surfaces of a fraction of the GC stars of material which have undergone H burning at temperatures of a few ten millions K (Prantzos *et al.* 2007; less for the C and N anticorrelation). This pattern is peculiar to GC stars; field stars show only changes in C and N abundances, expected from the typical evolution of low-mass stars (Sweigart & Mengel 1979); this pattern in GC stars is primordial, since it is observed in stars at all evolutionary phases (Gratton *et al.* 2001); and the whole star is involved (Cohen *et al.* 2002).

In addition, since the sixties (Sandage and Wildey 1967, van den Bergh 1967), we know that the horizontal branches (HB) of some GCs can be rather peculiar. This problem, usually known as the *the second parameter* problem, still lacks a comprehensive understanding: many mechanisms, and many parameters have been proposed to explain the HB peculiarities, but none apparently is able to explain the entire observational scenario. It is quite possible that a combination of parameters is responsible for the HB morphology

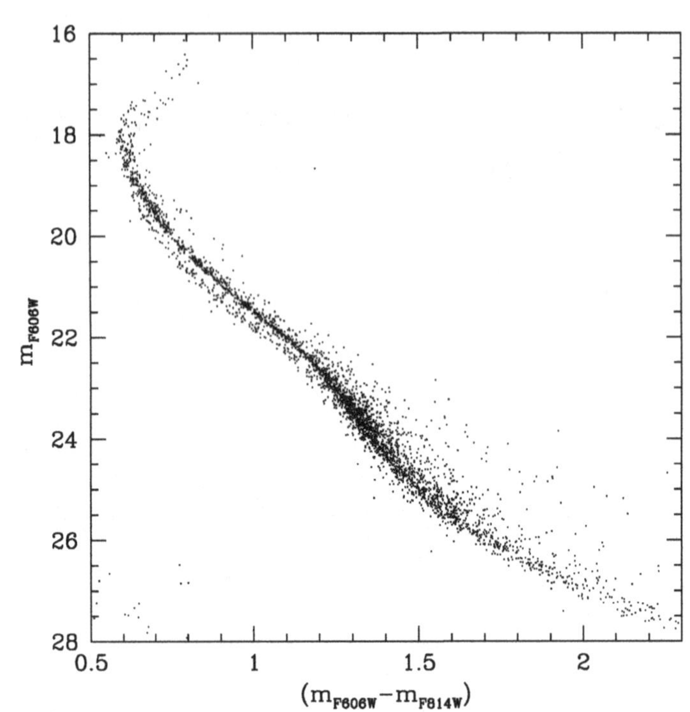

Figure 1. The multiple MS of Omega Centauri. This spectacular CMD (from Bedin *et al.* 2009, in preparation) comes from multi-epoch observations of a field at 17 arcmin from the cluster center. The plotted stars have been selected on the basis of their proper motion and all of them are cluster members. The split of the two MSs is clearly visible from the TO down to $m_{F606W} = 23.0$. Then the two MS seem to merge, due to the increased photometric error. Some stars are on the red side of the two main MSs; they are too far from the red MS to be binaries. These stars correspond to the third MS discussed in Villanova *et al.* (2007), which is likely related to the RGB-a of Pancino *et al.* (2002).

(Fusi Pecci *et al.* 1993). Surely, the total cluster mass seems to be a relevant parameter (Recio-Blanco *et al.* 2006).

It is tempting to relate the second parameter problem to the complex abundance pattern of GCs. Since high Na and low O abundances are signatures of material processed through hot H-burning, they should be accompanied by high He-contents (D'Antona & Caloi 2004). In most cases, small He excesses up to $\delta Y \sim 0.04$ (that is Y ~ 0.28, assuming the original He content was the Big Bang one) are expected. While this should have small impact on the colors and magnitudes of stars up to the tip of the RGB, a large impact is expected on the colors of the HB stars, since He-rich stars should be less massive. For example, in the case of GCs of intermediate metallicity ([Fe/H]~ -1.5), the progeny of He-rich, Na-rich, O-poor RGB stars should reside on the blue part of the HB, while those of the "normal" He-poor, Na-poor, O-rich stars should be within the instability strip, or redder than it. Actually, mean HB colors are influenced by the mass loss along the RGB and by small age differences of 2-3 Gyr. However, within a single GC a correlation is expected between the distribution of masses (i.e. colors) of the HB-stars and of Na and O abundances.

In summary, a number of apparently independent observational facts seems to suggest that, at least in some GCs, there are stars which have formed from material which must have been processed by a previous generation of stars.

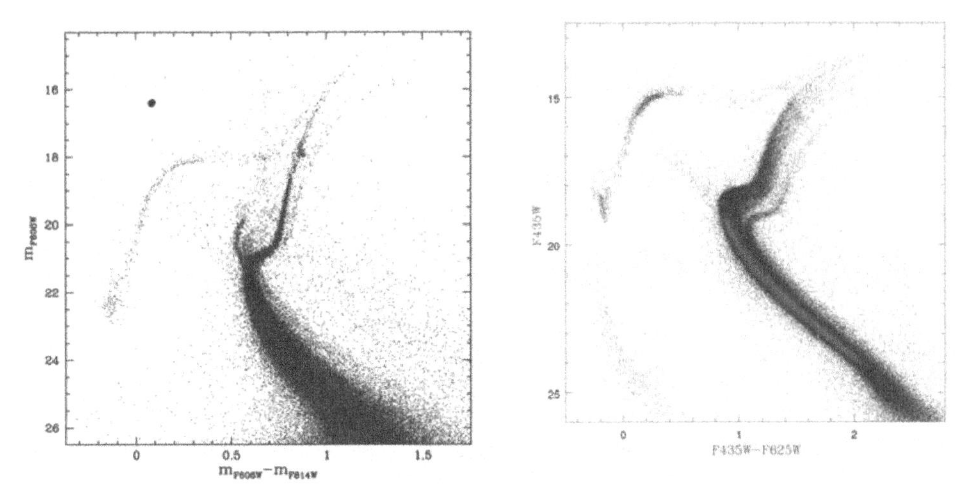

Figure 2. The CMD of M54 (left panel) resembles the CMD of ω Cen in many aspects. Do the two clusters have a similar origin?

The questions is: do we have some direct, observational evidence of the presence of multiple populations in GCs? Very recent discoveries, made possible by high accuracy photometry on deep HST images, allowed us to positively answer to this question. In this paper, we will summarize these new observational facts, and briefly discuss their link to the complex abundance pattern and to the anomalous HBs.

2. Observational evidence of multiple populations in GCs

2.1. *The first discoveries*

The first, direct observational evidence of the presence of more than one stellar population in a GC was published by Bedin *et al.* (2004). Bedin *et al.* found that, for a few magnitudes below the turn-off (TO), the main sequence (MS) of ω Centauri splits in two (Fig. 1). Indeed, the suspicion of a MS split in ω Cen was already raised by Jay Anderson in his PhD thesis, but the result was based on only one external WFPC2 field, and this finding was so unexpected that he decided to wait for more data and more accurate photometry to be sure of its reality. Bedin *et al.* (2004) confirmed the MS split in Anderson's field and in an additional ACS field located 17 arcmin from the cluster center. Now, we know that the multiple MS is present throughout the cluster, although the ratio of blue to red MS stars diminishes going from the cluster core to its envelope (Sollima *et al.* 2007, Bellini *et al.* 2009, in preparation). We also know that there is a third, redder MS (see Fig. 1), containing about 5% of the ω Cen MS stars (Villanova *et al.* 2007), probably related to the most metal rich RGB-a of Pancino *et al.* (2002).

The more shocking discovery concerning the multiple populations in ω Cen, however, came from a follow-up spectroscopic analysis that showed that the blue MS has twice the metal abundance of the dominant red branch of the MS (Piotto *et al.* 2005). The only isochrones that would fit this combination of color and metallicity were extremely enriched in helium ($Y \sim 0.38$) relative to the dominant old-population component, which presumably has primordial helium.

Indeed, the scenario in ω Cen is even more complex. As is already evident in the CMD of Bedin *et al.* (2004), the three MSs of ω Cen spread into a highly multiple sub-giant branch (SGB), with five distinct components, characterized by different metallicities and ages (Sollima *et al.* 2005, Villanova *et al.* 2007; the latter has a detailed discussion.)

These results reinforced the suspicion that the multiple MS of ω Cen could just be an additional peculiarity of an already anomalous object, which might not even be a GC, but a remnant of a dwarf galaxy instead.

On this respect, it might be instructive to compare the CMD of ω Cen (Fig 2, right panel), with the CMD of M54 (Fig. 2, left panel). The two CMDs look rather similar. We know that M54 almost coincides with the nucleus of the disrupting Sagittarius dwarf galaxy. And the complexity of the CMD of M54 of Fig. 2 is indeed due to the fact that we observe, in the same field, both M54 stars and background/foreground stars of the Sagittarius dwarf nucleus. M54 might have originated in the nucleus of its hosting galaxy, or ended there from elsewhere as a consequence of dynamical friction (Bellazzini *et al.* 2008). The important fact here is that both M54 and the Sagittarius nucleus now are located in the same place, in mutual dynamical interaction. It is very tempting to think that, a few Gyrs ago, ω Cen could have been exactly what we now find in the nucleus of the Sagittarius.

The spectacular case of ω Cen stimulated a number of investigations which showed that the multiple-population scenario is not a peculiarity of a single object. Piotto *et al.* (2007) showed that also the CMD of NGC 2808 is split into three MSs. Because of the negligible dispersion in Fe peak elements (Carretta *et al.* 2006), Piotto *et al.* (2007) proposed the presence of three groups of stars in NGC 2808, with three different He contents, in order to explain the triple MS. These groups may be associated to the three groups with different Oxygen content discovered by Carretta *et al.* (2006). These results are also consistent with the presence of a multiple, extended HB, as discussed thoroughly in D'Antona and Coloi (2004) and D'Antona *et al.* (2006).

Also NGC 1851 must have at least two, distinct stellar populations. In this case the observational evidence comes from the split of the SGB in the CMD of this cluster obtained from ACS/HST data (Milone *et al.* 2008a). More recently, Peter Stetson (see paper in the present Proceedings) has identified the SGB split also in the cluster envelope, thanks to spectacular photometry from ground-based wide field images. If the magnitude difference between the two SGBs were due only to an age difference, the two star-formation episodes would have to have been separated by at least 1 Gyr. However, as shown by Cassisi *et al.* (2007), the presence in NGC 1851 of two stellar populations, one with a normal α-enhanced chemical composition, and one characterized by a strong CNO-Na anticorrelation pattern, could reproduce the observed CMD split. In this case, the age spread between the two populations could be much smaller, possibly consistent with the small age spread implied by the narrow TO of NGC 2808. In other words, the SGB split would be mainly a consequence of the metallicity difference, and only negligibly affected by (a small) age dispersion. The Cassisi *et al.* (2007) hypothesis is supported by the presence of a group of CN-strong and a group of CN-weak stars discovered by Hesser *et al.* (1982), and by a recent study by Yong & Grundahl (2008) who find a Na-O anticorrelation among NGC 1851 giants.

NGC 1851 is considered a sort of prototype of bimodal-HB clusters. Milone *et al.* (2008a) note that the fraction of fainter/brighter SGB stars is remarkably similar to the fraction of bluer/redder HB stars. Therefore, it is tempting to associate the brighter SGB stars to the CN-normal, s-process element-normal stars and to the red HB, while the fainter SGB should be populated by CN-strong, s-process element-enhanced stars which should evolve into the blue HB. In this scenario, the faint SGB stars should be slightly younger (by a few 10^7 to a few 10^8 years) and should come from processed material

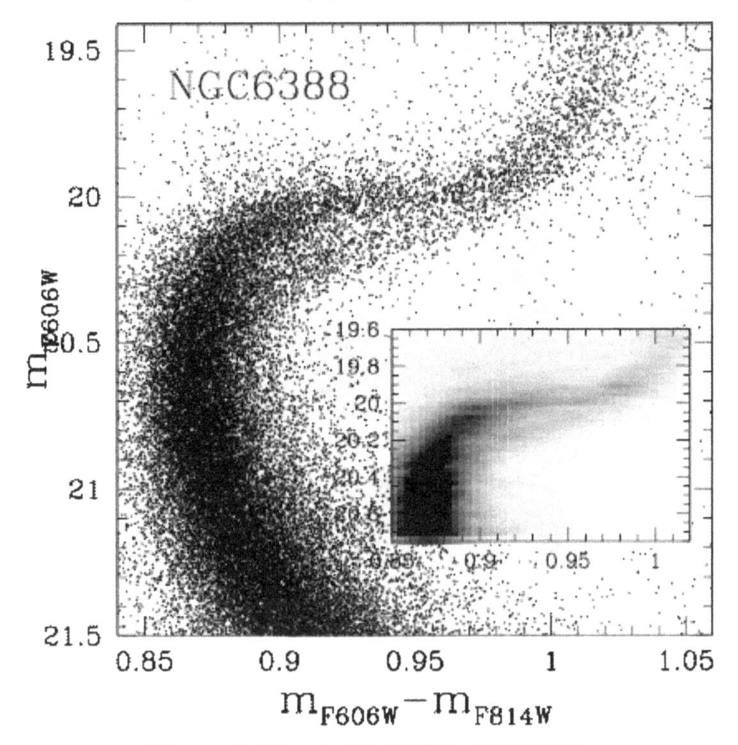

Figure 3. The double SGB in NGC 6388.

which might also be moderately He enriched, a fact that would help explain why they evolve into the blue HB. By studying the cluster MS, Milone *et al.* (2008a) exclude an He enrichment larger than $\Delta Y=0.03$, as expected also by the models of Cassisi *et al.* (2007). Nevertheless, this small He enrichment, coupled with an enhanced mass loss, would be sufficent to move stars from the red to the blue side of the RR Lyrae instability strip. Direct spectroscopic measurements of SGB and HB stars in NGC 1851 are badly needed.

2.2. *More recent findings. I. Galactic GCs*

Prompted by the results on ω Cen, NGC 2808, and NGC 1851 we used HST archive images and new proprietary data (GO10922 and GO11233, PI Piotto) to search for multiple populations. We are still working on the optimization of the software for the optimal extraction of high-accuracy photometry from the WFPC2 images (in particular for the new data acquired in the last months). For the moment, we did not find another cluster with multiple MSs as in ω Cen and NGC 2808, thought there are a couple of suspected cases.

However, we did find many clusters (at least seven, at the moment) with a double SGB (Piotto *et al.* 2009, in preparation). Among these, the most interesting cases are those of NGC 6388 (Fig. 3), M22 (Fig. 4), and M54 (Fig. 5).

Figure 3 shows that, even after correction for differential reddening, the SGB of NGC 6388 closely resembles the SGB of NGC 1851. The results, originally coming from HST data, have been recently confirmed also using near-IR, multi-conjugate adaptive optics images (Moretti *et al.* 2008) collected with MAD@VLT. NGC 6388, as well as its twin cluster NGC 6441, are two extremely peculiar clusters. Since Rich *et al.* (1997), we know

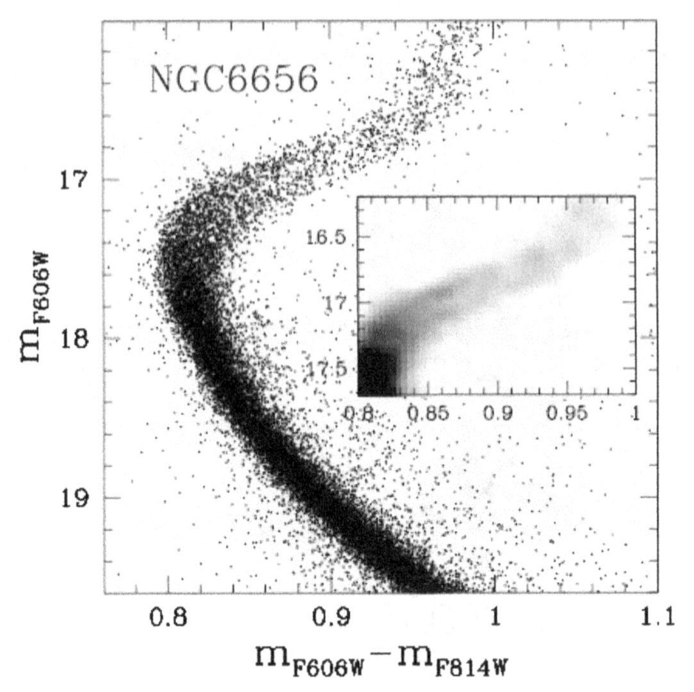

Figure 4. The double SGB in NGC 6656 (M22).

that, despite their high metal content – higher than in 47 Tucanae – they have a bimodal HB, which extends to extremely hot temperatures (Busso *et al.* 2007), totally un-expected for this metal rich cluster. NGC 6388 stars also display a Na-O anticorrelation (Carretta *et al.* 2007). Unfortunately, the available data do not allow us to study the MS of this cluster, to search for a MS split. In this context, it is worth noting that Caloi and D'Antona (2007), in order to reproduce the HB of NGC 6441, propose the presence of three populations, with three different He contents, one with an extreme He enhancement of Y=0.40. Such a strong enhancement should be visible in a MS split, as in the case of ω Cen and NGC 2808. A strong He enhancement and a consequent MS split may also apply to NGC 6388, because of the many similarities with NGC 6441. As soon as the new WF3 and the restored ACS instruments at HST will be available, we plan to test these predictions (GO11739, PI Piotto).

The case of M22 (Fig. 4) is a very interesting one. For decades this cluster has been suspected to have metallicity variations, including a spread in [Fe/H]. The iron spread has been controversial until very recently, when, thanks to the aquisition of UVES@VLT high-resolution spectra of RGB stars, we (Marino *et al.* 2009, in preparation) could demonstrate that, not only the [Fe/H] spread is confirmed, but that, indeed, there is a bimodal distribution in the iron content, and that this distribution is correlated with the abundance of s-process elements (as Y, Zr, Ba): Stars rich in Fe and Ca are also s-process element rich. The bimodal distribution of the SGB stars in M22 shown in Fig. 4 might be related to these two metallicity groups.

Also M54 shows a double SGB (Fig. 5). The fact that the stars populating the two SGBs are members of the M54 GC, and not field stars, is confirmed by the fact that they share exactly the same radial distribution within our ACS images centered on the cluster center.

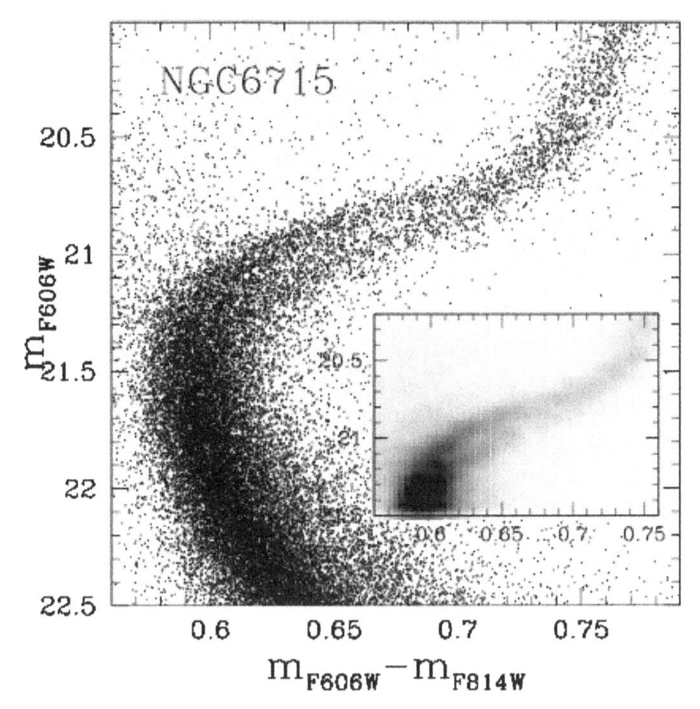

Figure 5. The double SGB in NGC 6715 (M54).

Piotto *et al.* (2009) shows that the double SGB phenomenon is quite common among massive GCs, though the fraction of stars in the two SGBs varies from cluster to cluster.

2.3. *More recent findings. II. Magellanic Cloud clusters*

The multiple-population phenomenon in star clusters is not confined to Galactic GCs only. The suspicion that some clusters in the Large Magellanic Cloud (LMC) could host more than one generation of stars has been raised in the past (e.g., Vallenari *et al.* 1994, Bertelli *et al.* 2003). However, only when high-precision photometry from ACS/HST images became available, could Mackey & Broby Nielsen (2007) clearly demonstrate the presence of two populations, with an age difference of ~300 Myr, in the 2 Gyr old cluster NGC 1846, in the LMC. In this case, the presence of the two populations is inferred by the presence of two TOs in the CMD. Mackey *et al.* (2008) identified two additional LMC clusters with multiple populations. More recently, Milone *et al.* (2008b), from the analysis of the CMDs of 16 intermediate age LMC clusters using HST archive data, showed that the multiple population phenomenon might be rather common among LMC clusters: 11 (70%!) have CMDs which are not consistent with the presence of a single, simple stellar population (see also Kozhurina-Platais *et al.* poster at this meeting). Also the Small Magellanic Cloud seems to host a cluster with a CMD that is not consistent with a single stellar population (Glatt *et al.* 2008).

3. An alternative approach to the search for multiple populations

The presence of abundance spreads in GC stars is well known. As already mentioned, in some cases more metallicity groups of stars can be isolated in the same cluster, as in the case of NGC 2808 (Carretta *et al.* 2006).

However, there is a recent finding for the GC M4 which is worth describing in some detail. Using high-resolution UVES@VLT spectra of more than 100 giants, Marino *et al.* (2008) have shown that also in this cluster, stars show a well defined Na-O anticorrelation (as in all GCs studied so far, searching for this phenomenon). The important result is that the distribution in Na (or O) content is clearly bimodal (see inset of Fig. 6), and this bimodal distribution is correlated with a bimodal distribution in CN strength among M4 stars. Stars that are Na-rich are also CN-strong.

The bimodality is also visible in the CMDs built using U-band images, as shown in Fig. 6. This is due to the strong effect of the CN bands on the U magnitude, as demonstrated by Marino *et al.* (2008). This is an interesting feature. Also in the case of M22, where Marino *et al.* (2009) have identified two groups of stars with different metal contents (see above), the U *vs* (U-V) CMD allows us to distinguish the two different populations. This RGB split or broadening in the U vs (U-B) CMD may be an alternative way to search for multiple stellar populations in GCs, in particular for clusters where large samples of high-resolution spectra are not available, or are not easily obtainable.

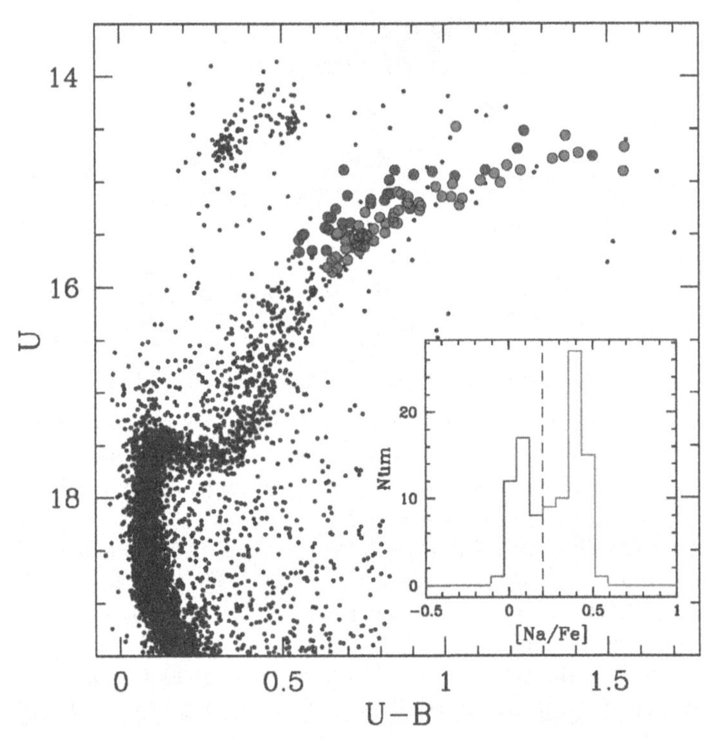

Figure 6. The Na and O distributions in M4 are bimodal (see inset). The bimodal distribution in these chemical elements is reflected also in a bimodal distribution of stars along the RGB.

4. Discussion

In the previous sections we have summarized the direct evidence we have of multiple generations of stars in star clusters. The observational scenario is rather complex. So far, multiple populations have been indentified by the presence of:

- multiple, distinct MSs, as in the case of ω Cen and NGC 2808;
- TO-SGB splits, in seven Galactic GCs, eleven intermediate-age LMC clusters, and at least one SMC cluster;

• bimodal or multimodal distributions of light elements, such as Na or O, or in the CN-strength, in some case associated with a broadening or multimodal distribution of RGB stars in the CMD involving the U-band.

In general, the multiple-population phenomenon differs from cluster to cluster, in the way it shows itself, in the ratio of the different populations present in the same cluster, and in the separation of the different sequences. An important property, shared by all the clusters where the phenomenon has been seen so far, is that the different populations are distinct. Only in some LMC clusters does there seem to be a broadened distribution, but it is not clear whether this is just due to photometric errors, which do not allow the separation of the single sequences, or is an intrinsic feature (Milone *et al.* 2008b).

At the moment, we cannot say whether the three different manifestations of the multimodality of cluster stellar populations reflect a single phenomenon. For example, it has been proposed (Bekki & Mackey 2008) that the origin of the bimodal populations in LMC clusters could come from an encounter of a young cluster with a giant molecular cloud, where the formation of a second generation of stars is triggered by the encounter itself. On the other hand, the multiple populations identified in the (generally more massive) Galactic globular clusters could be due to a second (or third) generation of stars which formed from material polluted by the ejecta from a variety of possible first-generation stars (see review by Yi in these proceedings). For sure, these GCs are clearly not simple, single-stellar-population objects. The emerging evidence is that the star-formation history can vary strongly from cluster to cluster, and that some GCs are able to produce very unusual objects, as no such He-rich MS stars have ever been found elsewhere. Reconstruction of this star-formation history requires a better understanding of the chemical-enrichment mechanisms, since not only the site of hot H-burning proposed to explain the He enhancement, but also the NaO anticorrelation, remains unclear. There are two requirements: (i) the temperature should be high enough; and (ii) the stars where the burning occur should be able to give back the processed material to the intracluster matter at a velocity low enough that it can be kept within the GC itself (a few tens of km/s). Candidates include: (i) Massive ($M > 10M_{\odot}$) rotating stars (Decressin *et al.* 2007); (ii) the most massive among the intermediate-mass stars undergoing hot-bottom burning during their AGB phase (Ventura *et al.* 2001), and possibly (iii) Population III stars (again see Yi's contribution in these proceedings). The first two mechanisms act on different timescales (10^7 and 10^8 yr, respectively), and each solution has its pros and cons (Renzini 2008). The massive-star scenario should avoid mixing the O-poor, Na-rich material with that material that is rich in heavy elements from SNe, while it is not clear how the chemically processed material could be retained by the proto-cluster in spite of the fast winds and SN explosions always associated with massive stars. Producing the right pattern of abundances from massive AGB stars seems to require considerable fine tuning. In addition, both scenarios require that either the IMF of GCs was very heavily weighted toward massive stars, or that some GCs should have lost a major fraction of their original population (Bekki & Norris 2006), and even then may be the remnants of tidally disrupted dwarf galaxies, as suggested by the complexity in the CMDs of ω Cen and of M54.

The observational scenario is becoming more complex, but, new results might have indicated the right track for a comprehensive understanding of the formation and early evolution of GCs. We are perhaps, for the first time, close to solving what has been for decades, and still, is a broken puzzle.

Acknowledgements. I wish to warmly thank Jay Anderson, Andrea Bellini, Luigi R. Bedin, Santi Cassisi, Ivan King, Antonino P. Milone, Alessia Moretti, and Sandro

Villanova, without whom most of the results presented in this review would not have been possible. A special thanks to Alvio Renzini and Raffaele Gratton for many enthusiastic discussions on the subject of multi-populations in star clusters. I acknowledge partial support by MIUR (PRIN2007) and ASI under contract ASI-INAF I/016/07/0.

References

Anderson, J. *et al.* 2008, *AJ*, 135, 2114

Bedin, L. R., Piotto, G., Anderson, J., Cassisi, S., King, I. R., Momany, Y., & Carraro, G. 2004, *ApJ*, 605, L125

Bekki, K. & Mackey, A. D. 2008, arXiv/0812.0631

Bekki, K. & Norris, J. E. 2006, *ApJ* (Letters), 637, L109

Bellazzini, M. *et al.* 2008, *AJ*, 136, 1147

Busso, G. *et al.* 2007, *A&A*, 474, 105

Caloi, V. & D'Antona, F. 2007, *A&A*, 463, 949

Carretta, E., Bragaglia, A., Gratton, R. G., Leone, F., Recio-Blanco, A., & Lucatello, S. 2006, *A&A*, 450, 523

Carretta, E. *et al.* 2007, *A&A*, 464, 957

Cassisi, S., Salaris, M., Pietrinferni, A., Piotto, G., Milone, A. P., Bedin, L. R., & Anderson, J. 2007, *ApJ*, 672, L115

Cohen, J. G., Briley, M. M., & Stetson, P. B. 2002, *AJ*, 123, 2525

D'Antona, F. & Caloi, V. 2004, *ApJ*, 611, 871

D'Antona, F., Bellazzini, M., Caloi, V., Pecci, F. Fusi, Galleti, S., & Rood, R. T. 2006, *ApJ*, 631, 868

Decressin, T., Meynet, G., Charbonnel, C., Prantzos, N., & Ekstrm, S. 2007, *A&A*, 464, 1029

Glatt *et al.* 2008, *AJ*, 136, 1703

Gratton, R. *et al.* 2001, *A&A*, 369, 87

Gratton, R., Sneden, C., & Carretta, E. 2004, *ARAA*, 42, 385

Hesser, J. E., Bell, R. A., Harris, G. L. H., & Cannon, R. D. 1982, *AJ*, 87, 1470

Mackey, A. D. & Broby Nielsen, P. 2007, *MNRAS*, 379, 151

Milone, A. P. *et al.* 2008a, *ApJ*, 673, 241

Milone, A. P. *et al.* 2008b, in press, arXiv0810.2558

Moretti, A. *et al.* 2008, in press, arXiv0810.2248

Piotto, G., *et al.* 2005, *ApJ*, 621, 777 (P05)

Piotto, G., *et al.* 2007, *ApJ*, 661, L53 (P07)

Prantzos, N., Charbonnel, C., & Iliadis, C. 2007, *A&A*, 470, 179

Recio-Blanco, A., Aparicio, A., Piotto, G., de Angeli, F., & Djorgovski, S. G. 2006, *A&A*, 452, 875

Renzini, A. 2008, *MNRAS*, 391, 354

Richer, H. *et al.* 2008, *AJ*, 135, 2141

Sandage, A. & Wildey, R. 1967, *ApJ*, 150, 469

Siegel *et al.* 2007, *ApJ*, 667, L57

Sollima, A., Pancino, E., Ferraro, F. R., Bellazzini, M., Straniero, O., & Pasquini, L. 2005, *ApJ*, 634, 332

Sollima, A., Ferraro, F. R., Bellazzini, M., Origlia, L., Straniero, O., & Pancino, E. 2007, *ApJ*, 654, 915

Sweigart, A. V. & Mengel, J. G. 1979, *ApJ*, 229, 624

Ventura, P., D'Antona, F., Mazzitelli, I., & Gratton, R. 2001, *ApJ* (Letters), 550, L65

Vallenari, A., Aparicio, A., Fagotto, F., Chiosi, C., Ortolani, S., & Meylan, G. 1994, *A&A*, 284, 447

Villanova, S., *et al.* 2007, *ApJ*, 663, 296

van den Bergh, S. 1967, *AJ*, 72, 70

Yong, D. & Grundahl, F. 2008, *ApJ*, 672, L29

Discussion

G. Meynet: 1) I would expect that He-rich stars are also those showing the greatest Na enrichment. Therefore stars on the blue main sequence of, e.g., NGC 2808, should show also the greatest Na enrichment. Is this the case? Is there any observational evidence of that? 2) The models which are used to derive the Na and O abundances, do they account for the fact that the star may be He- rich? If it is not the case, would this fact, if taken into account, change somewhat the results? Might it produce a less continuous Na- ratio distribution, reflecting the separated sequences derived in the CMD?

G. Piotto: 1) There is no direct measure of Na or O in MS stars in NGC 2808. However, I note that, accounting for the evolutionary times, the fraction of stars in the three MSs is such that we can link the reddest MS with the O-normal stars and the red sequence of the HB, the intermediate MS with the O-poor stars and the EBT1 part of the HB as defined by Bedin *et al.* (2000, A&A, 363, 159), and finally the bluest MS with the O-poorest group of stars and with the EBT2 HB. The last, hottest HB could be associated with the binary population; see Piotto *et al.* (2007, ApJ, 661, L53) for a more detailed discussion. 2) The models of Francesca DAntona and collaborators do take into account the He enhancement in the Na-O relation and in this way they can account for both the HB morphologies and the MS split observed, e.g., in NGC 2808 (see, e.g., DAntona *et al.* 2005, ApJ, 631, 868). If your question refers to the fact that strong He enhancement affects the precise models used to measure the abundances of the stars in the He-enhanced main sequences, I can answer that we did this check in Piotto *et al.* (2005 ApJ, 621, 777). We concluded that our He enhancement, up to $Y = 0.4$, is affecting the measured metallicity of the blue main sequence of ω Cen by 0.02-0.03 dex, well within our measurement error.

A. Dotter: You drew a correlation between M54 and ω Cen, but while ω Cen shows a small age dispersion and large He dispersion, M54 has a large age dispersion and no evidence for enhanced He.

G. Piotto: I do not think we can say that M54 has no evidence of He enhancement: simply it has not yet been properly observed to infer the presence of He enhancement. As for the second point, it is true that M54 shows a less-dispersed SGB than ω Cen, but what this means in terms of age dispersion remains to be established, for both clusters. However, if you take the M54 and the Sagitarius nucleus together, they do show a large dispersion along the SGB. And now, try to imagine how the CMD of the M54 field will look a few Gyr from now.

V. Poole: Have you noticed any trends about when multiple populations occur in globular clusters?

G. Piotto: No. However, note that this is a new field, and we know too few clusters with clear evidence of multiple populations. It is still too early to identify any trend. Also, note that the multiple population evidence is different in different clusters.

C. Deliyannis: You showed a population of dwarfs in ω Cen that have 40% helium. Do you have any suggestions as to how ω Cen can create such a population?

G. Piotto: I am as concerned as you are about the possible presence of a stellar population with $Y = 0.40$. Unfortunately, at the moment this solution seems to be the only one that can account for both the photometric and spectroscopic observations. As for the origin of such a high He abundance, see the review by Yi in these proceedings.

A. SARAJEDINI: A comment about the association of a ω Cen with M54, as nuclei of disrupted dwarfs: Bellazzini *et al.* (2008, arxiv:0807.0105) show that M54 is not likely to be the nucleus of Sgr. It moved to its current location from somewhere else.

G. PIOTTO: I am perfectly aware of the conclusions by Bellezini *et al.* (2008). My point in my talk is independent of whether or not M54 was born at the center of the Sagittarius dwarf galaxy or that it ended there because of dynamical friction. The fact is that, today, M54 and the nucleus of the Sgr dwarf galaxy are at the same place, and, therefore, they will dynamically evolve (mix) together. In a few billion years, M54 and the Sgr nucleus CMD will look very, very similar to the present day CMD of ω Cen, and, eventually the surrounding Sgr dwarf galaxy will have been completely stripped away by the Galactic tidal field.

H. RICHER: It would be nice to measure the proposed high He abundance in ω Cen directly. Have you tried to see it among hot stars on the horizontal branch?

G. PIOTTO: Direct measurement of He in HB stars can be performed only in a narrow temperature interval: stars hotter than the Grundahl jump ($\sim 11,000$ K) are affected by He sedimentation and metal levitation; stars cooler than $\sim 9,000$ K do not show He lines. We have recently proven the feasibility of this measurement for NGC 6752 stars in the 8,000 K to 10,000 K temperature interval, and surely the next application shall be in ω Cen stars. An alternative method would be to look for He-rich HB stars by analyzing their location on the $\log T$ vs. $\log g$ relation, and this project is ongoing.

Giampaolo Piotto

The Ages of Stars
Proceedings IAU Symposium No. 258, 2008
E.E. Mamajek, D.R. Soderblom & R.F.G. Wyse, eds.

doi:10.1017/S1743921309031895

Recovering the ages and metallicities of stars of a complex stellar population system

Sebastian L. Hidalgo[1], **Antonio Aparicio**[1,2], **and Carme Gallart**[1]

[1]Instituto de Astrofísica de Canarias, Vía Láctea, La Laguna, Tenerife, Canary Islands, Spain
[2]Departamento de Astrofísica, Universidad de La Laguna, Tenerife, Canary Island, Spain

Abstract. We present a new method to solve for the star-formation history (SFH) of a complex stellar population system from the analysis of the color-magnitude diagram (CMD). The SFH is obtained in four steps: i) computing a synthetic CMD, ii) simulating observational effects, iii) parameterization and sampling of the synthetic and observed CMDs, and iv) solving and averaging the solutions. The consistency and stability of the method have been tested using a mock stellar population.

The method has been used to solve the SFH of a set of six isolated Local Group dwarf galaxies observed with HST. The main goal is to probe the effects of cosmological processes, such as reionization in the early star formation, or the ability of SNe feedback to remove gas in small halos, in dwarf galaxies free from environmental effects due to the strong interaction with the host galaxy.

Keywords. Hertzsprung-Russell diagram, galaxies: stellar content, (galaxies:) Local Group, methods: numerical

---

1. Introduction

The color-magnitude diagram (CMD) is the best tool to derive the star formation history (SFH) of resolved galaxies. If the CMD is deep enough, stars born all over the life-time of the galaxy can be observed. However, deciphering an accurate SFH in a complex stellar population system requires some relatively sophisticated techniques. The most extended, and probably most powerful, technique is the one based on synthetic CMD analysis. We present a procedure and a set of algorithms designed for this task. The main code we present here is based in the same principle as that used by Aparicio, Gallart, & Bertelli (1997) for the solution convergence. Applied in the most general way, it derives the SFR of a system as a function of both time and metallicity, from the comparison of its CMD star distribution with the star distribution in a synthetic CMD.

2. Basic concepts and definitions

We will adopt here the following approach: considering that time and metallicity are the most important variables in the problem, we define the SFH as a function $\psi(t,z)$ such that $\psi(t,z)\mathrm{d}t\mathrm{d}z$ is the number of stars formed at time t' in the interval $t < t' \leqslant t+\mathrm{d}t$ and with metallicity z' in the interval $z < z' \leqslant z+\mathrm{d}z$, per unit time and metallicity. $\psi(t,z)$ is a distribution function and can be identified with the usual SFR, but as a function both of time and metallicity.

There are several other functions and parameters related to the SFH, that we will consider here as auxiliary. The initial mass function (IMF), $\phi(m)$, and a function accounting for the frequency and relative mass distribution of binary stars, $\beta(f,q)$, are the main ones. The solution found for the SFH depends on the assumptions made for these functions. The method we present here uses several choices of both with the aim of setting constraints on them.

Other parameters affecting the solution of $\psi(t, z)$ are distance and reddening, including differential reddening. But the strongest limitation on the observational information is produced by the *observational effects.* These include all the factors affecting and distorting the observational material, namely signal-to-noise limitations, defects on the detector, and crowding and blending of stars. The consequences are loss of stars, changes in measured stellar colors and magnitudes, and external errors larger and more difficult to control than internal ones.

3. Running the algorithms

The method is designed to obtain a SFH in four steps, each one, handled by an algorithm:

- Generating one (or several) global synthetic stellar population, which serves as a model (algorithm: IAC-star).
- Simulating observational effects on the model (obsersin).
- Parameterization and sampling the model and data (minniac).
- Solving the equations (IAC-pop).

3.1. *IAC-star*

IAC-star (Aparicio & Gallart 2004) is a code for synthetic CMD computation. In short, the algorithm is intended to be as general as possible and allows a variety of inputs for the initial mass function (IMF), star formation rate, metallicity law, and binarity. IAC-star is used to generate a global synthetic stellar population with a large number of stars with ages and metallicities with a constant distribution over the full interval of variation of $\psi(t, z)$ in time and metallicity. If functions $\phi(m)$, $\beta(f, q)$ are to be explored, a global synthetic stellar population must be generated for each choice of $\phi(m)$ and $\beta(f, q)$, as mentioned before. Among other quantities, IAC-star provides magnitudes in several filters, age, and metallicity for each synthetic star, and the integral of the SFH (i.e. total mass ever formed in the model), used to normalize the solutions. Synthetic magnitudes are used to plot the CMD (sCMD) associated with the global synthetic population.

3.2. *Obsersin*

Once the sCMD is generated, observational effects must be simulated. This is done by *obsersin* which makes use of the results of previously done completeness tests. The completeness tests are done by injecting a list of false stars in the observed images and recovering them using the same photometric procedure used for the photometry of real stars. From the list of the unrecovered stars and the differences between the injected (m_i) and recovered magnitudes (m_r) obsersin uses the following procedure to simulate the observational effects: for any star from sCMD (called a 'synthetic star') with magnitude m_s, a list of false stars with $|m_i - m_s| \leqslant \epsilon$ is created for each filter, with ϵ being a free input searching interval. From the stars in common to both filters, a single false star is selected by a simple random sampling. If m_i' and m_r' are the injected and recovered magnitudes of the selected false star in a given filter, then $m_s^e = m_s + m_i' - m_r'$ will be the magnitude of the synthetic star with observational effects simulated. The same is done for both filters.

With the procedure described above, the observational effects are simulated in sCMD star by star. The number of injected stars in the completeness tests, in comparison with the number of synthetic stars in sCMD, determines the quality of the simulation of the observational effects.

3.3. *Minniac*

IAC-pop uses a method introduced by Aparicio *et al.* (1997) to solve the SFH. The method assumes that any SFH can be given as a linear combination of simple populations (see §3.4). We define a simple stellar population by a number of stars with ages and metallicities within small intervals. Each simple population has a CMD associated that we call the partial model CMD (pCMD). With this definition, sCMD is formed for the sum of all pCMD. To compare pCMD with the observed CMD (oCMD) we create a set of boxes and count stars in them. This process is repeated for all the pCMD included in sCMD and gives the arrays M_i^j, containing the number of stars from partial model i populating the pCMD box j, and O^j, containing the number of observed stars in box j.

To minimize the dependency of the results on the selection of simple populations and on the size and position of the boxes, the process is repeated by introducing offsets in age and metallicity bins which define the simple populations. The age and metallicity interval for each simple population is fixed. For each new set of simple populations, the boxes can change in size and position.

In addition, for each set of simple populations and/or size and position of the sampling boxes, minniac can introduce an offset in color and magnitude in oCMD. For each new color-magnitude point, the process described above is repeated. This produces a manifold of input files which are used by IAC-pop (see below) to solve the SFH. We call *model parameterization* to each set of simple populations, position and size of boxes, and color-magnitude point. Each model parameterization has associated with it the arrays defined above, M_i^j and O^j.

3.4. *IAC-pop*

With the information provided by minniac, the distribution of stars in the CMDs can be calculated for any model SFH as a linear combination of the M_i^j calculated above:

$$M^j = A\sum_i \alpha_i M_i^j \tag{3.1}$$

It should be noted that $\alpha_i \geqslant 0$. A is a scaling constant.

The SFH best matching the distribution, O^j, of the observational CMD can be found using a merit function. In particular Mighell's χ^2_γ (Mighell 1999) is used:

$$\chi^2_\gamma = \sum_j \frac{(O^j + min(O^j,1) - M^j)^2}{O^j + 1} \tag{3.2}$$

We will use $\chi^2_\nu = \chi^2_\gamma/\nu$, where ν is the number of degrees of freedom. In our case $\nu = k - 1$, where k is the number of boxes defined in the CMD. Minimization of χ^2_ν provides the best solution as a set of α_i values. IAC-pop makes use of a genetic algorithm (Charbonneau 1995) for efficient searching of the χ^2_ν minimum.

The solution SFH can be written as:

$$\psi(t,z) = A\sum_i \alpha_i \psi_i \tag{3.3}$$

where ψ_i refers to partial model i, with i taking values from 1 to $n \times m$, and A is again a scaling constant.

All solutions given by IAC-pop for the same color-magnitude point are averaged. The solutions obtained for different color-magnitude points are used to calculate external uncertainties. To find the color-magnitude offset which gives the best χ^2_ν, several choices of color-magnitude points must be explored. This procedure minimizes the impact of the uncertainties from the reddening and distance determination and the differences between the evolutionary models and data.

4. Consistency test

To test the consistency and stability of the method, a mock stellar population has been computed using IAC-star and analyzed with the procedure described in §3 to obtain its SFH as if it were a real one. Results have been compared with the input used to compute the mock population. The IAC-star input parameters used to compute the mock population were as follows. The BaSTI stellar evolution library (Pietrinferni *et al.* 2004) and the Castelli & Kurucz (2003) bolometric correction library were used. The number of stars in the associated mock CMD (mCMD) was 10^5. The star formation ranges from 14 Gyr ago to the present, with a constant SFR, $\psi(t)$, for that period. The metallicity increases with time, with initial and final metallicities $z_0 = 0.0001$ and $z_f = 0.008$ and some metallicity dispersion at each time. Finally, no binary stars were considered and the IMF by Kroupa, Tout, & Gilmore (1993) was used. The integral of $\psi(t,z)$ (i.e. the total mass ever transformed into stars) for this system is $\Psi_T = 2.02 \times 10^6 M_\odot$. The SFH $\psi(t,z)$ of the mock population is shown in Fig. 1 on the left panel. The volume below the curved surface and over the age-metallicity plane gives the mass that has been ever transformed into stars within the considered age-metallicity interval. The SFR as a function of time only, $\psi(t)$, and of metallicity only, $\psi(z)$, are also shown. The associated mCMD is shown in the right panel.

Figure 2 shows on the left panel the $\psi(t,z)$ solution for the mock population given in Fig. 1. Agreement is good, being the differences between input and solution within the error bars, showing the algorithm's robustness against observational effects. The right panel shows the CMD corresponding to the solution.

Table 1 summarizes the consistency test. Column 1 identifies the CMD. Column 2 gives the χ^2_ν value of the solution. Columns 3, 4 and 5 give, respectively, the total mass (M_T), mean age ($< age >$), and mean metallicity ($< z >$) of the stars in the mock population (first row) and the solutions. The agreement of the integral and average values between the mock population and the solution is good in all cases.

Table 1. Results for the self-consistency test

CMD	χ^2_ν	M_T ($10^6 M_\odot$)	$< age >$ (Gyr)	$< z >$
Mock		2.02	7.00	0.0026
Solution	0.9	2.00 ± 0.02	6.91 ± 1.10	0.0026 ± 0.0005

5. SFHs of the LCID project

The procedure described in the previous sections have been applied to LCID (Local Cosmology from Isolated Dwarfs) project. LCID was granted 110 orbits on the HST@ACS to obtain deep photometry of five Local Group isolated dwarf galaxies, each with a different level of current star formation activity and gas fraction: IC1613, Leo A, Tucana, LGS 3 and Cetus. The Phoenix dwarf galaxy was observed previously with the HST@WFPC2

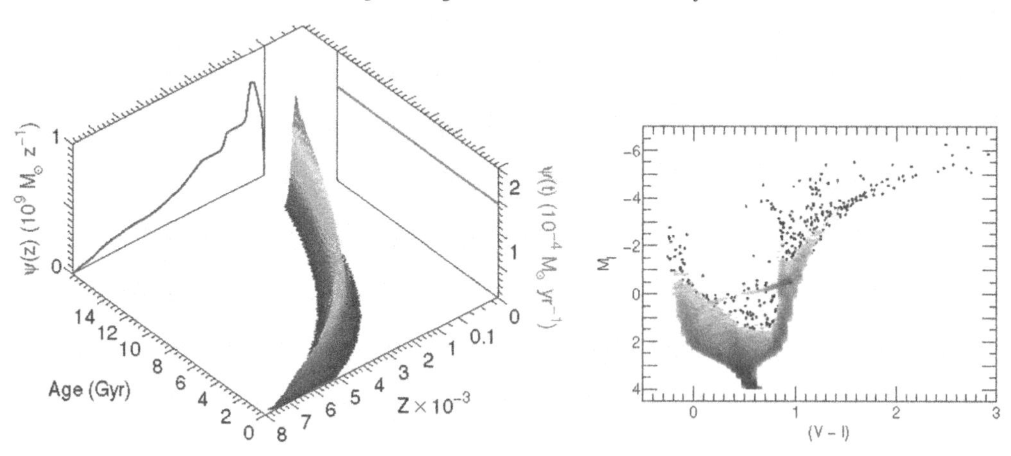

Figure 1. The SFH $\psi(t,z)$ of the mock population (left panel). Age and metallicity are given in the horizontal axis. The volume below the curved surface and over the age-metallicity plane gives the mass that has been ever transformed into stars within the considered age-metallicity interval. The mono-dimensional $\psi(t)$ and $\psi(z)$ are shown on the ψ-age plane (in red in the paper electronic version) and on the ψ-metallicity plane (in blue in the paper electronic version) respectively. The CMD of the mock population (mCMD)is shown on the right panel. Grey levels show the density of stars. A factor of two in density exists between each two successive gray levels.

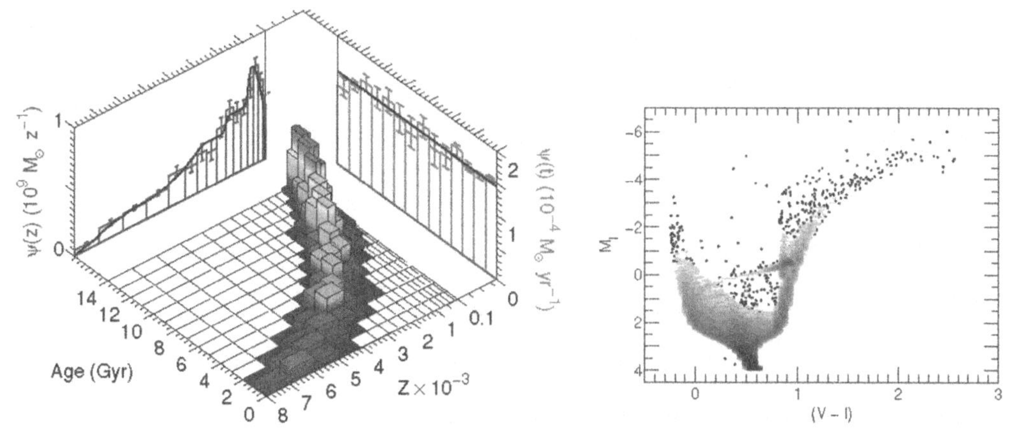

Figure 2. Solution of $\psi(t,z)$ obtained for the mock population (left panel). The caption is the same as in Figure 1. The CMD corresponding to the solution is shown on the right panel. To be compared with the mCMD shown in Figure 2.

over 24 orbits and also added to the LCID sample. The final objective is to analyze how star formation has proceeded in isolated dwarfs, free of strong tidal effects.

The galaxies were observed in F475W and F814W ACS filters (except Phoenix which was observed in F555W and F814W WFPC2 filters). The photometry list was obtained using DOLPHOT (Dolphin (2000)) (the photometry of Phoenix was obtained using DAOPHOT/ALLFRAME (Stetson, 1994)). The process described in this paper was used to obtain the preliminary SFHs of all galaxies, which are shown in Figure 3. Preliminary results show that all galaxies except LeoA have old (>10 Gyr) star formation. The first star formation even seems to be stronger in Tucana and Cetus than the other galaxies. LeoA shows a delay in the first star-formation event. However, in the case of LeoA, our

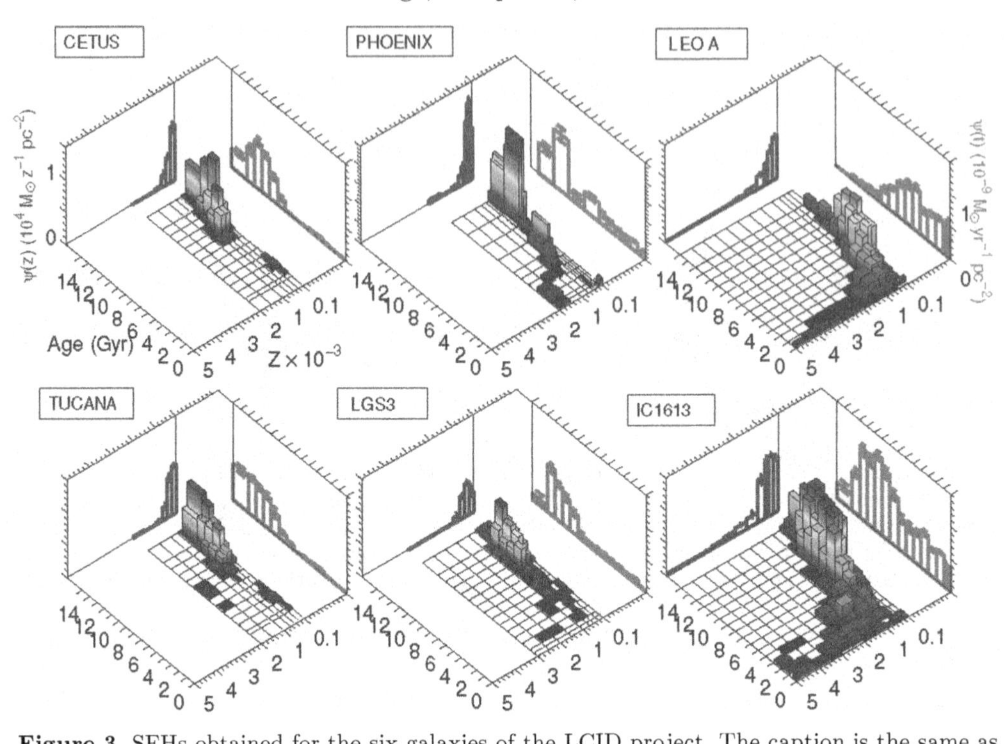

Figure 3. SFHs obtained for the six galaxies of the LCID project. The caption is the same as in Figure 1.

results show that the SFR for stars older than 10 Gyr depends on the distance and reddening assumed.

6. Conclusions

We have presented several algorithms which have been shown to be a useful tools to obtain the SFH from the CMD of resolved stellar systems. The procedure has been run through a consistency test showing the robustness of the method against the uncertainties introduced by realistic observational effects.

The method has been applied to obtain the SFHs of a set of six Local Group dwarf galaxies from the LCID project.

References

Aparicio, A. & Gallart, C. 2004, *AJ*, 128, 1465
Aparicio, A., Gallart, C., & Bertelli, G. 1997, *AJ*, 114, 669
Aparicio, A. & Hidalgo, S. L. 2008, *AJ*, submitted
Castelli, F. & Kurucz, R. L. 2003, *New Grids of ATLAS9 Model Atmospheres IAU Symposium*
Charbonneau, P. 1995, *ApJS*, 101, 309
Dolphin, A. E 2000, *PASP*, 112, 1383
Kroupa, P., Tout, C. A., & Gilmore, G. 1993, *MNRAS*, 262, 545
Mighell, K. J. 1999, *ApJ*, 518, 380
Pietrinferni, A., Cassisi, S., Salaris, M., & Castelli, F. 2004, *ApJ*, 612, 168
Stetson, P. B. 1994, *PASP*, 106, 250

Discussion

M. Salaris: Given our current inability to predict HB Morphologies, what is the role played by the observed HB stars in your SFH determinations?

S. Hidalgo: We have not used HB stars at all. We use MS and sub-giant branch stars only.

M. Tosi: Could you comment on why Cole does find a (secondary) SF peak in Leo A at the earliest epochs while you have very little (if any) SF at those epochs?

S. Hidalgo: The star formation history derived by Cole *et al.* (2007, ApJ, 659, L17) uses a different stellar evolution library than this solution, which could introduce some differences between them. But, moreover, the secondary peak found by Cole is, within the error bars, compatible with the solution that I present here.

R. Wyse: How robust is your result that star formation in Leo A did not start earlier than 10 Gyr ago? You mentioned that different stellar evolution models can give older ages.

S. Hidalgo: The solutions have been tested against several consistency tests. All the SFHs have been derived homogeneously. All other galaxies show evident SF older than 10 Gyr. Leo A presents a very low SF rate for that epoch, but there is something. We know that Leo A has an old population because there are RR-Lyrae stars in the galaxy. We tried to derive the SFH with two different stellar evolution libraries to look for the differences in the solutions.

Enrico Vesperini

Walter Maciel

The Ages of Stars
Proceedings IAU Symposium No. 258, 2008
E.E. Mamajek, D.R. Soderblom & R.F.G. Wyse, eds.

doi:10.1017/S1743921309031901

Multiple population theory: The extreme helium population problem

Sukyoung K. Yi

Yonsei University, Department of Astronomy, Seoul 120-749, Korea
email: yi@yonsei.ac.kr

Abstract. The spreads in chemical abundances inferred by recent precision observations suggest that some or possibly all globular clusters can no longer be considered as simple stellar populations. The most striking case is ω Cen in the sense that its bluest main-sequence, despite its high metallicity, demands an extreme helium abundance of $Y \approx 0.4$. I focus on this issue of "the extreme helium population problem" in this review.

Keywords. stars: abundances, AGB and post-AGB, chemically peculiar, evolution Hertzsprung-Russell diagram, Population II, Galaxy: general, abundances, globular clusters, ISM: abundances

1. Introduction

Globular clusters may not be a robust example for *simple* stellar populations any more. Perhaps there is no such thing as simple stellar populations from the beginning. The classic globular clusters, such as ω Cen, NGC 2808, and NGC 1851, are now suspected to be composed of heterogeneous populations, and recent data from space-based telescopes with unprecedented resolving accuracy are hinting at a great fraction of Milky Way globular clusters being composite populations, at least chemically.

At the centre of these debates is ω Cen. It has long been known as a mysterious object. To begin with, this spectacular southern cluster is the most massive in the Milky Way, with some million solar masses. Its unusually broad red-giant branch (RGB) was found to indicate *discrete* multiple populations by the magnificent effort and insight of Lee *et al.* (1999), using a mere 0.9m telescope. More recent work, with much superior instruments, unambiguously revealed the multiplicity of this giant cluster. Norris (2004) and Bedin *et al.* (2005) sequentially found that the multiplicity is evident not just on the red giant branch but also on the main sequence. To everyone's surprise, the bluest main sequence is too blue for the measured metallicity for ω Cen and is in fact more metal rich than the redder main-sequence stars (Piotto *et al.* 2005). If such a blue colour for such a metal-rich population is real, it unavoidably indicates the possibility of the scorchingly high helium abundance, $Y \approx 0.4$. The blue main-sequence population constitutes 30% by number (Bedin *et al.* 2004; Sollima *et al.* 2007) and thus is not something we can simply sweep under the carpet. If there is any good news in this apparent nightmare, the blue main-sequence population seems at least to be younger than the majority of the stars in this cluster, perhaps by a couple of billion years (Villanova *et al.* 2007).

Such an age range has significance for the horizontal branch morphology in this cluster. This and many other clusters exhibit an extended horizontal branch, and its origin has been a long-debated issue. Apparently, the same level of extreme helium inferred by the blue main-sequence can also explain the extreme blueness of the extended horizontal branch (Lee *et al.* 2005). If this prevails in other clusters as well, the hitherto mysterious origin for the extended horizontal branch may also be solved by the extreme helium content.

Apparently many more clusters show multiple sequences, either on the main sequence and/or on the sub-giant branch (Piotto 2008, this volume), even though it is not yet clear whether such multiplicities are also to be interpreted as originating from an extreme helium content. More massive clusters tend to show multiple sequences more often, and interestingly the same trend is found for the extended horizontal branch (Recio-Blanco *et al.* 2006; Lee *et al.* 2007).

Extragalactic counterparts to ω Cen and the like may have been found in the giant elliptical galaxy M87 in the Virgo cluster (Sohn *et al.* 2006; Kaviraj *et al.* 2007). Using the Hubble Space Telescope Sohn *et al.* (2006) found that most of the massive globular clusters in M87 are UV-bright, despite their likely old ages, as if they have an extremely hot horizontal branch. Through an extensive test using the UV-focused population synthesis models of Yi (2003), Kaviraj *et al.* (2007) concluded that the UV strengths (a tracer of the horizontal-branch morphology) of these clusters are even stronger than that of ω Cen, by more than a magnitude. Whatever is causing the multiplicity in ω Cen seems to affect the M87 clusters even more greatly.

The existence of massive clusters showing various anomalies seems to corroborate the idea of these clusters originally being something of a different nature, for example, nucleated dwarf spheroidal galaxies (Lee *et al.* 1999).

All things considered, there appears to be a huge conspiracy. It is not yet clear whether the cause of the multiplicity of the main sequence is the cause of that of the horizontal branch. However, they all fit in a very sensible storyline. Although it ruins the old and naive concept of "simple stellar populations", multiplicity by itself is perhaps not a huge problem. The extreme helium abundance inferred by the blue main-sequence population is an exciting discovery to observers but a desperate-to-forget nightmare to theorists. I now discuss why this is so.

2. Significance

The significance of this issue is immense. First, obtaining an understanding how such an extreme helium abundance could be possible is an interesting challenge. It also influences the current age-dating techniques that are based on precise main-sequence fitting and on detailed horizontal-branch analysis. The seemingly settled issue of the identity of the second parameter of horizontal-branch morphology may enter a new stage with the not-so-new idea of variations in helium abundance, with a clearer understanding ot the helium enrichment processes. The endless debate on the origin of the UV-upturn found in bright elliptical galaxies may also find a new and compelling explanation with helium. Obvious too is the impact on the issue of the age of the universe, as globular clusters and bright elliptical galaxies are often considered the oldest stellar populations in the Universe.

3. Observational facts and inferences

Finding a solution to the case of ω Cen is only a beginning step, since other clusters show different constraints, but it would still be a good start. So I attempt to find a solution, adopting some of the most widely discussed channels.

Our *simplified* constraints are as follows.

- The age separation: the blue main-sequence subpopulation is 1–3 billion years younger than the red main-sequence subpopulation; i.e., $t(bMS) \approx t(rMS) - 1$–$3$ Gyr (Lee *et al.* 2005; Stanford *et al.* 2007; Villanova 2007). I think the exact value is poorly constrained but for now adopt a value $\Delta t = 1\,\mathrm{Gyr}$.

• The mass fraction: the number (and mass) fraction of the blue main-sequence sub-population is roughly 30% (Bedin *et al.* 2004), i.e., $f(bMS) = 0.3$. I will try to aim to find a solution that satisfies this. However, this may not place as strong a constraint as I assume, if the mass evolution of sub-populations is complex. I will discuss this in detail in §6.

• Discrete sub-populations: the main sequence and horizontal branch splits appear very sharp and discrete. Hence, a stochastic element in a solution to the extreme helium abundance cannot be dominant. Instead, it has to offer a process that leads to a clear prediction in helium abundance.

• The metal abundance: $Z(rMS) = 0.001$ and $Z(bMS) = 0.002$ (Piotto *et al.* 2005). The metallicity of the blue main-sequence stars is difficult to pin down due to their faintness and so is still uncertain. But it seems clear that it is higher than that of the red main-sequence stars.

• The helium abundance: the helium abundance of the blue main sequence sub-population is 40% by mass, i.e., $Y = 0.4$. In reality, the observed colour-magnitude diagram shows up to 5 sub-populations. But it is impossible to make a model that pins down all the sub-populations found. Hence, I approximate them into 2 sub-populations: the red main sequence, with an ordinary helium abundance, and the blue main sequence, with an extreme helium abundance. As I will discuss at the end, it is perhaps very important to remind ourselves repeatedly that *the helium abundance has not been directly measured, but is only inferred from main-sequence fitting.* Despite this, I take the helium abundance as given.

• The helium enrichment parameter: the helium and metal abundances together lead to an incredible value for the helium enrichment parameter, $\Delta Y/\Delta Z \approx 70$. Ordinary populations with ordinary stars yield $\Delta Y/\Delta Z \approx 2$–$3$, even for a wide range of stellar initial mass functions. Hence, this poses the most challenging problem of all. I will focus most of my tests on this issue.

• Other elements: spectroscopic measurements of carbon and nitrogen are available, i.e., $[C/M] \sim 0$ and $[N/M] \sim 1$ for the blue main sequence population. However, their accuracy appears not to be as good as one might hope for, and estimated errors (i.e. measurement significance) are not provided. It is already a daunting task to reproduce just the helium properties, and so I will only use this information as a reference.

4. Asymptotic giant branch stars

The most obvious candidate origin for such an extreme helium abundance is asymptotic giant branch (AGB) stars (e.g., Izzard *et al.* 2004; D'Antona *et al.* 2005, among many papers). Although there is quite a scatter in the predictions of chemical yields from the AGB, there is a consensus that these stars generate a copious amount of helium, but only a small amount of metals (e.g., Maeder 1992). This is good for our present purposes, since we do not just want to produce a lot of helium but want also to achieve a very high value of the helium enrichment parameter $\Delta Y/\Delta Z$. Supernovae, for comparison, produce too high a mass in metals to satisfy this constraint, although they are also good producers of helium. This is such a basic point that it does not require much elaboration, but it has recently been discussed in a quantitative matter by Choi & Yi (2007).

AGB stars in a narrow mass range ($M \approx 5$–$6 M_\odot$) indeed release ejecta with a high value of the helium enrichment parameter, equal to that we aim to achieve. So if a population receives the stellar mass ejecta mainly from asymptotic giant stars and little else, it is in principle possible to achieve such a high value of helium enrichment parameter. More massive stars would produce both metals and helium. Hence, an *ad hoc* scenario can be set up to maximise the impact of the asymptotic giants in terms of the helium enrichment parameter, with all the mass ejecta from massive stars (say $M > M_{\rm esc}$ where $M_{\rm esc} \sim 5$–$10\ M_\odot$) posited to escape the gravitational potential where subsequent star formation occurs. The effectiveness of this *maximum AGB scenario* has been discussed by a few groups (e.g., Karakas *et al.* 2006; Bekki *et al.* 2007), and Choi & Yi (2008) performed a detailed calculation to check its viability.

Choi & Yi (2008) adopted a toy model where the original gas reservoir does not accept any new gas infall from outside and the material ejected from massive stars above $M_{\rm esc}$ escape it, supposedly via supernova explosions, hence maximising the helium enrichment effect from AGB stars. It is plausible that the kinetic energy of the material ejected from supernova explosions equals the escape energy from such a small potential well. It is assumed that some fraction (50 – 100%) of the initial gas is used to form the first population (the red main sequence) and the subsequent population (the blue main sequence) will be born from the remnant gas, mixed with the material ejected from the first stellar population. The abundance of the initial gas is assumed to be the abundance of the red main sequence population of ω Cen. If a higher fraction of the initial gas reservoir is used to build the first population, it would obviously result in a higher value of helium abundance and helium enrichment parameter for the second population, but only a small amount of gas becomes available for the second population to form out of.

If we do not adopt any constraint on the age difference between the red and blue main-sequence populations, we can achieve a very high helium abundance ($Y \approx 0.36$ which is almost as high as we aim to reach) and the maximum value of helium enrichment parameter of about 70, as we hoped for. In this case, the age difference is roughly 0.1 Gyr, and the second generation is virtually a pure recycling product of the first generation, consisting of material ejected from stars within a narrow mass range of 5–6 $M_\odot$. But in this case, the total mass ratio between the red and blue main-sequence populations becomes 99.3: 0.7; that is, only 0.7% of the total population in ω Cen can benefit from this scenario. Since the blue main sequence population is observed to be 30% instead 0.7%, there is a factor of 40 discrepancy! I call this "the mass deficit problem".

One may achieve somewhat different estimates by adopting different yields. For example, Renzini (2008) uses the recent yield for the so-called "super-AGB stars" to find that the mass discrepancy can be as small as 15 instead of 40.

If we take the age difference of roughly 1 Gyr as a valid constraint, the situation becomes dramatically worse. This is because, even if we assume the $M_{\rm esc}$ argument, the stars in the mass range 2–5 $M_\odot$ will now contribute to the gas reservoir through stellar mass ejecta, with this ejecta in general of substantially lower helium abundance (~ 0.3) and helium enrichment parameter (~ 2–5). Consequently, this scenario with 1 Gyr age separation can achieve only up to $Y \approx 0.3$ and $\Delta Y/\Delta Z \approx 10$, while the upper limit in the mass fraction of the second generation is just 7% (instead of 30%). Let alone the shortcoming in the helium properties, the mass fraction requirement cannot be met, either.

The verdict on the maximum AGB scenario and its variation can be summarised as follows. The extreme helium-related properties are almost impossible if the age difference is a meaningful constraint, hence making this scenario totally implausible. If the age separation constraint can be eased, then the extreme value of the helium enrichment parameter (*but not the helium abundance itself*) can be reproduced by the first generation of asymptotic giants, under the following conditions, and with the following criticisms.

- The stellar mass ejecta from massive stars of $M > 6M_\odot$ must all escape the gravitational potential well. If the 'super AGB' scenario (e.g. Siess 2007) is adopted, this mass limit can be as high as $10M_\odot$. If all supernova ejecta leaves the system in a high-velocity wind, this is not a bad assumption, but assuming that the supernova ejecta leaves completely without affecting the remaining gas in the reservoir is extreme and very unlikely.

• The blue main-sequence population must form exactly 0.1 Gyr after the red main sequence population, in disagreement with the 1 Gyr separation suggested by previous studies. I personally think the age separation constraint is not very strong and thus 0.1 Gyr is not particularly unappealing.

• The mass deficit of a factor of 40 (which can be somewhat smaller if 'super' AGB stars are adopted) is a serious threat and requires a rescue plan. A possible remedy may be found in the details of the cluster dynamical evolution, which is discussed in §6.

• An encouraging aspect of this scenario is that the discreteness of the separated populations is easy to explain. The second generation forms from the mass loss of the first generation, 0.1 Gyr later.

5. Fast-rotating massive stars

A totally different solution was put forward by massive-star evolution models. Maeder & Meynet (2006) suggested that metal-poor massive stars that are rotating nearly at their break-up speed may release a lot of helium via a *slow wind* before they start burning heavy elements and explode as a supernova. Their idea came from their earlier work (Maeder & Meynet 2001) that suggested (1) low-metallicity stars reach break-up rotational speed more easily by the combined effects of stellar (slow) winds and rotation, (2) they have efficient mixing of their core materials, that is, helium and other heavier elements (depending on the rotation speed) are mixed out to the surface, and (3) during their blue loop, after the red giant phase, a fast contraction leads to extensive mass loss from the helium- and nitrogen-enhanced surface material.

The elemental yields via slow (stellar) winds are sensitive to the rotational speed adopted. For example for a 60 $M_{\odot}$ star with $logZ = -5$, a fast rotating model at 85% of the break-up speed yields the helium abundance of 5.86 solar mass, the metal abundance of 0.09 solar mass, and thus the helium enrichment parameter of 63.3 (which is very close to our aim!). On the other hand, for a moderately fast rotating model at 35% of the break-up speed, the yields become $\Delta Y = 1.73$, $\Delta Z = 2.6e-5$, and $\Delta Y/\Delta Z \approx 10^5$. These extremely fast-rotating stars generate excessively high values of $\Delta Y/\Delta Z$ and too little of helium. The fast rotating stars overproduce carbon and nitrogen abundances compared to observation, while the moderate rotating stars reproduce the observation better. But we still select the fast-rotating models in our exercise (Choi & Yi 2007) because they produce much more helium and thus are more likely to satisfy our aim.

The toy model of Choi & Yi (2007), using the metal-poor massive rotating stars of Maeder & Meynet (2006), shows that a simple population based on an ordinary initial mass function can indeed achieve high values of both the helium abundance and $\Delta Y/\Delta Z$ in the stellar mass loss, as we aim to recover. These values are further elevated by the helium-dominant contribution from asymptotic giants, until lower-mass giants become the main source of chemical yields. Thus this phenomenon of high helium properties lasts only for a short period of time, of order 0.1–0.2 Gyr, just as in the AGB scenario. Once the population becomes older than that, its accumulated stellar mass loss will no longer have such high values of its helium properties.

We find, however, that the amount of gas with these high helium properties can be only roughly 1.4% of the total stellar mass of ω Cen, which is a factor of 20 too small for it to be the sole solution to this problem. This mass deficit of a factor of 20 is smaller (and thus better) than that of the asymptotic giant branch star scenario simply because this time we have helium contributions from massive rotating stars as well as from asymptotic giants. Here, we assume that only the slow-wind material (stellar mass loss) from the massive stars remains in the gravitating system and the fast-wind material (explosions) leaves the system without polluting the gas reservoir.

In conclusion, we could not find a solution if the age separation of 1 Gyr or so is a meaningful constraint. For a much smaller age separation, of order 0.1 Gyr, we could achieve high values of the helium parameters, but even in this case the mass available for the formation of the second generation is a factor of 20 smaller than the requirements in ω Cen. This problem has been noted also by a much more detailed dynamical simulation of Decressin, Baumgardt, & Kroupa (2008; see also the article by Decressin in this volume). We will discuss this further in §6.

Another serious problem in this scenario is the carbon abundance. While it depends strongly on the rotational speed adopted, the 60 $M_\odot$ model, with 85% of the break-up speed, suggests that the slow-wind mass loss will be highly enriched in carbon, which is not supported by the observational data (Piotto *et al.* 2005).

For this scenario to be appealing, we also need to understand how a specific rotation speed is determined for the stars. Why does it happen to some clusters (like ω Cen) but not to others? Is it randomly given to each cluster, and not to each forming star? That will be very odd. This scenario with fast-rotating massive stars certainly adds to what was already possible from the asymptotic giant stars and thus provides a positive contribution. However, it alone does not appear to provide a full solution to our problem.

6. Dynamical evolution

The blue main-sequence population seems to be more centrally concentrated than is the red main-sequence population. If this were true from the start, one would expect that the spatially more-extended red main-sequence population would lose more stars throughout its dynamical evolution. D'Ercole *et al.* (2008; and also the poster at this meeting) indeed suggested that a substantial fraction of the first generation of stars may escape the system if some conditions are met. For example, if the initial mass distribution within each globular cluster follows the King profile and *if its King radius is equal to its true tidal radius*, then it is very easy to shed some high-velocity stars into space. In this case, only 2-3% of the original first-generation stars may remain in the cluster, mainly due to kinetic energy injection by supernova explosions and two-body relaxation. If this is true, it makes both the AGB scenario and the massive fast-rotating star scenario viable.

Whether these conditions were easy to meet by the first generation clusters is not yet clear, however. More traditional studies (e.g., Fall & Zhang 2001) based on evaporation by two-body relaxation, gravitational shocks, and stellar mass loss suggest an order of magnitude milder mass evolution.

The mass evaporation is supposed to be sensitive to the mass of the cluster in the sense that *a more massive cluster would shed less mass.* So, if the dynamical evaporation was indeed the key to this extreme helium phenomenon, it would be very unlikely to happen preferentially to the most massive clusters. Unfortunately for this scenario, ω Cen is the Milky Way's most massive cluster and the other clusters showing multiplicity, NGC 2808 and M 54, are among the most massive, too. Besides, the extended horizontal-branch globular clusters in the Milky Way, and the UV-brightest clusters in M 87, all occupy the highest-mass end in the total cluster-mass distribution of the host galaxy. In this sense, the dynamical evaporation picture loses its charm.

If D'Ercole *et al.*'s dramatic mass evolution is applicable to all globular clusters, then it would have a significant impact on the cluster luminosity-function evolution. Typical clusters in the Milky Way are of a million solar masses presently, which is in the same order as the mass of the giant molecular clouds, the main site of star formation, and also as the mass of the star clusters forming in nearby merging galaxies. In this regard, I feel that this scenario of shedding 98% of the initial mass of the first population is likely a rare event. Perhaps this is why the main sequence splits are not a common feature. Otherwise, that is, if such a dramatic mass evaporation had been true to all clusters, then we should find our galactic stellar halo to have at least ten billion solar masses, which is an order or magnitude greater than the current estimate. I strongly feel that physical understanding of the dynamical process (when such conditions are met) is required, and detailed dynamical models, cross-checking with the observed cluster luminosity functions, are called for.

7. The first stars

There must have been stars that formed before population I and II stars. This is evident from several arguments. Theoretically, the mass of the first objects that experience dynamical instability is estimated to be stellar rather than galactic. This is consistent with the fact that reionisation is (although indirectly) observed through the cosmic microwave background radiation studies. Observationally, despite the fact that the big bang itself did not generate any appreciable amount of heavy elements, totally metal-deficient stars are not found anywhere. Even the most metal-deficient stars show $\log Z/Z_{\odot} \sim -5$ and more typical metal-poor stars have metallicities greater than a hundredth of the solar value. This means that the pregalactic gas reservoir must have been substantially enriched in metals. The most probable objects for this are the first stars, a.k.a. population III stars. The first stars are often thought to have been very massive, above a hundred solar mass, while other possibilities are also being considered.

The duration of the first star-formation episode is considered to be extended well into that of population II (Bromm & Loab 2006). If we are considering a proto-galactic scale system, the mixing timescale for the chemical elements may have been of order a hundred million years, and thus considering both the extended star formation timescale and varying mixing timescale, some *chemical inhomogeneity* in the gas reservoir for the population II star formation was inevitable.

Marigo *et al.* (2003) have computed the chemical yields for such metal-free stars of mass between 100 and 1000 $M_{\odot}$. Surprisingly, their models suggest that the first stars were very efficient in generating and releasing helium into space, but not metals. This is mainly because the first stars had such an enormous radiation pressure that the balance between mass accretion and radiative pressure was difficult to achieve; that is, the strong radiation pressure blew away the gas that was being accreted. So the first energy generation involving hydrogen burning was possible, but before the star reaches the next stage it would release much materials processed by then: i.e., helium. This results in a high helium to metal ratio, as we were looking for.

Choi & Yi (2007) have indeed investigated this effect on the helium enrichment in the gas cloud. They found that the the range between 100 and 1000 solar masses, a lower-mass first star produces a much larger value of $\Delta Y/\Delta Z \sim 10^{7-8}$. (No, this is not a typo.) The first stars of 1000 $M_{\odot}$ are predicted by this model to have $\Delta Y/\Delta Z \sim 10^{2}$, which is much closer to our aim. Adopting a Salpeter initial mass function†, we found that a first-star population with a mass range 100—1000 $M_{\odot}$ releases virtually no metals but

† As I type this part I just learned of Professor Salpeter's death. We have just lost one of the greatest astrophysicists of our time.

abundant helium, and thus reaches $\Delta Y/\Delta Z \sim 500$. A population with a higher value of the lower mass bound results in a gradually lower value. Eventually, a population purely made up of 1000 solar-mass stars would have $\Delta Y/\Delta Z \approx 70$.

After letting the first star population evolve for a billion years the remnant gas cloud (primordial gas left out of the first star formation plus the stellar mass loss mixed evenly) reaches the metallicity of the blue main sequence ($Z = 0.002$), the helium abundance ($Y = 0.4$), and so the helium enrichment parameter ($\Delta Y/\Delta Z \approx 70$), with no further free parameter.

This scenario is briefly investigated by Choi & Yi (2008) and can be chronologically described as follows.

(*a*) The majority of first stars form in the Universe at redshift roughly at 20 ($t \equiv 0$).

(*b*) These stars release much helium and some metals.

(*c*) The chemical mixing in the proto-galactic cloud took a long time, and after hundreds of millions of years, chemically mixed regions are more common than unmixed regions.

(*d*) From a chemically mixed region, the red main-sequence population of ω Cen forms ($t \sim 0.5$ Gyr).

(*e*) From the pristine (unmixed) gas in the vicinity a second generation of first stars forms ($t \sim 0.7$ Gyr).

(*f*) They generate abundant helium and little metals and enrich the remnant gas cloud to $Z \sim 0.002$, $Y \sim 0.4$ and thus $\Delta Y/\Delta Z \sim 70$.

(*g*) From this gas cloud, the blue main-sequence population of ω Cen forms ($t \sim 1.5$ Gyr).

(*h*) The blue main-sequence population merges into the more massive red main-sequence population soon after their formation.

This picture is very rough however and contains many caveats:

- The first-star chemical yields may be highly uncertain. A more robust understanding of the formation and evolution of the first stars will perhaps come in the near future, but, more importantly, independent calculations (besides Marigo *et al.*) are required immediately.
- In this scenario the first stars (at least the ones that led to the gas reservoir for the formation of the blue main-sequence population) should have very high mass, of order 1000 $M_{\odot}$. This is not supported by some recent first-star studies.
- We need not just a couple of first stars in this region but more than one hundred. How such material gathers up in this proto cloud is a mystery, especially when first stars are often believed to form isolated, rather than in multiplets.
- The physics, in terms of the chemical mixing and its timescale, is highly uncertain, as is the case for other scenarios.

Given all these uncertainties, it is difficult to argue that the first-star scenario is any more compelling than others. However, it is still a very exciting possibility. After all, we astronomers are always the first one to find something wrong, as well as new and important. This conjecture at least implores for more studies on the first stars.

8. Alternative theories

Alternative theories are also available. The velocity-dispersion dependent surface pollution of AGB ejecta scenario was put forward by Tsujimoto *et al.* (2007). A similar surface-pollution scenario was presented by Newsham & Terndrup (2007). While the channels for the pollution can be several, it provides an interesting possibility that the blue main-sequence stars are not truly so helium-rich as we believe, but pretend to be so by having unusually high helium abundance only on the stellar surface. Mass transfer of the surface material in binary stellar systems could be one channel, or if stars pass through the central region of the cluster, where helium-rich gas from the accumulated stellar mass loss is located, such stars may be polluted on the surface. However, it is very unlikely that 30% of the stars get contaminated like this. Besides, all these processes would occur in a random manner, so that the discreteness of the blue main-sequence would be unnatural.

The possibility of having primordial fluctuations in the helium density was presented by Chuzhoy (2006). In this study, the helium diffusion timescale for primordial gas of stellar mass was of order a hundred million years, and thus some of the birth clouds for the first stars were heavily enriched in helium. But again, the diffusion timescale must depend on the conditions in each birthcloud, which should be rather random, which again makes the main-sequence discreteness a tough problem to solve.

9. Conclusions

Theorists are often very optimistic, thinking that a tough problem to solve is challenging instead of mind-boggling. But I must admit that I am much more than puzzled by this "extreme helium population problem". The presence of multiple populations in globular-cluster size populations is surely a problem, given that numerical simulations of the kind performed by Bate, Bonnell, & Bromm (2008) suggest that the star formation in a cluster probably happens on the crossing time scale, which is only on the order of a million years. But we have seen other small populations that have a complex star formation history, e.g., the Carina dwarf galaxy (Smecker-Hane *et al.* 1994). A more critical issue is the extreme value of the helium properties. I do not believe that we have a compelling theory yet. Asymptotic giant branch stars are a familiar class and thus makes our mind susceptible. But I believe that I have shown that AGB-models still have a mass deficit problem, by a factor of at least 40, which is threateningly large even to astronomers. The same is true for the scenario of massive stars rotating nearly at the break-up speed. They alone cannot provide the full answer and suffer from a similar mass deficit problem. Its physical plausibility is also something to be worked out. The first-star scenario is fascinating because first stars are a mystery in general. We believe that they were once around but have never seen them, a bit like black holes. They provide a plausible solution, but just barely. It has so many caveats and uncertainties that cannot be clarified in the next few years. Hence it loses its charm, too.

I said at the end of my presentation at this conference that the enigmatic extreme helium population is so tough to theorists that I would almost feel happy if someone comes up to say "It was all a mistake from the start. There is no such extreme helium population". George Meynet disagreed with me. He instead said the problem is so enigmatic that we are greatly challenged and excited. I became humble at his constructive attitudes. I hope to see a more believable solution in the near future.

Acknowledgements

I thank Ena Choi for countless constructive discussions. Much of this review is based on several key papers written by Choi & Yi (2007, 2008), Decressin, Charbonnel & Meynet (2008), and by Renzini (2008). I am grateful to Young-Wook Lee, Suk-Jin Yoon, Ken Nomoto (during my visit to Tokyo University), and Enrico Vesperini for constructive discussions. Special thanks are due to Changbom Park for stimulating discussion on stellar collisions in clusters during my visit to the KIAS. This research has been supported by Korea Research Foundation (SKY).

References

Bedin, L. R., Piotto, G., Anderson, J., Cassisi, S., King, I. R., Momany, Y., & Carraro, G. 2004, *ApJ*, 605, L125

Bekki, K., Campbell, S. W., Lattanzio, J. C. & Norris, J. E., 2007, *MNRAS*, 337, 335

Choi, E. & Yi, S. K. 2007, *MNRAS*, 375, L1

Choi, E. & Yi, S. K. 2008, *MNRAS*, 386, 1332

Chuzhoy, L. 2006, *MNRAS*, 369, L52

D'Antona, F., Bellazzini, M., Caloi, V., Pecci, F. F., Galleti, S., & Rood, R. T. 2005, *ApJ*, 631, 868

Decressin, T., Charbonnel, C., & Meynet, G. 2007, *A&A*, 475, 859

Decressin, T., Baumgardt, H., & Kroupa, P. 2008, *A&A*, in press

Fall, S. M. & Zhang, Q. 2001, *ApJ*, 561, 751

Izzard, R. G., Tout, C. A., Karakas, A., & Pols, O. R. 2004, *MNRAS*, 350, 407

Karakas, A. I., Fenner, Y., Sills, A., Campbell, S. W., & Lattanzio, J. C. 2006, *ApJ*, 652, 1240

Kaviraj, S., Sohn, S. T., O'Connell, R. W., Yoon, S.-J, Lee, Y.-W., & Yi, S. K. 2007, *MNRAS*, 377, 987

Lee, Y.-W., Joo, J.-M., Sohn, Y.-J., Rey, S.-C., Lee, H.-C., & Walker, A. R. 1999, *Nature*, 402, 55

Lee, Y.-W., Joo, S.-J., Han, S.-I., Chung, C., Ree, C. H., Sohn, Y.-J., Kim, Y.-C., Yoon, S.-J., Yi, S. K., & Demarque, P. 2005, *ApJ*, 621, L57

Lee, Y.-W., Gim, H. B., & Casetti-Dinescu, D. I. 2007, *ApJ*, 661, L49

Maeder, A. 1992, *A&A*, 264, 105

Maeder, A. & Meynet G. 2001, *A&A*, 373, 555

Maeder, A. & Meynet G. 2006, *A&A*, 448, L37

Newsham, G. & Terndrop, D. M. 2007, *ApJ*, 664, 332

Norris, J. E. 2004, *ApJ*, 612, 25

Piotto *et al.* 2005, *ApJ*, 621, 777

Piotto, G., Bedin, L. R., Anderson, J., King, I. R., Cassisi, S., Milone, A. P., Villanova, S., Pietrinferni, A., & Renzini, A. 2007, *ApJ*, 661, 53

Recio-Blanco, A., Aparicio, A., Piotto, G., de Angeli, F., & Djorgovski, S. G. 2006, *A&A*, 452, 875

Renzini, A. 2008, *MNRAS*, 391, 354

Siess, L. 2007, *A&A*, 476, 893

Smecker-Hane, T. A., Stetson, P. B., Hesser, J. E., & Lehnert, M. D. 1994, *AJ*, 108, 507

Sohn, S. T., O'Connell, R. W., Jundu, A., Landsman, W. B., Burstein, D., Bohlin, R. C., Frogel, J. A., & Rose, J. A. 2006, *AJ*, 131, 866

Sollima, A., Ferraro, F. R., Ballazzini, M., Origlia, L., Straniero, O., & Pancino, E. 2007, *ApJ*, 654, 915

Stanford, L. M., Da Costa, G. S., Norris, J. E., & Cannon, R. D. 2007, *ApJ*, 667, 911

Tsujimoto, T., Shigeyama, T., & Suda, T. 2007, *ApJ*, 654, 139

Villanova, S., Piotto, G., King, I. R., Anderson, J., Bedin, L. R., Gratton, R. G., Cassisi, S., Momany, Y., Bellini, A., Cool, A. M. Recio-Blanco, A., & Renzini, A. 2007, *ApJ*, 663, 296

Yi, S. 2003, *ApJ*, 582, 202

Discussion

M. PINSONNEAULT: We don't see evidence of high helium in field populations. Is this consistent with the high He models, or do they pose constraints?

S. YI: That's a very interesting test. But, according to both the AGB scenario and the massive rotating star scenario, the age separation between the first and second generations must be in a very narrow window (i.e., $\sim$ 30 - 300 Myr). Unless this condition is met, the extreme helium population may not be possible, which would explain the paucity of them in the field.

R. WYSE: Does the assumption that the clusters can retain the ejecta of stars with $M \sim 6M_{\odot}$ (your escape mass) have implications for the gas and dust content of globular clusters (there are very low upper limits)?

S. YI: You're right. A more acceptable $M_{\rm esc}$ would be close to $10M_{\odot}$. A model with $M_{\rm esc} = 6$ is an extreme one, and it tells us how extreme the AGB scenarios must be to make them work.

J. MELBOURNE: I was wondering what is driving the blue MS colors, is it an age effect or is the helium playing a role?

S. YI: Extreme helium abundance.

G. MEYNET: The Na-O anticorrelation is not seen in halo field stars. In the framework of the Pop III scenario that you presented, how do you explain the absence in field stars? Why only in clusters?

S. YI: Yes, if there is a causal connection between the Na-O anti-correlation and the extreme helium, that would be a threat to the first-star scenario.

Carme Gallart and Sukyoung Yi

Michelle Cignoni

The Ages of Stars
Proceedings IAU Symposium No. 258, 2008
E.E. Mamajek, D.R. Soderblom & R.F.G. Wyse, eds.

doi:10.1017/S1743921309031913

The ages of Galactic globular clusters in the context of self-enrichment

T. Decressin[1], H. Baumgardt[1], P. Kroupa[1], G. Meynet[2] and C. Charbonnel[2,3]

[1]Argelander Institute for Astronomy (AIfA), Auf dem Hügel 71, D-53121 Bonn, Germany
email: decressin@astro.uni-bonn.de, holger@astro.uni-bonn.de, pavel@astro.uni-bonn.de,

[2]Geneva Observatory, University of Geneva, chemin des Maillettes 51, CH-1290 Sauverny, Switzerland
email: Georges.Meynet@unige.ch, Corinne.Charbonnel@unige.ch

[3]LATT, CNRS UMR 5572, Université de Toulouse, 14 avenue Edouard Belin, F-31400 Toulouse Cedex 04, France

Abstract. A significant fraction of stars in globular clusters (about 70%-85%) exhibit peculiar chemical patterns, with strong abundance variations in light elements along with constant abundances in heavy elements. These abundance anomalies can be created in the H-burning core of a first generation of fast-rotating massive stars, and the corresponding elements are conveyed to the stellar surface thanks to rotational induced mixing. If the rotation of the stars is fast enough, this material is ejected at low velocity through a mechanical wind at the equator. It then pollutes the interstellar medium (ISM) from which a second generation of chemically anomalous stars can be formed. The proportion of anomalous stars to normal stars observed today depends on at least two quantities : (1) the number of polluter stars; (2) the dynamical history of the cluster, which may lose different proportions of first- and second-generation stars during its lifetime. Here we estimate these proportions, based on dynamical models for globular clusters. When internal dynamical evolution and dissolution due to tidal forces are accounted for, starting from an initial fraction of anomalous stars of 10% produces a present-day fraction of about 25%, still too small with respect to the observed 70-85%. In the case of gas expulsion by supernovae, a much higher fraction is expected to be produced. In this paper we also address the question of the evolution of the second-generation stars that are He-rich, and deduce consequences for the age determination of globular clusters.

Keywords. globular clusters: general, stellar dynamics, stellar evolution

1. Introduction

It has long been known that globular-cluster stars present some striking anomalies in their content of light elements whereas their heavy elements (i.e., Fe-group, α-elements) remain fairly constant from star to star (with the notable exception of ω Cen). While in all the Galactic globular clusters studied so far one finds "normal" stars with detailed chemical compositions similar to those of field stars of the same metallicity (i.e., same [Fe/H]), one also observes numerous "anomalous" main-sequence and red-giant (RGB) stars that are simultaneously deficient (to various degrees) in C, O, and Mg, as well as enriched in N, Na, and Al (for reviews see Gratton *et al.* 2004; Charbonnel 2005).

These abundance variations are expected to result from H-burning nucleosynthesis at high temperatures, around 75×10^6 K (Denisenkov & Denisenkova 1989, 1990; Langer & Hoffman 1995; Prantzos *et al.* 2007). Such temperatures are not reached in the low-mass main-sequence and RGB stars that are chemically peculiar, meaning that the stars inherited their abundance anomalies at stellar birth.

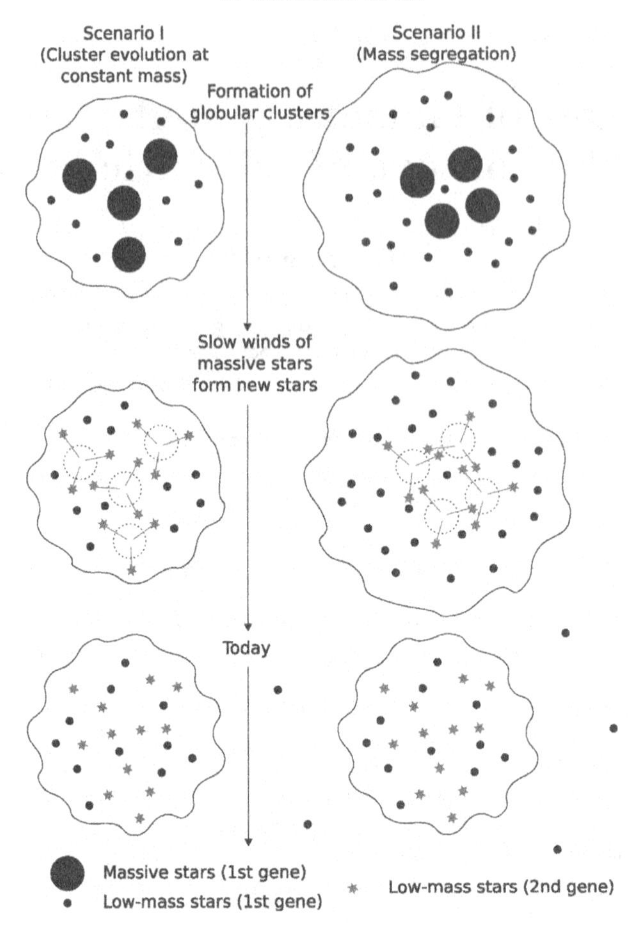

Figure 1. Schematic evolution of a globular cluster: a first generation of stars is born from a giant molecular cloud. Massive stars of the first generation evolve and give birth to a second generation of low-mass stars (dashed symbols in middle panels). Then the cluster evolves and today (lower panels) a mixture of first and second generation low-mass stars is present. In the right panel the cluster is initially mass segregated and massive stars (and hence stars of second generation) are concentrated toward the cluster centre.

Here we follow the work of Prantzos & Charbonnel (2006) and Decressin *et al.* (2007b) who propose that abundance anomalies are build up by fast rotating, fast evolving massive stars. During their main sequence evolution, rotationally induced mixing transports elements synthesised in the convective H-burning core to the stellar surface. For stars heavier than 20 $M_\odot$, the surface reaches break-up at the equator (i.e., the centrifugal equatorial force balances gravity), provided that their initial rotational velocity is high enough. In such a situation, a slow mechanical wind develops at the equator and forms a disc around the stars, similar to what happens in Be stars (Townsend *et al.* 2004; Ekström *et al.* 2008). The material in the discs is strongly enriched in H-burning products and has a slow velocity, allowing it to remain trapped in the potential well of the cluster. On the contrary, matter released later through radiatively driven winds during most of the He-burning phase and then through SN explosions has a very high velocity and is lost from the cluster. Therefore, new stars can form only from the matter available in discs, with the abundance patterns we observe today. Thus globular clusters can contain two populations of low-mass stars: a first generation which has the chemical composition of the material out of which the cluster formed (similar to field stars with similar metallicity); and a second generation of stars harbouring the abundance anomalies born from the ejecta of fast-rotating massive stars. This scenario is sketched in Fig. 1.

2. Dynamical issues

2.1. *The number ratio between the two populations in globular clusters*

Based on the determination of the compositions of giant stars in NGC 2808 by Carretta *et al.* (2006), Prantzos & Charbonnel (2006) determined that around 70% of stars show abundance anomalies in this specific cluster today. Decressin *et al.* (2007a) find similar results for NGC 6752 with their analysis of the data of Carretta *et al.* (2007): around 85% of the cluster stars (of the sample of 120 stars) present abundance anomalies. Therefore most stars still evolving in globular clusters seem to be second-generation stars.

How can one produce such a high fraction of chemically peculiar stars? The main problem is that, assuming a Salpeter (1955) IMF for the polluters, the accumulated mass of the slow winds ejected by the fast-rotating massive stars would only provide 10% of the total number of low-mass stars. To match the observations thus requires either (a) a flat IMF with a slope of 0.55 instead of 1.35 (Salpeter's value), or (b) that 95% of the first generation stars have escaped the cluster (Decressin *et al.* 2007a). Here we first investigate whether such a high loss of stars is possible, and what are the main processes that could drive it.

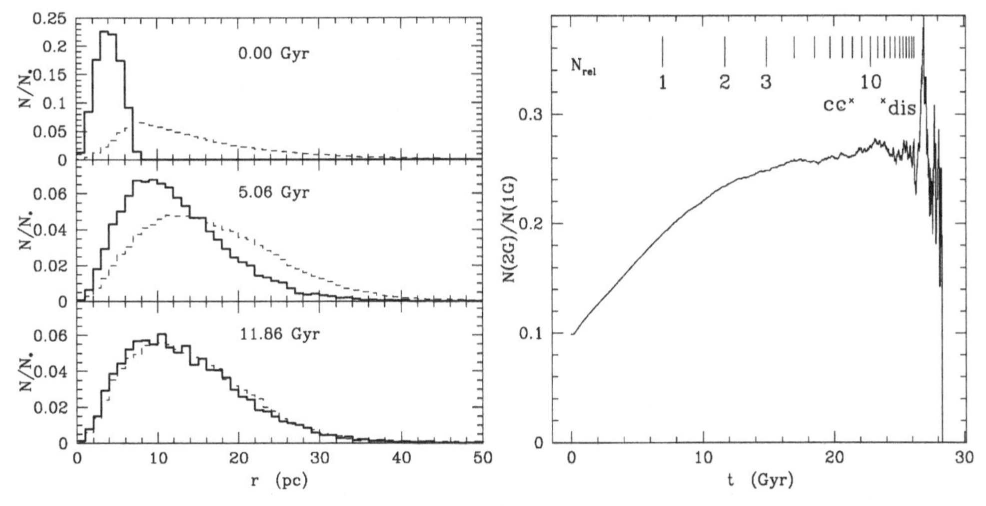

Figure 2. *Left:* radial distribution of the first (dashed lines) and second (full lines) generation of low-mass stars at three different times. Each histogram is normalised to the total number of stars in each population. *Right:* Number ratio between the second (with low initial specific energy) and first (with high initial specific energy) population of low-mass stars in a cluster with initially 128K stars as a function of time. At the top of each panel the number of elapsed relaxation times, with crosses indicating the times of core collapse and of cluster dissolution.

2.2. *Dynamical evolution of globular clusters*

First we assume that the globular clusters display primordial mass segregation so that the massive stars are located at their center. Since we expect that the formation of the second generation of low-mass stars happens locally around individual massive stars (see Decressin *et al.* 2007a for more details), the second generation of stars will also be initially more centrally concentrated than the first generation. In such a situation, two competitive processes act in the clusters: the loss of stars in the outer parts of the cluster will first reduce the number of bound first-generation stars; and the dynamical spread of the initially more concentrated second-generation stars will stop this differential loss when the two populations are dynamically mixed.

Our analysis, based on the N-body models computed by Baumgardt & Makino (2003) with the collisional Aarseth N-body code NBODY4 (Aarseth 1999), is presented in detail in Decressin *et al.* (2008).

As these models have been computed for a single stellar population, we apply the following process to mimic the formation of a cluster with two dynamically distinct populations: we sort all the low-mass stars ($M \leqslant 0.9$ $M_\odot$) according to their specific energy (i.e., their energy per unit mass). We define the second stellar generation as the stars with lowest specific energies, (i.e., those which are most tightly bound to the cluster due to their small central distance and low velocity). The number of second-generation stars is given by having their total number representing 10% of the total number of low-mass stars.

In Fig. 2 one can see the radial distribution of the two populations at various epochs. Initially, first-generation stars show an extended distribution up to 40 pc whereas the second-generation stars (with low specific energy) are concentrated within 6 pc around the centre.

Progressively the second-generation stars spread out due to dynamical encounters so their radial distribution extends. However this process operates on a long timescale: even after 5 Gyr of evolution the two populations still have different distributions. The bottom panel of Fig. 2 shows that after nearly 12 Gyr of evolution (slightly less than 3 elapsed relaxation times) the two populations have similar radial distributions and can no longer be distinguished owing to their dynamical properties.

As previously seen, the effect of the external potential of the Galaxy on the cluster is to strip away stars lying in the outer part of the cluster. Initially, as only stars of the first generation populate the outer part of the cluster owing to their high specific energy, only these first-generation stars are lost in the early times. This lasts until the second-generation stars migrate towards the outer part of the cluster. Depending on the cluster mass, it takes between 1 to 4 Gyr to start losing second-generation stars. Due to the time-delay to lose second-generation stars, their remaining fraction in the cluster is always higher than that of the first-generation stars, except during the final stage of cluster dissolution. Fig. 2 (right panel) quantifies this point by showing the time evolution of the number ratio of second to first-generation stars. As a direct consequence of our selection procedure, the initial ratio is 0.1; it then increases gradually with time and it tends to stay nearly constant as soon as the two distributions are similar. Finally, at the time of cluster dissolution (i.e., when the cluster has lost 95% of its initial mass, indicated by the label "dis" in Fig. 2), large variations occur due to the low number of low-mass stars present in the cluster. In Fig. 2 (right panel) we have also indicated the number of elapsed relaxation times, showing that the increase of the number ratio lasts only 3 relaxation times.

The fraction of second-generation stars relative to first-generation ones increases by a factor of 2.5 over the cluster history. Therefore, these second-generations stars can account for 25% of the low-mass stars present in the clusters. Compared to the observed ratios (70% and 85% in NGC 2808 and NGC 6752 respectively) the internal dynamical evolution and the dissolution due to the tidal forces of the host Galaxy are not efficient enough. An additional mechanism is thus needed to expel the first-generation stars more effectively.

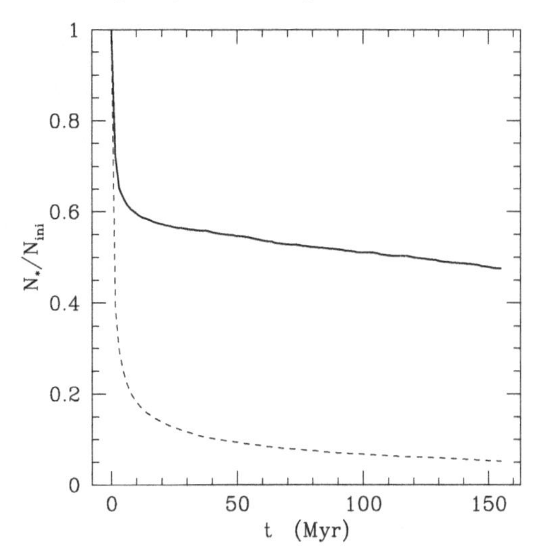

Figure 3. Evolution of the fraction of first- (dashed line) and second-generation (full line) stars still bound to the cluster with initial parameters: SFE of 0.33, $r_h/r_t = 0.06$ and $\tau_M/t_{\mathrm{Cross}} = 0.33$.

2.3. *Gas expulsion*

As it operates early in the cluster history (a few million years after cluster formation at the latest), initial gas expulsion by supernovae is an ideal candidate for such a process. As the gas still present after the star formation is quickly removed, it ensues a strong lowering of the potential well of the cluster so that the outer parts of the cluster can become unbound.

Baumgardt & Kroupa (2007) computed a grid of N-body models to study this process and its influence on cluster evolution by varying the free parameters: star formation efficiency, SFE, ratio between the half-mass and tidal radius, r_h/r_t, and the ratio between the timescale for gas expulsion to the crossing time, $\tau_M/t_{\mathrm{Cross}}$. They show in particular that, in some extreme cases, the complete disruption of the cluster can be induced by gas expulsion. This process has also been used successfully by Marks *et al.* (2008) to explain the challenging correlation between the central concentration and the mass function of globular clusters as found by De Marchi *et al.* (2007).

We have applied the same method as the one we used in § 2.2 to the models of Baumgardt & Kroupa (2007). Fig. 3 shows that in the case of a cluster which loses around 90% of its stars, the ejection of stars from the cluster mostly concerns first-generation ones. At the end of the computation only 5% of the first-generation stars remain bound to the cluster, along with around half of the second-generation stars. Therefore the number ratio of second to first-generation stars increases by a factor of 10: half of the population of low-mass stars still populating the cluster are second-generation stars. Further, the initial radial distribution is not totally erased by this mechanism, as the second-generation stars are still more centrally distributed. We can expect that this ratio will continue to increase during the long-term evolution of the cluster (see Decressin *et al.*, in preparation).

Thus if globular clusters are born mass segregated, dynamical processes (gas expulsion, tidal stripping and two-body relaxation) can explain the number fraction of second-generation stars with abundance anomalies. Similar conclusions have been reached by D'Ercole *et al.* (2008).

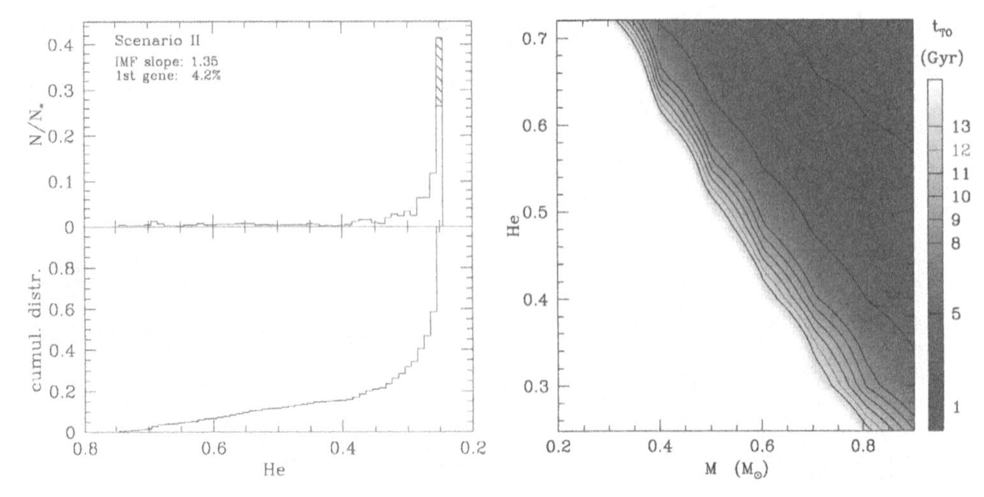

Figure 4. *Left:* distribution function of He for low-mass stars of second (white area) and first (hatched area) generation (top panel) and the cumulative distribution function of He in low mass-stars (bottom panel). *Right:* Age at the turn-off for low-mass stars as a function of their mass (0.2–0.9 $M_\odot$) and initial He value (0.245-0.72, mass fraction). White area indicates stars still on the main sequence after 15 Gyr of evolution.

3. He-rich stars

Since the abundance variations in light elements are expected to be due to H-burning whose direct product is Helium, we expect that second-generation stars are also enriched in He, to some degree. Unfortunately the He abundance cannot be directly measured in globular-cluster stars and we have only indirect evidence for an overabundance of He. The globular clusters ω Cen and NGC 2808 display multiple main sequences (Piotto *et al.* 2005, 2007); a double sub-giant branch is also found in NGC 1851 (Milone *et al.* 2008). Such features can be understood if the stars have a range of He contents. He enrichment is also a possible explanation for the appearance of extreme horizontal branches, as seen in several globular clusters (Caloi & D'Antona 2005, 2007).

3.1. *Evolution of He-rich stars*

As explained in Decressin *et al.* (2007a), material stored in the discs around massive stars is heavily enriched with He. Fig. 4 (left panel) gives our expected theoretical distribution function of the He-value in low-mass stars in NGC 6752. A main peak is present at $Y = 0.245$ and it extends up to 0.4. However a long tail toward higher Y-values is also present with around 12% of the stars with initial He value between 0.4 and 0.72.

To assess the implications for globular clusters induced by this population of He-rich stars we have computed a grid of low-mass stellar models from 0.2 to 0.9 $M_\odot$ at a metallicity of $Z = 0.0005$ (similar to the metal-poor globular cluster NGC 6752) for initial He mass fraction between 0.245 and 0.72 with the stellar evolution code STAREVOL V2.92 (see Siess *et al.* 2000; Siess 2006, for more details). These models have been computed without any kind of mixing except for an instantaneous mixing in convection zones. The adopted mass-loss rate follows the Reimers (1975) prescription (with the parameter value $\eta_R = 0.5$) with a $\sqrt{Z/Z_\odot}$ dependence. All models have been computed from the pre-main sequence phase to the end of the central He-burning phase.

For a given stellar mass, He-rich stars evolve faster on the main-sequence due to their lower initial H-content and to their higher luminosity. Figure 4 illustrates this point showing the turn-off age as a function of the initial mass and He mass fraction of stars.

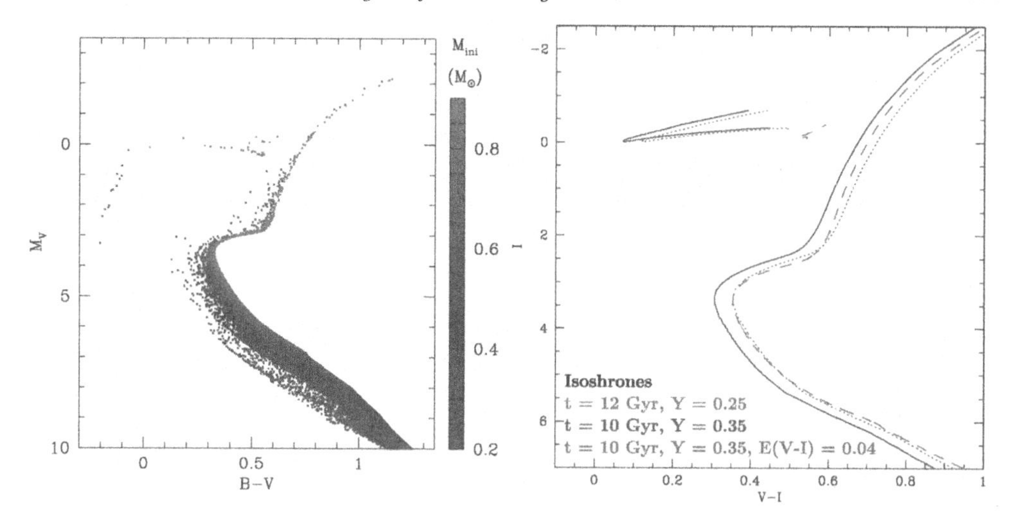

Figure 5. *Left:* Colour-magnitude diagram of a 12 Gyr old globular clusters with initial spread in He in stars similar to Fig. 4 (left panel). *Right:* isochrone of 12 Gyr and initial He of 0.25 (dashed lines) and isochrones of 10 Gyr and initial He of 0.35 without (full lines) and with (dotted lines) additional reddening.

After 12 Gyr, stars of 0.85 $M_\odot$ with standard helium ($Y = 0.245$), as well as He-rich stars of 0.4 $M_\odot$ ($Y = 0.6$), are leaving the main sequence.

3.2. *Effects on globular clusters*

Figure 5 (left panel) shows a synthetic colour-magnitude diagram (CMD) of a 12 Gyr old globular cluster with an initial spread in He as given by Fig. 4 (left panel). This CMD has been computed with a modified version of the program used by Meynet (1993) to investigate supergiant populations. The spread in He converts into a spread in mass at the turn-off. The luminosity increase of He-rich stars is mainly compensated by their shorter lifetime so that the turn-off luminosity is almost constant. In addition, due to their differences in opacity and to their compactness, they are also hotter. Thus the He-rich main-sequence and RGB stars are shifted to the left side of the CMD. They also occupy the blue part of the HB down to the extreme-HB location.

If we compare our theoretical CMD with the one observed by Brown *et al.* (2005) for NGC 6752 we note some discrepancies. First the theoretical width of the main sequence at the turn-off is too large for this cluster. Additionally NGC 6752 shows an extended blue HB, with no stars in the red part. This last discrepancy could be attributed to the low mass-loss rate used in the stellar models which do not remove enough mass along the RGB and hence produce stars on the HB that are too cool. As the theoretical spread of the initial He is strongly affected by the dilution of the disc ejected by fast rotating massive stars and the ISM, we plan to constrain this dilution with the observed sequences to check whether we are able to consistently reproduce both the abundance anomalies and the He-value inferred in globular clusters (see Decressin *et al.*, in preparation).

The uncertainties pertaining the ages of globular clusters are manifold. Among them, photometric uncertainties widen both sides of the main-sequence, unresolved binaries broaden the main-sequence towards cooler effective temperature, an increase of metallicity (as seen in ω Cen) induces redder sequences. The presence of He-rich stars can induce some additional uncertainties. Let us note that the He content is the only physical parameter which broadens the main-sequence only to its left (i.e., blue) side. In Fig. 5 (right panel) we evaluate uncertainties related to He-rich stars: we try to reproduce a

He-normal isochrone ($Y=0.25$) with a He-rich ($Y=0.35$) one. This could be done with an isochrone 2 Gyr younger along with an increase of the reddening of 0.04 magnitudes. The differences between the He-rich with reddening and He-normal isochrones are small around the TO, the main-sequence and the subgiant-branch. Discrepancies appear along the RGB and at the level of the HB, where the normal He-rich isochrone is much less extended toward the blue. Thus the age uncertainties due to a population of He-rich stars can be of the order of 1–2 Gyr.

Acknowledgements

T. D. and C. C. acknowledges financial support from the Swiss FNS.

References

Aarseth, S. J. 1999, *PASP*, 111, 1333
Baumgardt, H. & Kroupa, P. 2007, *MNRAS*, 380, 1589
Baumgardt, H. & Makino, J. 2003, *MNRAS*, 340, 227
Brown, T. M., Ferguson, H. C., Smith, E., *et al.* 2005, *AJ*, 130, 1693
Caloi, V. & D'Antona, F. 2005, *A&A*, 435, 987
Caloi, V. & D'Antona, F. 2007, *A&A*, 463, 949
Carretta, E., Bragaglia, A., Gratton, R. G., *et al.* 2006, *A&A*, 450, 523
Carretta, E., Bragaglia, A., Gratton, R. G., Lucatello, S., & Momany, Y. 2007, *A&A*, 464, 927
Charbonnel, C. 2005, in IAU Symposium 228, ed. V. Hill, P. François, & F. Primas, 347
De Marchi, G., Paresce, F., & Pulone, L. 2007, *ApJ*l, 656, L65
Decressin, T., Baumgardt, H., & Kroupa, P. 2008, *A&A*, in press
Decressin, T., Charbonnel, C., & Meynet, G. 2007a, *A&A*, 475, 859
Decressin, T., Meynet, G., Charbonnel, C., Prantzos, N., & Ekström, S. 2007b, *A&A*, 464, 1029
Denisenkov, P. A. & Denisenkova, S. N. 1989, *Astronomicheskij Tsirkulyar*, 1538, 11
Denisenkov, P. A. & Denisenkova, S. N. 1990, *Soviet Astronomy Letters*, 16, 275
D'Ercole, A., Vesperini, E., D'Antona, F., McMillan, S. L. W., & Recchi, S. 2008, *MNRAS*, 1228
Ekström, S., Meynet, G., Maeder, A., & Barblan, F. 2008, *A&A*, 478, 467
Gratton, R., Sneden, C., & Carretta, E. 2004, *ARA&A*, 42, 385
Langer, G. E. & Hoffman, R. D. 1995, *PASP*, 107, 1177
Marks, M., Kroupa, P., & Baumgardt, H. 2008, *MNRAS*, 386, 2047
Meynet, G. 1993, in The Feedback of Chemical Evolution on the Stellar Content of Galaxies, ed. D. Alloin & G. Stasińska, 40
Milone, A. P., Bedin, L. R., Piotto, G., *et al.* 2008, *ApJ*, 673, 241
Piotto, G., Bedin, L. R., Anderson, J., *et al.* 2007, *ApJ*l, 661, L53
Piotto, G., Villanova, S., Bedin, L. R., *et al.* 2005, *ApJ*, 621, 777
Prantzos, N. & Charbonnel, C. 2006, *A&A*, 458, 135
Prantzos, N., Charbonnel, C., & Iliadis, C. 2007, *A&A*, 470, 179
Reimers, D. 1975, Circumstellar envelopes and mass loss of red giant stars (Problems in stellar atmospheres and envelopes.), 229
Salpeter, E. E. 1955, *ApJ*, 121, 161
Siess, L. 2006, *A&A*, 448, 717
Siess, L., Dufour, E., & Forestini, M. 2000, *A&A*, 358, 593
Townsend, R. H. D., Owocki, S. P., & Howarth, I. D. 2004, *MNRAS*, 350, 189

Discussion

A. Bragaglia: Have you also tried to reproduce NGC 2808 with your models? You predict a continuous He enhancement, while from the three sequences we require discrete values.

T. Decressin: Actually I have used a local dilution of the slow winds produced by fast-rotating massive stars with the intra-cluster medium to be able to reproduce the continuous distribution of [O/Na] in NGC 6752. This scheme obviously leads to a continuous He-enrichment, contrary to what is observed in NGC 2808. In the future I will try to use this discrete He value to constrain the dilution process, which is presently poorly understood.

Mario Livio

Michal Simon

The Ages of Stars
Proceedings IAU Symposium No. 258, 2008
E.E. Mamajek, D.R. Soderblom & R.F.G. Wyse, eds.

doi:10.1017/S1743921309031925

The star clusters of the Magellanic Clouds

A. D. Mackey

Institute for Astronomy, University of Edinburgh,
Royal Observatory, Blackford Hill, Edinburgh, EH9 3HJ, UK
email: dmy@roe.ac.uk

Abstract. The Magellanic Clouds possess extensive systems of rich star clusters. These objects span a wide range in age and metal abundance, and are close enough to be fully resolved into individual stars. They represent the most accessible examples of such clusters and are therefore key to a wide variety of astronomical research. In this contribution I describe recent results from work on several problems in Magellanic Cloud cluster astronomy of relevance to *The Ages of Stars*. These include testing and constraining stellar evolution and simple stellar population models, investigating the formation and evolution of the Clouds themselves, and the discovery of several intermediate-age clusters which apparently possess more than one stellar population.

Keywords. Magellanic Clouds, galaxies: individual (Large Magellanic Cloud, Small Magellanic Cloud), galaxies: star clusters, galaxies: formation, galaxies: evolution

---

1. Introduction

The Large and Small Magellanic Clouds (LMC/SMC) are two dwarf irregular companions to the Milky Way, lying at distances of ~ 50 kpc and ~ 60 kpc, respectively. Both galaxies possess extensive systems of star clusters – the most recent census, by Bica *et al.* (2008), lists over 3700 such objects in total. A significant number of these (perhaps $\sim 100-200$) are rich star clusters with masses comparable to many of the globular clusters observed in the Milky Way. Unlike the Galactic globular clusters, however, Magellanic Cloud clusters have ages spanning the full range $\sim 10^6 - 10^{10}$ years: from very newly formed objects such as 30 Doradus, to clusters apparently coeval with the oldest Galactic globulars. In addition, the Magellanic Cloud clusters span roughly ~ 2 dex in metal abundance, and are sufficiently close that, with some effort, they may be studied on a star-by-star basis as fully-resolved systems. Taken together, these properties mean that Magellanic Cloud star clusters probe regions of parameter space (in age, metallicity and mass) that are not accessible in such detail anywhere else, and they are hence important to a surprisingly wide variety of astrophysical research (see e.g., Santiago 2009).

In the context of this meeting, there are two key questions we can ask – first, what can we learn from star clusters in the LMC and SMC about how to measure the ages of stars and star clusters; and second, assuming we are able to determine such ages with sufficient accuracy, what astrophysical questions do these two systems allow us to address? Covering the answers to these questions in detail would be worthy of a symposium (or, at the very least, a workshop) each, and is far beyond the scope of this contribution. Here I will merely touch on a few relevant topics which I, personally, find interesting and important, and which have seen significant recent work.

2. Testing stellar evolution models

Because they cover a wide variety of ages and metallicities, are generally considered as template single stellar populations, and are close enough to be studied as fully-resolved

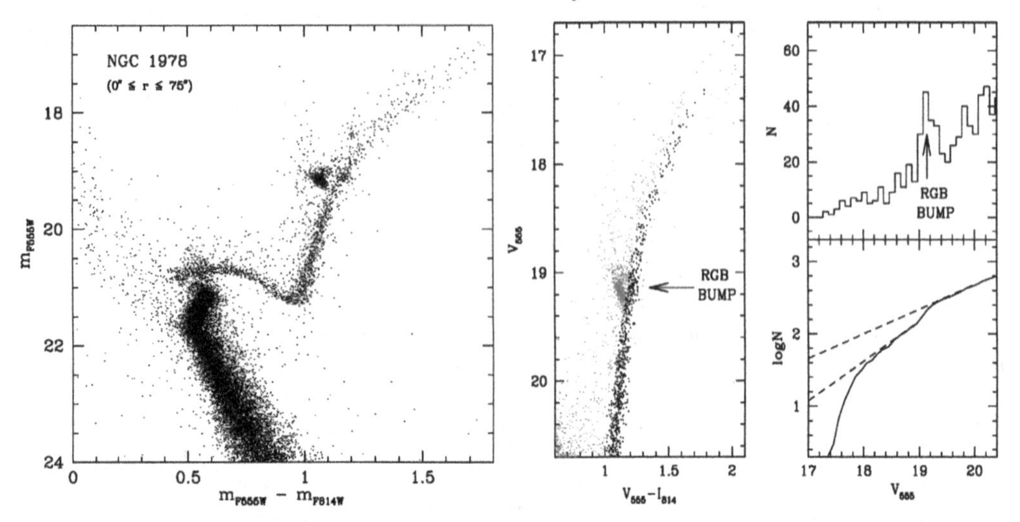

Figure 1. *Left:* CMD constructed from HST/ACS imaging of the rich intermediate-age LMC clusters NGC 1978, taken as part of our snapshot program #9891. The evolutionary sequences are narrow and very well defined. Note that stars brighter than $m_{F555W} \approx 18$ are saturated, so that the upper RGB is artificially broadened. *Centre and Right:* These plots, reproduced from Mucciarelli *et al.* (2007), show the presence of the RGB bump in the cluster CMD. This feature is particularly evident in the luminosity function of the RGB stars, shown at right.

systems, the massive star clusters of the Magellanic Clouds have long been used to progress various aspects of our understanding of stellar evolution, and to test and constrain stellar evolution models. Topics for study have been as wide-ranging as, for example, the properties of variable stars, including Cepheid variables (e.g., Bono & Marconi 1997); the evolutionary properties of high- and intermediate-mass stars (e.g., Massey & Hunter 1998; Barmina *et al.* 2002); the properties of stars undergoing very rapid phases of evolution such as occur on the AGB (e.g., Girardi & Marigo 2007; Lebzelter & Wood 2007; Lebzelter *et al.* 2008); and the properties of pre-main sequence stars (e.g., Nota *et al.* 2006; Carlson *et al.* 2007; Gouliermis *et al.* 2007; Hennekemper *et al.* 2008).

An interesting recent example has been the work done by Mucciarelli *et al.* (2007) on the very rich intermediate-age LMC cluster NGC 1978. This object was imaged with HST/ACS as part of our Cycle 12 snapshot survey of Magellanic Cloud clusters (program #9891, PI: G. Gilmore). The CMD for NGC 1978 is shown in the left panel of Fig. 1. Due to the high photometric accuracy of HST/ACS the evolutionary sequences are narrow and very well defined; further, because of the rather high mass of the cluster (2×10^5 M$_\odot$) these sequences are all very well populated – even for relatively rapid evolutionary phases such as the SGB. Mucciarelli *et al.* (2007) noted the presence of the so-called RGB bump – this occurs at $m_{F555W} \approx 19.10$ (Fig. 1, central panel). To emphasize this feature, Mucciarelli *et al.* (2007) constructed a luminosity function (LF) for the cluster RGB stars (Fig. 1, right panels) – the RGB bump shows up clearly in both the binned and cumulative forms of the LF. Although the RGB bump has been observed in several Galactic globulars, this is the first clear-cut detection of this feature in an intermediate-age cluster.

Mucciarelli *et al.* (2007) compare the observed morphology of the evolutionary sequences with the expectations of stellar evolution models from several groups. Because of their high quality photometry, the authors were able to incorporate both the shape of the cluster CMD and the observed ratios of stars on different evolutionary sequences – for example, the number of RGB stars to the number of SGB stars. These ratios provide

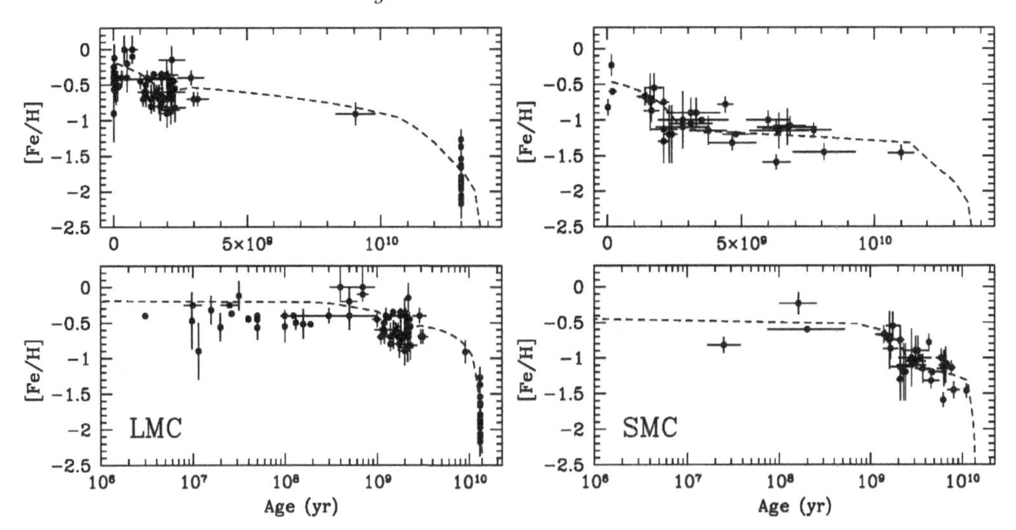

Figure 2. Age-metallicity relationships for clusters in the LMC (left panels) and SMC (right panels). In the upper panels the age axis is plotted linearly; in the lower panels it is plotted logarithmically. All data were assembled from compilations in the recent literature: Geisler *et al.* (2003); Glatt *et al.* (2008a,b); Kerber *et al.* (2007); Mackey & Gilmore (2003a,b); Mackey & Gilmore (2004); Mackey *et al.* (2006); Piatti *et al.* (2003); Piatti *et al.* (2005); Piatti *et al.* (2008); see also the references listed in these works. The dashed lines are chemical enrichment models by Pagel & Tautvaišienė (1998) for bursting star-formation histories.

direct constraints on the evolutionary time-scales. They find that the best-fitting models all require some degree of convective overshooting to properly reproduce both evolutionary sequence morphologies and star counts. Furthermore, none of the models performed particularly well in reproducing the location of the RGB bump. Future observations of this feature in additional clusters may therefore be useful for placing empirical constraints on the evolutionary tracks of intermediate-age stars.

3. Formation and evolution of the Magellanic Clouds

The star cluster systems of the Magellanic Clouds are of considerable importance to our understanding of the formation and evolution of the Clouds themselves. Fig. 2 presents age-metallicity relationships derived from LMC and SMC clusters. There are several notable features in these two plots. Both show a clear increase in metal abundance with time, indicating the process of chemical enrichment in the two Clouds. At any given age (except perhaps for the very oldest objects), the SMC star clusters are on average more metal-poor than those in the LMC, indicating that chemical enrichment processes have proceeded more slowly in the smaller galaxy. Over-plotted are the predictions of chemical enrichment models by Pagel & Tautvaišienė (1998) for bursting star-formation histories in the two Clouds. These do a good job of matching the observed cluster data – better, arguably, than do models for continuous star-formation histories (see also Da Costa & Hatzidimitriou 1998; Piatti *et al.* 2005). One important question is whether the age-metallicity relationships from star clusters match those derived from field stars – that is, can we reliably infer chemical enrichment histories in distant galaxies where the field populations cannot be resolved, by studying the integrated properties of individual star clusters? This is still an unsolved problem; however, the Magellanic Clouds are clearly excellent laboratories for its study, and significant progress is being made (see, e.g., the contribution by Gallart in these proceedings).

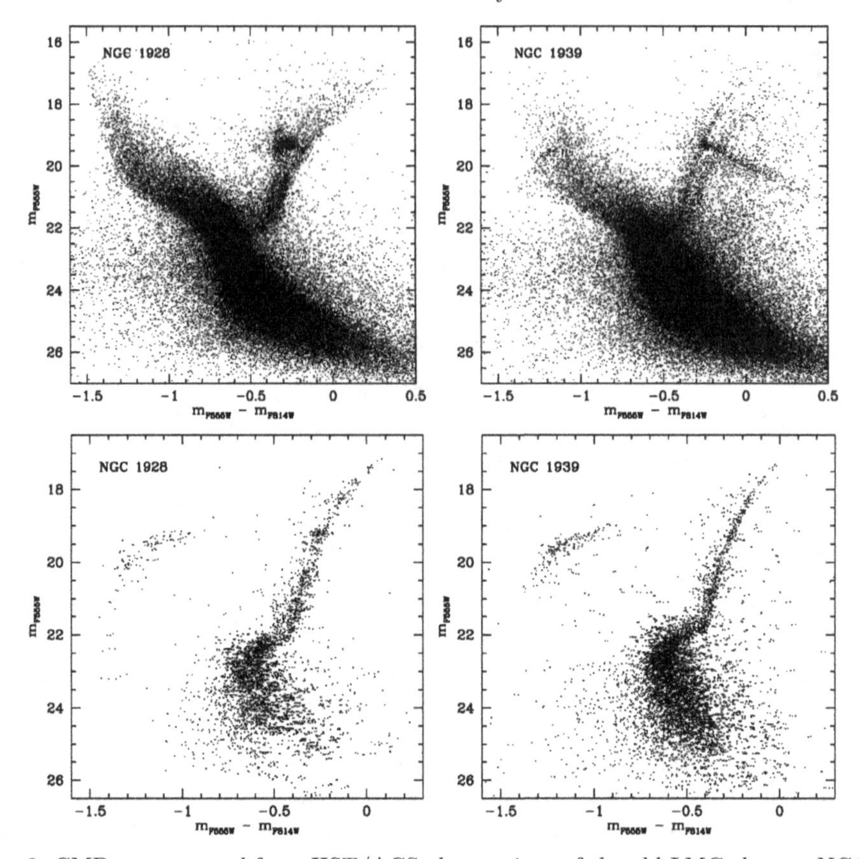

Figure 3. CMDs constructed from HST/ACS observations of the old LMC clusters NGC 1928 (left) and 1939 (right) taken as part of our snapshot program #9891 (Mackey & Gilmore 2004). The upper panels show all stellar detections in the two ACS observations – the LMC bar fields against which the clusters are projected are so densely populated that it is impossible to clearly discern the cluster sequences. Only after a careful statistical subtraction of the field has been made (lower panels) does the ancient nature of the two targets become evident. Comparison of the fiducial sequences for NGC 1928 and 1939 with those for the oldest Galactic and LMC clusters revealed the full group of objects to be coeval within ~ 2 Gyr.

What is certainly clear is that the observed cluster formation history in a galaxy does not necessarily accurately reflect its star formation history. A notable feature of the LMC cluster formation history is an almost complete dearth of clusters with ages in the range $\sim 3 - 13$ Gyr. This "age gap" is quite evident in the upper left panel of Fig. 2, and is a well-known feature of the LMC system (e.g., Jensen *et al.* 1988; Da Costa 1991; Sarajedini 1998; Rich *et al.* 2001). Barring one object (see below), extensive searches have failed to locate any clusters lying in the age gap (e.g., Geisler *et al.* 1997); however, the LMC star-formation rate was clearly non-zero during this period of time (e.g., Holtzman *et al.* 1999; Smecker-Hane *et al.* 2002; Carrera *et al.* 2008). Curiously, the SMC does not appear to have an age gap, as it possesses many intermediate-age star clusters (e.g., Mighell *et al.* 1998; Rich *et al.* 2000; Glatt *et al.* 2008b). The LMC age gap has been interpreted theoretically in terms of repeated tidal interactions between the LMC, SMC, and Milky Way (Bekki *et al.* 2004); however, it is not yet clear how such models can be reconciled with the recent large LMC and SMC proper motions measured with HST (Kallivayalil *et al.* 2006a,b; Besla *et al.* 2007).

The one LMC cluster known to have an age lying in the range 3 – 13 Gyr is ESO 121-SC03. This was first studied by Mateo *et al.* (1986), who identified it as an unusual, and possibly unique, LMC member. ESO 121-SC03 was imaged as part of our Cycle 12 HST snapshot survey of rich star clusters in the Magellanic Clouds (program #9891), which has allowed construction of a deep CMD, and precise constraints to be placed on its age. Mackey *et al.* (2006) showed that the CMD of ESO 121-SC03 is very similar to that of the young Galactic globular cluster Palomar 12, which is known to have been accreted from the Sagittarius dwarf galaxy. Mackey *et al.* (2006) further demonstrated that ESO 121-SC03 is $73 \pm 4\%$ as old as the well-studied Galactic globular cluster 47 Tuc (NGC 104). Together these results translate into an age of 9.0 ± 0.8 Gyr, confirming that ESO 121-SC03 is indeed the only known LMC cluster to fall in the age gap.

It is also of interest to consider the oldest clusters in the two Magellanic systems. The LMC in particular possesses a significant population of extremely old objects, with ages ~ 13 Gyr – these clusters form the upper boundary of the age gap. Deep, high-resolution imaging from HST has allowed detailed relative age comparisons between members of this population and the oldest Galactic globular clusters (e.g., Brocato *et al.* 1996; Olsen *et al.* 1998; Johnson *et al.* 1999; Mackey & Gilmore 2004). These studies have shown that, to within the precision of current techniques, the oldest LMC clusters are coeval with the oldest Galactic globulars. Furthermore, there is a dispersion of at most ~ 2 Gyr among the ensemble of ancient LMC clusters. These results imply that the LMC formed at the same time as the Milky Way. There is also a significant spread in iron abundance among the ancient LMC globular clusters (Fig. 2), indicating that, at least in some locations, there must have been rapid early chemical enrichment in this galaxy.

A relevant question is whether the census of ancient LMC clusters is now complete. Some indication of the difficulties involved in answering this question can be seen from Fig. 3, which shows CMDs for the most recently identified members of this population, NGC 1928 and 1939 (Mackey & Gilmore 2004). These two clusters are so compact and lie against such highly crowded LMC bar fields that it required HST/ACS to obtain CMDs definitively proving their age; even then, the task was not straightforward. It is conceivable other such clusters still remain unidentified. Furthermore, a number of the oldest LMC clusters are extremely remote objects – NGC 1841 lies more than 14° from the LMC centre. It is certainly possible that unknown faint, remote old LMC clusters exist; forthcoming surveys should facilitate the discovery of any such objects.

In the SMC the story is somewhat different. There is only one cluster known to be older than 10 Gyr, NGC 121, and several HST-based studies have shown this object to clearly be ~ 2 Gyr younger than the oldest Galactic and LMC globulars (Mighell *et al.* 1998; Shara *et al.* 1998; Glatt *et al.* 2008a). It is not clear what the somewhat young age of NGC 121 tells us about the formation of the SMC – it may be that it results from delayed cluster formation in this galaxy, or it may result from the random survival of only one object out of a small population of ancient clusters. NGC 121 is also not particularly metal-poor, indicating that in the SMC, as in the LMC, there must have been some significant chemical enrichment relatively early on.

4. Integrated light measurements

The massive star clusters of the Magellanic Clouds are extremely useful as calibrators of simple stellar population (SSP) models, because they span a wide variety of ages and metal abundances, and, in addition, because they are located at a very convenient distance where precise observations of both their integrated and resolved properties may be obtained. With such measurements in hand, accurate ages and metallicities derived from

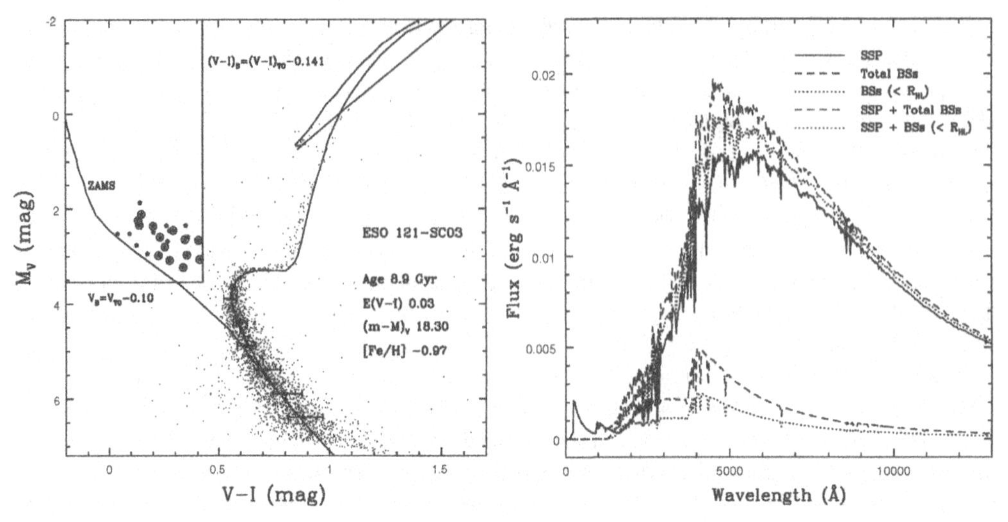

Figure 4. *Left:* A HST/ACS CMD for the LMC age gap cluster ESO 121-SC03. The significant population of blue straggler stars in this cluster is highlighted. *Right:* The effect of these stars on the cluster's integrated spectral energy distribution (ISED) is illustrated. The solid line is the ISED from the SSP model of appropriate age and metallicity. The dotted lines show the expected contribution to the ISED made by blue stragglers within the cluster half-light radius; while the dashed lines show the expected contribution made by all blue stragglers thought to be cluster members. The effect of the blue stragglers is quite significant, especially blueward of ~ 7000Å. These plots have been reproduced from Xin *et al.* (2008).

the resolved observations may be compared to the predictions of evolutionary synthesis models derived from, for example, the integrated cluster colours observed in different pass-bands. Providing empirical tests and constraints on SSP models in this way is of significant importance, because in galaxies beyond the Local Group it is at present not possible to resolve clusters into individual stars – therefore, information on cluster ages and metal abundances may only be obtained from integrated light measurements. An interesting example of this process is the recent work of Pessev *et al.* (2006, 2008), who assembled a database of near-infrared and optical integrated colours for a large sample of massive Magellanic Cloud clusters, and then used this information in combination with the known ages and metallicities of these objects (as derived from resolved photometry and spectroscopy) to test the performance of four popular SSP models in optical/near-IR colour-colour space.

One hitherto poorly-investigated aspect of the above problem is the effect of stars which are not well-incorporated into SSP models on the properties derived from integrated measurements. For example, blue stragglers are found in many different types of star cluster; however, such objects are not included in SSP models because they are generally formed as the result of binary star evolution or stellar dynamical processes. The presence of such stars in a cluster can add a significant contribution to the integrated light at blue and ultraviolet wavelengths, especially in older clusters where the stars are generally quite red, and this can lead to significant errors when deriving ages and metallicities from SSP models – specifically, the derived cluster age will be too young and the derived metal abundance too low†. As a demonstration of this problem, Xin *et al.* (2008) studied the LMC age gap cluster ESO 121-SC03, which possesses a significant population of blue stragglers (Fig. 4). They showed that these stars can make a significant alteration

† Note that a similar role can be played by extreme blue horizontal branch stars in very old clusters – the parameters controlling horizontal branch morphology are still not fully understood, and therefore not well-incorporated into SSP models.

to the derived broad-band integrated colours of the cluster, as well as to its integrated spectral energy distribution. These can lead to an age underestimate of $\sim 40-60\%$ and a metallicity underestimate of $\sim 30-60\%$. Ultimately, by assessing such effects over a wide variety of cluster ages, metallicities and dynamical environments, it should be possible to derive empirical corrections for SSP models to account for the effects of blue stragglers. Magellanic Cloud clusters will certainly play a significant role in achieving this goal.

5. Multiple stellar populations in Magellanic Cloud clusters

Recent photometric studies of massive intermediate-age star clusters in the Magellanic Clouds have discovered objects exhibiting extremely unusual main-sequence turn-offs (MSTOs) in their CMDs. The first indications that such clusters might exist date back several years to observations obtained with terrestrial facilities – Bertelli *et al.* (2003) demonstrated that the LMC cluster NGC 2173 apparently possesses an unusually large spread in colour about its MSTO, while Baume *et al.* (2007) obtained a similar result for the LMC cluster NGC 2154.

More recent studies based on deep precision photometry from HST/ACS have revealed the truly peculiar nature of many intermediate-age LMC clusters. The clearest example is NGC 1846, which was first imaged as part of our Cycle 12 snapshot survey of rich star clusters in the Magellanic Clouds (program #9891). The CMD for this cluster displays two distinct MSTO branches (Mackey & Broby Nielsen 2007), but is otherwise as expected for an intermediate-age cluster. In particular, it possesses a narrow red-giant branch, sub-giant branch and main sequence, and a compact, well-defined red clump (Fig. 5). This implies that the turn-off features are not due to significant line-of-sight depth or differential reddening in this cluster. The narrow sequences further suggest a minimal internal dispersion in [Fe/H]; however, they do not allow strong constraints to be placed on the possibility of internal variations in other chemical abundances – for example, CN, O, Na, or [α/Fe] – as are observed for several of the peculiar massive Galactic globular clusters (e.g., Piotto *et al.* 2008).

Our Cycle 12 snapshot data revealed several additional candidates possessing unusual MSTOs – for example, NGC 1806, 1783, 1852, and 2154, in decreasing order of significance. CMDs for these clusters are presented in Mackey *et al.* (2009); our result for NGC 2154 apparently confirms the observations of Baume *et al.* (2007). Three of the most populous clusters in our sample of candidates, NGC 1846, 1806, and 1783, have deeper ACS imaging available in the HST public archive – this is from GO program #10595 (PI: P. Goudfrooij). Several groups have independently recently published CMDs from these data (Mackey *et al.* 2008; Milone *et al.* 2008; Goudfrooij *et al.* 2009) – each group used different photometry software but the CMDs match quite closely (although differing in some fine details). In Fig. 6 I present the results from Mackey *et al.* (2008). Each of the three clusters unambiguously possesses a peculiar MSTO: NGC 1846 and 1806 possess two MSTO branches, while NGC 1783 possesses a turn-off covering a spread in colour much larger than can be explained by the photometric uncertainties (note the very narrow upper main sequence). This feature may represent a bifurcated turn-off in which the branches are unresolved on the CMD (see e.g., Milone *et al.* 2008), or it may represent a smooth spread of stars.

As with NGC 1846, both NGC 1806 and 1783 possess very narrow sequences across the remainder of their CMDs, implying no significant line-of-sight depth, differential reddening, or internal dispersion in [Fe/H] for either of these two systems. Mucciarelli *et al.* (2008) obtained high resolution spectra for 6 RGB stars in NGC 1783 and found no significant star-to-star dispersion in [α/Fe]; however similar measurements are not presently available for NGC 1846 or 1806. If, in the absence of this information, we assume

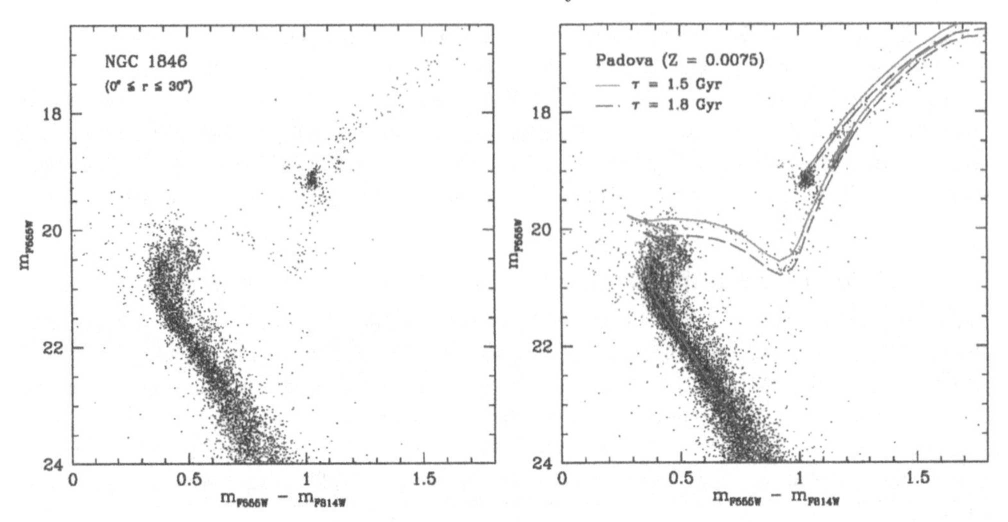

Figure 5. *Left:* HST/ACS CMDs from our HST snapshot imaging of NGC 1846. The double main-sequence turn-off is clearly visible (Mackey & Broby Nielsen 2007); otherwise the CMD is as expected for an intermediate-age cluster (see text). *Right:* Isochrone fit to the HST #9891 CMD for NGC 1846. Under the assumption of chemical homogeneity and a uniform distance and foreground extinction, an age spread of ~300 Myr within this ~1.8 Gyr old cluster can closely reproduce the observed CMD.

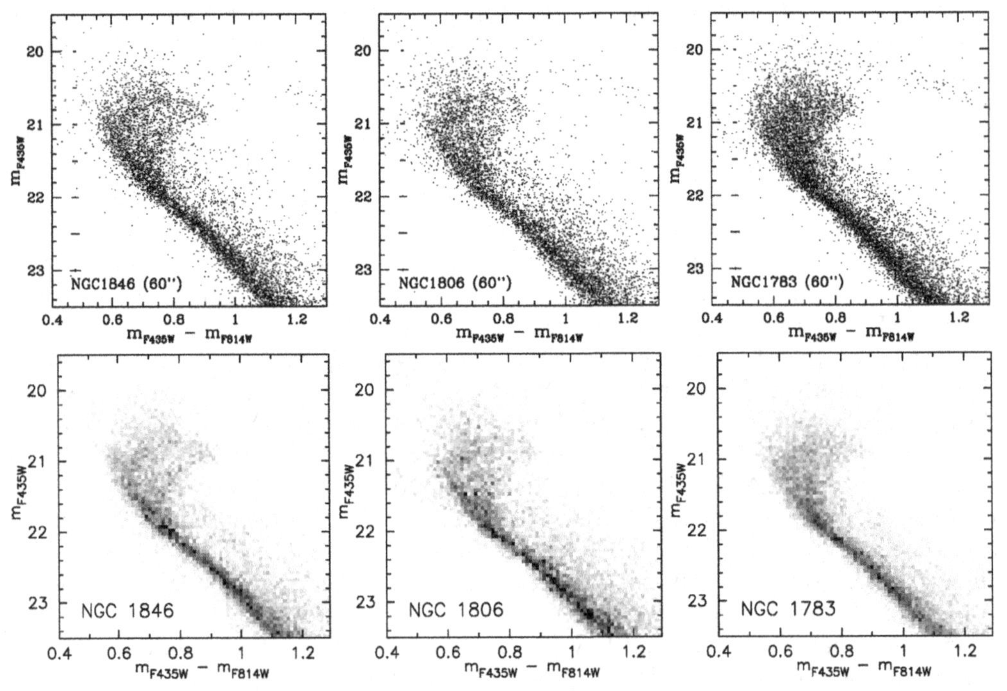

Figure 6. HST/ACS CMDs (upper panels) and Hess diagrams (lower panels) constructed from imaging taken under HST program #10595, obtained from the public archive. These deep CMDs confirm the peculiar nature of the clusters identified in our HST #9891 snapshot imaging. NGC 1846 and 1806 clearly display bifurcated main-sequence turn-offs, while NGC 1783 has a turn-off exhibiting a large spread in colour. Each cluster has a very narrow main sequence and a significant population of unresolved binary stars.

complete chemical homogeneity in each of the three systems, the simplest explanation for the peculiar MSTOs is that each cluster possesses at least two stellar populations of differing ages. Isochrone fitting (e.g., Fig. 5) results in implied age spreads of $\sim 200-300$ Myr within each cluster, with the oldest populations being $\sim 1.8-2.0$ Gyr old (Mackey & Broby Nielsen 2007; Mackey *et al.* 2008; Milone *et al.* 2008). Note that each of the three clusters possesses a significant population of unresolved binary stars (evident from their CMDs); however, these do not explain the MSTO morphologies (Mackey *et al.* 2008).

Milone *et al.* (2008) studied a large sample of 16 intermediate-age LMC clusters with HST/ACS imaging, and found that 11 of these possess CMDs exhibiting MSTOs which are not consistent with being simple, single stellar populations. Of these 11, four clearly show bifurcated MSTOs (the three described above plus NGC 1751) while the remainder possess more sparsely populated CMDs that make the precise morphologies of their peculiar MSTOs difficult to ascertain. All are consistent with being double TOs, or alternatively with possibly being more smoothly distributed intrinsic broadenings. Nonetheless, all 11 clusters again exhibit narrow sequences across the remainder of their CMDs, suggesting that the peculiar MSTO features are due to internal age dispersions of $\sim 150-300$ Myr. Of their three most populated clusters with double MSTOs, Milone *et al.* (2008) found that the inferred younger populations may comprise up to $\sim 70\%$ of the stars in the central regions of these systems. It is further worth noting that one intermediate-age SMC cluster, NGC 419, also likely possesses a peculiar MSTO (Glatt *et al.* 2008b).

The rapidly-growing number of intermediate-age Magellanic Cloud star clusters that are being found to possess peculiar MSTO morphologies suggests that such a feature may not be uncommon for this type of object. This is a surprising result since, as discussed above, Magellanic Cloud star clusters have long been treated as prototypical single stellar populations, used to test and calibrate stellar evolution and SSP models. A number of possible origins for these peculiar clusters have been proposed – for example, self-pollution by AGB stars; the merger of two (or more) star clusters formed in the same giant molecular cloud (GMC); or a scenario where clusters interact and merge with star-forming GMCs (Bekki & Mackey 2009). One alternative possibility is that the MSTO morphologies represent some hitherto poorly-understood aspect of the evolution of intermediate-age stars. In this case, Magellanic Cloud clusters will, once again, be teaching us something new about stellar evolution.

Acknowledgements

ADM is supported by the Marie Curie Excellence Grant of Annette Ferguson, which comes from the European Commission under contract MCEXT-CT-2005-025869. ADM is grateful to Alessio Mucciarelli and Yu Xin for their kind help with Figures 1 and 4.

References

Barmina, R., Girardi, L., & Chiosi, C. 2002, *A&A*, 385, 847

Baume, G., Carraro, G., Costa, E., Mendez, R. A., & Girardi, L. 2007, *MNRAS*, 375, 1077

Bekki, K., Couch, W., Beasley, M., Forbes, D., Chiba, M., & Da Costa, G. 2004, *ApJ*, 610, L93

Bekki, K. & Mackey, A. D. 2009, *MNRAS*, in press

Bertelli, G., Nasi, E., Girardi, L., Chiosi, C., Zoccali, M., & Gallart, C. 2003, *AJ*, 125, 770

Besla, G., *et al.* 2007, *ApJ*, 668, 949

Bica, E., Bonatto, C., Dutra, C. M., & Santos Jr., J. F. C. 2008, *MNRAS*, 389, 678

Brocato, E., Castellani, V., Ferraro, F. R., Piersimoni, A. M., & Testa, V. 1996, *MNRAS*, 282, 614

Carlson, L. R., *et al.* 2007, *ApJ*, 665, L109

Carrera, R., Gallart, C., Hardy, E., Aparicio, A., & Zinn, R. 2008, *AJ*, 135, 836

Da Costa, G. S. 1991, in R. Haynes & D. Milne (eds.), *Proceedings IAU Symposium 148: The Magellanic Clouds* (Kluwer Academic Publishers: Dordrecht), 183
Da Costa, G. S. & Hatzidimitriou, D. 1998, *AJ*, 115, 1934
Geisler, D., Bica, E., Dottori, H., Clariá, J. J., Piatti, A. E., & Santos Jr., J. F. C. 1997, *AJ*, 114, 1920
Geisler, D., Piatti, A. E., Bica, E., & Clariá, J. J. 2003, *MNRAS*, 341, 771
Girardi, L. & Marigo, P. 2007, *A&A*, 462, 237
Glatt, K., *et al.* 2008a, *AJ*, 135, 1106
Glatt, K., *et al.* 2008b, *AJ*, 136, 1703
Goudfrooij, P., Puzia, T. H., Kozhurina-Platais, V., & Chandar, R. 2009, *AJ*, submitted
Gouliermis, D. A., *et al.* 2007, *ApJ*, 665, L27
Hennekemper, E., *et al.* 2008, *ApJ*, 672, 914
Holtzman, J. A., *et al.* 1999, *AJ*, 118, 2262
Jensen, J., Mould, J., & Reid, N. 1988, *ApJS*, 67, 77
Johnson, J. A., Bolte, M., Stetson, P. B., Hesser, J. E., & Somerville, R. S. 1998, *ApJ*, 527, 199
Kallivayalil, N., *et al.* 2006a, *ApJ*, 638, 772
Kallivayalil, N., van der Marel, R. P., & Alcock, C. 2006b, *ApJ*, 652, 1213
Kerber, L.O., Santiago, B. X., & Brocato, E. 2007, *A&A*, 462, 139
Lebzelter, T. & Wood, P. R. 2007, *A&A*, 475, 643
Lebzelter, T., Lederer, M. T., Cristallo, S., Hinkle, K. H., Straniero, O., & Aringer, B. 2008, *A&A*, 486, 511
Mackey, A. D. & Gilmore, G. F. 2003a, *MNRAS*, 338, 85
Mackey, A. D. & Gilmore, G. F. 2003b, *MNRAS*, 338, 120
Mackey, A. D. & Gilmore, G. F. 2004, *MNRAS*, 352, 153
Mackey, A. D., Payne, M. J., & Gilmore, G. F. 2006, *MNRAS*, 369, 921
Mackey, A. D. & Broby Nielsen, P. 2007, *MNRAS*, 379, 151
Mackey, A. D., Broby Nielsen, P., Ferguson, A. M. N., & Richardson, J. C. 2008, *ApJ*, 681, L17
Mackey, A. D., Broby Nielsen, P., Ferguson, A. M. N., & Richardson, J. C. 2009, in J.Th. van Loon & J. M. Oliveira (eds.), *Proceedings IAU Symposium 256: The Magellanic System: Stars, Gas, and Galaxies* (CUP: Cambridge), in press
Massey, P. & Hunter, D. A. 1998, *ApJ*, 493, 180
Mateo, M., Hodge, P., & Schommer, R. A. 1986, *ApJ*, 311, 113
Mighell, K. J., Sarajedini, A., & French, R. S. 1998, *AJ*, 116, 2395
Milone, A. P., Bedin, L. R., Piotto, G., & Anderson J. 2008, *A&A*, in press
Mucciarelli, A., Ferraro, F. R., Origlia, L., & Fusi Pecci, F. 2007, *AJ*, 133, 2053
Mucciarelli, A., Carretta, E., Origlia, L., & Ferraro, F. R. 2008, *AJ*, 136, 375
Nota, A., *et al.* 2006, *ApJ*, 640, L29
Olsen, K. A. G., Hodge, P. W., Mateo, M., Olszewski, E. W., Schommer, R. A., Suntzeff, N. B., & Walker, A. R. 1998, *MNRAS*, 300, 665
Pagel, B. E. J., & Tautvaišienė, G. 1998, *MNRAS*, 299, 535
Pessev, P. M., Goudfrooij, P., Puzia, T. H., & Chandar, R. 2006, *AJ*, 132, 781
Pessev, P. M., Goudfrooij, P., Puzia, T. H., & Chandar, R. 2008, *MNRAS*, 385, 1535
Piatti, A. E., Bica, E., Geisler, D., & Clariá, J. J. 2003, *MNRAS*, 344, 965
Piatti, A. E., Sarajedini, A., Geisler, D., Seguel, J., & Clark, D. 2005, *MNRAS*, 358, 1215
Piatti, A. E., Geisler, D., Sarajedini, A., Gallart, C., & Wischnjewsky, M. 2008, *MNRAS*, 389, 429
Piotto, G. 2008, *Mem.S.A.It.*, 79, 3
Rich, R. M., Shara, M. M., Fall, S. M., & Zurek, D. 2000, *AJ*, 119, 197
Rich, R. M., Shara, M. M., & Zurek, D. 2001, *AJ*, 122, 842
Santiago, B. X. 2009, in J.Th. van Loon & J. M. Oliveira (eds.), *Proceedings IAU Symposium 256: The Magellanic System: Stars, Gas, and Galaxies* (CUP: Cambridge), in press
Sarajedini, A. 1998, *AJ*, 116, 738
Shara, M. M., Fall, S. M., Rich, R. M., & Zurek, D. 1998, *ApJ*, 508, 570
Smecker-Hane, T. A., Cole, A. A., Gallagher III, J. S., & Stetson, P. B. 2002, *ApJ*, 566, 239
Xin, Y., Deng, L., de Grijs, R., Mackey, A. D., & Han, Z. 2008, *MNRAS*, 384, 410

Discussion

S. Yi: The AGB scenarios produce extreme helium populations best if the age separation between the old and young populations is roughly 100 Myr, which is the case with NGC 1846. Do you see a bluer MS for the younger population?

D. Mackey: I've often wondered about that but unfortunately the data we presently have in hand are not sufficiently deep to resolve any split in the main sequence. However, there's no reason why we can't look for this in the future; the required observations are not too tough and it's certainly a vital problem to address.

J. Kalirai: Is there any spatial segregation between the stars in the brighter vs. fainter turnoffs of the LMC clusters? Do you see evidence for mass segregation between the two populations?

D. Mackey: In my original paper on NGC 1846 (Mackey & Broby Nielsen 2007, MNRAS, 379, 151), I looked at this and saw no segregation between the two populations ("young" and "old" turn-off stars). However, another recent study is claiming to see some spatial distinction (P. Goudfrooij, private communication). I'm not sure yet how to reconcile the two results but it would certainly help constrain formation scenarios if we could get a definitive answer.

Doug Mackey

Eric Jensen and Steven Margheim

The Ages of Stars
Proceedings IAU Symposium No. 258, 2008
E.E. Mamajek, D.R. Soderblom & R.F.G. Wyse, eds.

doi:10.1017/S1743921309031937

White dwarf cosmochronology: Techniques and uncertainties

Maurizio Salaris

Astrophysics Research Institute, Liverpool John Moores University,
12 Quays House, Birkenhead, CH41, 1LD, UK
email: ms@astro.livjm.ac.uk

Abstract. White dwarfs represent the endpoint of the evolution of the large majority of stars formed in the Galaxy. In the last two decades observations and theory have improved to a level that makes possible to employ white dwarfs for determining ages of the stellar populations in the solar neighborhood, and in the nearest star clusters. This review is centered on the theory behind the methods for white dwarf age-dating, and the related uncertainties, with particular attention paid to the problem of the CO stratification, envelope thickness and chemical composition, and the white dwarf initial-final-mass relationship.

Keywords. convection, dense matter, stars: atmospheres, stars: interiors, white dwarfs

1. Introduction

White dwarfs (WDs) are the last evolutionary phase of stars with initial masses smaller than about 10-11 $M_{\odot}$. The large majority of WDs, i.e. with progenitor masses below $\sim 6 - 7 M_{\odot}$, have masses between ~ 0.55 and ~ 1.0 $M_{\odot}$ and are made of an electron degenerate core of carbon and oxygen. Higher mass WDs have an oxygen and neon core. Some WDs originated by low-mass progenitors stripped of their envelope during the Red Giant Branch phase (e.g. in a binary system) have masses below $0.5 M_{\odot}$ and are made of helium. In all cases, the electron degenerate core is surrounded by a layer of pure He with mass of the order of $M_{He} \sim 10^{-2}$ M_{WD} or less. This He-layer can be, in turn , surrounded by a H-layer with mass of the order of $M_H \sim 10^{-4}$ M_{WD} or less. Due to very efficient atomic diffusion, all metals in the envelope during the progenitor's evolution have settled at the bottom of the He-layer. WDs are spectroscopically labelled as DA if they have essentially a pure-H outer envelope, or non-DA in case of no hydrogen in the outer layers. Among the non-DA WDs the DB subclass corresponds to a pure He-envelope. The number ratio non-DA/DA appears to change with temperature for WDs in the solar neighborhood. For the range of luminosities of interest here, Tremblay & Bergeron (2008) find empirically an increase of the non-DA/DA ratio when T_{eff} decreases below ~ 10000 K, which is ascribed to mixing of a thin H-envelope with the more massive underlying He-layers. Other types of non-DA objects, which typically appear at low T_{eff}, arise probably from convective mixing of the He-layer with underlying metals, and/or accretion of metals from the interstellar medium. This review will be focused on WDs with a carbon oxygen (CO) core and will consider DB models as templates for the evolution of non-DA WDs.

The basic structure of a WD is simple. Almost all WD mass is in the CO core, that is nearly isothermal, because of the high electron conductivity. The WD evolution is a cooling process, whereby the core acts as energy reservoir (the energy available to be radiated away is the internal energy of the CO ions), the outer non-degenerate layers control the rate of energy outflow, and the luminosity and core temperature decrease

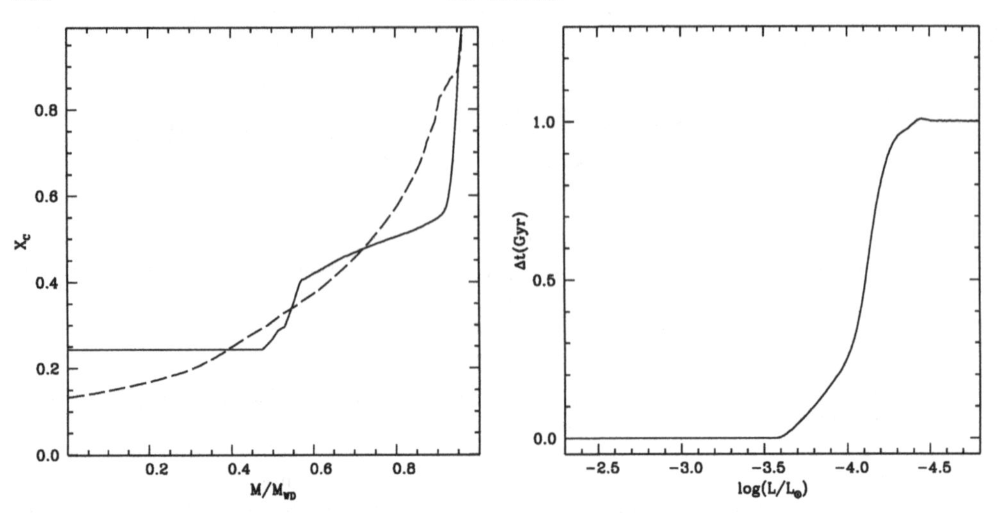

Figure 1. *Left* : Initial carbon abundance profile in the core of a $0.61M_{\odot}$ WD model from Salaris *et al.* (1997) calculations (solid line), and after chemical redistribution upon crystallization (dashed line). *Right* : Time delay due to the chemical redistribution, from the models by Salaris *et al.* (2000).

with time. Given that most stars are or will become WDs, plus the existence of a well defined relationship between cooling time and luminosity, and the long cooling timescales, WDs are very attractive candidates to unveil the history of star formation in the Galaxy. During the last two decades observations and theory have improved to a level that has made finally possible to employ WDs for determining ages of the stellar populations in the solar neighborhood, and in the nearest star clusters.

2. White Dwarf evolution

The main phases of WD cooling evolution are briefly sketched below. The figures for the luminosity ranges are just indicative, for more precise values depend on the exact chemical stratification in the core and chemical composition of the non-degenerate envelope.

$\log(L/L_{\odot}) > -1.5$. The brightest stages of WD evolution are dominated by neutrino emission (mainly plasma-emission). By the end of this phase of neutrino cooling, differences in the thermal structures at fixed mass, due to different approaches to the WD cooling sequence disappear. If the mass in the H-envelope is above a threshold of $\approx 10^{-4} M_{WD}$ (the exact value depending on the WD mass), hydrogen burning through the *pp* chain becomes effective. Pulsational studies seem to constrain the M_H to values generally below this threshold, see, e.g., Castanheira & Kepler (2008).

$-3 < \log(L/L_{\odot}) < -1.5$. The main source of energy is the internal energy of the ions. The Coulomb parameter Γ in the core is above unity, i.e. the ions are in the liquid phase.

$\log(L/L_{\odot}) < -3.0$. The Coulomb parameter Γ reaches the critical value $\Gamma_{cryst} \sim 180$, and the ions in the core undergo a phase transition from liquid to solid. This introduces two new energy sources. The first one is the latent heat of crystallization, of the order $K_B T$ per crystallized ion, where K_B is the Boltzmann constant. The second source is related to the phase diagram of the CO binary mixture, see, e.g., Isern *et al.* (2000). In brief, the equilibrium composition of the CO mixture of the solid and liquid phases are not the same. The net effect is a migration of oxygen towards the central regions with the consequent release of gravitational energy (see, Fig. 1).

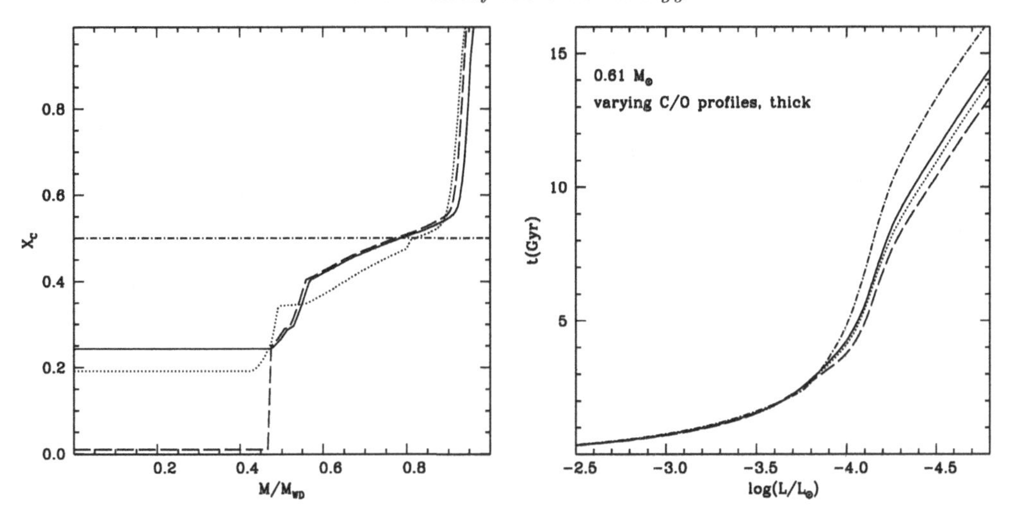

Figure 2. *Left* : Test carbon profiles in the core of a $0.61 M_\odot$ model. *Right* : Corresponding cooling timescales (see text for details).

Debye cooling. When crystallization of the CO core is essentially complete, the specific heat follows the Debye law, decreasing as T^3. At this stage, the energy provided by the compression of the non-degenerate envelope (through the virial theorem) is enough to prevent the sudden disappearance of WDs with sufficiently thick envelopes.

The local detailed energy budget of a WD is given by, as shown in Isern *et al.* (1997):

$$-\left(\frac{dL_{\rm r}}{dm}+\epsilon_\nu\right)=C_{\rm v}\frac{dT}{dt}+T\left(\frac{\partial P}{\partial T}\right)_{V,X_0}\frac{dV}{dt}-l_{\rm s}\frac{dM_{\rm s}}{dt}\delta(m-M_{\rm s})+\left(\frac{\partial E}{\partial X_0}\right)_{T,V}\frac{dX_0}{dt}$$

where E is the internal energy per unit mass, L the local luminosity, ϵ_ν the neutrino energy losses per unit mass, $V=1/\rho$ is the specific volume. The first term in the right-hand side is the heat capacity of the star, the second one denotes the energy contribution due to changes in volume, which is usually negligible, apart from the latest stages of cooling. The third term represents the energy contribution due to the latent heat release upon crystallization ($l_{\rm s}$ is the latent heat of crystallization and dM_s/dt is the rate of growth of the solid core due to crystallization). The delta function specifies that the latent heat is released at the solidification front. The last term denotes the energy released by the change of chemical abundances (chemical separation) due to the phase diagram of the CO mixture. X_0 here represents the oxygen mass fraction. Figure 1 displays the effect of chemical separation upon crystallization on a $0.61 M_\odot$ model, with $M_H=10^{-4} M_{WD}$ and $M_{He}=10^{-2} M_{WD}$. The delay in the cooling process caused by this phenomenon is comparable to the effect of latent heat release.

Apart from uncertainties in the input physics (equation of state, opacities, CO phase diagrams), the detailed energy budget is strongly dependent on the chemical stratification of the degenerate core and of the envelope. Varying the chemical composition of the core affects the internal energy available to be radiated away, the latent heat release and chemical redistribution upon crystallization. Varying the composition and mass thickness of the envelope affects the rate of energy release because of the change of opacity. In general, more transparent envelopes (e.g. thinner or absent H-layers) speed up the cooling process. To date, predictions from stellar evolution calculations are subject to sizable uncertainties regarding the CO profiles and envelope stratification at the onset of WD

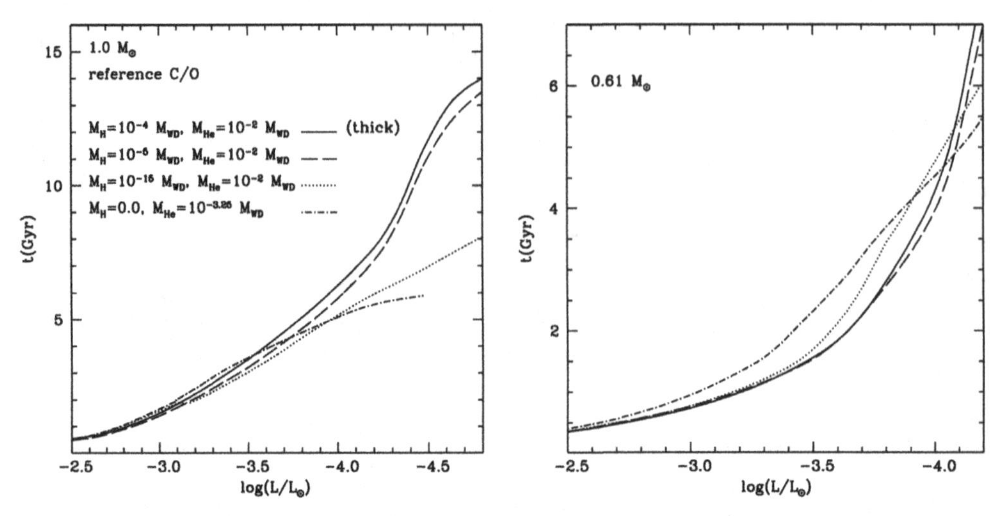

Figure 3. *Left* : Cooling timescales for a $1.0M_{\odot}$ WD model with varying envelope properties. *Right* : As in the left panel but for a $0.61M_{\odot}$ WD model.

cooling. Regarding the CO profile, the main reason is not only the uncertainty in the $^{12}C+\alpha$ reaction rate – estimated to be of the order $1\sigma=\pm 30$ % according to the latest determination by Kunz *et al.* (2002) – but also, and probably even more importantly, the treatment of core convection during the central He-burning phase. As discussed in detail by Straniero *et al.* (2003), for a given stellar mass, whether or not semiconvection is included, whether or not breathing pulses are suppressed and the way in which they are suppressed can alter substantially the final CO profile. There are good observational reasons why breathing pulses should be suppressed in model computations – e.g. Cassisi *et al.* (2003) – but the way in which they are suppressed can affect substantially the CO abundances. As for the envelopes, uncertainties in the mass-loss efficiency during the Asymptotic Giant Branch (AGB) phase do not allow a firm prediction of their chemical composition and thickness at the start of WD cooling. The uncertainty in the mass-loss along the AGB affects also the CO stratification in a more subtle way. Changing the mass-loss efficiency affects the initial (main sequence)-final (WD) mass relationship (IFMR), hence the relationship between the mass of the CO core (and associated chemical profile) at the end of central He-burning and its final value at the end of the AGB.

Figure 2 shows numerical tests that give an order-of-magnitude idea of uncertainties on cooling times due to alternative CO profiles. The cooling times of a $0.61M_{\odot}$ WD with $M_H=10^{-4}M_{WD}$ and $M_{He}=10^{-2}M_{WD}$ are calculated assuming 4 different chemical profiles in the core. The solid line represents the reference profile of Fig. 1, the dashed-dotted line shows a flat CO profile with equal abundances (that maximizes the effect of chemical separation), the dashed line shows the effect of an alternative treatment of He-burning core convection following the results by Straniero *et al.* (2003), the dotted line accounts for a different IFMR (progenitor with a smaller initial mass). The largest effect compared to the reference model is obtained with the flat CO profile, that causes an age increase up to 1.5-2 Gyr for the oldest WDs. The other profiles cause generally a reduction of cooling times by at most ~ 1 Gyr.

Figure 3 displays the cooling times of a $1.0M_{\odot}$ with the CO stratification by Salaris *et al.* (1997) and different envelopes. The model without H-layer is representative of a non-DA WD, and has the fastest cooling time, because of a lower envelope opacity.

Age differences up to ~ 8 Gyr appear for the faintest observed WDs ($\log(L/L_{\odot}) \sim -4.5$). The case with $M_H = 10^{-15} M_{WD}$ deserves a further discussion. In this model, surface convection (efficient below a certain effective temperature) has been inhibited, so that the outermost H-layer is never mixed with the underlying, much thicker He-layer. If mixing is allowed, the H-layer is engulfed by the He envelope at temperatures well above 10000 K, when the cooling timescales of DA and non-DA models are essentially the same. From that moment on the model behaves as the non-DA calculations shown in the figure, because the hydrogen is completely diluted within the He-envelope – see also the discussion in D'Antona & Mazzitelli (1987). Thicker H-layers escape mixing with the underlaying He-envelope (for example $M_H = 10^{-4} - 10^{-5} M_{WD}$), or mix at lower luminosities. The right-hand panel of the same figure shows the case of a $0.61 M_{\odot}$. Notice how for cooling ages between ~ 1 and ~ 4 Gyr the model with pure He envelope is sizably brighter than models with an external H-layer, whereas this effect is smaller for the $1 M_{\odot}$ model. A surface H-layer with $M_H \approx 10^{-10} M_{\odot}$ would mix at $T_{eff} \sim 9000$ K, i.e. $\log(L/L_{\odot}) \sim -3.0$, and after the mixing episode the cooling timescales will run parallel to the case of the non-DA model.

3. White dwarf cosmochronology

Fundamental working tools for WD age dating are WD isochrones, i.e. the CMD of WDs born from a single-burst, single-metallicity population (see, Fig. 4). They are routinely used to study the cooling sequences of WDs in star clusters, see, e.g., Hansen *et al.* (2007), Bedin *et al.* (2008a). Computations of WD isochrones require a grid of WD models for different masses, an IFMR and evolutionary timescales of the WD progenitors plus appropriate bolometric corrections. From a WD isochrone, the luminosity function (LF) in a given passband can be calculated after assuming an initial mass function (IMF) for the WD progenitors. Here the reference WD isochrones are computed using the cooling models by Salaris *et al.* (2000) that make use of Salaris *et al.* (1997) CO profiles, envelopes with $M_H = 10^{-4} M_{WD}$ and $M_{He} = 10^{-2} M_{WD}$, and progenitor lifetimes for a metal mass fraction Z = 0.0198 (solar) from Pietrinferni *et al.* (2004). The adopted IFMR is displayed in Fig. 5 as a solid line. The reference LFs are computed using a Salpeter IMF for the WD progenitors. Figure 4 displays, as an example, four isochrones and the corresponding LFs in the F606W ACS filter. The age indicator for a WD population is the faint end of the isochrones, that corresponds to a cut-off in the luminosity function. Due to the finite age of the stellar population, the more massive WDs formed from higher-mass and shorter-lived progenitors pile up at the bottom of the cooling sequence, producing the turn to the blue (i.e. a turn towards lower radii) visible at the faint end of the isochrones. Increasing the age obviously makes the bottom end of the isochrones fainter, because of the longer cooling times. The F606W magnitude of the LF cut-off changes with age by ~ 0.2 mag/Gyr at old ages, and ~ 0.5 mag/Gyr at intermediate ages. This in principle makes WDs better suited for age determinations compared to the main sequence turn off (the corresponding changes of F606W for the turn off are ~ 0.1 mag/Gyr and ~ 0.4 mag/Gyr, respectively), once the cluster distance modulus is fixed.

Figure 5 shows the sensitivity of the LF to changes in the adopted IFMR. The two IFMRs used in the test are well within the scatter of points that defines semiempirical estimates for objects belonging to the Galactic Disk system (field stars and open clusters). The full IFMR (especially the high-mass end) of old star clusters is, from an empirical point of view, largely unknown. The shape of the resulting LFs is severely affected

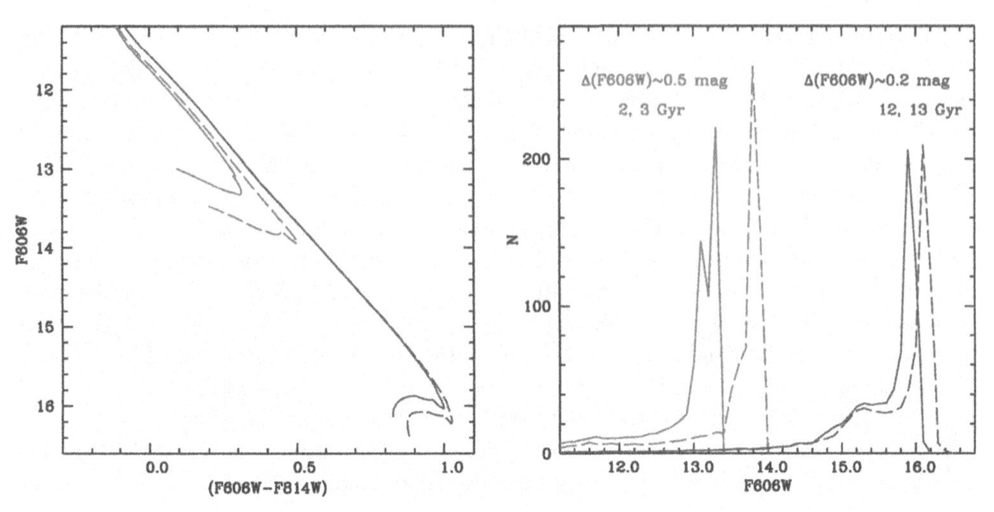

Figure 4. *Left* : WD isochrones for 2, 3, 12, 13 Gyr. *Right* : Luminosity functions in the F606W ACS filter, for the isochrones in the left panel.

by the choice of IFMR. Bumps and plateaus do appear, depending on the age of the population. The magnitude of the cut-off is only moderately affected for the older ages. If one compares just the observed magnitude of the LF cut-off with theory, these two different IFMRs cause an age uncertainty below 10% for the two oldest ages displayed. The effect for the youngest LF is larger. Taking into account the overall shape of the LF would put firm constraint on the form of the IFMR and this should mitigate this source of indetermination for the WD ages. The matter is however complicated by the fact that a change of the progenitor IMF also alters the overall shape of the LF, as well as photometric errors and the presence of unresolved WD+WD binaries, this latter point being discussed in some detail by Bedin *et al.* (2008b). Additionally, the transition of a fraction of DA objects to non-DA could also cause some local change in the shape of the LF, because of the change of evolutionary speed if the mixing happens at T_{eff} below ≈10000 K.

The assumed CO stratification and envelope chemical composition are more troublesome from the point of view of age estimates, as shown in Fig. 6, that displays the effect of changing the CO profile from the reference case to a flat one. At old ages, LFs with 1.5 Gyr age differences are close to indistinguishable, provided that the progenitor IMF is a power law with exponent $\alpha = -1.7$, instead of $\alpha = -2.35$ (Salpeter). Given that the progenitor IMF – or, more precisely, the progenitor mass function MF that could be different from the IMF because of the dynamical evolution of the stellar system – is at some level an unknown quantity in the observed cluster fields, this degeneracy between theoretical LFs with different ages is worrying. An analogous result is found when comparing the reference case with the LF computed from models with a thinner H-layer ($M_H = 10^{-5} M_{WD}$). Figure 6 shows how at old ages a systematic error of at least 0.8 Gyr seems to be unavoidable given our inability to predict the envelope thickness and composition of WDs.

One could try – if allowed by the quality of the photometric data at hand, see e.g. Hansen *et al.* (2007) – to use the full CMD to break these age degeneracies. Figure 7 compares reference isochrones with the case of a flat CO profile and a thinner H-layer, respectively. Notice how the degeneracy of the LF is mirrored by a near degeneracy in the CMD. To assess whether it is possible to use the distribution of points in the

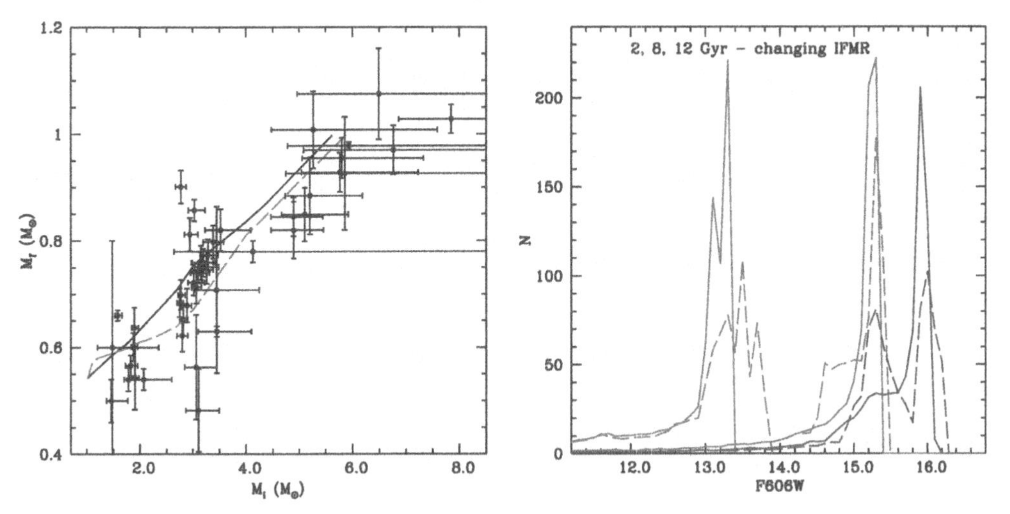

Figure 5. *Left* : IFMRs adopted in the calculations presented in this paper. The solid line is the reference IFMR. Datapoints with error bars are semiempirical estimates from Salaris *et al.* (2009) and, at low progenitors masses, from Catalan *et al.* (2008a), for WDs in open clusters and in the field. *Right* : Effect of varying the IFMR on the LF of star clusters with the labelled ages. The different line-styles correspond to the two IFMRs in the left panel.

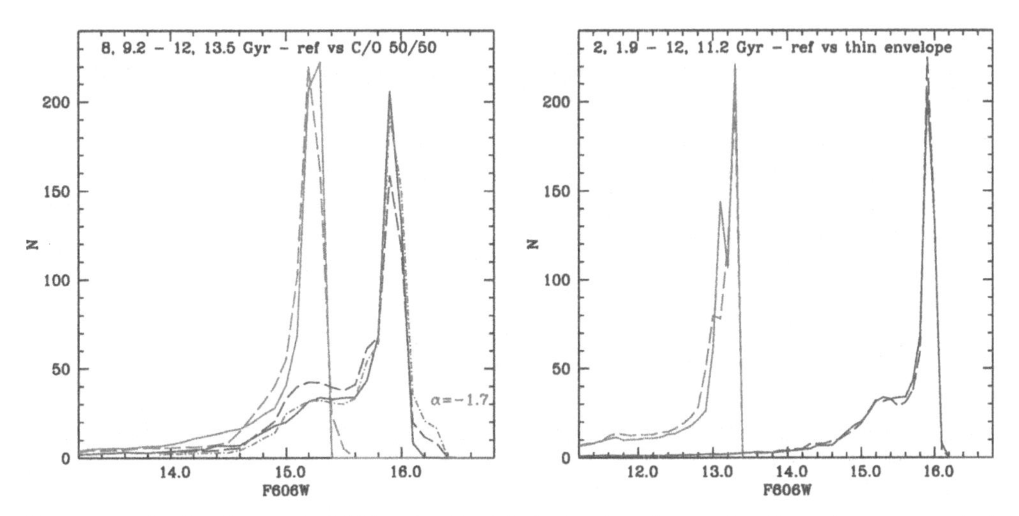

Figure 6. *Left* : Theoretical LFs with the labelled pairs of ages, for the reference case (solid lines) and for models with a flat CO profile (dashed lines). The case of a flat CO profile and a progenitor mass function with exponent $\alpha = -1.7$ is displayed as a dashed-dotted line. *Right* : As for the left panel, but for the case of reference (solid line) versus a thinner ($M_H = 10^{-5} M_{WD}$) H-layer (dashed line).

CMD to break the age degeneracy, the following test has been performed. Photometric errors and completeness as a function of F606W from the (F606W, F814W) photometry of NGC 6791 by Bedin *et al.* (2008a) have been considered. The reference F606W magnitudes have been then shifted so that the 50% completeness level is reached at F606W = 16, i.e. a luminosity on the descending branch of the LF cut-off in Fig. 6. This is a reasonable assumption for the existing photometries of old clusters. Monte-Carlo simulations have been then employed to produce synthetic samples of WDs from the pairs of old isochrones in Fig. 7, that include the prescribed photometric error law. A

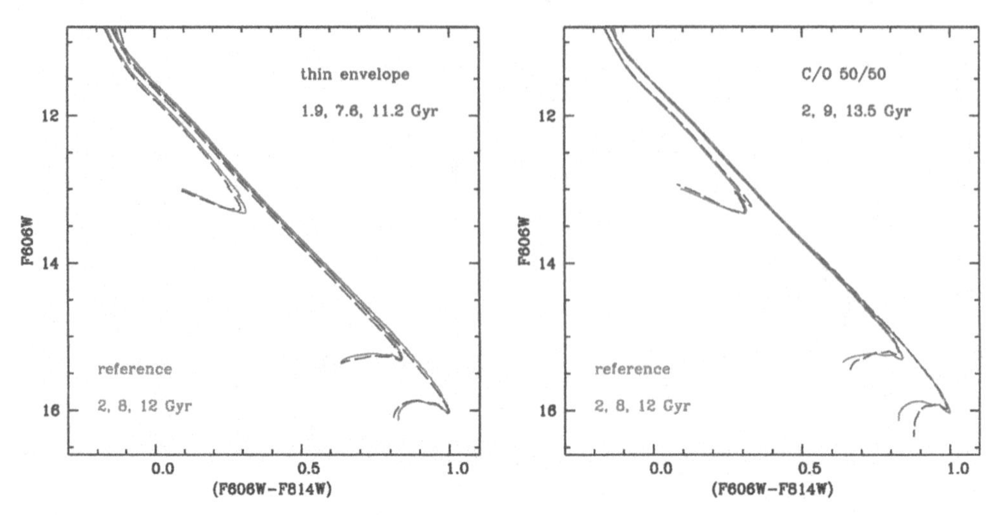

Figure 7. *Left* : Isochrone for the reference case and the labelled ages (solid lines) versus isochrones with a thinner ($M_H = 10^{-5} M_{WD}$) H-layer (dashed lines) and the three labelled ages. *Right* : As in the left panel but for the case of reference (solid lines) versus flat CO profiles (dashed lines).

2-dimensional KS-test applied to several realizations of the two pairs of synthetic WDs (sample size equal to 400 objects, but the precise value is not critical) for F606W$\leqslant$16 gives probabilities well below 95% (assumed as threshold probability) that the WD populations in each pair are actually generated by two different distributions. Halving the errors at a given magnitude would produce again the same result for the case of thinner H-layers. A fainter (by about 0.1-0.2 mag) magnitude limit would allow a discrimination between the reference and flat CO case.

The presence of a fraction of non-DA objects (e.g. up to 30-40%) modelled with pure-He envelopes does not affect sensibly the LF of old populations, because at those ages non-DA objects would be mainly located at much fainter magnitudes, well below detectability. However, their effect on the LF may be significant for ages of 2-3 Gyr, that correspond to luminosities where the cooling timescales of WDs with pure He-envelopes are longer than the case of H-envelopes. The situation may become more complicated in case a sizable fraction of DA objects have a range of H-layer masses, and undergo mixing with the He-layer at luminosities where the evolutionary speed is very different for H- and He-envelopes. In this case, additional features in the LF (and stellar distribution on the CMD) can appear, related to the change in evolutionary timescales.

Regarding the age of field WDs, investigations are traditionally focused on the LF of WDs in the solar neighborhood, see, e.g., Leggett *et al.* (1998). Figure 8 displays an observational LF from Catalan *et al.* (2008b), compared to the theoretical counterpart calculated from the reference models described above, solar metallicity progenitors, a Salpeter IMF and a continuous star formation rate. All theoretical LFs are normalized to the observed star counts at $\log(L/L_{\odot}) = -3.2$. The position of the cut-off in the empirical LF provides an age $t = 11.5 \pm 1.0$ Gyr for the onset of star formation in the local Disk. The right panel of Figure 8 shows also the result for models with a flat CO profile and a thinner H-layer ($M_H = 10^{-5} M_{WD}$) respectively. The systematic change in age is essentially the same as for the case of the star counts from isochrones discussed above. The left panel displays also the case of a substantial primordial fraction of DB objects, taken as representative of the whole non-DA population. The effect is at the level

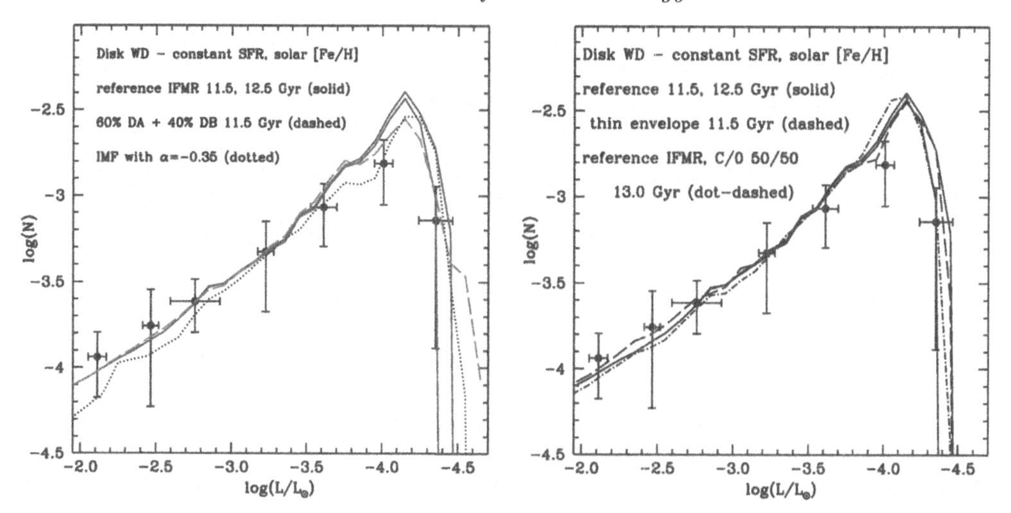

Figure 8. *Left* : Observed LF for WDs in the solar neighborhood from Catalan *et al.* (2008b) compared, respectively, to the reference DA theoretical counterparts for the labelled ages, a LF computed allowing for a 40% fraction of DB objects, and a LF with DA models but an IMF with exponent $\alpha = -0.35$. *Right* : As in the left panel, but with additional LFs obtained from models with a flat CO profile and with a thinner H-layer ($M_H = 10^{-5} M_{WD}$) respectively (see text for details).

of only a few 10^8 yr, because of a different shape of the LF at the lower end. The faster cooling times of He-envelope models tend to distribute the fainter DB objects below the observed magnitude of the cut-off. Tremblay & Bergeron (2008) data show that the transformation of DA into non-DA objects due to surface convection – that changes the observed non-DA/DA observed ratio from ~ 0.25 to ~ 0.4-0.5 – happens in local Disk objects at $T_{eff} \sim 10000$ K, i.e. at a temperature where evolutionary timescales of DA and DB models are virtually the same. This means that an overall 40% primordial fraction of DB models is an adequate assumption, as far as the effect on the Disk LF is concerned. Changing the exponent of the IMF has also an effect on the age because of a different shape of the LF. The change applied to the theoretical LF in Fig. 8 causes an age change by $\approx$0.5 Gyr. Finally, the effect of the alternative IFMR displayed in Fig. 5 causes a decrease of the derived age by less than 1 Gyr.

4. Conclusions

As discussed in some details in the previous sections, uncertainties in the CO stratification and envelope composition give probably the largest contribution to the systematic errors in age estimates from the WD LF. The analysis of the WD distribution across the CMD may not always help in removing these sources of uncertainties. Also the choice of the IFMR may play a non-negligible role in the final error budget on WD ages. Together with these systematics arising from the modelling of core convection and mass-loss during the previous evolutionary phases, uncertainties in the adopted input physics introduce additional sizable systematic differences in the cooling times calculated by different authors. Comparisons of a $0.6 M_{\odot}$ pure-C WD model by Fontaine *et al.* (2001) with results from the Salaris *et al.* (2000) code, using a simple gray $T(\tau)$ relationship for the surface boundary conditions, $M_H = 10^{-4} M_{WD}$ and $M_{He} = 10^{-2} M_{WD}$, reveal differences of ~ 1 Gyr for cooling times of the order of 12-13 Gyr. Additional physics not yet included in age estimates of WD populations may be potentially responsible for other systematic effects. Deloye & Bildsten (2002) and García-Berro *et al.* (2008) have explored the role played by ^{22}Ne diffusion in the liquid phase, concluding that for progenitor metallicities solar or supersolar and old ages ($\approx$10 Gyr), a $\approx$1 Gyr age increase in estimates from the LF has to be accounted for. There are however still sizable uncertainties in the determination of the appropriate Ne diffusion coefficient. The effect of minor chemical species on the phase diagram of the WD core and on the time delays due to chemical separation is still not well established. Calculations by Segretain (1996) for a CONe ternary mixture and solar progenitor metallicity show a negligible effect compared to results for the CO binary mixture. The effect of the chemical separation of Fe has not been yet assessed from the appropriate multibody phase diagram, although, as summarized in Isern *et al.* (2002), estimates from an effective N/Fe binary mixture (whereby N has the average charge of a flat CO profile) give a delay of the order of 1 Gyr for a $0.6M_{\odot}$ WD, that is probably an upper limit to the true effect. Taken all together, these confirmed and potential sources of uncertainties show that WD evolution is not just a simple cooling problem that naturally provides ages more accurate than, e.g. main sequence turn off ages. Instead, it is testing the models and – as long as ab-initio solid theoretical predictions are still lacking – tuning the uncertain inputs against turn off ages in star clusters, that should be the most fruitful approach to firmly establish WD cosmochronology as a useful tool on its own.

References

Bedin, L. R., King, I. R., Anderson, J., Piotto, G., Salaris, M., Cassisi, S., & Serenelli, A. 2008, *ApJ*, 678, 1279

Bedin, L. R., Salaris, M., Piotto, G., Cassisi, S., Milone, A. P., Anderson, J., & King, I. R. 2008, *ApJ*, 679, L29

Castanheira, B. G. & Kepler, S. O. 2008, *MNRAS*, 385, 430

Cassisi, S., Salaris, M., & Irwin, A. W. 2003, *ApJ*, 588, 862

Catalán, S., Isern, J., García-Berro, E., Ribas, I., Allende Prieto, C., & Bonanos, A. Z. 2008a, *A&A*, 477, 213

Catalán, S., Isern, J., García-Berro, E., & Ribas, I. 2008b, *MNRAS*, 387, 1693

D'Antona, F. & Mazzitelli, I. 1987, *in The Second Conference on Faint Blue Stars*, D.S. Hayes and J.W. Liebert (eds.), L. Davis Press, p. 635

Deloye, C. J. & Bildsten, L. 2002, *ApJ*, 580, 1077

Fontaine, G., Brassard, P., & Bergeron, P. 2001, *PASP*, 113, 409

García-Berro, E., Althaus, L. G., Córsico, A. H., & Isern, J. 2008, *A&A*, 677, 473

Hansen, B. M. S. & Liebert, J. 2003, *ARA&A*, 41, 465

Hansen, B. M. S., *et al.* 2007, *ApJ*, 671, 380

Kunz, R., *et al.* 2002, *ApJ*, 567, 643

Isern, J., Mochkovitch, R., García-Berro, E., & Hernanz, M. 1997, *ApJ*, 485, 308

Isern, J., García-Berro, E., Hernanz, M., & Chabrier, G. 2000, *ApJ*, 528, 397

Isern, J., García-Berro, E. , Hernanz, M., & Salaris, M 2002, *Contributions to Science*, 2, 237

Leggett S. K., Ruiz M. T., & Bergeron P. 1998, *ApJ*, 497, 294

Pietrinferni, A., Cassisi, S., Salaris, M., & Castelli, F. 2004, *ApJ*, 612, 168

Salaris, M., Dominguez, I., García-Berro, E., Hernanz, M., Isern, J., & Mochkovitch, R. 1997, *ApJ*, 486, 413

Salaris, M., García-Berro, E., Hernanz, M., Isern, J., & Saumon, D. *ApJ*, 544, 1036

Salaris, M., Serenelli, A., Weiss, A., & Miller Bertolami, M. 2009 *ApJ*, in press

Segretain, L. 1996, *A&A*, 310, 485

Straniero, O., Dominguez, I., Imbriani, G., & Piersanti, L. 2003, *ApJ*, 583, 878

Tremblay, P.-E. & Bergeron, P 2008, *ApJ*, 672, 1144

Discussion

G. Piotto: You have listed six sources of uncertainty in the age we can derive from the WD cooling sequence. How do they size up? Can we leave this room with an idea of the uncertainty in the cluster age derived from white dwarf cooling sequences?

M. Salaris: It is difficult to say with some precision. $10^{-4} M_{\rm WD}$ is an upper limit to the H-layer thickness. So, an uncertainty there goes always in the direction of reducing cooling timescales. The effect of CO profiles can go both ways. ^{22}Ne diffusion goes in the direction of always increasing $t_{\rm cool}$. Uncertainties due to different input physics may probably go both ways. I would say that globally we have something of the order of +2-3 to −2 Gyr at old ages. This is of course just a very rough estimate.

J. Melbourne: So you point out all the variations in the models, but what about the large potential variations in the stars themselves. Variations in He envelopes and CO profiles? How do you account for these uncertainties in your methods?

M. Salaris: If the mixing treatment, the initial-final mass relationship and the metallicities of the progenitors are specified, the CO profiles of WDs with different masses are uniquely defined. The problem is the envelope; that depends on the details of the mass loss at the level of 10^{-2} to $10^{-4} M_{\odot}$. One has to make assumptions, treating the H and He layer thickness essentially as free parameters (of course within certain limits).

S. Leggett: What is the importance of star formation history or the age of the disk as implied by the luminosity function?

M. Salaris: This is a very good point. The somewhat standard assumption is a constant SFR with solar metallicity progenitors. I suspect the effect is less important than uncertainties in the CO profiles and envelope thickness. Matt Wood (1992, ApJ, 386, 539) has investigated this issue. From his Figures 13 and 15 it transpires that the age of the disk derived from the luminosity function cut-off is not greatly affected by the choice of the SFR, at least for the cases analyzed in his paper.

P. Demarque: What are the chances of learning about the core helium buring phase from WD observations? How does diffusion complicate things?

M. Salaris: Comparisons between cooling ages and turn- off ages of clusters may give indications about the internal CO stratification, hence about He-burning core mixing. Also, studies of pulsating WDs can put constraints on the core C/O ratios. Diffusion should not affect the CO profile but it is important to explain the lack of metals in the envelopes of WDs. Also, ^{22}Ne diffusion in the liquid phase alters the ^{22}Ne profile in the core.

Maurizio Salaris and Brian Chaboyer

Jason Melbourne

The Ages of Stars
Proceedings IAU Symposium No. 258, 2008
E.E. Mamajek, D.R. Soderblom & R.F.G. Wyse, eds.

doi:10.1017/S1743921309031949

White dwarfs as astrophysical probes

Jasonjot S. Kalirai

Space Telescope Science Institute, 3700 San Martin Drive, Baltimore MD, 21231
email: jkalirai@stsci.edu

Abstract. Much of our knowledge regarding the *ages of stars* derives from our understanding of the Hertzsprung-Russell Diagram. The diagram is typically dominated by hydrogen burning main-sequence stars, which historically, have been used to establish our most fundamental knowledge of stellar ages and evolution. In this brief article, I highlight how deep ground and space based imaging can uncover the stellar remnants of these hydrogen burning stars, white dwarfs. We have followed up our initial discovery of several large white dwarf populations in nearby star clusters with multiobject spectrographs. The spectroscopy allows us to characterize the properties of the remnant stars (e.g., mass, temperature, and age), which are in turn used to shed new light on fundamental astrophysical problems. Specifically, we estimate the ages of the Milky Way disk and halo, provide the inputs needed to calculate the chemical evolution of galaxies, and re-iterate the important role of HB stars in producing the UV-upturn seen in elliptical galaxies.

Keywords. Galaxy: open clusters and associations: individual (NGC 7789, NGC 6819, and NGC 6791) – stars: evolution – techniques: photometric, spectroscopic – white dwarfs

1. Introduction

Ejnar Hertzsprung, in 1905, found direct evidence that stars of the same temperature, and with the same parallax (and therefore at the same distance), could have very different luminosities. His observations led him to suggest at least two broad bins that characterize stars, bright "giants" and fainter "dwarfs". The first Hertzsprung-Russell (H-R) Diagram was presented by him in 1911, consisting of a small number of nearby stars. Henry Norris Russell quickly improved the quantity and quality of data on the H-R Diagram (Russell 1913; 1914), adding more accurate observations of both field stars and nearby co-moving groups (e.g., the Hyades and the Pleiades). Russell's analysis suggested clear evidence for groupings and sequences on the H-R diagram, and even hinted at a mass-luminosity relation. For example, Russell noted that "one corner of the diagram is vacant...There do not seem to be any faint white stars".

The H-R diagram represents one of the most widely used plots in astrophysics. Much of our knowledge on the ages of stars, and our grasp of fundamental stellar evolution, is based on our ability to understand and model observables in this plane. Historically, this work has largely involved the abundant hydrogen burning stars and bright giant stars, which are both easy to see in the nearby Galaxy. With the construction of sensitive wide-field imagers on 4-m and 8-m telescopes, as well as the launch of the HST, astronomers have recently been able to probe the H-R diagram to unprecedented depths. In addition to many other discoveries, the "faint white stars" that Russell noted as being absent in the diagram, have been abundantly discovered. These stars, white dwarfs, represent the end products of low and intermediate mass hydrogen-burning stars.

White dwarfs are extremely useful probes in several avenues of stellar and galactic astrophysics. First, almost all stars will eventually form white dwarfs given the bias in the initial mass function, and therefore white dwarf represent a unique link to the distribution

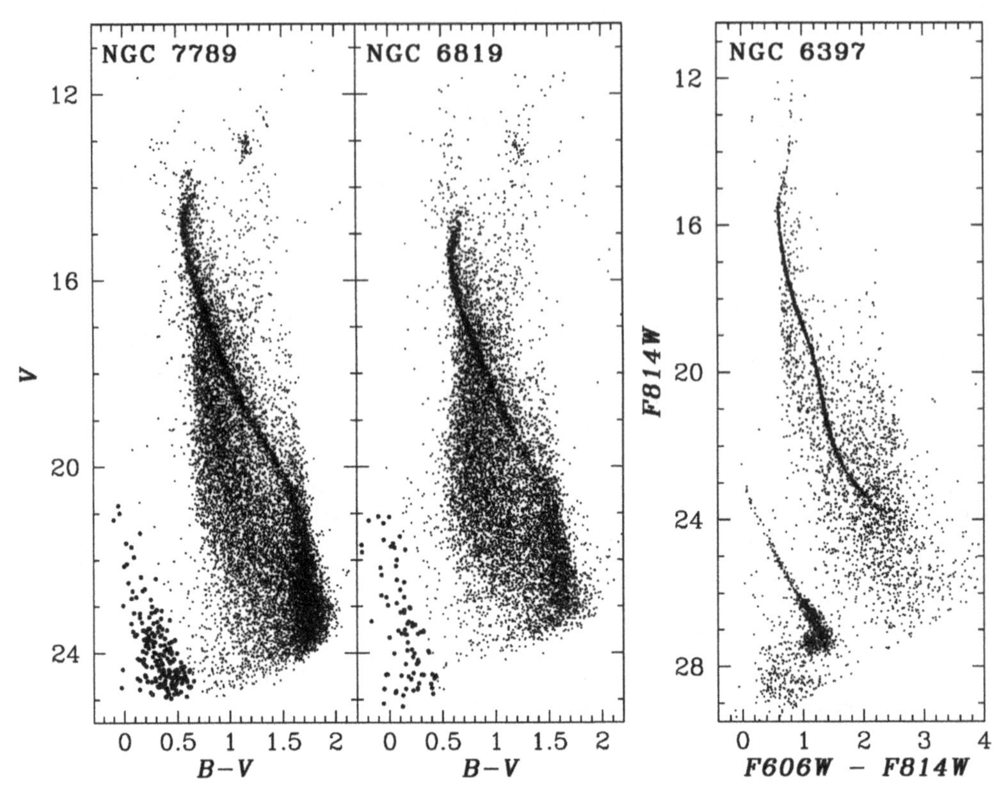

Figure 1. CFHT/CFH12K H-R diagrams of the star clusters NGC 7789 ($t = 1.4$ Gyr) and NGC 6819 ($t = 2.5$ Gyr), and an HST/ACS H-R diagram of NGC 6397 ($t = 12$ Gyr). The diagrams show both a scattered distribution of stars that are in the foreground/background of each cluster, as well as organized sequences representing the cluster stars. The main-sequences can be tracked to low mass stars (to the hydrogen burning limit in the case of NGC 6397), and the brighter giants are also detected. The faint-blue parts of the H-R diagrams show abundant populations of white dwarfs in each cluster. In the case of NGC 6397, the most massive star on the main-sequence currently has a mass of 0.8 $M_{\odot}$, and therefore the white dwarfs represent the end products of stars similar to the Sun.

and properties of first generation stars in old stellar populations. Specifically, the numbers and masses of white dwarfs can be used to place constraints on the initial-mass function of a stellar population as well as an estimation of how much mass stars lose through their evolution (Reimers 1975; Renzini & Fusi Pecci 1988; Weidemann 2000; Kalirai *et al.* 2008). White dwarfs also contain no nuclear energy sources and therefore simply cool and fade with time, making them excellent cosmic clocks to date stellar populations. Finally, the properties of these stars (e.g., mass and temperature) in different environments can provide detailed clues to enhance our understanding of fundamental stellar evolution.

2. Measuring Stellar Mass Loss

In Figure 1 we present observervational H-R diagrams of two rich open star clusters, NGC 7789 and NGC 6819, and one globular cluster, NGC 6397. The open clusters were observed with the CFH12K mosaic CCD camera on the 4-meter Canada-France-Hawaii Telescope (CFHT) as a part of the CFHT Open Star Cluster Survey (Kalirai *et al.* 2001a; 2001b). The globular cluster was observed as a part of a very large (123 orbit) Hubble Space Telescope (HST) allocation with the Advanced Camera for Surveys (ACS),

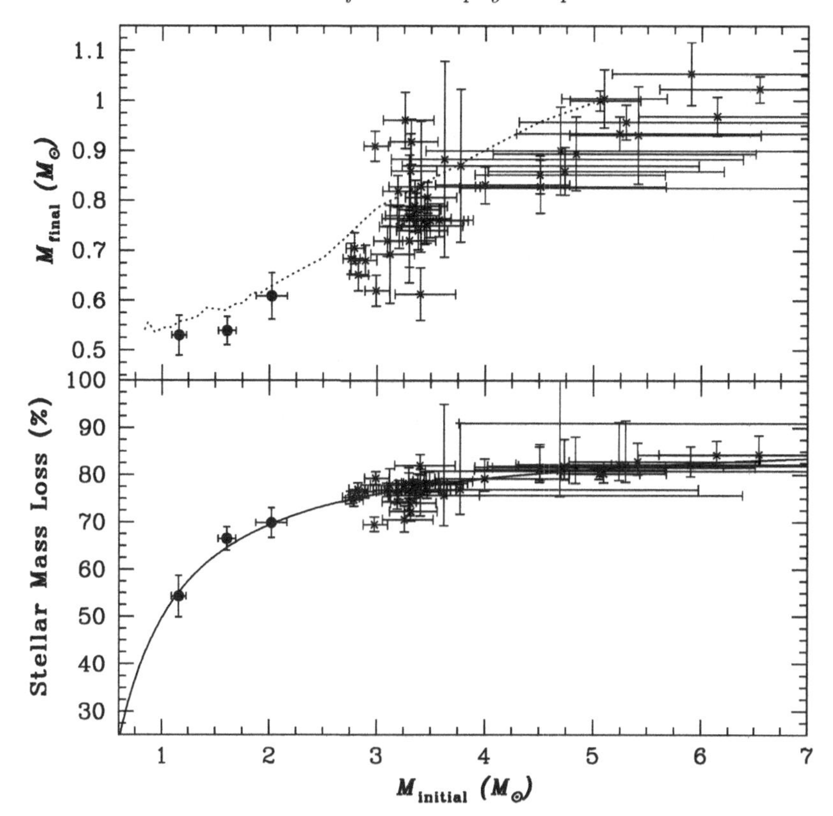

Figure 2. *Top* – The initial-final mass relation including all previous data (crosses) and the results from NGC 7789, NGC 6819, and NGC 6791 (larger data points). For these clusters, we have binned all white dwarfs in a given cluster and plotted a single data point indicating the weighted mean mass and 2σ uncertainty in the mean. The dotted curve indicates the theoretical initial-final mass relation from Marigo (2001) for Solar metallicity. *Bottom* – The total integrated stellar mass as a function of initial mass, along with our best fit linear relation. These results suggest that the initial and final mass of Solar metallicity stars are connected by $M_{\text{final}} = (0.109 \pm 0.007)\ M_{\text{initial}}\ +\ 0.394 \pm 0.025\ M_\odot$ (solid curve in Figure 2).

Richer *et al.* (2006). The color-magnitude diagrams of all three clusters show rich main-sequences and stars in post-main sequence evolutionary stages. The white dwarfs that represent the end products of these stars, after they suffer mass loss along the red giant branch and asymptotic giant branch, form a tight sequence in the faint-blue part of each H-R diagram.

We can use these white dwarfs to directly estimate how much mass normal main-sequence stars will lose through their evolution and inject back into the interstellar medium. This relation, the initial-final mass relation, is critical to our understanding of chemical evolution of galaxies, the star formation efficiency in these systems (Somerville & Primack 1999), and is also necessary in order to use white dwarfs to date the Galactic disk and halo (e.g., Ferrario *et al.* 2005 and Hansen *et al.* 2007). We construct the relation by combining the results of the two open star clusters here with similar observations of four other clusters with different ages (e.g., Kalirai *et al.* 2008). A cluster with a younger age will lead to constraints on the higher mass end of the relation (since higher mass stars will evolve to white dwarfs faster), whereas an older cluster constrains the low mass end of the relation.

The masses of the white dwarfs in each cluster are calculated by observing these stars spectroscopically (observations made with LRIS on Keck I). The spectrum of the white dwarfs contains broadened Balmer lines which are modelled (as discussed in Bergeron, Saffer, & Liebert 1992 and Bergeron, Liebert, & Fulbright 1995) to yield the effective temperature and gravity of each star, and therefore the mass. In the younger clusters of age ~100 Myrs, we find that the white dwarfs have masses of $\sim 1\ M_{\odot}$, whereas in older 10 Gyr clusters, the stars have masses of $\sim 0.5\ M_{\odot}$ (see Kalirai et al. 2005; 2007; 2008 for more information). These final remnant masses are connected with initial masses that represent the progenitor star of each white dwarf using knowledge of the cluster age and main-sequence lifetimes. For example, the difference between the cluster age and the white dwarf cooling age represents the lifetime of the progenitor star, from which we can calculate its mass. The result of this calculation, for a dozen clusters in the literature combined with our results, is illustrated in Figure 2.

The initial-final mass relation illustrates that Solar type stars will lose ~50% of their mass through stellar evolution, stars with 2.5 – 3 $M_{\odot}$ will lose 70 – 75% of their mass, and stars with 5 – 6 $M_{\odot}$ will lose ~80% of their mass. In Kalirai *et al.* (2008), we compare this relation to theoretical expectations and briefly explore the implications of the derived mass loss for chemical evolution studies. In Kalirai *et al.* (2007), we also find direct evidence that the masses of white dwarfs in the old, metal-rich cluster NGC 6791 are well below the expected value based on an extrapolation of the relation. Likely, this suggests that mass loss was more efficient in the higher metallicity environment. As discussed in that paper, this result naturally explains the anomalously low white dwarf cooling age measured for NGC 6791 by Bedin *et al.* (1995) since a large fraction of the white dwarfs have helium cores and cool slower than normal carbon-oxygen core white dwarfs.

Increased mass loss in higher metallicity environments is also likely to produce large populations of extreme horizontal branch stars, as seen in NGC 6791. The progenitors of these stars can be singly evolved red giants that lost enough mass on the red giant branch to avoid the helium flash. If the stars are well below the critical mass, they will form helium core white dwarfs. However, some of the stars that were close to the critical mass can also ignite the helium at a later stage while cooling to the white dwarf track, and subsequently populate the extreme horizontal branch (see e.g., Castellani & Castellani 1993). This result has important implications for our understanding of the integrated colors of elliptical galaxies. These old, metal-rich systems are in some ways similar to NGC 6791, and often show UV-excesses that are easy to explain through blue horizontal branch morphologies.

3. The Age of the Galactic Disk and Halo

The initial-final mass relation derived above represents an important ingredient to properly use white dwarfs as chronomoters to date stellar populations, such as the Milky Way disk and halo. As illustrated in the right panel of Figure 1, deep HST observations can uncover even the detailed white dwarf cooling sequences in the nearest globular clusters, such as M4 (Richer *et al.* 2004; Hansen *et al.* 2004), NGC 6397 (Richer *et al.* 2006; Hansen *et al.* 2007), and 47 Tuc (pending cycle 17 HST/ACS program). Modelling the luminosity and color functions of the white dwarfs in these halo clusters, through simulations that propogate main-sequence stars with a given mass function through the initial-final mass relation to the white dwarf stage, suggest an age of ~12 Gyr for the clusters (see e.g., Hansen *et al.* 2007). For the field white dwarf population of the Milky Way disk, a similar analysis suggests a significantly younger age of ~8 Gyr (see Figure 3).

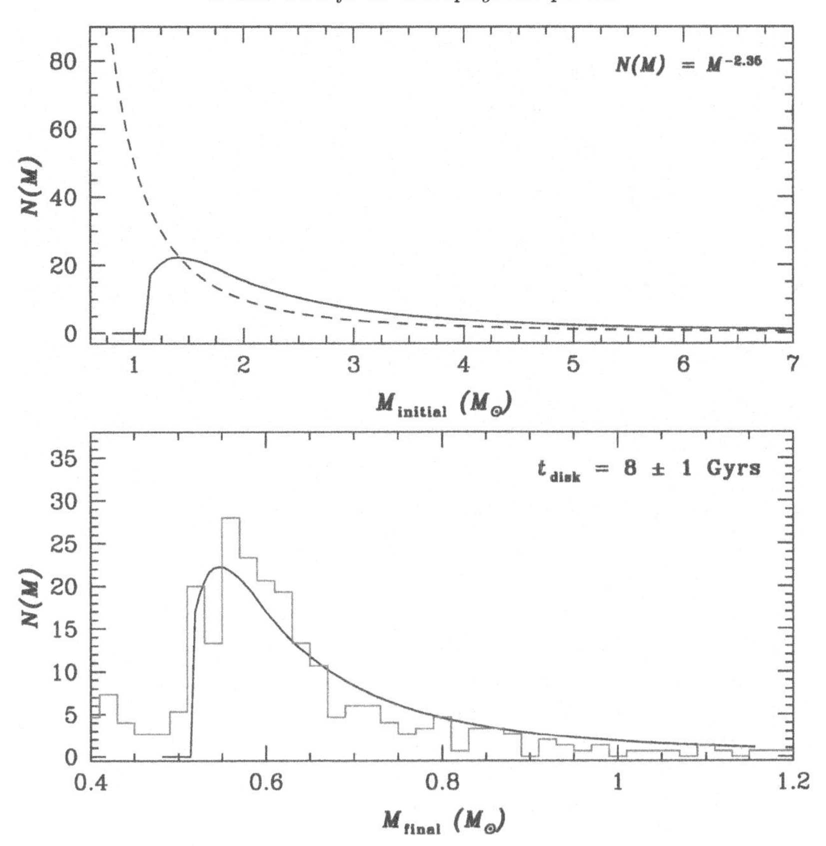

Figure 3. A Salpeter mass function (dashed curve) illustrates the large number of low mass stars that will be formed over any stellar population's lifetime (top panel). The solid curve shows a cutoff imposed by the star formation history of the population (i.e., we assume that the population has formed stars continously over the past 8 Gyr, and only those stars that would have evolved to white dwarfs in that time are shown). Propogating this function through the initial-final mass relation yields a nice fit to the observed field white dwarf luminosity function of the Milky Way disk (bottom panel).

4. Conclusions

White dwarfs are extremely interesting stellar probes to improve our understanding of several astrophysical problems. Although difficult to observe given their faintness, high quality imaging and spectroscopic observations of these stars can be obtained with both ground and space based telescopes. These data have provided new constraints on our understanding of stellar mass loss and evolution, and have been used to provide independent age estimates for the Galactic disk and halo. The properties of white dwarfs in different environments can also improve our general understanding of chemical evolution in galaxies and the stellar content of elliptical galaxies.

References

Bedin, L. R., Salaris, M., Piotto, G., King, I. R., Anderson, J., Cassisi, S., & Momany, Y. 2005, *ApJL*, 624, 45

Bergeron, P., Saffer, R. A., & Liebert, J. 1992, *ApJ*, 394, 228

Bergeron, P., Liebert, J., & Fulbright, M. S. 1995, *AJ*, 444, 810

Castellani, M. & Castellani, V. 1993, *ApJ*, 407, 649

Ferrario, L., Wickramasinghe, D., Liebert, J., & Williams, K. A. 2005, *MNRAS*, 361, 1131

Hansen, B. M. S., *et al.* 2004, *ApJS*, 155, 551

Hansen, B. M. S., *et al.* 2007, *ApJ*, 671, 380

Hertzsprung, E. 1905, *Zeitschrift fur Wissenschaftliche Photographie*, 3, 442

Hertzsprung, E. 1911, *Publ. Astrophys. Observ. Potsdam*, 22, 1

Kalirai, J. S., Richer, H. B., Fahlman, G. G., Cuillandre, J., Ventura, P., D'Antona, F., Bertin, E., Marconi, G., & Durrell, P. 2001a, *AJ*, 122, 257

Kalirai, J. S., Richer, H. B., Fahlman, G. G., Cuillandre, J., Ventura, P., D'Antona, F., Bertin, E., Marconi, G., & Durrell, P. 2001b, *AJ*, 122, 266

Kalirai, J. S., Richer, H. B., Reitzel, D., Hansen, B. M. S., Rich, R. M., Fahlman, G. G., Gibson, B. K., & von Hippel, T. 2005, *ApJL*, 618, L123

Kalirai, J. S., Bergeron, P., Hansen, B. M. S., Kelson, D. D., Reitzel, D. B., Rich, R. M., & Richer, H. B. 2007, *ApJ*, 671, 748

Kalirai, J. S., Hansen, B. M. S., Kelson, D. D., Reitzel, D. B., Rich, R. M., & Richer, H. B. 2008, *ApJ*, 676, 594

Marigo, P. 2001, *A&A*, 370, 194

Reimers, D. 1975, *Societe Royale des Sciences de Liege, Memoires*, 8, 369

Renzini, A. & Fusi Pecci, F. 1988, *ARAA*, 26, 199

Richer, H. B. *et al.* 2004, *AJ*, 127, 2771

Richer, H. B. *et al.* 2006, *Science*, 313, 936

Russell, H. N. 1913, *The Observatory*, 36, 324

Russell, H. N. 1913, *The Observatory*, 37, 165

Somerville, R. S. & Primack, J. R. 1999, *MNRAS*, 310, 1087

Weidemann, V. 2000, *A&A*, 363, 647

Discussion

M. PINSONNEAULT: Can you comment on the potential implications of the high He WD fraction on the age of NGC 6791?

J. KALIRAI: This can likely explain the anomalously- low WD cooling age of NGC 6791. He-core WDs cool a factor of 3 times slower than C-O core WDs, and therefore there is an expectation for a brighter WD peak in the cluster luminosity function, as detailed by Bedin *et al.* The result also naturally explains the large number of EHB stars in this cluster.

S. YI: If $M_{\rm WD} \sim 0.5$, as you find, we expect to see a lot more PAGB stars than found in GCs, because low-M PAGB stars live long. This was also a problem in Brown *et al.*'s M32 observation. Any comment?

J. KALIRAI: Changing the mass could be one of several explanations for this. The nuclear timescales may not be the dominant mechanism, and the TP-AGB models are very uncertain. The result is consistent with Brown *et al.*'s M32 observations given that the lifetimes may not scale strongly with mass.

K. WILLIAMS: A comment that the initial-final mass relation, especially at the high-mass end, is very sensitive to the input cluster ages (see Salaris *et al.* 2008). We need to be cautious to avoid circular arguments in using WDs to get ages in younger open clusters.

J. KALIRAI: I fully agree. This is especially true at young ages where the MS turnoff morphology is vertical in an optical CMD. Alternate age measurements, such as lithium or MS turn- on ages will be required.

Jasonjot Kalirai

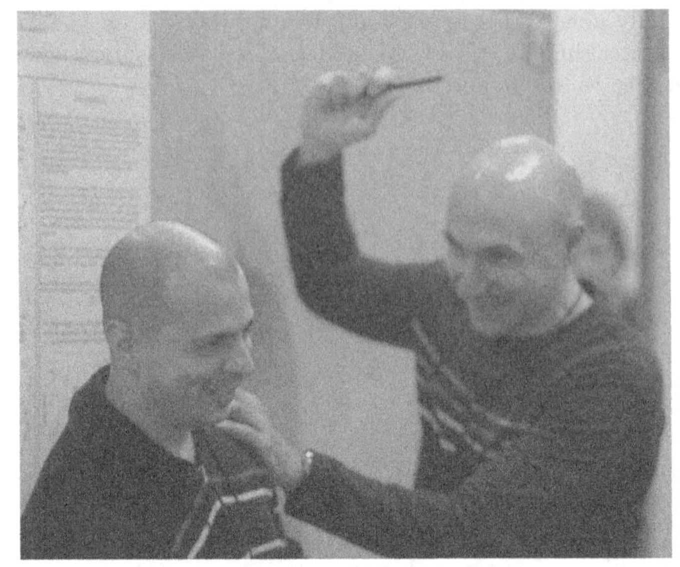

Andrea Bellini and Maurizio Salaris

Andreas Kaufer

The Ages of Stars
Proceedings IAU Symposium No. 258, 2008
E.E. Mamajek, D.R. Soderblom & R.F.G. Wyse, eds.

doi:10.1017/S1743921309031950

Stellar chronology with white dwarfs in wide binaries

S. Catalán[1,2], A. Garcés[1,2], I. Ribas[1,2], J. Isern[1,2] and E. García–Berro[1,3]

[1]Institut d'Estudis Espacials de Catalunya, c/ Gran Capità 2–4, 08034 Barcelona, Spain
email: catalan@ieec.uab.es

[2]Institut de Ciències de l'Espai, CSIC, Facultat de Ciències, UAB, 08193 Bellaterra, Spain

[3]Departament de Física Aplicada, Escola Politècnica Superior de Castelldefels, Universitat Politècnica de Catalunya, Avda. del Canal Olímpic s/n, 08860 Castelldefels, Spain

Abstract. White dwarfs are the evolutionary end product of stars with low and intermediate masses. The evolution of white dwarfs can be understood as a cooling process, which is relatively well known at the moment. For this reason, wide binaries containing white dwarfs are a powerful tool to constrain stellar ages. We have studied several wide binaries containing white dwarfs with two different purposes: when the age of the companion of the white dwarf can be determined with accuracy, we use the binary to improve the knowledge about the white dwarf member. On the contrary, if the companion is a low-mass star with no age indicator available, the white dwarf member itself is used to calibrate the age of the system. In this contribution we present some results using both methodologies to constrain the ages of wide binaries.

Keywords. stars: low-mass, stars: white dwarfs, stars: binaries: visual, stars: activity, stars: evolution

1. Introduction

It is sound to assume that the members of a common proper motion pair, or a wide binary, were born simultaneously and with the same chemical composition. Since the components are well separated (from 100 to 1000 AU), mass exchange between them is unlikely and it can be considered that they have evolved as isolated stars (Oswalt *et al.* 1988). Thus, these kind of systems can be used to perform tests on evolutionary models or even to infer the properties of a less known member from the study of the companion.

In our case, we have studied wide binaries containing white dwarfs. White dwarfs are the final remnants of low- and intermediate-mass stars. About 95% of main-sequence stars will end their evolutionary pathways as white dwarfs and, hence, the study of the white dwarf population provides details about the late stages of the life of the vast majority of stars. Since white dwarfs are long-lived objects, they also constitute useful objects to study the structure and evolution of our Galaxy (Liebert *et al.* 2005; Isern *et al.* 1998). The evolution of white dwarfs can be described as a cooling process, which is relatively well understood at the present moment (Salaris *et al.* 2000). The total age of a white dwarf can be expressed as the sum of its cooling time and the pre-white dwarf lifetime of its progenitor. So, if some external constraints are available, white dwarfs can be used as age calibrators, as it will be shown in the next sections.

White dwarfs in wide binaries have been studied for years. Wegner (1973) carried out a survey of Southern Hemisphere wide binaries containing degenerate members. Later, Oswalt (1981) performed a spectroscopic study of pairs containing white dwarfs, and continued with this work obtaining different catalogs (e.g. Oswalt *et al.* 1988), which have been very useful for more recent works. In this contribution we study wide binaries containing white dwarfs with two main objectives. If the age of the companion of the white dwarf can be determined with relative accuracy, we use the binary system to better understand the white dwarf member, in our case, to improve the initial-final mass relationship of white dwarfs, specially at the low-mass domain. On the contrary, if the companion of the white dwarf is a low-mass star and no age indicators are available, the white dwarf member itself is used to constrain the age of the system.

2. Application I. The initial-final mass relationship of white dwarfs.

The initial-final mass relationship of white dwarfs connects the mass of a white dwarf with that of its progenitor in the main-sequence. This function is of fundamental importance for several aspects of modern astrophysics since it is required as an input for the study of the chemical evolution of galaxies, the determination of the ages of globular clusters and their distances, and also to understand the properties of the Galactic population of white dwarfs. Despite its importance, an accurate measurement of this relationship is still not available and, thus, more efforts are needed from both the theoretical and the observational perspectives to improve it (Weidemann 2001).

From an observational point of view, the final mass of a white dwarf can be determined from spectroscopic observations and the use of cooling sequences (Salaris *et al.* 2000). On the other hand, the initial mass can be derived by considering its progenitor lifetime and some stellar tracks (Domínguez *et al.* 1999). As previously pointed out, the total age of a white dwarf has two components, the cooling time (that can be easily obtained from cooling sequences) and the main-sequence lifetime of its progenitor, which depends on the metallicity. So, in order to obtain the progenitor lifetime, the knowledge of the total age of the white dwarf is necessary. For this reason, white dwarfs for which some external constraints are available are typically studied to be able to derive the initial and final masses. This is the case of white dwarfs belonging to open clusters, since the total ages and metallicities of the progenitors can be inferred from the cluster properties. However, until recently, open clusters data only covered the region of initial masses above 2.5 $M_{\odot}$, since they are in general young and the resulting white dwarfs are massive.

Another way to derive the initial and final masses of white dwarfs is to study white dwarfs belonging to wide binaries. Wide binaries can be relatively old, so, we expected to extend the observational data of the initial-final mass relationship to the low-mass domain. In this case, the total age and metallicity of the progenitor of the white dwarf can be inferred from the study of the companion star, taking into account that the stars were born at the same time and with the same chemical composition. We selected a sample list of wide binaries containing a DA white dwarf (i.e. with the only presence of hydrogen absorption lines) and a FGK star companion (see Catalán *et al.* 2008a, and references therein). The ages of the FGK companions were derived from isochrone fitting, if the star was relatively evolved, or using our preliminary X-ray luminosity-age calibration, which is well established for ages below 1 Gyr (see next section), if the star was close to the ZAMS. In Fig. 1 we group the data we obtained for 6 white dwarfs in wide binaries with our results from a recalculation of the final and initial masses of several open clusters' white dwarfs. The observational data contains now 62 white dwarfs. It is important to emphasize that all the values below 2.5 $M_{\odot}$ correspond to our data obtained from wide binaries (WBs — Catalán *et al.* 2008a) and some recent data based on old open clusters (K08 — Kalirai *et al.* 2008). The coverage of the low-mass end of the initial-final mass relationship is specially important since it is the most populated bin according to the Salpeter's initial mass function and at the same time it guarantees,

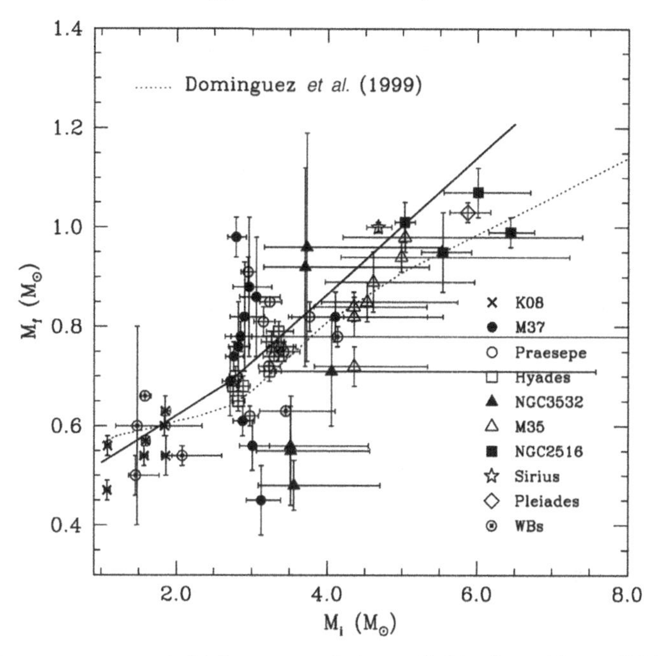

Figure 1. Final masses versus initial masses of the available data. The solid lines correspond to the analytical expressions derived from a weighted least squares linear fit of these data. The dotted line represents the theoretical relation for solar composition of Domínguez *et al.* (1999).

according to the theory of stellar evolution, the study of white dwarfs with masses near the most common value, $M \sim 0.57\,M_\odot$, which is representative of the 90 per cent of the white dwarf population (Kepler *et al.* 2007). Thus, these new data increase considerably the statistical significance of the semi-empirical initial-final mass relationship.

2.1. *An analytical relationship*

Following some recent works on this subject (Ferrario *et al.* 2005, Williams 2007), we assume that the initial-final mass relationship can be described as a linear function. As can be seen in Fig. 1 the theoretical initial-final mass relationship of Domínguez *et al.* (1999) (dotted line) can be divided in two different linear functions, each one above and below 2.7 $M_\odot$, with a shallower slope for small masses probably due to the smaller efficiency of mass loss. Taking this into account we have performed a weighted least-squares linear fit for each region, obtaining for $M_{\rm i} < 2.7\,M_\odot$:

$$M_{\rm f} = 0.096 M_{\rm i} + 0.429 \qquad \Delta M_{\rm f} = 0.05\ M_\odot \tag{2.1}$$

and for $M_{\rm i} > 2.7\,M_\odot$:

$$M_{\rm f} = 0.137 M_{\rm i} + 0.318 \qquad \Delta M_{\rm f} = 0.12\ M_\odot \tag{2.2}$$

These expressions have been overplotted in Fig. 1 as two solid lines. Taking into account the scatter of the data and the values of the reduced χ^2 of these fits (7.1 and 4.4, respectively) we have computed the dispersion of the derived final masses, obtaining 0.05 $M_\odot$ and 0.12 $M_\odot$, respectively. In past works, it was necessary to include in the fit a fictitious anchor point to represent the canonical white dwarf mass, $M_{\rm f} \sim 0.57\ M_\odot$ (Kepler *et al.* 2007), since no data were available at the low-mass domain. In our case,

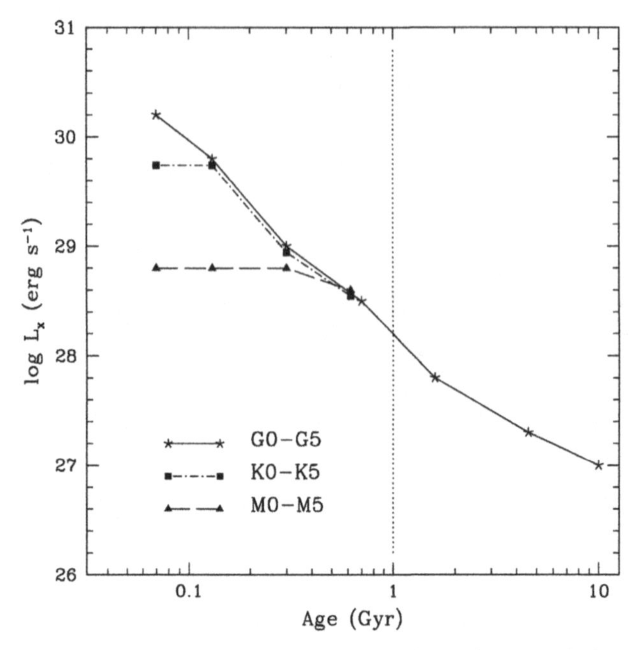

Figure 2. Calibration between the X-ray luminosity and age for stars belonging to the Sun in Time program (G type) covering up to 10 Gyr (solid line). The dashed-dotted and dashed lines correspond to our preliminary extension of this calibration to other spectral types (K and M, respectively) with data up to 1 Gyr.

this is not necessary since we are including new data that cover this typical mass value of white dwarfs.

It is worth mentioning that as an indirect test we have computed the theoretical white dwarf mass distribution considering different initial-final mass relationships, and the best fit to the recent observational data based on the SDSS (Degennaro *et al.* 2008) is obtained when our two analytical expressions are used (see Catalán *et al.* 2008b).

3. Application II. The determination of ages of low-mass stars.

When a wide binary is composed of a white dwarf and a low-mass star for which no age indicator is available, the white dwarf component can be used as the age calibrator (Silvestri *et al.* 2005). The determination of the ages of low-mass stars has many applications in Astrophysics such as, for instance, the calibration of the decrease of high-energy emissions. Magnetic activity in the Sun, and low-mass stars in general, manifests itself in the form of high-energy and particle emissions. Besides an associated strong level of variability over various timescales (rotational modulation, flares, cycles), the overall activity decreases very rapidly with time. This is related to the rotational spin down and the subsequent loss of efficiency of energy generation mechanisms.

The case of the Sun has been studied specifically within the Sun in Time program, using stars with the same spectral type but different ages, in order to estimate its past history and future evolution (Ribas *et al.* 2005). In Fig. 2 we show our calibration of the X-ray luminosity and age for G stars (solid line), covering up to 10 Gyr. We also show our preliminary results from the extension of this study to K and M stars (dashed-dotted and dashed lines, respectively). For this purpose we used stars with well known ages, for instance, stars belonging to open clusters or moving groups (NGC2547, IC2391, IC2602,

α Per, Pleiades, UMa moving group, Hyades). As can be seen in Fig. 2, high-energy emissions decrease monotonically with age. The flat part of these diagrams (i.e., constant $\log L_x$) corresponds to the saturation phase, which is longer in lower-mass stars. It can be noticed as well that no accurate data above 1 Gyr is available for K and M spectral types. By studying wide binaries containing white dwarfs we will be able to extend these calibrations to older ages, and at the same time estimate the decrease of the rotational speed of stars with age. At the present moment, there are some relations between the rotational period and the activity of stars, but the link with age is still problematic although some improvements have been achieved recently (Mamajek & Hillenbrand 2008, Messina *et al.* 2001). A major application of this method will be to obtain Age-Rotation-Activity relations for different spectral types (GKM) and perform a cross-check with ages derived from the gyrochronology method (Barnes *et al.* 2007).

The understanding of the past and current evolution of stars is also essential for characterizing the atmospheres of their planets (if any) since the particle emissions may have a relevant influence on them (e.g. Scalo *et al.* 2007). Moreover, having a method to derive the ages of low-mass stars above 1 Gyr will have many applications in dynamic and kinematic studies of our Galaxy (Famaey *et al.* 2005).

3.1. *Method*

We have selected a sample of 27 wide binaries containing a DA white dwarf and a GKM star (2 G, 6 K and 19 M) from the recent revision of the NLTT Catalogue (Chanamé & Gould 2004). Since a spectroscopic analysis of these white dwarfs to obtain their atmospheric parameters at the present moment is still not possible, we have used the available photometry to obtain an effective temperature to be able to derive a rough estimation of the ages. We have used the photometric effective temperature and assumed the canonical value for the surface gravity of white dwarfs, $\log g = 8.0$, in order to obtain the masses and cooling times considering the cooling sequences of Salaris *et al.* (2000). Then, we have considered the analytical expressions given in the previous section (Catalán *et al.* 2008b) to obtain their initial masses. Finally, using the stellar tracks of Domínguez *et al.* (1999) we have derived the progenitor lifetimes. Adding these to the cooling times, we have obtained preliminary values for the total ages of the white dwarfs, i.e. the ages of the low-mass companions. The values obtained are generally larger than 1 Gyr, so, with this method we will be able to derive the ages of old low-mass stars for which no age indicator is currently available.

At the present moment, we have been granted time at several observatories to obtain spectroscopic observations for the white dwarf members. Once these data are available, a more precise value for the effective temperature and surface gravity will be obtained from performing a fit of the Balmer lines to synthetic spectra. Then, an accurate value for the masses of the white dwarfs will be derived straightforwardly using cooling sequences. It should be taken into account that when white dwarfs are very cool ($T_{\rm eff} < 12000$ K) the spectroscopic masses are not completely reliable, since their atmospheres could be well enriched in helium, while retaining its DA appearance, and this would mimic a larger surface gravity (Bergeron *et al.* 1992). In those cases we will use other methods, like, for instance, the derivation of the gravitational redshift of the white dwarf using the radial velocity of the GKM companion.

Considering the dependency of this method on the cooling sequences, stellar models and the initial-final mass relationship assumed, an accuracy of 20-40% is expected to be achieved in the ages derived. Thus, although in some cases the uncertainty in the age can be large, with this method we will be placing age constraints to stars that did not have any age indicator up to now.

The next step in this work will be the characterization of the low-mass companions. A photometric and spectroscopic analysis of these stars will be performed to determine their effective temperatures and metallicities. Moreover, the activity of these stars will be studied as well using X-ray data (e.g. ROSAT), if available, or some flux-flux relationships, for instance, those of Montes *et al.* (1996). This will allow us to obtain the necessary data to accurately define the activity-age relations.

4. Conclusions

In this contribution we have explored two different methods to constrain the ages of wide binaries composed of a white dwarf and a low-mass star. When the age of the companion of the white dwarf can be determined with accuracy we have used the wide binary to improve the initial-final mass relationship of white dwarfs. This study has allowed us to cover the low-mass domain of this relationship, which was poorly studied until now. This has a great importance since the low-mass domain is the most populated bin of masses according to the Salpeter's initial mass function. On the other hand, when no age indicators are available for the companion, we have used the white dwarf member itself to constrain the age of the system. Our preliminary study of 27 wide binaries from the NLTT Catalogue has shown that wide binaries are a powerful tool to derive the ages of low-mass stars. This method has many applications in stellar and planetary physics such as the calibration of Age-Rotation-Activity relations for different spectral types (GKM), the understanding of the evolution of the high-energy emissions of exoplanet host stars and dynamic and kinematic Galactic studies, among others.

References

Barnes, S. A. 2007, *ApJ*, 669, 1167
Bergeron, P., Wesemael, F., & Fontaine G. 1992, *ApJ*, 387, 288
Catalán, S., Isern, J., García–Berro, E., *et al.* 2008a, *A&A*, 477, 213
Catalán, S., Isern, J., García–Berro, E., & Ribas, I. 2008b, *MNRAS*, 387, 1693
Chanamé, J. & Gould, A. 2004, *ApJ*,601, 289
Domínguez, I., Chieffi, A., Limongi, M., & Straniero, O. 1999, *ApJ*, 524, 226
Famaey, B., Jorissen, A., Luri, X., *et al.* 2005, *ESASP*, 576, 129
Ferrario, L., Wickramasinghe, D. T., Liebert, J., & Williams, K. A. 2005, *MNRAS*, 361, 1131
Messina, S., Rodono, M., & Guinan, E. F. 2001, *A&A*, 366, 215
Isern, J., García-Berro, E., Hernanz, M., Mochkovitch, R., & Torres, S. 1998, *ApJ*, 503, 239
Kalirai, J. S., Hansen, M. S. B., Kelson, D. D., *et al.* 2008, *ApJ*, 676, 594
Kepler, S. O., Kleinman, S. J., Nitta, A., *et al.* 2007, *MNRAS*, 375, 1315
Liebert, J., Young, P. A., Arnett, D., Holberg, J. B., & Williams, K. A. 2005, *ApJ*, 630, L69
Mamajek, E. E. & Hillenbrand, L. A. 2008, *ApJ*, 687, 1264
Montes, D., Fernandez-Figueroa, M. J., Cornide, M., & de Castro, E. 1996, *ASPC*, 109, 657
Oswalt, T. 1981, Ph. D. thesis, AA (Ohio State Univ., Columbus)
Oswalt, T., Hintzen, P., & Luyten, W. 1988, *ApJS*, 66, 391
Ribas, I., Guinan, E. F., Güdel, M., & Audard, M. 2005, *ApJ*, 622, 680
Salaris, M., García–Berro, E., Hernanz, M., Isern, J., & Saumon D. 2000, *ApJ*, 544, 1036
Scalo, J., Kaltenegger, L., Segura, A. *et al.* 2007, *AsBio*, 7, 85
Silvestri, N. M., Hawley, S. L., & Oswalt, T. D. 2005, *AJ*, 129, 2428
Wegner, G. 1973, *MNRAS*, 165, 271
Weidemann, V. 2000, *A&A*, 363, 647
Williams, K. A. 2007, *ASPCS*, 372, 85

Discussion

R. JEFFRIES: You have made an implicit assumption that the components are non-interacting. But components with separations up to $\sim$ 100 AU may have interacted in the past: the slow wind of the AGB progenitor of the WD can be accreted by the low-mass companion, resulting in spin up and higher x-ray activity (Jeffries and Stevens 1996, MNRAS, 279, 180).

S. CATALAN: The typical separations of this type of systems is 100 AU or usually more (up to 1,000 AU), so I don't think that is probable that they have been interacting in the past.

L. HILLENBRAND: Just a question of clarification: the sample is nearby enough for spatially-resolved data on the two components, correct? What is the distance range of your pairs?

S. CATALAN: A few tens of parsecs, out to 100 pc.

Silvia Catalán

The Ages of Stars
Proceedings IAU Symposium No. 258, 2008
E.E. Mamajek, D.R. Soderblom & R.F.G. Wyse, eds.

doi:10.1017/S1743921309031962

Towards a precise white dwarf cooling age of a globular cluster

Harvey B. Richer[1], Saul Davis[1], Jason Kalirai[2], Aaron Dotter[3] and R. Michael Rich[4]

[1]Department of Physics and Astronomy, University of British Columbia, Vancouver, BC, Canada
email: richer@astro.ubc.ca

[2]Space Telescope Science Institute, 3700 San Martin Drive, Baltimore, MD 21218
email: jkalirai@stsci.edu

[3]Department of Physics and Astronomy, University of Victoria, Victoria, BC, Canada
email: dotter@uvic.ca

[4]Department of Physics and Astronomy, UCLA, Los Angeles, Ca 90095
email: rmr@astro.ucla.edu

Abstract. The white dwarf cooling age of a globular star cluster provides a potentially precise method of determining the ages of these ancient systems. This age-dating technique should be viewed as one distinct from that of turn-off ages, with a largely different set of input physics and problems. As such the ages produced by these two methods are complimentary and we seek convergent to the same value. In addition to deep photometry and astrometry of cluster stars, precise distances to the clusters and their reddenings are required. Theoretical models of both main sequence stars and cooling white dwarfs are also needed as well as the masses of the white dwarfs and an initial-final mass relationship. In this contribution I discuss a potentially precise approach to cluster distances via a geometric technique (comparing the internal proper motion dispersion of cluster stars with their radial velocity dispersion) and spectroscopically determined masses of M4 white dwarfs at the top of the cooling sequence. These latter data extend the initial-final mass relationship down to the lowest mass stars that are currently forming white dwarfs.

Keywords. astrometry, stars: white dwarfs, Galaxy: globular clusters: individual (M 4)

Discussion

M. LIU: 1) For your proper motion measurements, do you use only the two (closely spaced) epochs from Gemini AO, or include the older CFHT archival data? 2) Given the concerns about the metallicity dependence of the initial-find mass relation, what will the fundamental limitations be in determining the ages, since the clusters used to calibrate the initial-final mass relation have different metallicities?

H. RICHER: 1) The proper motions themselves are derived only from the Gemini data but the older (and much less precise) CFHT data were useful in assessing the errors in the proper motions. 2) Except for the data I just showed on M4, all the clusters used in the plot are about solar metallicity. It is interesting however, that this point for [Fe/H] near -1.3 seems to fit the same relationship quite well.

I. KING: When your distance from comparing dispersions of radial velocities and proper motions is still at the 20% level of accuracy, the straightforward comparison works fine, but when you aim at higher accuracies you will need a better dynamical model, which takes into account anisotropy of dispersions and rotation.

H. Richer: I certainly agree with this. As the data improve, particularly the radial velocity dispersion which now dominates the error, a more sophisticated model will be required. We are not there yet, but soon I hope to be.

Harvey Richer

The Ages of Stars
Proceedings IAU Symposium No. 258, 2008
E.E. Mamajek, D.R. Soderblom & R.F.G. Wyse, eds.

doi:10.1017/S1743921309031974

Brown dwarfs as Galactic chronometers

Adam J. Burgasser

Massachusetts Institute of Technology, Kavli Institute for Astrophysics and Space Research, Building 37, Room 664B, 77 Massachusetts Avenue, Cambridge, MA 02139, USA
email: ajb@mit.edu

Abstract. Brown dwarfs are natural clocks, cooling and dimming over time due to insufficient core fusion. They are also numerous and present in nearly all Galactic environments, making them potentially useful chronometers for a variety of Galactic studies. For this potential to be realized, however, precise and accurate ages for individual sources are required, a prospect made difficult by the complex atmospheres and spectra of low-temperature brown dwarfs; degeneracy between mass, age and luminosity; and the lack of useful age trends in magnetic activity and rotation. In this contribution, I review five ways in which ages for brown dwarfs are uniquely determined, discuss their applicability and limitations, and give current empirical precisions.

Keywords. stars: binaries, stars: fundamental parameters, stars: kinematics, stars: late-type, stars: low-mass, brown dwarfs.

1. Introduction

Brown dwarfs are very low-mass stars whose masses ($M \lesssim 0.075\ M_{\odot}$) are insufficient to sustain the core hydrogen fusion reactions that balance radiative energy losses (Kumar 1963; Hayashi & Nakano 1963). Supported from further gravitational contraction by electron degeneracy pressure, evolved brown dwarfs continually cool and dim over time as they radiate away their initial contraction energy, ultimately achieving photospheric conditions that can be similar to those of giant planets. The first examples of brown dwarfs were identified as recently as 1995 (Nakajima *et al.* 1995; Rebolo *et al.* 1995). Today, there are hundreds known in nearly all Galactic environments, identified largely in wide-field, red and near-infrared imaging surveys such as 2MASS, DENIS, SDSS and UKIDSS. The known population of brown dwarfs encompasses the late-type M ($T_{eff} \approx$ 2500–3500 K), L ($T_{eff} \approx$1400–2500 K) and T spectral classes ($T_{eff} \approx$600–1400 K; e.g., Vrba *et al.* 2004), while efforts are currently underway to find even colder members of the putative Y dwarf class (see review by Kirkpatrick 2005).

Because brown dwarfs cool over time, their spectral properties are inherently time-dependent, making them potentially useful chronometers for Galactic studies (much like white dwarfs; see contributions by M. Salaris, J. Kalirai, S. Catalán and H. Richer). However, the primary observables of a brown dwarf—temperature, luminosity and spectral type—depend on both mass and age (and weakly on metallicity). This degeneracy complicates characterizations of individual sources and mixed populations. Unfortunately, traditional stellar age-dating methods do not appear to be applicable for brown dwarfs. Magnetic activity metrics, such as the frequency and strength of optical or X-ray non-thermal emission, appear to be largely age-invariant (e.g., Stelzer *et al.* 2006) and quiescent emission drops off precipitously in the early L dwarfs as cool photospheres are decoupled from field lines † (e.g., Mohanty *et al.* 2002; see contribution by A. West). Long-term angular momentum loss in brown dwarfs is far more muted than in stars,

† Interestingly, radio emission does not drop off for cooler brown dwarfs, and may even increase relative to bolometric flux, although there are currently few detections (e.g., Berger 2006).

Table 1. Age-dating Methods for Brown Dwarfs.

Technique	Applicable for	Pros	Cons	Precision	Examples in the Literature
Cluster members & companions	nearby clusters; companions to age-dated stars	precise ages based on stellar/cluster work; calibration for other techniques and evolutionary models	generally restricted to young clusters; wide (resolvable) & close (RV) companions rare; must verify coevality/ association; atmospheres variable for $t \lesssim 10$ Myr	~10% clusters; ~50–100% for companions	1,2,3,4
6708 Å Li I absorption	$t \sim 10$–200 Myr; resolved binaries; individual field sources with $T_{eff} > 1500$ K and $t \lesssim 2$ Gyr (limits only)	consistent predictions from different models; straightforward test; largely insensitive to atmospheric properties	requires high sensitivity, high resolution spectra; low brightness region; not useful for T dwarfs or for $t \lesssim 10$ Myr; relies on accurate evolutionary models	~10% for young clusters; upper/lower limits for all others	5,6,7
Mass standards	astrometric/RV binaries	precise masses yield precise ages; weakly sensitive to atmospheric properties	suitable systems are rare; long-term follow-up required; relies on accurate evolutionary models	~10–20%	8,9,10,11
Surface gravities	any source with a well-measured spectrum	applicable to individual sources; particularly useful for T dwarfs	low precision; other factors (e.g., metallicity) complicate analysis; relies on accurate evolutionary and atmospheric models	~50–100%	12,13,14
Kinematics	well-defined groups or populations	useful statistical test for various subclasses; insensitive to evolutionary or atmospheric models	very low precision; large groups required to beat statistical noise; susceptible to selection biases	~100–300%	15,16,17

References: (1) Bouvier *et al.* (1998); (2) Luhman *et al.* (2003); (3) Geballe *et al.* (2001); (4) Burgasser (2007); (5) Stauffer *et al.* (1998); (6) Jeffries & Oliveira (2005); (7) Liu & Leggett (2005); (8) Zapatero Osorio *et al.* (2004); (9) Stassun *et al.* (2006); (10) Liu *et al.* (2008); (11) Dupuy *et al.* (2008); (12) Mohanty *et al.* (2004); (13) Burgasser *et al.* (2006a); (14) Saumon *et al.* (2007); (15) Schmidt *et al.* (2007); (16) Zapatero Osorio *et al.* (2007); (17) Faherty *et al.* (2008).

and there is no clear rotation-activity relation for L dwarfs (e.g., Reiners & Basri 2008). Exploitation of the cooling properties of brown dwarfs is therefore a favorable approach for determining their ages.

In this contribution, I review five methods currently employed to age-date brown dwarfs and summarize their applicability, inherent limitations and current (typical) precisions. Table 1 provides a summary of the methods discussed in detail below.

2. Cluster members and companions

The most straightforward way to age-date an individual brown dwarf is to borrow from its environment, a tactic that is suitable for members of coeval clusters/associations and companions to age-dated stars. Brown dwarfs are well-sampled down to and below the deuterium fusing mass limit ($M \lesssim 0.013\ M_{\odot}$) in the youngest nearby clusters ($t \lesssim 5$ Myr), as their luminosities are greater at early ages. Brown dwarfs have also been identified in somewhat older (10–50 Myr) "loose associations" in the vicinity of the Sun ($d \lesssim 50$–100 pc; e.g., Kirkpatrick *et al.* 2008). For older and more distant clusters ($d \gtrsim 1$ kpc), decreasing surface temperatures and compact radii exacerbate the sensitivity issues that plague low-mass stellar studies (see contribution by G. Piotto). There are as yet no known brown dwarfs in globular clusters, despite detection of the end of the stellar main sequence in systems such as NGC 6397 (Richer *et al.* 2008).

For brown dwarfs in young clusters, numerous studies have examined age-related trends in colors (e.g., Jameson *et al.* 2008), spectral characteristics (e.g., Allers *et al.* 2007), accretion timescales (e.g., Mohanty *et al.* 2005) and circum(sub)stellar disk evolution (e.g., Scholz *et al.* 2007). These have been coarsely quantified, and appear to be most useful at very young ages ($t \lesssim 10$ Myr). Surface properties and luminosities are highly variable at these ages due to sensitivity to formation conditions (e.g., Baraffe *et al.* 2002), ongoing accretion, complex magnetic effects (e.g., Reiners *et al.* 2007) and age spreads within a cluster (see contribution by R. Jeffries). Hence, while brown dwarfs in clusters with ages spanning ~1–650 Myr are now well-documented, with age uncertainties as good as 10% (for the LDB technique; see § 3), useful predictive trends are probably limited to ages of $\gtrsim$10 Myr.

Known brown dwarf companions to main sequence stars now number a few dozen, spanning a wide range of separation, age and composition. Many of these systems are widely-separated so that their brown dwarf companions can be directly studied. Age uncertainties for companion brown dwarfs depend on stellar dating methods which are generally more uncertain (50–100%; e.g., Liu *et al.* 2008) than cluster ages. Searches for substellar companions to more precisely age-dated white dwarfs (e.g. Day-Jones *et al.* 2008; Farihi *et al.* 2008) and subgiants (Pinfield *et al.* 2006) have so far met with limited success. Nevertheless, brown dwarf companions to age-dated stars serve as important benchmarks for calibrating other age-dating methods at late ages ($\gtrsim$500 Myr) and are fundamental for testing evolutionary models (see contribution by T. Dupuy).

3. 6708 Å Li I Absorption

Lithium is fused at a lower temperature than hydrogen (2.5×10^6 K), resulting in a somewhat lower fusing mass limit (0.065 $M_\odot$; Bildsten *et al.* 1997). Because the interiors of low mass stars and brown dwarfs are fully convective at early ages, an object with a mass above this limit will fully deplete its initial reservoir of lithium within ~200 Myr. Hence, any system older than this which exhibits Li I absorption has a mass less than 0.065 $M_\odot$ and is therefore a brown dwarf (e.g., Rebolo *et al.* 1992). With a mass limit, one can use evolutionary models to determine an age limit.

In the age range ~10–200 Myr, the degree of lithium depletion in low-mass stars and brown dwarfs is itself mass-dependent, occurring earlier in more massive stars which first achieve the necessary core temperatures. Hence, over this range the age of an individual source can be precisely constrained if its mass is known. A more practical approach, however, is to ascertain the age of a group of coeval low-mass objects based on which sources do or do not exhibit Li I absorption; this is the lithium depletion boundary (LDB) technique. Different evolutionary models yield remarkably similar predictions for the location of the LDB over a broad range of ages (Burke *et al.* 2004), and the boundary itself is readily identifiable in color-magnitude diagrams. As such, this technique has been used to age-date several nearby young clusters and associations (e.g., Stauffer *et al.* 1998; Barrado y Navascués *et al.* 1999; Jeffries & Oliveira 2005; Mentuch *et al.* 2008). LDB studies have also provided independent confirmation of other cluster-dating methods such as isochrone fitting (Jeffries & Oliveira 2005). A variant of the LDB technique for coeval binary systems has been proposed by Liu & Leggett (2005), in which a system that exhibits Li I absorption in the secondary but not in the primary can have both lower and upper age limits assigned to it (note that the presence/absence of Li I in both components simply sets a single upper/lower age limit). This technique requires resolved optical spectroscopy of both components and can be pursued only for a few (rare) wide low-mass pairs (e.g., Burgasser *et al.*, in prep.) or using high spatial-resolution spectroscopy (e.g. Martín *et al.* 2006). No single brown dwarf pair straddling the LDB has yet been identified.

Despite its utility, the detection of LI I absorption in brown dwarf spectra has limitations. The 6708 Å line lies in an relatively faint spectral region for cool L-type dwarfs, so sensitive spectral observations on large telescopes are typically required to detect (or convincingly rule out) this feature. For optically-brighter M-type brown dwarfs, high-resolution observations are generally required to distinguish Li I absorption from overlapping molecular absorption features. Young brown dwarfs ($t \lesssim 50$ Myr) with low surface gravities show weakened alkali line absorption (see § 5), including Li I, making it again necessary to obtain sensitive, high-resolution observations (Kirkpatrick *et al.* 2008). For brown dwarfs cooler than ~1500 K (i.e., the T dwarfs), lithium is chemically depleted in the photosphere through its conversion to LiCl, LiF or LiOH (Lodders 1999). As such, practical age constraints using Li I can only be made for systems younger than ~2 Gyr.

4. Mass standards

One way of breaking the mass/age/luminosity degeneracy for an individual brown dwarf is to explicitly measure its mass. This is feasible for sufficiently tight brown dwarf binaries for which radial velocity (RV) and/or astrometric orbits can be measured. Of the ~100 very low mass (M_1,M_2 $\leqslant$ 0.1 $M_\odot$) binary systems now known, only a handful have sufficiently short periods that large portions of their RV orbits (e.g., Joergens & Müller 2007; Blake *et al.* 2008), astrometric orbits (e.g., Lane *et al.* 2001; Bouy *et al.* 2004; Liu *et al.* 2008; Dupuy *et al.* 2008), or both (Zapatero Osorio *et al.* 2004; Stassun *et al.* 2006) have been measured. With measurable total system masses or mass functions, individual masses can be estimated from relative photometry or directly determined from recoil motion in both components (e.g., Stassun *et al.* 2006). The individual masses and component luminosities can then be compared to evolutionary models to determine ages.

Liu *et al.* (2008) have suggested that such "mass standards" provide more precise constraints on the physical properties (including ages) of brown dwarfs as compared to "age standards", namely companions to main sequence stars. Orbital masses can currently be constrained to roughly 5–10% precision, translating into 10–20% uncertainties in ages based on evolutionary tracks (versus 50–100% for main-sequence stars). More importantly, brown dwarf binaries with mass measurements and independent age determinations—i.e., companions to age-dated stars and cluster members—can provide specific tests of the evolutionary models themselves. Further details are provided in the contribution by T. Dupuy.

5. Surface gravity

Only 10–20% of brown dwarfs are found to be multiple (e.g., Burgasser *et al.* 2006b) and few of these are suitable for orbital mass measurements. A proxy for mass is surface gravity, which can be determined directly from a brown dwarf's spectrum. For $T_{eff} \lesssim$ 2500 K and ages greater than ~50 Myr, evolutionary models predict that brown dwarf surface gravities ($g \propto M/R^2$) are roughly proportional to mass due to near-constant radii (roughly equal to Jupiter's radius). Surface gravity is also proportional to photospheric pressure ($P_{ph} \propto g/\kappa_R$, where κ_R is the Rosseland mean opacity), which in turn influences the chemistry, line broadening and (in some cases) opacities of absorbing species in the photosphere. Hence, "gravity-sensitive" features in a brown dwarf's spectrum can be used to infer its mass and, through evolutionary models, its age.

Examples of gravity-sensitive features include the optical and near-infrared VO bands and alkali lines in late-type M and L dwarfs, all of which evolve considerably between field dwarfs ($\log g \approx 5$ cgs), young cluster dwarfs ($\log g \approx 3$–4) and giant stars ($\log g \approx 0$; e.g., Luhman 1999). Enhanced VO absorption and weakened alkali line absorption is a characteristic trait of young brown dwarf spectra (e.g., Gorlova *et al.* 2003; Allers *et al.* 2007). Alkali features in particular are useful for cooler brown dwarfs as VO condenses out of the photosphere. Quantitative analyses of these features have been used to distinguish "young" ($\gtrsim$10 Myr) from "very young" sources thus far ($\lesssim$5 Myr; e.g., Cruz *et al.* 2007). More robust metrics await larger and more fully-characterized samples.

Another important surface gravity diagnostic is collision-induced H_2 absorption, a smooth opacity source spanning a broad swath of the infrared (e.g., Linsky 1969; Borysow *et al.* 1997). H_2 absorption is weakened in the low-pressure atmospheres of young cluster brown dwarfs, resulting in reddened near-infrared spectral energy distributions and colors; in particular, a characteristic, triangular-shaped H-band (1.7 μm) flux peak (e.g., Lucas *et al.* 2001; Kirkpatrick *et al.* 2006). The proximity of many young and reddened brown dwarfs ($<$ 100 pc) rules out ISM absorption as a primary source for this reddening. Jameson *et al.* (2008) have exploited this trend by using a proper-motion selected sample of nearby young cluster candidate members to infer an age/color/luminosity relation for brown dwarfs younger than 0.7 Gyr, with a stated accuracy of $\pm$0.2 dex in log(age), or about 60% fractional uncertainty. Kinematically older low-mass stars and brown dwarfs in the Galactic disk (e.g., Faherty *et al.* 2008) and halo populations (e.g., Burgasser *et al.* 2003) exhibit unusually blue near-infrared colors due to enhanced H_2 absorption. However, differences in metallicities and condensate cloud properties can muddle surface gravity determinations in these sources by modulating the photospheric pressure through opacity effects (changing κ_R; e.g., Leggett *et al.* 2000; Looper *et al.* 2008).

The use of H_2 absorption as a surface gravity indicator is particularly useful for T dwarfs, as H_2 dominates the K-band (2.1 μm) opacity and significantly influences near-infrared colors (e.g., Burgasser *et al.* 2002; Knapp *et al.* 2004). Several groups now employ this feature to estimate the atmospheric properties of individual T dwarfs (e.g., Burgasser *et al.* 2006a; Warren *et al.* 2007; Burningham *et al.* 2008), typically through the use of spectral indices that separately sample surface gravity (e.g., the K-band) and temperature variations (e.g., H_2O or CH_4 bands). These indices are compared to atmospheric models calibrated by one or more benchmarks (e.g., a companion to a precisely age-dated star), and evolutionary models are used to determine individual masses and ages. Typical uncertainties of $\log g$ ~0.3 dex translate into 50-100% uncertainties in age, comparable to uncertainties for main sequence stars. Again, variations in metallicity can mimic variations in surface gravity, although a third diagnostic such as luminosity can break this degeneracy (e.g., Burgasser 2007). As atmosphere models improve in fidelity, parameters are increasingly inferred from direct fits to spectral data, with comparable uncertainties (e.g., Saumon *et al.* 2007; Cushing *et al.* 2008).

6. Kinematics

When a sufficiently large enough sample of brown dwarfs is assembled, one can apply standard kinematic analyses, building from the assumption that gravitational perturbations lead to increased velocity dispersions with time (e.g., Spitzer & Schwarzschild 1953; see contribution by B. Nordström). Velocity dispersions are typically characterized by a time-dependent power-law form, i.e., $\sigma \propto (1 + t/\tau)^{\alpha}$ (e.g., Wielen 1977; Hänninen

& Flynn 2002). Other statistics, such as Galactic scale height, can also be tied to age through kinematic simulations (e.g., West *et al.* 2008) to calibrate secondary age diagnostics such as magnetic activity (see contribution by A. West).

Samples of very low mass stars and brown dwarfs have only recently become large enough that kinematic studies are feasible. The largest samples (over 800 sources) have been based on proper motion measurements (e.g., Schmidt *et al.* 2007; Casewell *et al.* 2008; Faherty *et al.* 2008). For field dwarfs, dispersion in tangential velocities for both magnitude- and volume-limited samples indicate a mean age in the range 2–8 Gyr, largely invariant with spectral type. This is consistent with the ages of more massive field stars and population synthesis models (e.g., Burgasser 2004). However, when field samples are broken down by color (Faherty *et al.* 2008) or presence of magnetic activity (Schmidt *et al.* 2007), distinct age groupings are inferred, indicating that both very old (i.e., thick disk or halo) and very young (i.e., thin disk or young association) brown dwarf populations coexist in the immediate vicinity of the Sun. Indeed, one of the major results from these studies is the identification of widely-dispersed brown dwarf members of nearby, young moving groups such as the Hyades (e.g., Bannister & Jameson 2007).

With only two dimensions of motion measured, proper motion samples may produce biased dispersion measurements depending on the area of sky covered by a sample. Full 3D velocities require RV measurements which are more expensive and have thus far been obtained only for a small fraction of the known brown dwarf population (e.g., Basri & Reiners 2006; Blake *et al.* 2007). A seminal study by Zapatero Osorio *et al.* (2007) of 21 nearby, late-type dwarfs with parallax, proper motion and RV measurements found considerably smaller 3D velocity dispersions for L and T dwarfs than GKM stars, suggesting that local brown dwarfs are young ($t \sim 0.5$–4 Gyr). The discrepancy between this result and the proper motion studies may be attributable to small number statistics and/or contamination by young moving groups; ~40% of the brown dwarfs in the Zapatero Osorio *et al.* (2007) sample appear to be associated with the Hyades. Resolving this discrepancy requires larger RV samples, which has the side benefit of potentially uncovering RV variables that can be used as mass standards (e.g., Basri & Reiners 2006).

7. Improvements and future work

With several methods for age-dating brown dwarfs over a broad range of phase space now available, opportunities to use these objects as chronometers for various Galactic studies look to be increasingly promising; e.g., age-dating planetary systems, examining cluster age spreads, testing Galactic disk dynamical heating models, and direct measures of the substellar mass function and birthrate in the field and other populations. However, there are areas where improvements in uncertainties are needed and basic assumptions tested. Surface gravity determinations in particular require better constraints, since these enable age-dating of individual sources. In the short term, improvements in spectral models can help in this endeavor; however, a sufficiently sampled grid of benchmark sources may obviate the need for models entirely. Benchmarks should increasingly arise from mass standards, for which age constraints are more precise; these additionally provide necessary tests of evolutionary models upon which most of the age-dating techniques hinge. Improved angular resolution and sensitivity with *JWST* and the next generation of large (> 25m) telescopes will increase resolved binary sample sizes by expanding the volume in which they can be found and monitored. These facilities will also aid searches for substellar cluster members in old open field and globular clusters and, perhaps more importantly, mass standards in these clusters to facilitate more stringent tests of evolutionary models (T. Dupuy, priv. comm.). Finally, a larger, more complete sample of brown dwarfs with precise RV measurements will both improve our statistical constraints on the age of the local brown dwarf population (and subpopulations) while additionally aiding in the search for mass standards.

The author thanks T. Dupuy, S. Leggett, M. Liu, & A. West for helpful comments.

References

Allers, K. N., Jaffe, D. T., Luhman, K. L., *et al.* 2007, *ApJ*, 657, 511
Bannister, N. P. & Jameson, R. F. 2007, *MNRAS*, 378, L24
Baraffe, I., Chabrier, G., Allard, F., & Hauschildt, P. H. 2002, *A&A*, 382, 563
Barrado y Navascués, D., Stauffer, J. R., & Patten, B. M. 1999, *ApJL*, 522, L53
Basri, G. & Reiners, A. 2006, *AJ*, 132, 663
Berger, E. 2006, *ApJ*, 648, 629
Bildsten, L., Brown, E. F., Matzner, C. D., & Ushomirsky, G. 1997, *ApJ*, 482, 442
Blake, C. H., Charbonneau, D., White, R. J., Marley, M. S., & Saumon, D. 2007, *ApJ*, 666, 1198
Blake, C. H., Charbonneau, D., White, R. J., Torres, G., Marley, M. S., & Saumon, D. 2008, *ApJL*, 678, L125
Borysow, A., Jorgensen, U. G., & Zheng, C. 1997, *A&A*, 324, 185
Bouvier, J., Stauffer, J. R., Martin, E. L., Barrado y Navascues, D., Wallace, B., & Bejar, V. J. S. 1998, *A&A*, 336, 490
Bouy, H., Duchêne, G., Köhler, R., *et al.* 2004, *A&A*, 423, 341
Burgasser, A. J. 2004, *ApJS*, 155, 191
—. 2007, *ApJ*, 658, 617
Burgasser, A. J., Burrows, A., & Kirkpatrick, J. D. 2006a, *ApJ*, 639, 1095
Burgasser, A. J., Kirkpatrick, J. D., Brown, M. E., *et al.* 2002, *ApJ*, 564, 421
Burgasser, A. J., Kirkpatrick, J. D., Burrows, A., Liebert, J., Reid, I. N., Gizis, J. E., McGovern, M. R., Prato, L., & McLean, I. S. 2003, *ApJ*, 592, 1186
Burgasser, A. J., Kirkpatrick, J. D., Cruz, K. L., Reid, I. N., Leggett, S. K., Liebert, J., Burrows, A., & Brown, M. E. 2006b, *ApJS*, 166, 585
Burke, C. J., Pinsonneault, M. H., & Sills, A. 2004, *ApJ*, 604, 272
Burningham, B., Pinfield, D. J., Leggett, S. K., *et al.* 2008, *MNRAS*, 1183
Casewell, S. L., Jameson, R. F., & Burleigh, M. R. 2008, *MNRAS*, 390, 1517
Cruz, K. L., Reid, I. N., Kirkpatrick, J. D., Burgasser, A. J., Liebert, J., Solomon, A. R., Schmidt, S. J., Allen, P. R., Hawley, S. L., & Covey, K. R. 2007, *AJ*, 133, 439
Cushing, M. C., Marley, M. S., Saumon, D., Kelly, B. C., Vacca, W. D., Rayner, J. T., Freedman, R. S., Lodders, K., & Roellig, T. L. 2008, *ApJ*, 678, 1372
Day-Jones, A. C., Pinfield, D. J., Napiwotzki, R., Burningham, B., Jenkins, J. S., Jones, H. R. A., Folkes, S. L., Weights, D. J., & Clarke, J. R. A. 2008, *MNRAS*, 388, 838
Dupuy, T. J., Liu, M. C., & Ireland, M. J. 2008, ArXiv e-prints
Faherty, J. K., Burgasser, A. J., Cruz, K. L., Shara, M. M., Walter, F. M., & Gelino, C. R. 2008, ArXiv e-prints
Farihi, J., Becklin, E. E., & Zuckerman, B. 2008, *ApJ*, 681, 1470
Geballe, T. R., Saumon, D., Leggett, S. K., Knapp, G. R., Marley, M. S., & Lodders, K. 2001, *ApJ*, 556, 373
Gorlova, N. I., Meyer, M. R., Rieke, G. H., & Liebert, J. 2003, *ApJ*, 593, 1074
Hänninen, J. & Flynn, C. 2002, *MNRAS*, 337, 731
Hayashi, C. & Nakano, T. 1963, Progress of Theoretical Physics, 30, 460
Jameson, R. F., Lodieu, N., Casewell, S. L., Bannister, N. P., & Dobbie, P. D. 2008, *MNRAS*, 385, 1771
Jeffries, R. D. & Oliveira, J. M. 2005, *MNRAS*, 358, 13
Joergens, V. & Müller, A. 2007, *ApJL*, 666, L113
Kirkpatrick, J. D. 2005, *ARAA*, 43, 195

Kirkpatrick, J. D., Barman, T. S., Burgasser, A. J., McGovern, M. R., McLean, I. S., Tinney, C. G., & Lowrance, P. J. 2006, *ApJ*, 639, 1120

Kirkpatrick, J. D., Cruz, K. L., Barman, T. S., *et al.* 2008, ArXiv e-prints, 808

Knapp, G. R., Leggett, S. K., Fan, X., *et al.* 2004, *AJ*, 127, 3553

Kumar, S. S. 1963, *ApJ*, 137, 1121

Lane, B. F., Zapatero Osorio, M. R., Britton, M. C., Martín, E. L., & Kulkarni, S. R. 2001, *ApJ*, 560, 390

Leggett, S. K., Allard, F., Dahn, C., Hauschildt, P. H., Kerr, T. H., & Rayner, J. 2000, *ApJ*, 535, 965

Linsky, J. L. 1969, *ApJ*, 156, 989

Liu, M. C., Dupuy, T. J., & Ireland, M. J. 2008, ArXiv e-prints

Liu, M. C. & Leggett, S. K. 2005, *ApJ*, 634, 616

Lodders, K. 1999, *ApJ*, 519, 793

Looper, D. L., Kirkpatrick, J. D., Cutri, R. M., *et al.* 2008, *ApJ*, 686, 528

Lucas, P. W., Roche, P. F., Allard, F., & Hauschildt, P. H. 2001, *MNRAS*, 326, 695

Luhman, K. L. 1999, *ApJ*, 525, 466

Luhman, K. L., Stauffer, J. R., Muench, A. A., Rieke, G. H., Lada, E. A., Bouvier, J., & Lada, C. J. 2003, *ApJ*, 593, 1093

Martín, E. L., Brandner, W., Bouy, H., Basri, G., Davis, J., Deshpande, R., & Montgomery, M. M. 2006, *A&A*, 456, 253

Mentuch, E., Brandeker, A., van Kerkwijk, M. H., Jayawardhana, R., & Hauschildt, P. H. 2008, ArXiv e-prints

Mohanty, S., Basri, G., Jayawardhana, R., Allard, F., Hauschildt, P., & Ardila, D. 2004, *ApJ*, 609, 854

Mohanty, S., Basri, G., Shu, F., Allard, F., & Chabrier, G. 2002, *ApJ*, 571, 469

Mohanty, S., Jayawardhana, R., & Basri, G. 2005, *ApJ*, 626, 498

Nakajima, T., Oppenheimer, B. R., Kulkarni, S. R., Golimowski, D. A., Matthews, K., & Durrance, S. T. 1995, *Nature*, 378, 463

Pinfield, D. J., Jones, H. R. A., Lucas, P. W., Kendall, T. R., Folkes, S. L., Day-Jones, A. C., Chappelle, R. J., & Steele, I. A. 2006, *MNRAS*, 368, 1281

Rebolo, R., Martin, E. L., & Magazzu, A. 1992, *ApJL*, 389, L83

Rebolo, R., Zapatero-Osorio, M. R., & Martin, E. L. 1995, *Nature*, 377, 129

Reiners, A. & Basri, G. 2008, *ApJ*, 684, 1390

Reiners, A., Seifahrt, A., Stassun, K. G., Melo, C., & Mathieu, R. D. 2007, *ApJL*, 671, L149

Richer, H. B., Dotter, A., Hurley, J., *et al.* 2008, *AJ*, 135, 2141

Saumon, D., Marley, M. S., Leggett, S. K., *et al.* 2007, *ApJ*, 656, 1136

Schmidt, S. J., Cruz, K. L., Bongiorno, B. J., Liebert, J., & Reid, I. N. 2007, *AJ*, 133, 2258

Scholz, A., Jayawardhana, R., Wood, K., Meeus, G., Stelzer, B., Walker, C., & O'Sullivan, M. 2007, *ApJ*, 660, 1517

Spitzer, L. J. & Schwarzschild, M. 1953, *ApJ*, 118, 106

Stassun, K. G., Mathieu, R. D., & Valenti, J. A. 2006, *Nature*, 440, 311

Stauffer, J. R., Schultz, G., & Kirkpatrick, J. D. 1998, *ApJL*, 499, L199+

Stelzer, B., Micela, G., Flaccomio, E., Neuhäuser, R., & Jayawardhana, R. 2006, *A&A*, 448, 293

Vrba, F. J., Henden, A. A., Luginbuhl, C. B., *et al.* 2004, *AJ*, 127, 2948

Warren, S. J., Mortlock, D. J., Leggett, S. K., *et al.* 2007, *MNRAS*, 381, 1400

West, A. A., Hawley, S. L., Bochanski, J. J., Covey, K. R., Reid, I. N., Dhital, S., Hilton, E. J., & Masuda, M. 2008, *AJ*, 135, 785

Wielen, R. 1977, *A&A*, 60, 263

Zapatero Osorio, M. R., Lane, B. F., Pavlenko, Y., Martín, E. L., Britton, M., & Kulkarni, S. R. 2004, *ApJ*, 615, 958

Zapatero Osorio, M. R., Martín, E. L., Béjar, V. J. S., Bouy, H., Deshpande, R., & Wainscoat, R. J. 2007, *ApJ*, 666, 1205

Discussion

J. MELBOURNE: I was wondering how the brown dwarfs fit into the initial mass function?

A. BURGASSER: There has been considerable work on this in young clusters, for example work by Luhman and Hillenbrand, and they find the IMF is definitely declining in the brown dwarf regime, likely peaking above 0.1 $M_{\odot}$ (the Hyades may be somewhat different, however). In the field it is more difficult as one must invert the luminosity junction statiscally (again, part of the mass/ age/temperature degeneracy), but results there still indicate a decline (see question by Leggett). I would estimate that brown dwarfs are roughly equal in number to stars within a factor of a few.

S. LEGGETT: Recent UKIDSS results (Pinfield, Chin) imply the MF is flat or declining at very low masses. Your thoughts?

A. BURGASSER: The UKIDSS result is intriquing as it disagrees strongly with the 2MASS results. I worry a little bit about selection effects this early in the UKIDSS survey. It would be very useful to have an independent measure, for example from SDSS.

H. RICHER: Can you comment on the use of the gap between the end of the hydrogen-burning main sequence and brown dwarfs as a chronometer? We might see such a feature in 47 Tuc where we have a large program using WFC3 on HST coming up.

A. BURGASSER: Detecting such a gap would be very interesting primarily from its constraint on the physics of "low" temperature H fusion in the cores of these objects. The evolutionary models show their largest differences near the H-burning mass limit, and in that sense ages based around that limit may be less reliable. However, this would be one of those important empirical checks on the models to increase our trust in their use.

M. LIU: 1) Regarding the discrepancy in the typical ages of field objects from the Zaptero-Osorio and Faherty results, how much do selection effect of the two samples matter? 2) How reliable are the cluster/group memberships derived from proper motions (above)?

A. BURGASSER: 1) I would say selction bias is very possible in the Zapetero-Osorio results given the small size of their otherwise well-characterized sample. The Faherty study specifically looked at a large sample limited to 20 pc; volume limited, even if not volume complete. There is of course a chance that our Sun lies in a special region surrounded by young low-mass stars in general, but that would be surprising. 2) The Jameson study has some problems with the older clusters where candidates are exclusively proper-motion selected. It would be very useful to confirm those sources spectroscopically.

J. STAUFFER: If brown dwarfs do not spin down, then what you see is their birth angular momentum distribution (modulo how contraction or age is modifying this). Can't you use this information to then inform models of brown dwarf formation?

A. BURGASSER: Possibly, although most models focus on much earlier times, well before BDs have fully contracted. An additional problem is actually measuring rotation periods, which is proving challenging as these objects are less variable than desired (largely due to lack of spots). But I think the large rotation velocities may have implications on the frequency of disks or close binaries among brown dwarfs.

Kevin Covey and Andrew West

Kelle Cruz and Adam Burgasser

The Ages of Stars
Proceedings IAU Symposium No. 258, 2008
E.E. Mamajek, D.R. Soderblom & R.F.G. Wyse, eds.

doi:10.1017/S1743921309031986

Using magnetic activity and Galactic dynamics to constrain the ages of M dwarfs

Andrew A. West[1], Suzanne L. Hawley[2], John J. Bochanski[1], Kevin R. Covey[3] and Adam J. Burgasser[1]

[1]MIT Kavli Institute for Astrophysics and Space Research,
77 Massachusetts Ave, Cambridge, MA 02139-4307
email: aaw@mit.edu, jjb@mit.edu, ajb@mit.edu

[2]Department of Astronomy, University of Washington,
Box 351580, Seattle, WA 98195
email: slh@astro.washington.edu

[3]Harvard-Smithsonian Center for Astrophysics,
60 Garden Street, Cambridge MA 02138
email: kcovey@cfa.harvard.edu

Abstract. We present a study of the dynamics and magnetic activity of M dwarfs using the largest spectroscopic sample of low-mass stars ever assembled. The age at which strong surface magnetic activity (as traced by Hα) ceases in M dwarfs has been inferred to have a strong dependence on mass (spectral type, surface temperature) and explains previous results showing a large increase in the fraction of active stars at later spectral types. Using spectral observations of more than 40000 M dwarfs from the Sloan Digital Sky Survey, we show that the fraction of active stars decreases as a function of vertical distance from the Galactic plane (a statistical proxy for age), and that the magnitude of this decrease changes significantly for different M spectral types. Adopting a simple dynamical model for thin disk vertical heating, we assign an age for the activity decline at each spectral type, and thus determine the activity lifetimes for M dwarfs. In addition, we derive a statistical age-activity relation for each spectral type using the dynamical model, the vertical distance from the Plane and the Hα emission line luminosity of each star (the latter of which also decreases with vertical height above the Galactic plane).

Keywords. stars: activity, stars: kinematics, stars: late-type, stars: low-mass, brown dwarfs

---

1. Introduction

M dwarfs are the smallest and least luminous stars on the main sequence, yet they are the most numerous of all stellar constituents and have lifetimes longer than the current age of the Universe. Determining the age of an individual main sequence, field M dwarf is quite challenging. Stellar clusters would seem to provide the ideal environments for calibrating the ages of M dwarfs (because the age of the cluster is determined). However, the large distances to clusters older than a few Gyr (and resulting faintness of cluster M dwarfs), preclude any detailed observations of the intermediate or oldest cluster resident M dwarfs. Age estimates for field M dwarfs must therefore rely on a number of techniques that piggyback on other age dating methods. Although several methods exist for estimating the ages of M dwarfs (e.g. contribution by S. Catalan), we focus this contribution on what the magnetic activity of M dwarfs can tell us about their age.

M dwarfs are host to intense magnetics dynamos that give rise to chromospheric and coronal heating, producing emission from the X-ray to the radio. This magnetic activity has long been thought to be tied to the age of the host star. Almost 40 years ago, Wilson

& Woolley (1970) found a link between magnetic activity (as traced by the Ca II H & K emission lines) and the orbital elements (namely the inclination and eccentricity) of more than 300 late-type dwarfs; stars in near-circular orbits with small inclinations had the strongest activity. They concluded that as stars age, their orbits get more inclined and more eccentric (due to dynamical interactions), and thus, magnetic activity must also decline with age. Subsequent studies over the following decades have found similar connections between age and activity in low-mass stars (Wielen 1977; Giampapa & Liebert 1986; Soderblom *et al.* 1991; Hawley *et al.* 1996, Hawley *et al.* 1999, Hawley *et al.* 2000).

Although activity has been observed for decades, the exact mechanism that gives rise to the chromospheric heating is still not well-understood. In the Sun, magnetic field generation and subsequent heating is closely tied to the Sun's rotation. From helioseismology, we know that there is a rotational boundary between the inner solid-body rotating radiative zone and the outer differentially rotating convective zone (Parker 1993; Ossendrijver 2003; Thompson *et al.* 2003). This boundary, dubbed the tachocline, creates a rotational sheer that likely allows magnetic fields to be generated, preserved and eventually rise to the surface where they emerge as magnetic loops. These loops bring heat to the Sun's chromosphere and corona, driving both large stellar flares, as well as lower-level quiescent (or persistent) magnetic activity. The faster a star (with a tachocline) rotates, the stronger its magnetic heating and surface activity. Therefore, the angular momentum evolution of solar-type stars should play an important role in determining the magnitude of the observed activity.

Angular momentum loss from magnetized winds has been shown to slow rotation in solar-type stars; as a result magnetic activity decreases. Skumanich (1972) found that both activity and rotation decrease over time as a power law ($t^{-0.5}$). Subsequent studies confirmed the Skumanich results and empirically demonstrated a strong link between age, rotation and activity in solar-type stars (Barry 1988; Soderblom *et al.* 1991; Pizzolato *et al.* 2003; Mamajek & Hillenbrand 2008; see contribution by E. Mamajek).

There is strong evidence that the rotation-activity (and presumably age) relation extends from stars more massive than the Sun to smaller dwarfs (Pizzolato *et al.* 2003; Mohanty & Basri 2003; Kiraga & Stepien 2007). Therefore, rotation derived ages (using gyrochronology) may provide a useful independent estimate of age for M dwarfs when large enough calibration samples can be acquired (see contributions by S. Meibom, J. Irwin, and S. Barnes). However, at a spectral type of ~M3 (0.35 $M_\odot$; Reid & Hawley 2005; Chabrier & Baraffe 1997), stars become fully convective and the tachocline presumably disappears. Despite this changes, magnetic activity persists in late-type M dwarfs; the fraction of active M dwarfs peaks around a spectral type of M7 before decreasing into the brown dwarf regime (Hawley *et al.* 1996; Gizis *et al.* 2000; West *et al.* 2004).

It is unclear if the rotation-age-activity relation extends to the fully convective M dwarfs. A few empirical studies have uncovered evidence that activity and rotation might be linked in late-type M dwarfs (Delfosse *et al.* 1998; Mohanty & Basri 2003; Reiners & Basri 2007). In addition, recent simulations of fully convective stars find that rotation may play a role in magnetic dynamo generation (Dobler *et al.* 2006; Browning 2008). However, a recent study found detectable rotation in a few inactive M7 dwarfs, indicating that the situation may in fact be more complicated (West & Basri 2009).

Irrespective of rotation, many studies have found evidence that the age-activity relation extends into the M dwarf regime. Eggen (1990) observed a Skumanich-type decrease in activity as a function of age. Larger samples of M dwarfs have added further evidence that magnetic activity decreases with age (Fleming *et al.* 1995; Gizis *et al.* 2002). There is also evidence that M dwarfs may have finite active lifetimes. Stauffer *et al.* (1994) suggested that activity may not be present in the most massive M dwarfs in the Pleiades. Hawley *et al.* (2000) confirmed the Stauffer *et al.* (1994) claim by observing a a sample of clusters that spanned several Gyr in age. They were able to calibrate the lifetimes for early-type M dwarfs by observing the color at which activity (traced by Hα emission) was no longer present. Because of the small sample size, the derived Hawley *et al.* (2000) activity ages are lower limits at a given color or spectral type. As mentioned above, the Hawley *et al.* (2000) study was limited to younger, nearby clusters due to the intrinsic faintness of M dwarfs; thus, it could only probe ages of a few Gyr and spectral types as late as ~M3. Larger samples of M dwarfs are required to statistically derive age-activity relations that extend both in the ages they probe and the range of spectral types they cover.

In this contribution, we review recent studies that have used large spectroscopic samples to investigate the relationship between age and activity in M dwarfs. In addition, we derive a statistical Hα age-activity relation for M2-M7 dwarfs.

2. SDSS DR5 low-mass star spectroscopic sample

Large surveys such as the Sloan Digital Sky Survey (SDSS; York *et al.* 2000) have created optical and infrared catalogs of tens of millions of M dwarfs (Bochanski *et al.* 2009 in prep). In addition to the photometric data, the SDSS has obtained spectra for tens of thousands of M dwarfs. Recently, these surveys have been utilized to examine the statistical properties of M dwarfs, the dynamics of the Milky Way and detailed studies of magnetic activity (Hawley *et al.* 2002; West *et al.* 2004; West *et al.* 2006; Covey *et al.* 2007; Bochanski *et al.* 2007a, 2007b; West *et al.* 2008; hereafter W08).

Recently, the SDSS Data Release 5 (DR5) low-mass star spectroscopic sample was released (W08)†, containing over 44000 M and early L-type dwarfs. Radial velocities (accurate to within ~5 km s^{-1}) and photometric distances (from the $r - z$ colors; see Bochanski *et al.* 2009 in prep) were measured for all of the stars. Proper motions (matches to USNO-B; Munn *et al.* 2004) were obtained for over 27000 of the stars, allowing for full U, V, W 3-D space motions for over 27000 low-mass dwarfs, the largest such sample ever assembled. In addition, all of the spectra were run through the Hammer spectral typing facility (Covey *et al.* 2007) to assign spectral types and measure the strength of the Hα emission line. One of the hallmarks of the W08 sample is that it probes the entirety of the Galactic thin disk and extends well into the thick disk (~50-2500 pc above the Plane). This is due to the fact that the majority of SDSS fields are in the North Galactic Cap and that SDSS was designed to study the distant Universe; intrinsically faint Galactic M dwarfs are bright in a deep extragalactic survey.

3. Results

3.1. *Hα activity*

Activity in M dwarfs varies as a function of spectral type. Figure 1 shows the fraction of active M dwarfs as a function of spectral type from West *et al.* (2004). The active fraction is very small for early-type M dwarfs, peaks at M7-M8 and declines into the L dwarf regime. While some of the morphology of this relation may be due to the ability (or inability) to host a strong dynamo (the activity fraction decrease in the late-type M and L dwarfs is likely caused by the atmospheres becoming neutral), a large part of the shape may be due to age effects. If activity has a finite lifetime that changes as a function

† Measured quantities can be obtained electronically using the CDS Vizier database http://vizier.u.strasbg.fr/vis-bin/VizieR

of spectral type, and we observe a range of stellar ages, then stars with longer lifetimes will appear to have higher activity fractions.

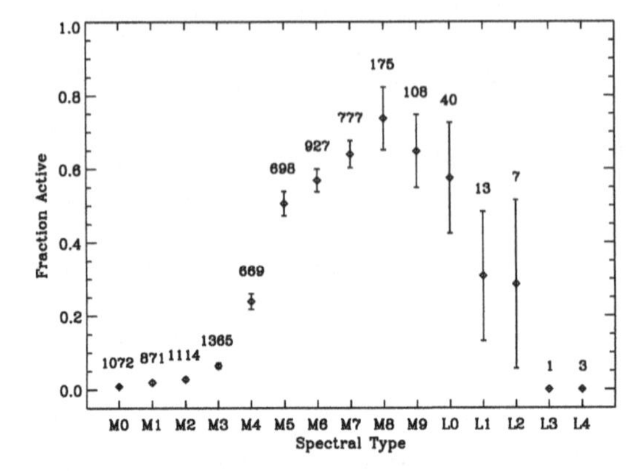

Figure 1. Fraction of active stars as a function of spectral type (reproduced from West *et al.* 2004). Numbers above each point represent the number of stars used to compute the fraction.

To test this hypothesis, West *et al.* (2004, 2006) showed that the fraction of active M7 stars decreases as a function of vertical distance from the Galactic Plane (see Figure 2). Stars are born in dynamically cold molecular gas near the midplane of the Galaxy. Over time they undergo dynamical interactions with molecular clouds, which add energy to the stellar orbits in all dimensions (see contribution by B. Nordström). It is this process that gives thickness to the Milky Way disk and allows us to use vertical distance as a proxy for age; stars further from the Plane are statistically older (they have to be dynamically heated to reach those heights), while stars near the Plane are statistically younger. Figure 2 shows that the younger stars, near the plane of the Milky Way are almost all active, whereas the fraction of active stars at larger distances above (and below) the Galaxy is significantly reduced; older stars have ceased being active. The decrease of active fractions as a function of vertical height above the Plane is seen for all M dwarf spectral types (W08).

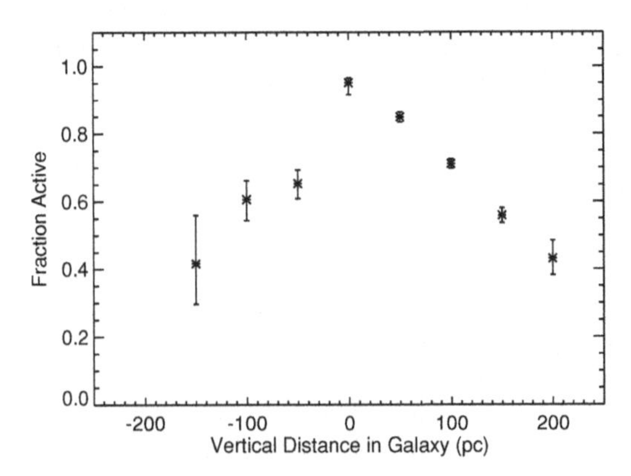

Figure 2. The fraction of active M7 stars as a function of vertical distance from the Galactic plane (reproduced from West *et al.* 2006). There is a significant decrease in the active fraction as a function of Galactic height, which can be used as a proxy for age. Younger stars near the Plane are almost all active, whereas older stars, further from the Plane have ceased being active.

W08 also showed that the amount of activity, quantified by $L_{\mathrm{H}\alpha}/L_{\mathrm{bol}}$ (the ratio of the luminosity in the Hα emission line to the bolometric flux of the star) decreases over time. Figure 3 shows the median $\log(L_{\mathrm{H}\alpha}/L_{\mathrm{bol}})$ as a function of vertical distance from the Plane. The narrow error bars represent the spread of the values and the wide error bars indicate the uncertainty of the median relation in each bin. The decrease as a function of height is a statistically significant over most of the spectral types. Some of the spectral types were omitted from Figure 3 because they lacked a sufficient number of active stars for a robust study.

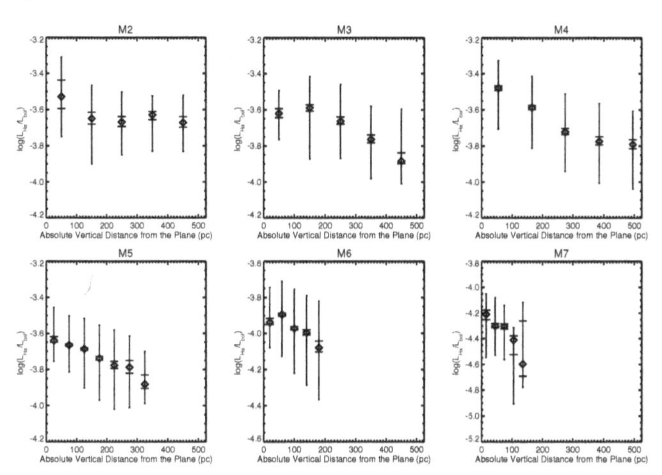

Figure 3. Median $\log(L_{\mathrm{H}\alpha}/L_{\mathrm{bol}})$ as a function of vertical distance from the Plane (reproduced from W08). The narrow error bars represent the spread of the values and the wide error bars indicate the uncertainty of the median relation in each bin. The decrease as a function of height is a statistically significant over most of the spectral types.

3.2. *1-D dynamical model*

West *et al.* (2006, 2008) devolved a 1D dynamical model that traces the vertical dynamics of stars as a function of time. The model assumes a constant star formation rate and adds a new population of 50 stars every 200 Myr for a total simulation time of 10 Gyr. Each new population of stars begins with a randomly drawn velocity dispersion 8 km s^{-1} (Binney *et al.* 2000). Orbits are integrated using a "leap frog" integration technique (Press *et al.* 1992) and the vertical Galactic potential from Siebert *et al.* (2003).

Dynamical heating was simulated by altering the velocities of stars such that their velocity dispersions as a function of age match a $\sigma_W \propto t^{0.5}$ relation (Wielen 1977; Fuchs *et al.* 2001; Hänninen & Flynn 2002; Nordström *et al.* 2004). Energy was added to stars that were within a certain distance from the Plane. This "region of influence" is a way to parameterize the cross section of interaction with molecular clouds during a Plane crossing (see West *et al.* 2006 for more information). The majority of the molecular gas is constrained to a small range of Galactic heights and the dynamical interaction depends on the proximity to the cloud, the density of the gas and the velocity of the star, all of which are absorbed in the "region of influence". The size of the "region of influence" was varied from ± 0.5 to ± 5 pc in intervals of 0.5 pc. Seventy simulations were run for each spectral type. Each simulation recorded the velocity, position, and age of each star.

3.3. *Age-activity*

W08 introduced a binary activity state to the dynamical models; stars started their lives as active and after a finite "lifetime", the activity turned off. The active lifetimes were varied from 0.0 to 9.0 Gyr in 0.5 Gyr intervals for each spectral type. The resulting model active fractions (as a function of vertical distance from the Plane) were compared to the empirical relations using a chi-squared minimization technique and a best-fit model for each spectral type was determined. The resulting Hα activity lifetimes are shown in Figure 4. The results from the Hawley *et al.* (2000) cluster study are overplotted for comparison (dotted). W08 found that there is a significant increase in activity lifetimes between spectral types M3 and M5, possibly indicating a physical change in the production of magnetic fields as the stellar interiors become fully convective.

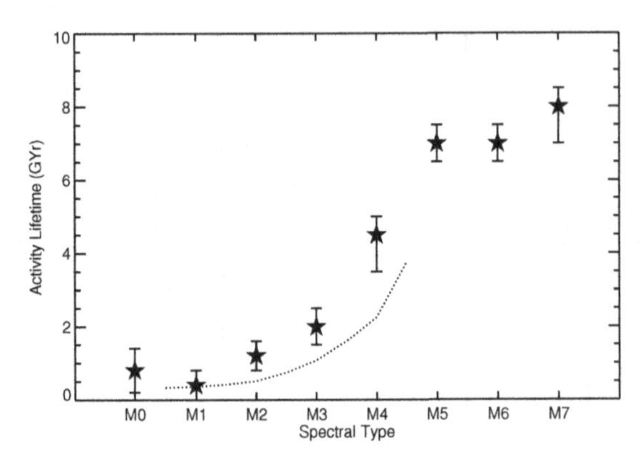

Figure 4. The Hα activity lifetimes of M dwarfs as determined by comparing the SDSS DR5 spectroscopic data to 1D dynamical simulations (stars; reproduced from W08). The Hawley *et al.* (2000) activity lifetime relation is overplotted for comparison. As predicted, the Hawley relation provides a lower limit to the ages. W08 found that there is a significant increase in activity lifetimes between spectral types M3 and M5, possibly indicating a physical change in the production of magnetic fields as full convection sets in.

Figure 3 suggests that activity is not simply a binary process, but rather that the amount of magnetic activity may decrease over time. To quantify this decrease, we compared the median $\log(L_{\mathrm{H}\alpha}/L_{\mathrm{bol}})$ from Figure 3 to the median ages in the same vertical distance bins from the W08 1D dynamical models. The result, shown in Figure 5, gives a statistical, quantitative relationship between the Hα activity of an M dwarf and it's age. Horizontal error bars represent the spread in age at a given bin. The age-activity relations are consistent with a smooth Skumanich-like activity decrease until the star reaches its activity lifetime, after which it rapidly falls to its eternal inactive state.

We fit the activity-age relations in Figure 5 with a function of the form:

$$\log(L_{\mathrm{H}\alpha}/L_{\mathrm{bol}}) = \frac{a}{l^n - t^n} - b, \tag{3.1}$$

where a, b and n are coefficients, l is the active lifetime (Gyr), and t is the age measured in Gyr. The exponent n was forced to be the same for all spectral types. The lifetime l coefficient was allowed to vary within the uncertainties of the derived active lifetimes from W08 (Figure 4). The resulting best-fit coefficients can be found in Table 1. We caution that these relations are purely statistical and may not be appropriate for use with individual stars. Additional discretion should be used with the functional fits, which may not be justifiably extrapolated beyond where data exist.

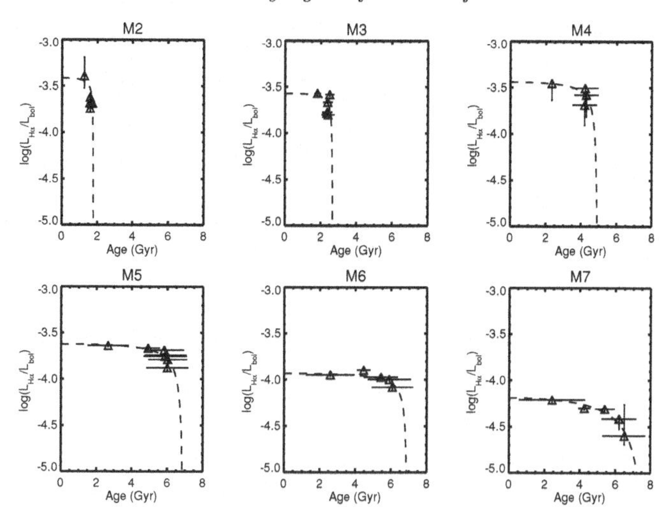

Figure 5. Median $\log(L_{\mathrm{H}\alpha}/L_{\mathrm{bol}})$ as a function of age (as derived from the median age of the same bins in the W08 dynamical models). The error bars represent the spread in the distributions. The best-fit function is overplotted for each spectral type (dashed).

Table 1. Coefficients for best-fit activity-age relation

Spectral Type	a	b	n	l (Gyr)
M2	0.106	3.38	2.0	1.8
M3	0.213	3.54	2.0	2.7
M4	1.41	3.39	2.0	5.0
M5	2.85	3.57	2.0	7.0
M6	1.78	3.90	2.0	7.0
M7	11.8	4.01	2.0	8.0

4. Summary

Large astronomical surveys have produced M dwarf samples of unprecedented size. We have shown that using the statistical foothold of the largest low-mass spectroscopic sample ever assembled, a strong tie between magnetic activity (as traced by Hα) and age has been established and that M dwarfs have finite active lifetimes. These lifetimes are a strong function of spectral type, dramatically increasing as the M dwarf interiors become fully convective. We derived age-activity relations for M2-M7 dwarfs based on the average activity as a function of height above (or below) the Galactic plane. Future studies will extend these relations to other spectral types. In addition, these relations can be confirmed and calibrated using both white dwarf cooling ages in wide binary systems that have both a white dwarf and an M dwarf (see contributions by S. Catalan, M. Salaris, and J. Kalirai; Silvestri *et al.* 2006), and deep cluster observations of M dwarfs (when such observations become possible).

References

Barry, D. C. 1988, *ApJ*, 334, 436

Binney, J., Dehnen, W., & Bertelli, G. 2000, *MNRAS*, 318, 658

Bochanski, J. J., Munn, J. A., Hawley, S. L., West, A. A., Covey, K. R., & Schneider, D. P. 2007a, *AJ*, 134, 2418

Bochanski, J. J., West, A. A., Hawley, S. L., & Covey, K. R. 2007b, *AJ*, 133, 531

Browning, M. K. 2008, *ApJ*, 676, 1262

Chabrier, G. & Baraffe, I. 1997, *A&A*, 327, 1039
Covey, K. R., *et al.* 2007, *AJ*, 134, 2398
Delfosse, X., Forveille, T., Perrier, C., & Mayor, M. 1998, *A&A*, 331, 581
Dobler, W., Stix, M., & Brandenburg, A. 2006, *ApJ*, 638, 336
Eggen, O. J. 1990, *PASP*, 102, 166
Fleming, T. A., Schmitt, J. H. M. M., & Giampapa, M. S. 1995, *ApJ*, 450, 401
Fuchs, B., Dettbarn, C., Jahreiß, H., & Wielen, R. 2001, in ASP Conf. Ser. 228: Dynamics of Star Clusters and the Milky Way, ed. S. Deiters, B. Fuchs, A. Just, R. Spurzem, & R. Wielen, 235
Giampapa, M. S. & Liebert, J. 1986, *ApJ*, 305, 784
Gizis, J. E., Monet, D. G., Reid, I. N., Kirkpatrick, J. D., Liebert, J., & Williams, R. J. 2000, *AJ*, 120, 1085
Gizis, J. E., Reid, I. N., & Hawley, S. L. 2002, *AJ*, 123, 3356
Gray, D. F. 1992, Science, 257, 1978
Hänninen, J. & Flynn, C. 2002, *MNRAS*, 337, 731
Hawley, S. L., Covey, K. R., *et al.* 2002, *AJ*, 123, 3409
Hawley, S. L., Gizis, J. E., & Reid, I. N. 1996, *AJ*, 112, 2799
Hawley, S. L., Reid, I. N., & Tourtellot, J. G. 2000, in Very Low-mass Stars and Brown Dwarfs, Edited by R. Rebolo and M. R. Zapatero-Osorio. Published by the Cambridge University Press, UK, 2000., p.109, ed. R. Rebolo & M. R. Zapatero-Osorio, 109–+
Hawley, S. L., Tourtellot, J. G., & Reid, I. N. 1999, *AJ*, 117, 1341
Kiraga, M. & Stepien, K. 2007, Acta Astronomica, 57, 149
Mamajek, E. E. & Hillenbrand, L. A. 2008, *ApJ*, 687, 1264
Mohanty, S. & Basri, G. 2003, *ApJ*, 583, 451
Munn, J. A., Monet, D. G., Levine, S. E., Canzian, B., Pier, J. R., Harris, H. C., Lupton, R. H., Ivezić, Ž., Hindsley, R. B., Hennessy, G. S., Schneider, D. P., & Brinkmann, J. 2004, *AJ*, 127, 3034
Nordström, B., Mayor, M., Andersen, J., Holmberg, J., Pont, F., Jørgensen, B. R., Olsen, E. H., Udry, S., & Mowlavi, N. 2004, *A&A*, 418, 989
Ossendrijver, M. 2003, *A&A Rev.*, 11, 287
Parker, E. N. 1993, *ApJ*, 408, 707
Pizzolato, N., Maggio, A., Micela, G., Sciortino, S., & Ventura, P. 2003, *A&A*, 397, 147
Press, W. H., Teukolsky, S. A., Vetterling, W. T., & Flannery, B. P. 1992, Numerical recipes in C. The art of scientific computing (Cambridge: University Press, —c1992, 2nd ed.)
Reid, N. & Hawley, S. L., eds. 2005, New light on dark stars : red dwarfs, low mass stars, brown dwarfs, ed. N. Reid & S. L. Hawley
Reiners, A. & Basri, G. 2007, *ApJ*, 656, 1121
Siebert, A., Bienaymé, O., & Soubiran, C. 2003, *A&A*, 399, 531
Silvestri, N. M., *et al.* 2006, *AJ*, 131, 1674
Skumanich, A. 1972, *ApJ*, 171, 565
Soderblom, D. R., Duncan, D. K., & Johnson, D. R. H. 1991, *ApJ*, 375, 722
Stauffer, J. R., Liebert, J., Giampapa, M., Macintosh, B., Reid, N., & Hamilton, D. 1994, *AJ*, 108, 160
Thompson, M. J., Christensen-Dalsgaard, J., Miesch, M. S., & Toomre, J. 2003, *ARAA*, 41, 599
West, A. A. & Basri, G. 2009, *ApJ*, submitted
West, A. A., Bochanski, J. J., Hawley, S. L., Cruz, K. L., Covey, K. R., Silvestri, N. M., Reid, I. N., & Liebert, J. 2006, *AJ*, 132, 2507
West, A. A., Hawley, S. L., Bochanski, J. J., Covey, K. R., Reid, I. N., Dhital, S., Hilton, E. J., & Masuda, M. 2008, *AJ*, 135, 785
West, A. A., *et al.* 2004, *AJ*, 128, 426
Wielen, R. 1977, *A&A*, 60, 263
Wilson, O. & Woolley, R. 1970, *MNRAS*, 148, 463
York, D. G., *et al.* 2000, *AJ*, 120, 1579

Discussion

R. SCHIAVON: How do you measure Hα? (A boring question.)

A. WEST: That's not a boring question. We have to be careful about how we measure Hα and the pseudo-continuum because of the strong molecular features. We also require that an active star have 1 Å equivalent width, which alleviates many problems. I would like to say we went through all 40,000 stars by eye, but we didn't. We have gone through a subset by eye and find we get Hα correct 96% of the time.

I. KING: It will be interesting to see your M-dwarf scale heights as a function of age. The empirical results will be so valuable because the theory is in such bad shape. Twenty years ago, when I was writing the dynamics chapter in *The Milky Way As A Galaxy*, I concluded that the theory of dynamical heating of the disk was in a very confused state, and I am not aware that it has been clarified in the two decades since.

A. WEST: Yes, we have compared our results to the empirical data of Nordström *et al.* (2004), West *et al.* (2006, 2008) and find excellent agreement. We have not however compared to theoretical predictions from star formation and dynamical theory.

B. WEAVER: With all these stars that you've measured, have you been able to determine a mass function for dwarf M stars.

A. WEST: Yes, our group at the University of Washington has derived two different luminosity and mass functions as part of the PhD theses of Kevin Covey (Covey *et al.* 2008, AJ, 136, 1778) and John Bochanski (PhD thesis, 2008; Bochanski *et al.* 2008, Proceedings of Cool Stars 15).

F. WALTER: What is the relation between $L_{\mathrm{H}\alpha}/L_{\mathrm{bol}}$ and age for the active stars, to the left of where the activity plummets. Is it a Skumanich-like power law? Is there any dependence of the slope on the spectral type?

A. WEST: There appears to be a Skumanich-like decrease before the rapid decrease. This is difficult to see in the early types because they are probing ages near the active lifetimes but at M5 and M6, a plateau is quite clear.

Andrew West

Trent Dupuy

The Ages of Stars
Proceedings IAU Symposium No. 258, 2008
E.E. Mamajek, D.R. Soderblom & R.F.G. Wyse, eds.

doi:10.1017/S1743921309031998

Confronting substellar theoretical models with stellar ages

Trent J. Dupuy[1], Michael C. Liu[1], and Michael J. Ireland[2]

[1]Institute for Astronomy, University of Hawai'i,
2680 Woodlawn Drive, Honolulu, HI 96822 USA
e-mail: tdupuy@ifa.hawaii.edu

[2]School of Physics, University of Sydney
NSW2006, Australia

Abstract. By definition, brown dwarfs never reach the main-sequence, cooling and dimming over their entire lifetime, thus making substellar models challenging to test because of the strong dependence on age. Currently, most brown dwarfs with independently determined ages are companions to nearby stars, so stellar ages are at the heart of the effort to test substellar models. However, these models are only *fully* constrained if both the mass and age are known. We have used the Keck adaptive optics system to monitor the orbit of HD 130948BC, a brown dwarf binary that is a companion to the young solar analog HD 130948A. The total dynamical mass of 0.109 ± 0.003 $M_\odot$ is the most precise mass measurement (3%) for any brown dwarf binary to date and shows that both components are substellar for any plausible mass ratio. The ensemble of available age indicators from the primary star suggests an age comparable to the Hyades, with the most precise age being $0.79^{+0.22}_{-0.15}$ Gyr based on gyrochronology. Therefore, HD 130948BC is unique among field L and T dwarfs as it possesses a well-determined mass, luminosity, and age. Our results indicate that substellar evolutionary models may underpredict the luminosity of brown dwarfs by as much as a factor of $\approx$2–3$\times$. The implications of such a systematic error in evolutionary models would be far-reaching, for example, affecting determinations of the initial mass function and predictions of the radii of extrasolar gas-giant planets. This result is largely based on the reliability of stellar age estimates, and the case study of HD 130948A highlights the difficulties in determining the age of an arbitrary field star, even with the most up-to-date chromospheric activity and gyrochronology relations. In order to better assess the potential systematic errors present in substellar models, more refined age estimates for HD 130948A and other stars with binary brown dwarf companions (e.g., ϵ Ind Bab) are critically needed.

Keywords. stars: brown dwarfs; techniques: high angular resolution; binaries: close, visual; infrared: stars

---

1. Introduction

Theoretical models of objects below the substellar limit ($M < 0.075$ $M_\odot$) are essential for characterizing the several hundred brown dwarfs and extrasolar gas-giant planets discovered to date. Thus, these models have become ubiquitous in the literature, even though empirical tests of their ability to accurately predict the properties of brown dwarfs has been limited to only a handful of relatively warm objects. To test substellar evolutionary models, the input parameters of mass and age must be determined. For young brown dwarfs, the M6.5 eclipsing binary 2MASS J05352184−0546085 in the Orion Nebula provides a unique benchmark (Stassun *et al.* 2006). Prior to this year, only three binaries provided dynamical mass measurements for field objects at or below the substellar limit: the M8.5 + M9 binary LHS 1070BC (Leinert *et al.* 2001; Seifahrt *et al.* 2008)); the M8.5 + M9 binary Gl 569Bab (Zapatero Osorio *et al.* 2004; Simon *et al.* 2006); and the L0.5 + L1 binary 2MASS J0746 + 2000AB (Bouy *et al.* 2004). Recent work has contributed several more dynamical masses for objects lower in both temperature and mass than previously studied: the mid-L dwarf GJ 802B (Ireland *et al.* 2008); the T5 + T5.5 dwarf binary 2MASS J1534-2952AB (Liu *et al.* 2008); and the L4 + L4 binary HD 130948BC (Dupuy *et al.* 2008). While mass measurements alone can provide very stringent tests of theoretical models (e.g., see Liu *et al.* 2008), substellar evolutionary models are only fully constrained when both the mass and age can be determined. In fact, precise ages are critical for such tests because brown dwarfs – unlike stars – never reach a main-sequence, so their properties depend very sensitively on their age.

Of the substellar field dwarfs with measured masses, only HD 130948BC has a precisely determined age. These nearly-identical L dwarfs were discovered by Potter *et al.* (2002) as companions to the young solar analog HD 130948A (G2V, [Fe/H] = 0.05). *Hipparcos* measured a distance of 18.17 ± 0.11 pc (van Leeuwen 2007) for the primary star, which enables a very precise dynamical mass measurement when paired with our well-determined orbital solution.

2. The mass of HD 130948BC

We have used Keck adaptive optics (AO) imaging to monitor the relative orbit of the two components of HD 130948BC (Figure 1). Combined with archival *Hubble Space Telescope* (*HST*) imaging and a re-analysis of the Gemini discovery data, our data span ≈7 years (≈70% of the orbital period). We fit a simple analytic PSF model to derive astrometry from the Keck and Gemini images, while TinyTim model PSFs were fit to the *HST* images. An individually tailored Monte Carlo simulation was used to determine the astrometric uncertainty for each observation epoch. The resulting astrometry is extremely precise with typical Keck errors of 300 μas, corresponding to ≈1 $R_{\odot}$ at the distance of this system, while the orbit is roughly the size of the asteroid belt. We determined the binary's orbital parameters and their confidence limits using a Markov Chain Monte Carlo

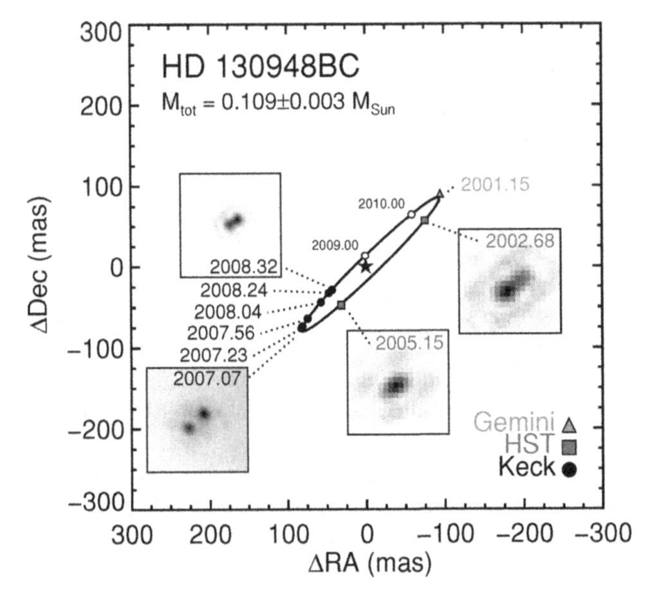

Figure 1. Keck (circles), *HST* (squares), and Gemini (triangle) relative astrometry for HD 130948BC along with the best-fit orbit. Error bars are smaller than the plotting symbols. The empty circles are the predicted positions in 2009 and 2010.

(MCMC) technique. The best-fit orbit has a reduced χ^2 of 1.06 (9 degrees-of-freedom), thus validating our astrometric error estimates. Applying Kepler's Law to the MCMC-derived orbital period ($P = 9.9^{+0.7}_{-0.6}$ yr) and semimajor axis ($a = 121 \pm 6$ mas) yields a dynamical mass of $0.1089^{+0.0020}_{-0.0017}$ $M_\odot$. Accounting for the additional uncertainty in the *Hipparcos* distance results in a dynamical mass of 0.109 ± 0.003 $M_\odot$ (114 ± 3 M_{Jup}).

In the following analysis, we apportion the total mass between the two components by converting the measured luminosity ratio into a mass ratio using evolutionary models. The resulting individual masses are very insensitive to the models used because the flux ratio is so close to unity (the steepness of the mass–luminosity relation means that even small differences in mass result in large differences in luminosity). Regardless, we are careful to conduct our analysis in a self-consistent manner free of circular logic.

3. The age of HD 130948A

As a young solar analog, multiple indicators are available to assess the age of HD 130948A:

• *Rotation/Gyrochronology* — Gaidos *et al.* (2000) measured two rotation periods of 7.69 and 7.99 days for HD 130948A. Thus, we adopt a rotation period of 7.84 ± 0.21 days and a $B-V$ color of 0.576 ± 0.016 mag from the *Hipparcos* catalog. We employ the Mamajek & Hillenbrand (2008) calibration of the "gyrochronology" relation originally introduced by Barnes (2007). The age we derive is $0.79^{+0.22}_{-0.15}$ Gyr, where the confidence limits are determined through a Monte Carlo approach in which the period, color, and empirical coefficients are drawn from normal distributions corresponding to their uncertainties.

• *Chromospheric Activity* — Henry *et al.* (1996) and Wright *et al.* (2004) measure $\log(R'_{HK})$ values of -4.45 and -4.50 for HD 130948A, respectively. Using the activity–age relation of Mamajek & Hillenbrand (2008), we derive ages of 0.4 and 0.6 Gyr from these $\log(R'_{HK})$ values. The empirical relation is expected to gives ages with an uncertainty of $\approx$0.25 dex, so we adopt a mean age of 0.5 ± 0.3 Gyr from this method.

• *X-ray Activity* — HD 130948A was detected by *ROSAT*, and Hünsch *et al.* (1999) measure $\log(L_X) = 29.01$ dex (cgs), which gives $\log(R_X) = -4.70$. Using the empirical relation of Gaidos (1998), this corresponds to an age of 0.1–0.3 Gyr, depending on whether we adopt α of 0.5 or 1/exp. The X-ray relation of Mamajek & Hillenbrand (2008), derived by combining their $\log(R_X)$–$\log(R'_{HK})$ and $\log(R'_{HK})$–age relations, gives an age of 0.5 Gyr. The X-ray luminosity of HD 130948A is in agreement with single G stars in the Pleiades and Hyades (28.9–29.0; Stern *et al.* (1995); Stelzer & Neuhäuser (2001)).

• *Isochrones* — Using high resolution spectroscopic data combined with a bolometric luminosity and model isochrones, Valenti & Fischer (2005) derived an age estimate of 1.8 Gyr, with a possible age range of 0.4–3.2 Gyr. From the same data and with more detailed analysis, Takeda *et al.* (2007) found a median age of 0.72 Gyr, with a 95% confidence range of 0.32–2.48 Gyr.

• *Lithium* — Measurements by Duncan (1981), Hobbs (1985), and Chen *et al.* (2001) give lithium equivalent widths of 95 ± 14, 96 ± 3, and 103 ± 3 mÅ, respectively, for HD 130948A. Compared to stars of similar color, these values are slightly lower than the mean for the Pleiades and slightly higher than for UMa and the Hyades, though consistent with the scatter in each cluster's measurements Soderblom *et al.* (1993a,b,c)

In summary, the most precise age estimate available for HD 130948A comes from gyrochronology, which gives an age of $0.79^{+0.22}_{-0.15}$ Gyr. All other age indicators agree with this estimate, though this is due to their large uncertainties rather than a true consensus.

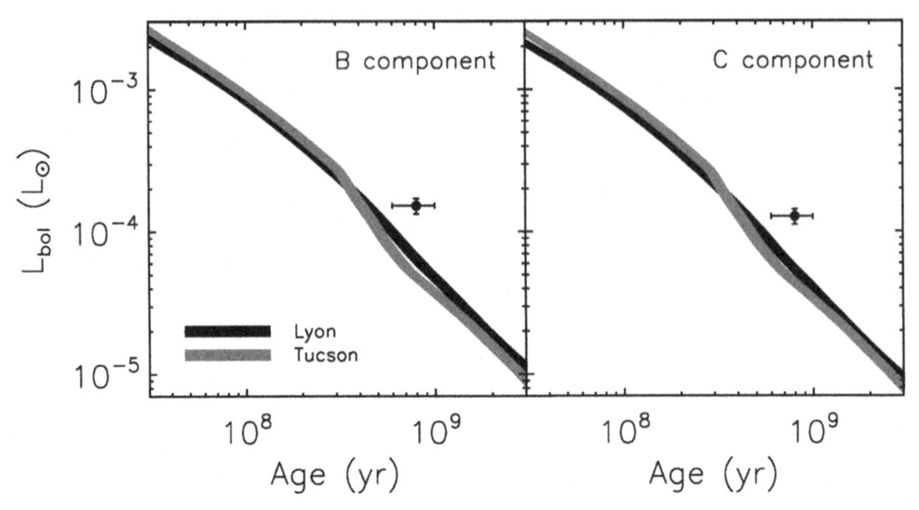

Figure 2. The filled circles mark the measured luminosities of HD 130948B and C at the age we derive for HD 130948A. The thick shaded lines are isomass lines from evolutionary models, where the line thicknesses encompass the 1σ errors in the individual masses of HD 130948BC. Although the two independent sets of models agree very well with one another, they underpredict the luminosities of HD 130948BC by a factor of $\approx$2–3×.

4. Substellar evolutionary models fully constrained

With a measured mass, luminosity, and age, HD 130948BC provides the first direct test of the luminosity evolution predicted by theoretical models for substellar field dwarfs. Both the Tucson models (Burrows *et al.* 1997) and Lyon models (DUSTY; Chabrier *et al.* 2000) underpredict the luminosities of HD 130948B and C given their masses and age. The discrepancy is quite large, about a factor of 2 for the Lyon models and a factor of 3 for the Tucson models (Figure 2). If the age and luminosities of HD 130948B and C had been used to infer their masses, the resulting estimates would have been too large by 20–30%. In order to explain this discrepancy entirely, model radii would have to be underpredicted by 30–40%. Alternatively, the age of HD 130948A would need to be $\approx$0.4 Gyr in order to resolve this discrepancy. Although such a young age is marginally consistent with the various age indicators; it is on the extreme young end of two independent, well-calibrated age estimates (gyrochronology and stellar isochrones). In order to better assess this discrepancy between models and data, a more refined age estimate for HD 130948A (e.g., from asteroseismology) is critically needed.

5. Lithium depletion in HD 130948BC

Since brown dwarfs are fully convective objects, they can rapidly deplete their initial lithium if their core temperature is ever high enough to do so. This threshold is reached around 0.065 $M_\odot$, and since this is below the hydrogen-burning mass-limit, this fact has been exploited to identify sufficiently old objects bearing lithium as substellar. In fact, the exact mass-limit for lithium burning is slightly different depending on which sets of theoretical models are used, and the masses of HD 130948B and C happen to be very close to these theoretically predicted mass-limits (Figure 3). According to the Tucson models, neither component is massive enough to have ever depleted a significant amount of its initial lithium. The Lyon models, on the other hand, predict that HD 130948B is massive enough to have depleted most of its lithium, while HD 130948C is not. Thus, resolved optical spectroscopy designed to detect the lithium doublet at 6708 Å would provide a

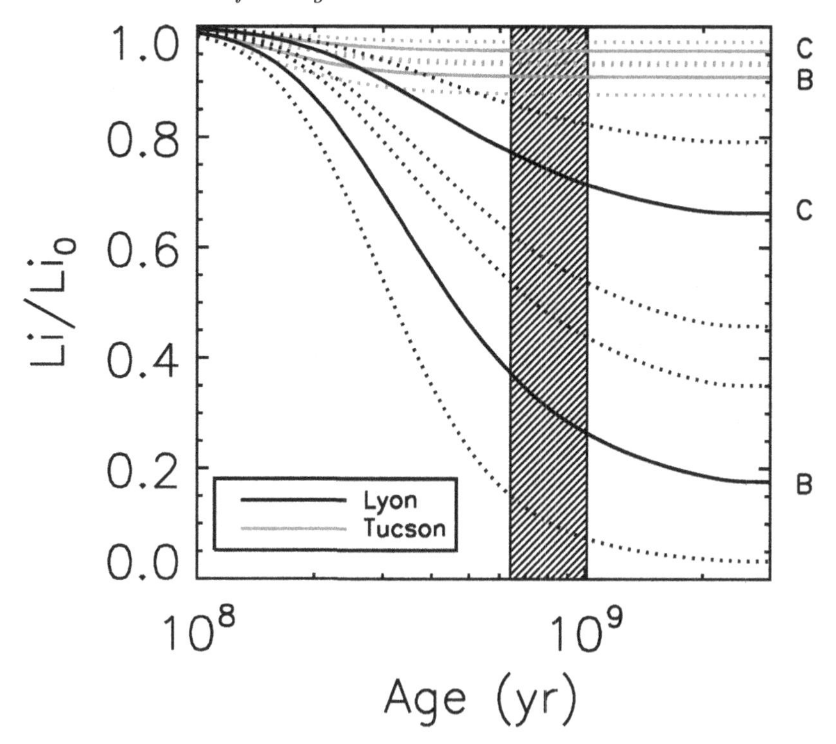

Figure 3. Lithium depletion as a function of age as predicted by evolutionary models. The solid lines correspond to the individual masses of HD 130948B and C. These lines are bracketed by dotted lines that correspond to the 1σ uncertainties in the individual masses. The ordinate is the fraction of initial lithium remaining. The hatched black box indicates the constraint from the age of HD 130948A estimated from gyrochronology.

very discriminating test of substellar evolutionary models, which are otherwise nearly indistinguishable (e.g., see Figure 2). This experiment can currently only be conducted with *HST*/STIS given the very small binary separation (< 130 mas).

6. Future prospects

Brown dwarfs hold the potential to address many astrophysical problems. For example, they are excellent laboratories in which to study ultracool atmospheres under a variety of conditions, and they may eventually be useful as Galactic chronometers given how sensitively their properties depend on their age (see contribution by A. Burgasser). However, the theoretical models we rely upon to characterize brown dwarfs have only begun to be rigorously tested by benchmark systems such as HD 130948BC. More results are expected to be forthcoming over the next several years for other brown dwarf binaries with stellar companions: ϵ Ind Bab (McCaughrean *et al.* 2004); Gl 417BC (Bouy *et al.* 2003); and GJ 1001BC (Golimowski *et al.* 2004). However, the utility of these systems as benchmarks critically depends on the confidence in the age estimates for their primary stars. Therefore, these stars deserve special attention so that state-of-the-art age-dating techniques (e.g., asteroseismology and gyrochronology) may be applied to them. Also, extending the empirical relations between age, stellar rotation, and chromospheric activity to include objects with as late a spectral type as possible will enable many more systems to be used as benchmarks for testing models. These relations are currently only calibrated for stars as late as early-K, but about half of the stars with brown dwarf companions have spectral types between early-K and early-M.

References

Barnes, S. A. 2007, *ApJ*, 669, 1167

Bouy, H., Brandner, W., Martín, E. L., Delfosse, X., Allard, F., & Basri, G. 2003, *AJ*, 126, 1526

Bouy, H., *et al.* 2004, *A&A*, 423, 341

Burrows, A., Marley, M., Hubbard, W. B., Lunine, J. I., Guillot, T., Saumon, D., Freedman, R., Sudarsky, D., & Sharp, C. 1997, *ApJ*, 491, 856

Chabrier, G., Baraffe, I., Allard, F., & Hauschildt, P. 2000, *ApJ*, 542, 464

Chen, Y. Q., Nissen, P. E., Benoni, T., & Zhao, G. 2001, *A&A*, 371, 943

Duncan, D. K. 1981, *ApJ*, 248, 651

Dupuy, T. J., Liu, M. C., & Ireland, M. J. 2008, ApJ, in press (astro-ph/0807.2450)

Gaidos, E. J. 1998, *PASP*, 110, 1259

Gaidos, E. J., Henry, G. W., & Henry, S. M. 2000, *AJ*, 120, 1006

Golimowski, D. A., *et al.* 2004, *AJ*, 128, 1733

Henry, T. J., Soderblom, D. R., Donahue, R. A., & Baliunas, S. L. 1996, *AJ*, 111, 439

Hobbs, L. M. 1985, *ApJ*, 290, 284

Hünsch, M., Schmitt, J. H. M. M., Sterzik, M. F., & Voges, W. 1999, *A&AS*, 135, 319

Ireland, M. J., Kraus, A., Martinache, F., Lloyd, J. P., & Tuthill, P. G. 2008, *ApJ*, 678, 463

Leinert, C., Jahreiß, H., Woitas, J., Zucker, S., Mazeh, T., Eckart, A., & Köhler, R. 2001, *A&A*, 367, 183

Liu, M. C., Dupuy, T. J., & Ireland, M. J. 2008, ApJ, in press (astro-ph/0807.0238)

Mamajek, E., & Hillenbrand, L. 2008, *ApJ*, 687, 1264

McCaughrean, M. J., Close, L. M., Scholz, R.-D., Lenzen, R., Biller, B., Brandner, W., Hartung, M., & Lodieu, N. 2004, *A&A*, 413, 1029

Potter, D., Martín, E. L., Cushing, M. C., Baudoz, P., Brandner, W., Guyon, O., & Neuhäuser, R. 2002, *ApJ*, 567, L133

Seifahrt, A., Röll, T., Neuhäuser, R., Reiners, A., Kerber, F., Käufl, H. U., Siebenmorgen, R., & Smette, A. 2008, *A&A*, 484, 429

Simon, M., Bender, C., & Prato, L. 2006, *ApJ*, 644, 1183

Soderblom, D. R., Fedele, S. B., Jones, B. F., Stauffer, J. R., & Prosser, C. F. 1993a, *AJ*, 106, 1080

Soderblom, D. R., Jones, B. F., Balachandran, S., Stauffer, J. R., Duncan, D. K., Fedele, S. B., & Hudon, J. D. 1993b, *AJ*, 106, 1059

Soderblom, D. R., Pilachowski, C. A., Fedele, S. B., & Jones, B. F. 1993c, *AJ*, 105, 2299

Stassun, K. G., Mathieu, R. D., & Valenti, J. A. 2006, *Nature*, 440, 311

Stelzer, B., & Neuhäuser, R. 2001, *A&A*, 377, 538

Stern, R. A., Schmitt, J. H. M. M., & Kahabka, P. T. 1995, *ApJ*, 448, 683

Takeda, G., Ford, E. B., Sills, A., Rasio, F. A., Fischer, D. A., & Valenti, J. A. 2007, *ApJS*, 168, 297

Valenti, J. A. & Fischer, D. A. 2005, *ApJS*, 159, 141

van Leeuwen, F. 2007, Hipparcos, the New Reduction of the Raw Data

Wright, J. T., Marcy, G. W., Butler, R. P., & Vogt, S. S. 2004, *ApJS*, 152, 261

Zapatero Osorio, M. R., Lane, B. F., Pavlenko, Y., Martín, E. L., Britton, M., & Kulkarni, S. R. 2004, *ApJ*, 615, 958

Discussion

F. WALTER: It is always risky to attempt to pin down the age of a field star, even using multiple techniques that may agree. How much would the age have to be changed to place the L dwarfs on the proper evolutionary tracks?

T. DUPUY: I agree and would really like to see another independent measurement of the age, such as from asteroseismology. The age of the system would have to be about 0.4 Gyr to bring the models into agreement with the data.

E. JENSEN: Is there a measured metallicity for HD 130948A? The metallicity will affect the evolutionary models, both in the HR diagram and for Li depletion.

T. DUPUY: That's exactly right and is a detail I didn't go into. The metallicity of HD 130948A is basically solar, which means we can use the standard models. This is another reason why having brown dwarfs with stellar companions is great: you can make sure you're not being confused by metallicity effects like Adam talked about.

A. WEST: Does the fact that this system is a close binary affect the measured luminosity (because the radii are affected)?

T. DUPUY: The binary separation is about 2.2 AU, so it's unlikely that tidal effects are at work in this system. Also, it turns out that the two components receive about as much flux from each other as they do from the primary star, so irradiation shouldn't be affecting them much.

J. FERNANDEZ: The next main source of benchmarks for brown dwarfs will be Kepler and Corot. The precise determination of ages for the primary stars will be crucial.

Sydney Barnes

The Ages of Stars
Proceedings IAU Symposium No. 258, 2008
E.E. Mamajek, D.R. Soderblom & R.F.G. Wyse, eds.

doi:10.1017/S1743921309032001

Gyrochronology and its usage for main sequence field star ages

Sydney A. Barnes
Lowell Observatory,
1400 W. Mars Hill Road, Flagstaff, AZ 86001, USA
email: barnes@lowell.edu

Abstract. The construction of all age indicators consists of certain basic steps which lead to the identification of the properties desirable for stellar age indicators. Prior age indicators for main sequence field stars possess only some of these properties. The measured rotation periods of cool stars are particularly useful in this respect because they have well-defined dependencies that allow stellar ages to be determined with ~20% errors. This method, called gyrochronology, is explained informally in this talk, shown to have the desired properties, compared to prior methods, and used to derive ages for samples of main sequence field stars.

Keywords. stars: activity, stars: binaries, stars: evolution, stars: fundamental parameters, stars: individual (ξ Boo, 61 Cyg, α Cen, 36 Oph), stars: late-type, stars: rotation

---

1. Motivations

Things in our world come into being, exist in time, and eventually cease to be. The properties of these things usually change over time, so that specifying the age of something, whether it be a tree, a human being, or a star, immediately gives us a good idea of what some of its other properties might be. In galactic astronomy, the ages of individual stars assume a particular importance, because they constitute the ticks of the abstract cosmic clock that tells us how various astronomical phenomena change over time.

Sandage (1962, and earlier) first noted that the morphology of a cluster of stars in the Hertzsprung-Russell diagram might be used to derive its age. Demarque & Larson (1964) improved it substantially, and named it the isochrone method. Although venerable, this method is not very effective for main sequence stars because its principal variable, a star's luminosity, changes only slowly on the main sequence. Furthermore, for a field star, it also requires an excellent distance measurement, not easily accomplished. Consequently, isochrone ages for main sequence field stars have errors approaching ~100%.

Of the prior distance-independent methods, the most consistent relies on the declining chromospheric activity of a cool star (Wilson, 1963; Skumanich 1972; Noyes *et al.* 1984; Soderblom *et al.* 1991; Donahue 1998). However, chromospheric emission varies with a star's rotation phase, activity cycle phase, and possibly other variables, limiting the precision of such ages to ~50%.

All activity-related age indicators are ultimately related to the rotation rate of a star. However, attempts to harness rotation to derive ages were hindered by the ambiguity of $v \sin i$ measurements, and what we now know to be a dynamo-related bimodality, and associated transition, in very young stars. The $v \sin i$ ambiguity can be entirely circumvented by (precisely) measuring a star's (mass-dependent) rotation period instead, and the early bimodality can be identified, and related stars excised.

This method, named gyrochronology (and parsed *gyros-chronos-logos*) allows the derivation of a significantly more precise age than previously available, of a cool main sequence field star, from its measured color and rotation period. The period is typically determined from time-series measurements of the spot-related photometric modulation of starlight. This talk is an informal summary of this method, as detailed in Barnes (2007). Some results and terminology derive from the 'CgI scenario' for stellar rotation presented in Barnes (2003).

2. Background for the construction of all age indicators

Many of the issues in constructing age indicators are so obvious that they are routinely ignored! Let us therefore proceed by first stepping back, and considering the main steps in the construction of any age indicator. One needs to:

(*a*) **Find an observable, v, that changes 'well' with age;** ('Well' means that it works for single objects rather than for an ensemble, and also has the properties listed in Table 1.)

(*b*) **Determine the ages of suitable calibrators independently;** (This means measuring both the variable, v, for the calibrating objects, and the most trustworthy prior variables so that v can be related to earth rotations, pendulum swings, etc.)

(*c*) **Measure the functional form of the variable:** $v = v(t, w, x, ...)$; (t represents the age, and w, x, ... additional dependencies. Variables with the fewest dependencies are the most desirable.)

(*d*) **Invert that functional form to find** $t = t(v, w, x, ...)$; (Analytic inversions provide insight, but numerical inversions are usually necessary.)

(*e*) **Calculate the error:** $\delta t = \delta t(t, v, w, x, ...)$; (Although necessary, non-linearities and other complexities often make this final step difficult.)

Table 1. Characteristics of the three major age indicators for field stars

Property⇓ Method⇒	Isochrone Age	Chromospheric Age	Gyrochronology
Measurable easily?	? (Distance reqd.)	? (Repetition reqd.)	? (Repetition reqd.)
Sensitive to age?	No (on MS)	Yes	Yes
Insensitive to other parameters?	No	Yes	Yes
Technique calibrable?	Yes (Sun)	? (Sun?)	Yes (Sun)
Invertible easily?	No	Yes	Yes
Errors calculable/provided?	? (Difficult)	Yes?	Yes
Coeval stars yield the same age?	No (Field binaries)	?	Yes

The foregoing, and other practical considerations, suggest that the following properties are desirable for *stellar* age indicators.

(*a*) **Measurability for single stars:** The indicator should be properly defined, measurable easily itself, and preferably should not require many additional quantities to be measured, otherwise it cannot be used routinely.

(*b*) **Sensitivity to age:** The indicator should change substantially (and preferably regularly) with age, otherwise the errors will be inherently large.

(*c*) **Insensitivity to other parameters:** The indicator should have insensitive (or separable) dependencies on other parameters that affect the measured quantity, otherwise there is the potential for ambiguity. In particular, distance-independent methods are preferred.

(*d*) **Calibration:** The technique should be calibrable using an object (or set of objects) whose age(s) we know very well, otherwise systematic errors will be introduced.

(*e*) **Invertibility:** The functional dependence determined above should be properly invertible to yield the age as a function of the measured variables.

(*f*) **Error analysis:** The errors on the age derived using the technique ought to be calculable, otherwise no confidence can be attached to the ages.

(*g*) **Test of coeval stars:** The technique should yield the same ages for stars expected to be coeval, otherwise the validity of the technique itself must be questioned.

Table 1 summarizes the extent to which these properties are satisfied for the three major field star age indicators now available.

3. Introduction to rotational ages

Skumanich (1972) seems to have been the first to identify a relationship between the rotation rate of a star and its age. He used the averaged $v \sin i$ values of stars in selected open clusters, and that of the Sun, all of whose ages are known independently. It was not clear then that such a relationship could be used in any more than a statistical sense, partly due to the inherent ambiguity in $v \sin i$ measurements. Observations of $v \sin i$ values, and later, rotation periods in young open clusters revealed a wide dispersion in the rotation rates of coeval stars that discouraged the use of rotation as an age indicator.

A prescient attempt was made by Kawaler (1989) to use rotation to derive ages based on the Hyades rotation period sequence, but its reliance on various theoretically motivated assumptions, the poor fit to the warm Hyades stars and the rotational dispersion in young open clusters cast doubts on its viability.

However, the availability of large numbers of rotation periods in open clusters allowed the resolution of this 'dispersion' into distinct rotational sequences, C & I, in color-period diagrams, each with its own set of dependencies (Barnes 2003). This resolution shows that the (largely slower-rotating) I sequence does indeed spin down similar to Skumanich's initial suggestion, but the (largely faster-rotating) C sequence does not. However, C sequence stars change into I sequence stars within a couple of 100 Myr, so that all older cool stars must be of the I type, and spin down predictably. Furthermore, the spindown is convergent, in the sense that initial variations become increasingly unimportant with the passage of time.

These facts allow one to identify the principal dependencies of stellar rotation, which turn out to be stellar color/mass and age, and to identify empirically the tight relationship between them. This relationship must be true for all cool stars on the main sequence. Therefore, measuring a field star's color/mass and rotation period at once allow the age to be determined.

Furthermore, this method of determining the age is such that most of the properties considered desirable for an age indicator, as listed above, can be shown to be satisfied. Therefore, it seems appropriate to name the method 'gyrochronology.'

4. Color-period diagrams

Gyrochronology is ultimately based on color-period diagrams of open clusters, such as those shown in Fig. 1. (More such diagrams can be found in the papers by Meibom and Irwin in these proceedings.) The older (~600 Myr-old) Hyades cluster shows a distinct diagonal sequence, called I, of faster-rotating warmer stars and slower-rotating cooler stars, marked with circles. The way to understand the color-period diagram of the younger (~300 Myr-old) cluster NGC 3532 is to realize that *not only is this I sequence also present in this cluster, but another sequence, C, of faster-rotating stars*, marked with asterisks.

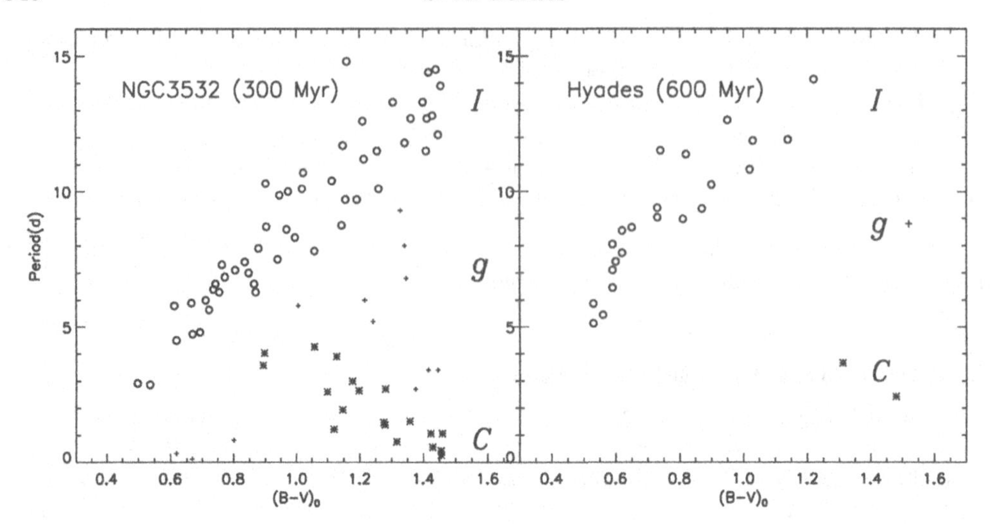

Figure 1. Color-period diagrams of the 300 Myr-old NGC3532 (Barnes, 1998) and the 600 Myr-old Hyades open clusters (primarily Radick *et al.* 1987). We see that the C-type stars in NGC 3532 (asterisks) have changed into I-type stars (circles) by Hyades age.

[The sequences are striking in the richer M 35 cluster (Meibom *et al.* 2008).] Comparing the two color-period diagrams tells us that almost all the C-type stars change into I-type stars by Hyades age. The stars in the rotational gap, g, between the two sequences can now be interpreted as stars in transition from the C- to the I-sequence.

These color-period diagrams also suggest that *the color/mass dependence of the I sequence is the same for both clusters.* This implies that the rotation period, P, of a star on this I sequence is expressible as the separable product of this mass dependence and of other variables, of which we might guess that the most important is the age, t, because stars spin down over time. Thus, we write $P = f(B - V).g(t)$.

5. The dependencies of I sequence stars

What might the age dependence, $g(t)$, be? A very good guess would simply be $g(t) = \sqrt{t}$, in agreement with the original suggestion by Skumanich (1972). Indeed, when the rotation periods, P, of stars in all measured open clusters are divided by $g(t) = \sqrt{t}$, the I sequences are brought into coincidence, as shown in Fig. 2, from Barnes (2007). (These early data include binaries, possibly aliased periods and other pathologies, hence the scatter.)

Fig. 3 shows a similar coincidence for field stars, the single unevolved set of Mt. Wilson stars from Baliunas *et al.* (1996). We have used individual chromospheric ages calculated using the prescription of Donahue (1998). It is obvious that the C sequence stars in the younger open clusters have all changed into I sequence stars in the older Mt. Wilson sample. Furthermore, by guessing the age dependence, $g(t)$, using Skumanich (1972), the mass dependence of the I sequence has been made manifest.

Indeed, one can fit this dependence using a function of the form

$$f(B - V) = a(B - V - c)^b \quad \text{giving} \quad a = 0.773 \pm 0.011, b = 0.601 \pm 0.024. \qquad (5.1)$$

The translational term, c, was simply equated to 0.4 in Barnes (2007) and to 0.5 in Barnes (2003). A subsequent fit by Meibom *et al.* (2008), using both a large sample of

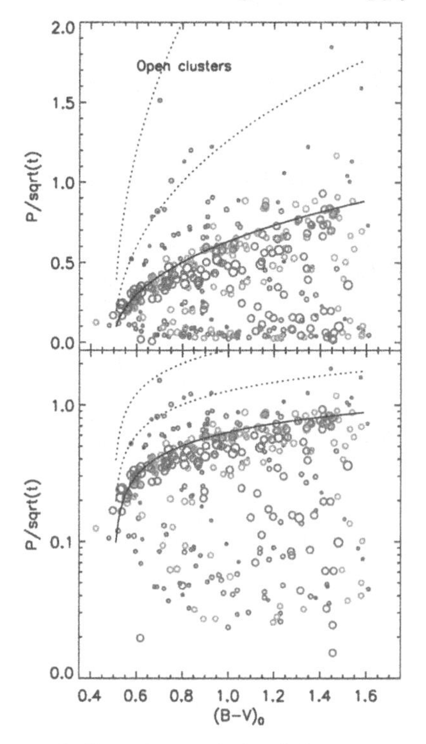

Figure 2. Rotation periods of cluster stars divided by the square roots of the cluster ages. Note the presence of both I sequence stars near the solid line, and C sequence stars below. (Figure from Barnes, 2007)

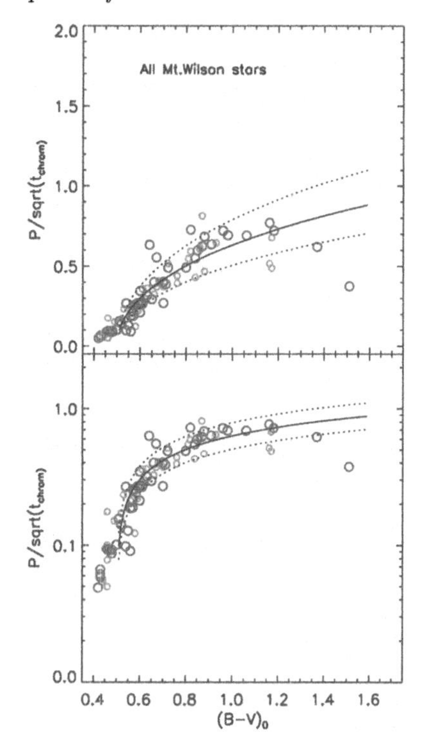

Figure 3. Rotation periods of single, main sequence Mt. Wilson stars, divided by the square roots of their chromospheric ages. Note that only I sequence stars are present. (Figure from Barnes, 2007)

rotation periods in the open cluster M 35, and spectroscopic membership information, gives

$$a = 0.770 \pm 0.014, b = 0.553 \pm 0.052, c = 0.472 \pm 0.027. \tag{5.2}$$

The point is that *regardless of the exact functional form chosen, a 2-3 parameter fit will suffice, and those parameters will be determined with small errors.*

One final step of the construction remains. Having specified $f(B-V)$, we now return to $g(t)$. It is reasonable to seek a power law dependence: $g(t) = t^n$. This allows us to calibrate the method using the Sun by ensuring that the above mass dependence gives the Solar rotation period at Solar age. This calibration gives $n = 0.519 \pm 0.007$.

Thus, the age of a star (in Myr) is simply given by inverting $P = f(B-V).g(t)$ to get

$$\log(t_{\rm gyro}) = \frac{1}{n}\{\log P - \log a - b \times \log{(B-V-c)}\} \tag{5.3}$$

where the constants a, b, c, n are as specified above, and base 10 logarithms are used.

6. Age error analysis

A virtue of the above formulation is that the age error can be simply calculated, and the various contributing error terms seen in perspective. The expression for the fractional age error, as calculated in Barnes (2007), is:

$$\frac{\delta t}{t} = 2\% \times \sqrt{3 + \frac{1}{2}(\ln t)^2 + 2P^{0.6} + \left(\frac{0.6}{x}\right)^2 + (2.4\ln x)^2} \tag{6.1}$$

where $x = B - V - 0.4$. For 1 Gyr-old stars of spectral types late F, early G, mid K and early M respectively, we get

$$\frac{\delta t}{t} = 2\% \times \begin{cases} \sqrt{26.9 + 6.4 + 66.5} & \text{when } B - V = 0.5\ (P = 7d); \\ \sqrt{26.9 + 8.9 + 16.9} & \text{when } B - V = 0.65\ (P = 12d); \\ \sqrt{26.9 + 12.1 + 2.5} & \text{when } B - V = 1.0\ (P = 20d); \\ \sqrt{26.9 + 15.4 + 0.35} & \text{when } B - V = 1.5\ (P = 30d). \end{cases} \tag{6.2}$$

which shows the relative contributions of the period and color errors (second and third terms, respectively). Color errors and differential rotation are the significant contributors for bluer and redder stars respectively.

The expression above evaluates to fractional age errors of 13-20% for 1 Gyr-old early M-late F stars, suggesting that relatively precise ages may indeed be derived for field stars, provided that the observable inputs, color and rotation period, are measured well.

7. Application to field star samples

Expressions (5.3) above and (6.1), for the gyro age and its error, respectively, are true for all I-type main sequence late F-early M stars, whether in clusters or in the field. We can therefore apply them to field star samples with measured rotation periods to derive ages where none were available before. The field star sample of Strassmeier *et al.* (2000) is an example.

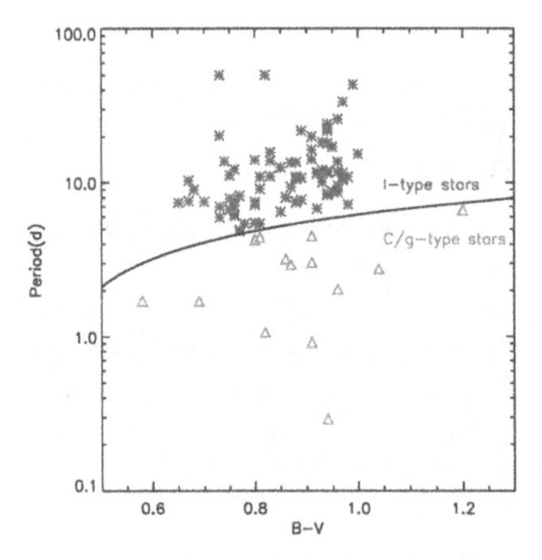

Figure 4. Color-period diagram for the Strassmeier *et al.* (2000) stars, showing the 100 Myr-old (gyro) isochrone used to discard possible C/g-type stars. Only the I-type stars above are retained for gyrochronology. (Figure from Barnes, 2007)

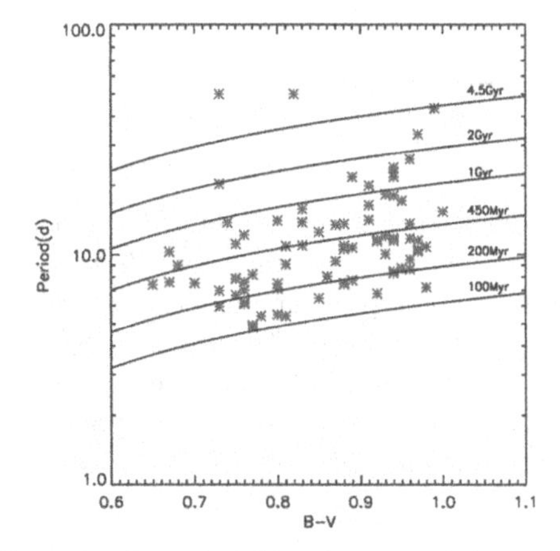

Figure 5. Rotational isochrones are shown for ages ranging from 100 Myr to 4.5 Gyr for the I-type stars in the Strassmeier *et al.* (2000) sample. Note the relative youth of the sample, in keeping with its active pedigree. (Figure from Barnes, 2007)

Fig. 4 displays the color-period diagram for this sample, and the 100 Myr (gyro) isochrone used to choose only the I-type stars for age analysis. Fig. 5 displays isochrones spanning the age range of the sample, showing its relative youth. Indeed, the median age is only 365 Myr, in agreement with the selection of the original sample by activity. The stars are individually tabulated in Barnes (2007), where the technique is also applied to the older sample of stars (median age of 1.2 Gyr) assembled by Pizzolato *et al.* (2003). For both samples, activity indicators like R_{HK} and L_X/L_{bol} are found to decline as expected with increasing gyro age.

8. Comparison with chromospheric ages

The best age indicator for nearby field stars over the past couple of decades has been the decline of chromospheric emission with age. The calibrations regularly used are those of Soderblom *et al.* (1991) and Donahue (1998), but see Mamajek & Hillenbrand (2008) and Mamajek's article in these proceeding for a recalibration including gyrochronology. It would therefore be appropriate to compare the new gyro ages with these older chromospheric ages. The best sample for this comparison is the Mt. Wilson sample of cool stars, one studied intensively for decades for chromospheric activity, and for which measured rotation periods are also available.

Fig. 6 shows this comparison, the cross indicating representative errors. The basic result to note is that there is reasonable agreement between the two ages because the upper left and lower right corners are unoccupied. A closer inspection shows that the chromospheric ages used here (Donahue 1998) are somewhat longer than the gyro ages, as the dashed median line shows. Dividing the stars into blue ($B-V < 0.6$), green ($0.6 > B-V > 0.8$), and red ($B-V > 0.8$), shows that the discrepancy relates mostly to the blue F stars, whose lifetime of 5 Gyr is indicated in the figure, as is the age of the universe.

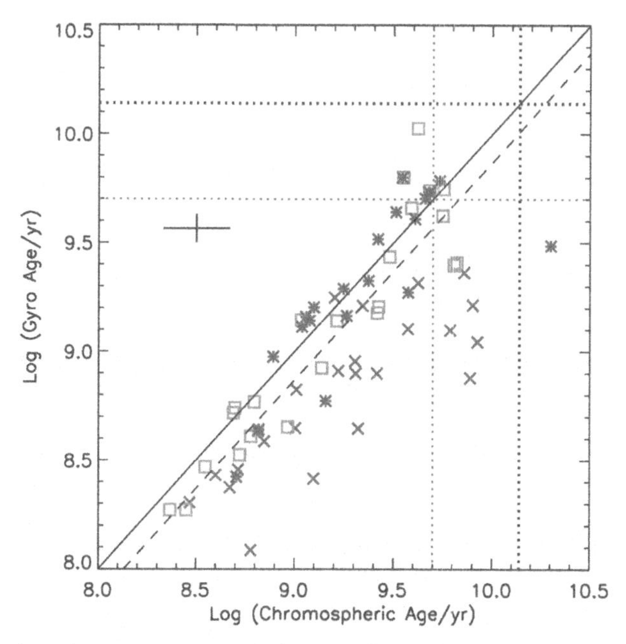

Figure 6. Comparison between gyro- and chromospheric ages for the Mt. Wilson star sample. Two conclusions can be drawn: (1) The two types of ages are in rough agreement, but the chromospheric ages are somewhat larger, and (2) The discrepancy relates mostly to the blue F stars (crosses) whose 5 Gyr lifetime is marked, rather than the green G stars (squares) or the red K stars (asterisks). (Figure from Barnes, 2007)

9. Comparison with isochrone ages

An equivalent comparison of gyro- and isochrone ages demonstrates the difficulty of deriving isochrone ages for field stars.

The most modern and homogeneous field star isochrone ages available are those for the SPOCS star sample of Takeda *et al.* (2007), who have undertaken a Bayesian age analysis, based on the method of Pont & Eyer (1994), and a prior uniform spectroscopic study of these stars by Valenti & Fischer (2005). The stars in common with those in Barnes (2007) are displayed in Fig. 7.

Despite the Bayesian technique's admirable attempt to account for the asymmetric error distribution in color-magnitude diagrams, the isochrone ages are still on average a factor of ~2.7 larger than the gyro ages. Some upper and lower limits are included when they represent wide binary stars with measured rotation periods. Related components are connected with dashed lines. These only serve to underscore the difficulty of deriving isochrone ages for non-cluster main sequence stars.

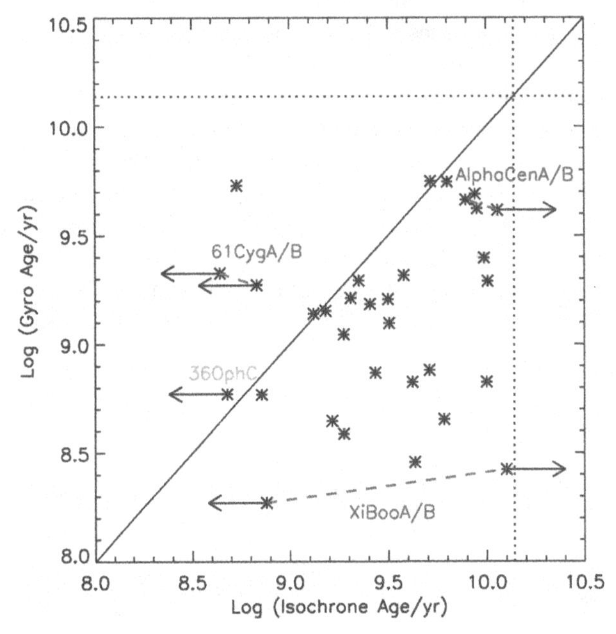

Figure 7. Comparison between gyro- and Bayesian isochrone ages for stars in common with the Takeda *et al.* (2007) SPOCS sample. The isochrone ages are on average ~2.7 times the gyro ages. Upper- and lower limits are included if they concern wide binaries with rotation periods. The components are connected by dashed lines. (Figure from Barnes, 2007)

10. Ages for wide binaries

Finally, we arrive at that very desirable property that an age determination method yield the same age for stars that we believe to be coeval. Indeed, there are a handful of wide binaries where rotation periods for both components have been measured. Thus, their ages may be determined independently. (We use wide binaries to be sure that there has been no tidal or magnetic hanky-panky between the components.)

The color-period diagram for the three available systems is shown in Fig. 8, along with the mean gyro isochrones and their errors for each pair, with details in Table 2 (from Barnes 2007). The ages for the components appear to be in agreement within the errors.

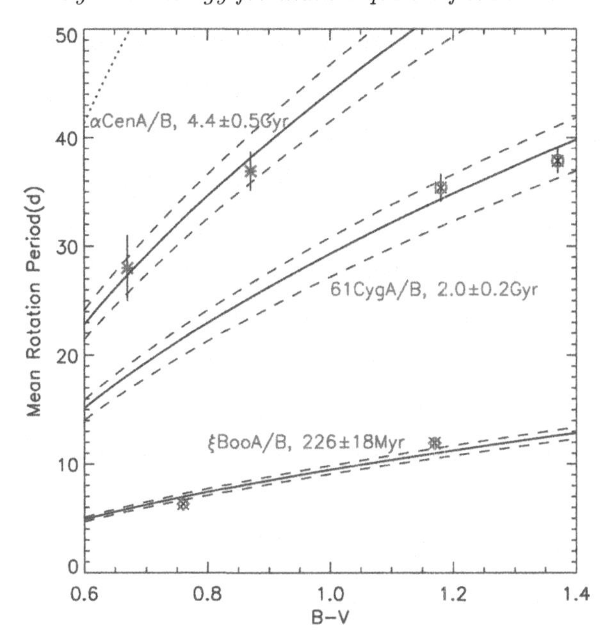

Figure 8. The components of the wide binaries ξ Boo, 61 Cyg, and α Cen appear to give the same gyro ages. (Figure from Barnes, 2007)

The age of the 36 Oph triple system is taken to be that of the presumably non-interacting tertiary component.

11. Conclusions

In summary, we have constructed an improved method of determining the age of a main sequence star from its measured rotation period, calibrated it using the Sun, and shown that the associated errors are smaller than those from prior methods.

The key steps of the construction are:

- All late F-early M stars become I-type rotators within a couple of 100 Myr,
- Their rotation periods, P, are describable as a product of two separable functions, f, and g, of the $B-V$ color and age, t, respectively: $P(B-V,t) = f(B-V).g(t)$,

Table 2. Ages for wide binary systems

System	Star	$B-V$	$\bar{P}_{rot}$	Age_{chromo}	Age_{iso}	Age_{gyro}
	HD131156A	0.76	6.31(0.05)	232 Myr	<760 Myr	187±21 Myr
ξ Boo	HD131156B	1.17	11.94(0.22)	508 Myr	>12600 Myr	265±28 Myr
	Mean					**226±18 Myr**
	HD201091	1.18	35.37(1.3)	2.36 Gyr	<0.44 Gyr	2.12±0.3 Gyr
61 Cyg	HD201092	1.37	37.84(1.1)	3.75 Gyr	<0.68 Gyr	1.87±0.3 Gyr
	Mean					**2.0±0.2 Gyr**
	HD128620	0.67	28(3)	5.62 Gyr	7.84 Gyr	4.6±0.8 Gyr
α Cen	HD128621	0.87	36.9(1.8)	4.24 Gyr	>11.36 Gyr	4.1±0.7 Gyr
	Mean					**4.4±0.5 Gyr**
	HD155886	0.85	20.69(0.4)	1.1 Gyr	......	1.42±0.19 Gyr
36 Oph	HD155885	0.86	21.11(0.4)	1.2 Gyr	......	1.44±0.20 Gyr
	HD156026	1.16	18.0(1.0)	1.4 Gyr	<0.48 Gyr	**0.59±0.07 Gyr**

- $f(B-V)$ and $g(t)$ can be determined empirically with small errors,
- $g(t)$ is such that initial variations become increasingly irrelevant with time,
- The functional dependence is easily inverted to get $t = t(P, B-V)$, and
- The age error, $\delta t = \delta t(t, P, B-V)$, is calculated.

The technique compares favorably with prior methods, which it complements, and passes some important tests. Precise time-series photometry is increasingly available from the ground and from space, making stellar rotation periods routinely measurable. Consequently, we recommend measuring rotation periods for appropriate main sequence cool stars where precise ages are desired, and using gyrochronology to derive them.

References

Baliunas, S., Sokoloff, D., & Soon, W. *et al.* 1996, *ApJL*, 457, 99
Barnes, S. A. 1998, *PhD Thesis*, Yale University
Barnes, S. A. 2003, *ApJ*, 586, 464
Barnes, S. A. 2007, *ApJ*, 669, 1167
Demarque, P. D. & Larson, R. B. 1964 *ApJ*, 140, 544
Donahue, R. A. 1998, in: R. A. Donahue & Bookbinder, J. A. (eds.), *Tenth Cambridge Workshop on Cool Stars, Stellar Systems and the Sun* (San Francisco: ASP), p. 1235
Kawaler, S. D. 1989, *ApJL*, 343, 65
Mamajek, E. E. & Hillenbrand, L. A. 2008, *ApJ*, 687, 1264
Meibom, S., Mathieu, R. D., & Stassun, K. G. 2008, *ApJ*, in press (arXiv/astroph: 0805:1040)
Noyes, R. W., Hartmann, L. W., Baliunas, S. L., Duncan, D. K., & Vaughan, A. H. 1984, *ApJ*, 279, 763
Pizzolato, N., Maggio, A., Micela, G., Sciortino, S., & Ventura, P. 2003, *A&A*, 397, 147
Pont, F. & Eyer, L. 1994, *MNRAS*, 351, 487
Radick, R. R., Thompson, D. T., Lockwood, G. W., Duncan, D. K. & Baggett, W. E. 1987, *ApJ*, 321, 459
Sandage, A. 1962, *ApJ*, 135, 349
Skumanich, A. 1972, *ApJ*, 171, 565
Soderblom, D. R., Duncan, D. K., & Johnson, D. R. H. 1991, *ApJ*, 375, 722
Strassmeier, K. G. Washuettl, A., Granzer, Th., Scheck, M., & Weber, M. 2000, *A&AS*, 142, 275
Takeda, G., Ford, E. B., Sills, A., Rasio, F. A., Fischer, D. A., & Valenti, J.A. 2007, *ApJS*, 168, 297
Valenti, J. A. & Fischer, D. A. 2005, *ApJS*, 159, 141
Wilson, O. C. 1963, *ApJ*, 138, 832

Discussion

P. Goudfrooij: Interesting method. However, it seems to me that for this method to work for any given star, one needs to know its foreground reddening very well, especially for the earlier- type stars where the "curvature" in the period-color relation is largest. I was wondering whether this issue could play a role in the apparent disagreement with ages from chromospheric lines.

S. Barnes: The disagreement with chromospheric ages is probably of a different origin. This is because reddening is not an issue for the (very close) Mount Wilson field stars. That said, there could be a residual effect in the calibration of gyrochronology if the cluster reddenings used are not quite correct.

E. Mamajek: How long does it take for a C-sequence star to "jump the gap" to the I-sequence? Is there a way to flag a star in the field as "gap" vs. "interface"?

S. BARNES: It varies from zero (late-F stars in the youngest clusters are already on the I sequence) to a few 100 Myr for early M stars. This includes a phase on the C-sequence. The transition from C to I is shorter because fewer stars are in the gap region in color-period diagrams, reminiscent of the Hertzsprung gap. Because of this timescale, any star above the 100 Myr isochrone in the color-period diagram is basically guaranteed to be an I-sequence star. Eventually we may have other ways of distinguishing the two, perhaps from magnetic field considerations.

M. PINSONNEAULT: There is a theoretical context here: One expects a tight relationship between rotation and age to exist only after stars lose memory of their initial conditions. I also find it very worrying that the models disagree with the data where the isochrone ages can be tested. Your thoughts?

S. BARNES: It is correct that the relationship will be tight after stars lose memory of their initial conditions. This happens as $1/t$. The data are what they are. The method is empirical. For the clusters, it does not really matter if their ages are slightly modified since they are all young. Any age error in there will simply show up as an increased spread in $f(B - V)$, which shows up as an increased error in the gyro age. The index n depends on the solar calibration. As for the models, I do not think that they get the rotational evolution correct, or even that the physics included is completely correct.

Theresa Antoja

Guido De Marchi

The Ages of Stars
Proceedings IAU Symposium No. 258, 2008
E.E. Mamajek, D.R. Soderblom & R.F.G. Wyse, eds.

doi:10.1017/S1743921309032013

Stellar ages from stellar rotation

Søren Meibom

Harvard-Smithsonian Center for Astrophysics
60 Garden Street, Cambridge, MA, 02138, USA
email: smeibom@cfa.harvard.edu

Abstract. Our ability to determine stellar ages from measurements of stellar rotation, hinges on how well we can measure the dependence of rotation on age for stars of different masses. Rotation periods for stars in open clusters are essential to determine the relations between stellar age, rotation, and mass. Until recently, ambiguities in $v \sin i$ data and lack of cluster membership information, prevented a clear empirical definition of the dependence of rotation on color. Direct measurements of stellar rotation periods for members in young clusters have now revealed a well-defined period-color relation. We show new results for the open clusters M35 and M34. However, rotation periods based on ground-based observations are limited to young clusters. The Hyades represent the oldest coeval population of stars with measured rotation periods. Measurements of rotation periods for older stars are needed to properly constrain the dependence of stellar rotation on age. We present our plans to use the Kepler space telescope to measure rotation periods in clusters as old as and older than the Sun.

Keywords. stars: rotation, stars: evolution, Galaxy: open clusters and associations: general

1. Introduction

Knowing stellar ages is fundamental to understanding the time-evolution of various astronomical phenomena related to stars and their companions. Accordingly, over the past decades much work has been focused on identifying the properties of a star that best reveal its age. For coeval populations of stars in clusters, the most reliable ages are determined by fitting model isochrones to single cluster members in the color-magnitude diagram. However, for the vast majority of stars not in clusters (unevolved late-type field stars), ages determined using the isochrone method are highly uncertain because the primary age indicators are nearly constant throughout their main-sequence lifetimes, and because their distances and thus luminosities are poorly known. Therefore, finding a distance-independent property of individual stars that can act as a reliable determinant of their ages will be of great value.

Stellar rotation (and the related measure of chromospheric activity - see paper by Mamajek, this volume) has emerged as a promising and distance- independent indicator of age (e.g. Skumanich 1972; Kawaler 1989; Barnes 2003, 2007). Skumanich (1972) first established stellar rotation as an astronomical clock by relating the average projected rotation velocity in young open clusters to their ages via the expression $\overline{v \sin i} \propto t^{-0.5}$. The Skumanich relation is limited in mass (color) to early G dwarfs and suffers from the ambiguity (due to the unknown inclination angle) of the $v \sin i$ data. Furthermore, for ages beyond that of the Hyades cluster ($\sim$ 625Myr), the Skumanich relationship is constrained only by a single G2 dwarf - the Sun.

Modern photometric time-series surveys in young open clusters can provide precisely measured stellar rotation periods (free of the $\sin i$ ambiguity) for F, G, K, and M dwarfs. Based on such new data and emerging empirical relationships between stellar rotation, color, and age, a method was proposed by Barnes (2003) to derive ages for late-type dwarfs from observations of their colors and rotation periods alone. We refer the reader to the paper in this volume by Barnes for a motivation and description of the method of *gyrochronology*. However, our ability to determine stellar ages from stellar rotation, hinges on how well we can measure the dependence of rotation on age for stars of different masses.

2. The key role of open clusters

As coeval populations of stars with a range of masses and well determined ages, open cluster fulfill a critically important role in calibrating the relations between stellar age, rotation, and color. Indeed, *open clusters can define a surface in the 3-dimensional space of stellar rotation period, color, and age, from which the latter can be determined from measurements of the former two* (see Figure 2 below).

This inherent quality of open clusters can only be fully exploited if precise stellar rotation periods (free of the $\sin i$ ambiguity) are measured for cluster members. Accordingly, the time base-line and frequency of time-series photometric observations must be long enough and high enough, respectively, to avoid a bias against detecting periods of more slowly rotating stars, and to avoid detection of false rotation periods due to aliases and a strong "window-function" in the data. Furthermore, measured rotation periods should be combined with information about cluster membership and multiplicity. Removing non-members and stars in close binaries affected by tidal synchronization will allow a better definition of the relationship between rotation period and color at the age of the cluster. Finally, identification of single cluster members will enable a better cluster age to be determined from isochrone fitting. The new results for the open clusters M35 and M34 shown in Figure 1, reflect the powerful combination of decade-long time-series spectroscopy for cluster membership and time-series photometry over 5 months for stellar rotation periods.

3. New observations in the open clusters M35 and M34

We carried out photometric monitoring campaigns over 5 consecutive months for rotational periods, and nearly decade-long radial-velocity surveys for cluster membership and binarity, on the ~150 Myr and ~200 Myr open clusters M35 and M34. For detailed descriptions of the observations, data-reduction, and data-analysis, see Meibom and Mathieu (2005); Meibom *et al.* (2006, 2008), and Braden *et al.* (2009).

Time-Series Photometric Observations: We surveyed, over a timespan of 143 days, a region of 40×40 arc minutes centered on each cluster. Images were acquired at a frequency of once a night both before and after a central block of 16 full nights with observations at a frequency of once per hour. The data were obtained in the Johnson V band with the WIYN 0.9m telescope on Kitt Peak. Instrumental magnitudes were determined from Point Spread Function photometry. Light curves were produced for more than 14,000 stars with $12 < V < 19.5$. Rotational periods were determined for 441 and 120 stars in the fields of M35 and M34, respectively (see Figure 1).

The spectroscopic surveys: M35 and M34 have been included in the WIYN Open Cluster Study (WOCS; Mathieu (2000)) since 1997 and 2001. As part of WOCS, 1-3 radial-velocity measurements per year were obtained on both clusters within the 1-degree field of the WIYN 3.5m telescope with the multi-object fiber positioner (Hydra) feeding a bench-mounted echelle spectrograph. Observations were done at central wavelengths of 5130Å or 6385Å with a wavelength range of ~200Å . From this spectral region with many narrow absorption lines, radial velocities were determined with a precision of < 0.4 km/s (Geller *et al.* 2008; Meibom *et al.* 2001). Of the stars with measured rotational periods in M35 and M34, 203 and 56, respectively, are radial-velocity members of the clusters (dark blue symbols in Figure 1). Including photometric members (light blue symbols in Figure 1), the total number of stars with measured rotational periods in M35 and M34, are 310 and 79.

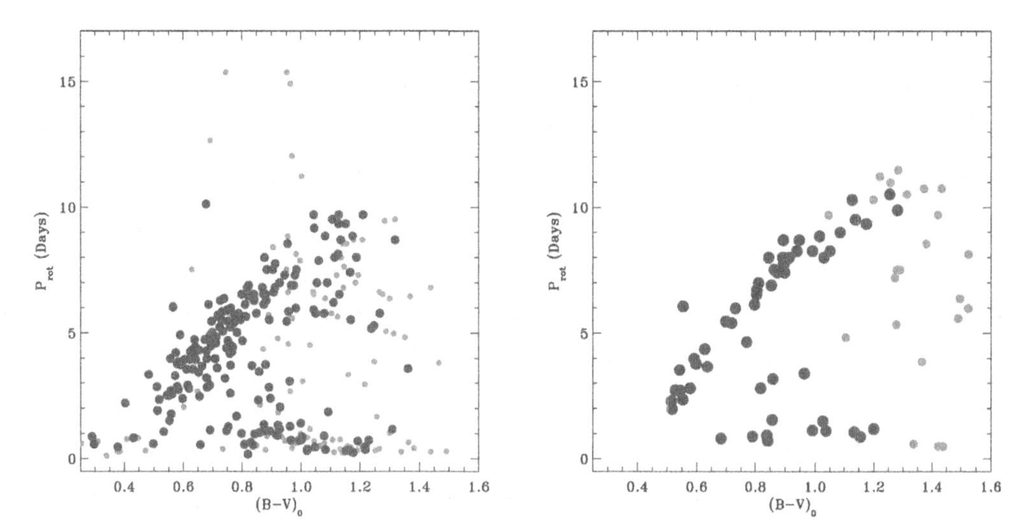

Figure 1. The distribution of stellar rotation periods with (B-V) color index for 310 members of M35 (*left*; (Meibom *et al.* 2008)) and 79 members of M34 (*right*). Dark blue (black) plotting symbols are used for radial-velocity members and light blue (grey) for photometric members.

4. The color-period diagram

Figure 1 shows the rotational periods for members in M35 and M34 plotted against their dereddened $B - V$ colors. The coeval stars fall along two well-defined sequences representing two different rotational states. One sequence displays a clear correlation between rotation period and color, and forms a diagonal band of stars whose periods are increasing with increasing color index (decreasing mass). The second sequence consists of rapidly rotating stars and shows little mass dependence. A small subset of stars is distributed between the two sequences. The distribution of stars in the color-period diagrams suggests that the rotational evolution is slow where we see the sequences and fast in the gap between them. Other areas of the color-period plane are either unlikely or "forbidden".

5. The dependence of stellar rotation period on color

For our purpose of determining the dependence of stellar rotation on stellar color, we can focus on the diagonal sequence of more slowly rotating stars in Figure 1. We can do so because surveys for stellar rotation in the older clusters M37 (550 Myr) and the Hyades (625 Myr) show that F, G, and K dwarfs spin down over a few hundred million years and converge onto this sequence (Hartman *et al.* 2008; Radick *et al.* 1987).

Barnes (2003, 2007, and this volume) refer to the diagonal sequence as the Interface (I) sequence and propose a 2 parameter function ($f(B - V)$) to represent it (Barnes 2007):

$$P(t, B-V) = g(t) \times f(B-V) \tag{5.1}$$

where

$$f(B-V) = a((B-V)-b)^c \tag{5.2}$$

with $a = 0.77$ and $c = 0.60$. Barnes (2007) fix b at a value of 0.4, and determine $g(t) = t^{0.52}$.

From the method of gyrochronology (Barnes 2003, 2007), the functional dependence between stellar color and rotation period ($f(B-V)$) will directly affect the derived ages, and will, if not accurately determined, introduce a systematic error. It is therefore important to constrain and test the color-rotation relation for stars on the I sequence as new data of sufficiently high quality becomes available. Meibom *et al.* (2008) fit $f(B-V)$ as given in equation [5.2] to the I sequence stars in M35, leaving all 3 coefficients (a, b, c) as free parameters. They get the same value of 0.77 for a, but a slightly different value of 0.55 for c. By leaving the translational term b free, a value of 0.47 was found. This value for b is interesting because it corresponds to the approximate $B-V$ color for F-type stars at the transition from a radiative to a convective envelope. This transition is also associated with the onset of effective magnetic wind breaking (e.g. Schatzman 1962), and known as the break in the Kraft curve (Kraft 1967). The value of 0.47 for the b coefficient therefore suggest that, for M35, the blue (high-mass) end of the I sequence begins at the break in the Kraft curve.

The I sequence in M34 is particularly well-defined and will be used to further constrain the dependence between rotation and color in a forthcoming paper (Meibom *et al.*, in preparation).

6. The dependence of stellar rotation on age

With well-defined color-rotation relations (I sequences) for clusters of different ages, we are able to constrain the dependence of stellar rotation on age for stars of different masses. Comparison of the rotation periods for F-, G-, and K-type I sequence dwarfs at different ages enable a direct test of the Skumanich relationship for early G dwarfs and for dwarfs of higher and lower masses.

Initial comparisons in Meibom *et al.* (2008) between the rotation periods of G and K dwarfs on the I sequences in M35 and the Hyades, suggest that the Skumanich time-dependence ($P_{rot} \propto t^{0.5}$) can account for the evolution in rotation periods between M35 and the Hyades for G dwarfs. However, the time-dependence for spin-down of K dwarfs is different and slower than Skumanich. In a more in depth analysis (preliminary results) Meibom *et al.* (2009; in preparation), calculate the mean rotation periods for late-F, G, early K, and late-K I sequence dwarfs in M35, M34, NGC3532 (Barnes 2003; 300 Myr), M37, and the Hyades. They find that the increase in the mean rotation period with age is consistent with Skumanich spin-down for the late-F and G dwarfs, whereas K dwarfs spin down significantly slower. The deviations from the Skumanich spin-down for K dwarfs, suggest that the rotation period for late-type stars cannot be expressed as the product of separable functions of time and color (Eq. [5.1]). Skumanich spin-down was assumed for late-F through early M stars in Barnes (2003, 2007) and in Kawaler (1989).

Eventually, when rotation periods of sufficient quality is available for a larger number of clusters, the effects on the rotational evolution of other stellar parameters, e.g. metallicity, and of the cluster environment, should be considered. Irwin, this volume, give a more complete list of published rotation data in clusters.

7. The Kepler mission - a unique opportunity

At the present time, the Hyades represent the oldest coeval population of stars with measured rotation periods. Measurements of rotation periods for older late-type dwarfs is needed to properly constrain the dependence of stellar rotation on age and mass and to calibrate the technique of gyrochronology. Figure 2 shows a schematic of the surface in the 3-dimensional space of rotation, color, and age. At the present time this surface is defined solely by color-period data in young clusters and for the Sun. The solid black curves represent the ages and color-ranges of FGK dwarfs in M35, M34, NGC3532, M37, and the Hyades. The color and age of the Sun is marked as a solid dot. The figure demonstrates clearly the need for observations of stellar rotation periods beyond the age of the Hyades.

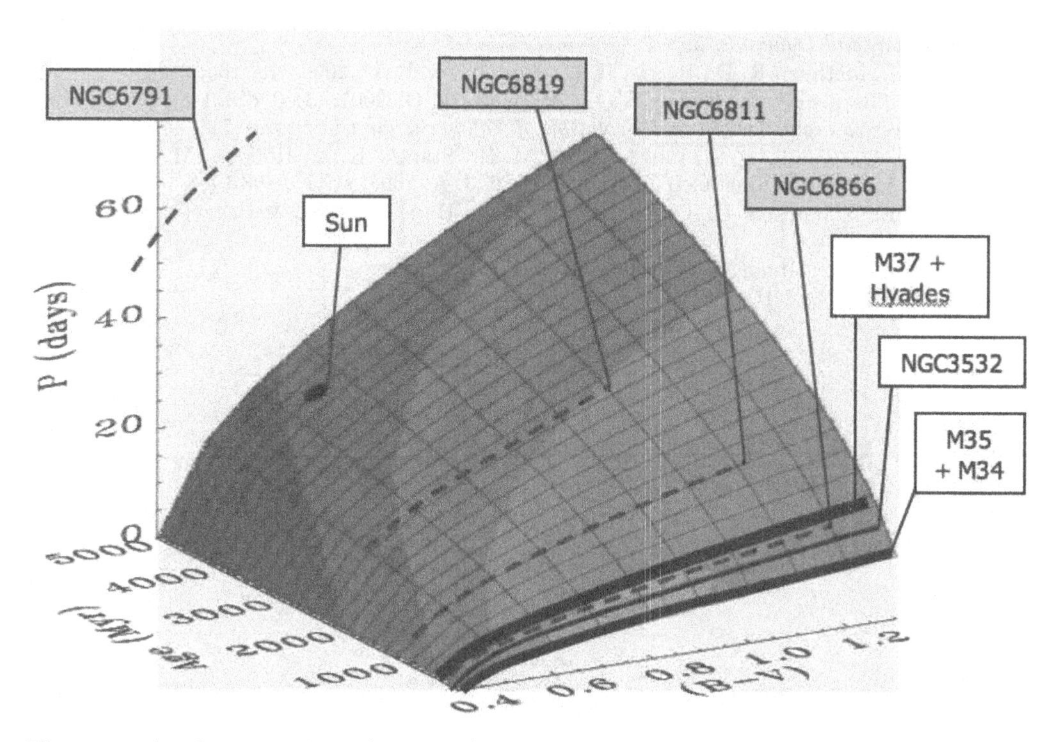

Figure 2. A schematic of the (presumed) empirical surface in the 3-dimensional parameter space of stellar age (Myr), color, and rotation period. The surface is currently defined *only* by stars in young open clusters (black solid lines), and by the Sun (black dot). The dashed blue lines mark the ages and approximate color ranges of FGK dwarfs in the 4 open clusters within the Kepler field.

The lack of periods for older stars (with the exception of the Sun) reflects the challenging task of measuring - from the ground - photometric variability for slowly rotating stars with ages of ∼1Gyr or more. However, the Kepler space telescope (scheduled for a 2009 launch), will provide photometric measurements with a precision, cadence, and duration, sufficient to measure stellar rotation periods from brightness modulations for stars as old as and older than the Sun. Four open clusters are located within the Kepler target region: NGC 6866 (∼0.5 Gyr), NGC 6811 (∼1 Gyr), NGC 6819 (∼2.5 Gyr), and NGC 6791 (∼10 Gyr). With Kepler we therefore have a unique opportunity to extend the age-rotation-color relationships beyond the age of the Hyades and the Sun. The dashed blue curves in Figure 2 mark the ages and approximate color ranges of FGK dwarfs in the 4 clusters.

References

Skumanich, A. 1972, *ApJ*, 171, 565

Kawaler, S. D. 1989, *ApJL*, 343, L65

Barnes, S. A. 2003, *ApJ*, 586, 464

Barnes, S. A., 2007, *ApJ*, 669, 1167

Meibom, S. & Mathieu, R. D. 2005, *ApJ*, 620, 970

Meibom, S., Mathieu, R. D., & Stassun, K. G. 2006, *ApJ*, 653, 621

Meibom, S., Mathieu, R. D., & Stassun, K. G. 2008, arXiv:0805.1040

Braden, E., Mathieu, R. D., & Meibom, S., in prep.

Mathieu, R. D. 2000, *ASP Conf. Ser. 198: Stellar Clusters and Associations: Convection, Rotation, and Dynamos*, p. 517.

Geller, A. M., Mathieu, R. D., Harris, H. C., & McClure, R. D. 2008, *AJ*, 135, 2264

Meibom, S., Barnes, S. A., Dolan, C., & Mathieu, R. D. 2001, *ASP Conf. Ser. 243: From Darkness to Light: Origin and Evolution of Young Stellar Clusters*, p. 711.

Hartman, J. D., Gaudi, B. S., Pinsonneault, M. H., Stanek, K. Z., Holman, M. J., McLeod, B. A., Meibom, S., Barranco, J. A., & Kalirai, J. A. 2008, arXiv:0803.1488

Radick, R. R., Thompson, D. T., Lockwood, G. W., Duncan, D. K., & Baggett, W. E. 1987, *ApJ*, 321, 459

Schatzman E. 1962, *Annales d'Astrophysique* 25, 18

Kraft R. P. 1967, *ApJ*, 150, 551

The Ages of Stars
Proceedings IAU Symposium No. 258, 2008
E.E. Mamajek, D.R. Soderblom & R.F.G. Wyse, eds.

doi:10.1017/S1743921309032025

The rotational evolution of low-mass stars

Jonathan Irwin[1] **and Jerome Bouvier**[2]

[1]Harvard-Smithsonian Center for Astrophysics,
60 Garden Street MS-16, Cambridge, MA 02138, USA
email: jirwin@cfa.harvard.edu
[2]Laboratoire d'Astrophysique, Observatoire de Grenoble,
BP 53, F-38041 Grenoble Cédex 9, France

Abstract. We summarise recent progress in the understanding of the rotational evolution of low-mass stars (here defined as solar mass down to the hydrogen burning limit) both in terms of observations and modelling. Wide-field imaging surveys on moderate-size telescopes can now efficiently derive rotation periods for hundreds to thousands of open cluster members, providing unprecedented sample sizes which are ripe for exploration. We summarise the available measurements, and provide simple phenomenological and model-based interpretations of the presently-available data, while highlighting regions of parameter space where more observations are required, particularly at the lowest masses and ages $\gtrsim 500$ Myr.

Keywords. stars: late-type, stars: low-mass, stars: pre–main-sequence, stars: rotation

---

1. Rotation period data

A compilation of most of the available rotation period (and some $v \sin i$) measurements in open clusters for stars with masses $M \lesssim 1.2\ \mathrm{M}_\odot$ is shown in Figure 1 (a list of references is included in Table 1). We plot rotation period as a function of stellar mass, rather than using the more conventional quantities of colour or spectral type on the horizontal axis, since the diagram spans ages from the early pre–main-sequence (PMS), at ~ 1 Myr in the ONC, to the main sequence (MS), and neither colour nor spectral type are invariant for a given star over this age range. The masses are, of course, model-dependent, but the majority of binning in mass and morphological examination used in this work require only that the mass scale is *approximately* correct, which should be reasonably well-satisfied by the PMS stellar evolution tracks. We use those of the Lyon group, from Baraffe *et al.* (1998) and I. Baraffe (private communication), throughout.

An expanded version of Figure 1, omitting many of the more sparsely sampled clusters and all the $v \sin i$ observations, is shown in Figure 2. By examining the evolution of the morphologies of these diagrams, we can already draw some (model-independent) conclusions regarding the evolution of the rotation periods in these clusters. For simplicity, we discuss two broad mass ranges: stars close to solar mass (e.g. $0.9 \lesssim M/\mathrm{M}_\odot \lesssim 1.1$), and fully-convective stars ($M \lesssim 0.4\ \mathrm{M}_\odot$), which represent the "tail" of the distribution which emerges especially in the NGC 2547 and NGC 2516/M35 panels of Figure 2.

Considering the solar mass stars first, it is clear that the behaviour on the PMS (which lasts to ~ 30 Myr for a solar mass star, until it reaches the zero-age main sequence, hereafter ZAMS) and on the MS up to the age of the Hyades, varies as a function of rotation rate. The slowest rotators in the ONC have periods of ~ 10 days. The most basic prediction we can make for the evolution of the rotation rate is to assume angular momentum is conserved, in which case as the star contracts towards the ZAMS, it should spin up. However, for these slowly-rotating stars, the period remains approximately constant all the way to NGC 2362, and has spun up to ~ 8 days by the age of NGC 2547 (~ 40 Myr),

Table 1. List of references for the panels in Figure 1.

Cluster	Age (Myr)	Source(s)
ONC	1	Herbst *et al.* (2001,2002) Stassun *et al.* (1999)
NGC 2264	2	Lamm *et al.* (2005) Makidon *et al.* (2004)
IC 348	3	Cohen *et al.* (2004) Littlefair *et al.* (2005) Cieza & Baliber (2006)
σ Ori	5	Scholz & Eislöffel (2004a)
ϵ Ori	5	Scholz & Eislöffel (2005)
NGC 2362	5	Irwin *et al.* (2008b)
IC 2391 IC 2602	30	Patten & Simon (1996) Barnes *et al.* (1999)
NGC 2547	40	Irwin *et al.* (2008a)
α Per	50	Stauffer *et al.* (1985, 1989) Prosser (1991) O'Dell & Collier Cameron (1993) O'Dell *et al.* (1994, 1996) Allain *et al.* (1996) Martín & Zapatero Osorio (1997) Prosser & Randich (1998) Prosser *et al.* (1998) Barnes *et al.* (1998)
Pleiades	100	van Leeuwen *et al.* (1987) Stauffer *et al.* (1987) Magnitskii (1987) Prosser *et al.* (1993a,b,1995) Krishnamurthi *et al.* (1998) Terndrup *et al.* (1999) Scholz & Eislöffel (2004b)
Pleiades $v \sin i$	100	Stauffer *et al.* (1984) Stauffer & Hartmann (1987) Soderblom *et al.* (1993) Jones *et al.* (1996) Queloz *et al.* (1998) Terndrup *et al.* (2000)
M50	130	Irwin *et al.* (2009)
NGC 2516 $v \sin i$ $v \sin i$	150	Irwin *et al.* (2007) Terndrup *et al.* (2002) Jeffries *et al.* (1998)
M35	150	Meibom *et al.* (2008)
M34	200	Irwin *et al.* (2006)
M37	485	Hartman *et al.* (2008)
Hyades	625	Radick *et al.* (1987) Prosser *et al.* (1995)
Praesepe	650	Scholz & Eislöffel (2007)

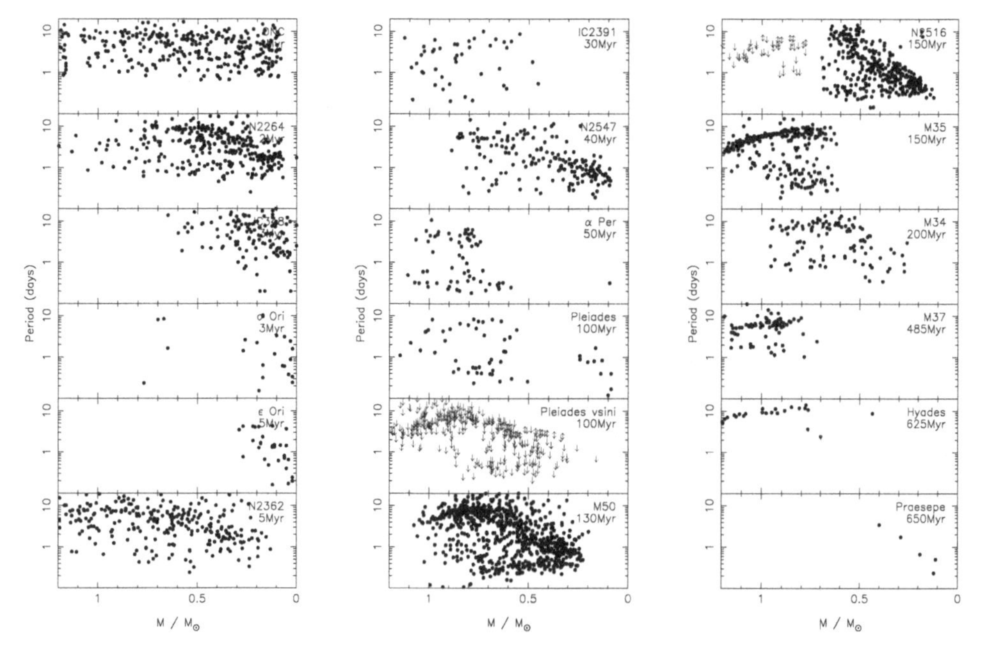

Figure 1. Compilation of 3100 rotation periods (and some $v \sin i$ measurements) for stars with masses $M \lesssim 1.2\ \mathrm{M}_\odot$ in young ($\lesssim 1$ Gyr) open clusters, from the literature. Plotted in each panel is rotation period as a function of stellar mass for a single cluster. The appropriate references for each panel are given in Table 1. All the masses used in this contribution were computed using the I-band luminosities of the sources and the models of Baraffe *et al.* (1998), assuming values of the age, distance modulus and reddening for the clusters taken from the literature.

at which point the contraction ceases. Over the same age range, the most rapid rotators gradually evolve from a period of ~ 1 day at the ONC, to ~ 0.6 days at NGC 2362 (~ 5 Myr), and ~ 0.2 days at NGC 2547. This follows closely the prediction from stellar contraction. We therefore conclude that there is some mechanism removing angular momentum from only the slow rotators, to an age of ~ 5 Myr, and a short time after this the angular momentum losses cease, leaving the star free to spin up for a short time until it reaches the ZAMS. The net result of this is to yield a wide range of rotation rates in the early-MS clusters such as the Pleiades and M35.

By the age of the Hyades however, the rotation rates converge onto a single well-defined sequence. It is clear that this process is well underway by the age of M37 ($\sim$ 485 Myr). There must therefore be another rotation-rate-dependent angular momentum loss mechanism operating on the early-MS, such that more rapid rotators lose more angular momentum, to drive all the stars toward the same rotation rate at a given mass.

In contrast, the behaviour of the lowest mass stars is clearly rather different. On the PMS, these stars all appear to spin up rapidly, and reach very rapid rotation rates at the ZAMS (for a $0.4\ \mathrm{M}_\odot$ star, the ZAMS is reached at ~ 150 Myr), and furthermore, the maximum rotation period seen is a very strong function of mass (this is most clear in the NGC 2516 diagram in Figure 2). This indicates that there is little rotation rate dependence in this mass domain due to the similarities between the morphologies of the diagrams especially from NGC 2264 to NGC 2516, and furthermore, that these stars lose little angular momentum on the PMS, in contrast to the slowly rotating solar mass stars. Moreover, between NGC 2516 and Praesepe, the limited quantity of data available suggest that there is essentially no evolution of the rotation period, so little angular momentum appears to be lost on the early-MS, again in contrast to the solar-type stars.

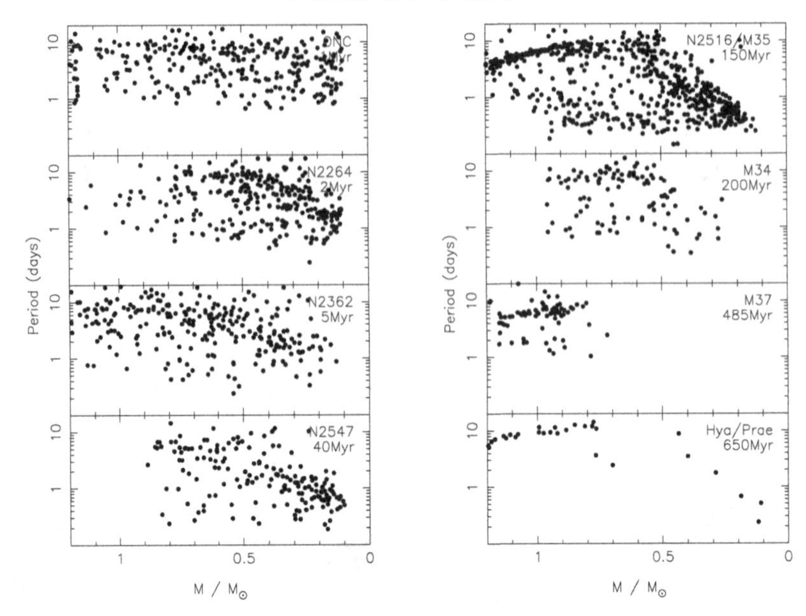

Figure 2. Enlarged version of Figure 1, showing a few selected clusters which have large samples of rotation periods available. NGC 2516 and M35 have been combined into a single panel, since they have essentially identical ages, as have the Hyades and Praesepe.

We proceed in §2 to examine the physical interpretation of these observational results.

2. Models of rotational evolution

Recall from §1 that we needed to invoke two mechanisms of angular momentum removal: one operating for $\sim 5 - 10$ Myr on the PMS, and one operating on the main sequence, with the property that it must produce a convergence in the rotation rates between ~ 100 and 600 Myr.

2.1. *Pre–main-sequence angular momentum losses*

For the first of these, the PMS angular momentum loss, the $\sim 5 - 10$ Myr timescale suggests an obvious candidate: the circumstellar discs that surround these stars in the earliest stages of their evolution dissipate on these timescales (e.g. Haisch *et al.* 2001). The precise mechanism by which angular momentum is removed due to the presence of a disc is still a matter of debate, with the currently-favoured hypothesis being an accretion-driven stellar wind (e.g. Matt & Pudritz 2005), or "disc locking" (e.g. Königl 1991; Collier Cameron *et al.* 1995). For our purposes, we shall simply treat the effect of the disc as holding the angular velocity of the star constant (by removing angular momentum) for the lifetime of the disc.

In reality, there will be a range of disc lifetimes, and thus the stars will be "released" from this angular velocity regulation at a range of times, giving rise to a spread of rotation rates on the PMS and ZAMS, as required by the observations. Moreover, examining one of the early-PMS clusters shows us a "snapshot" of this process in action. The best-observed example is the ONC, shown in Figure 3.

If we presuppose that stars begin with a tight distribution of initial rotation periods, centred around $8 - 10$ days, in a histogram of rotation period we should expect to see a population of stars at this period, and then a tail of faster-rotating stars, corresponding to the ones now uncoupled from their discs, with faster rotating stars having

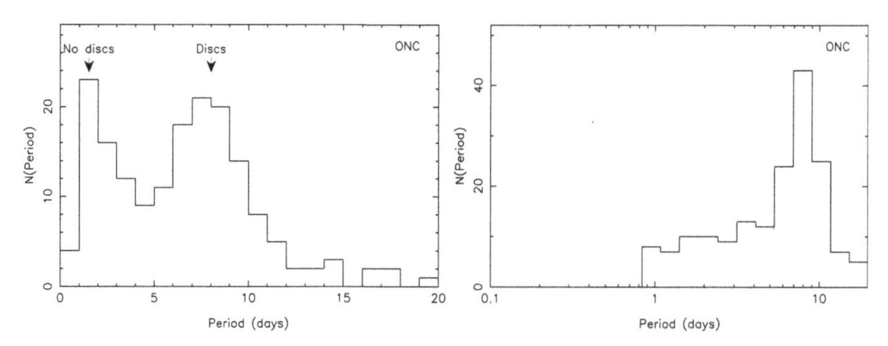

Figure 3. Distribution of rotation periods for ONC stars with masses $M \gtrsim 0.25\ \mathrm{M}_\odot$ (using the D'Antona & Mazzitelli 1998 models; this corresponds to $M \gtrsim 0.4\ \mathrm{M}_\odot$ for the Baraffe *et al.* 1998 models used in this contribution), replotted from Herbst *et al.* (2001). The two panels show the same distribution binned in linear period (left) and log period (right).

uncoupled earlier. This is indeed what is seen in Figure 3, where it is instructive to plot the distribution in log period rather than the more conventional linear period.

Moreover, this hypothesis makes an observationally-testable prediction: the population of slowly-rotating stars should show measurable indications of the presence of a disc (e.g. mid-IR excess) or of active accretion (i.e. be classical T-Tauri stars, CTTS), whereas the rapidly-rotating stars should not have discs, or have recently-dissipated discs, and not be active accretors (i.e. weak-lined T-Tauri stars, WTTS).

In reality, extracting this observational signal from the cluster data has proved to be difficult (see Rebull *et al.* 2004 for a detailed discussion of this issue). Rebull *et al.* (2006) presented the first statistically significant detection of this effect, using mid-IR excess measurements from the IRAC instrument aboard the *Spitzer* space telescope, the current method-of-choice, finding that slowly-rotating stars are indeed more likely to posses discs than rapidly-rotating stars. However, in doing so this study revealed a wrinkle in the argument: they also found a puzzling population of slowly-rotating stars without discs. Several subsequent studies (e.g. Cieza & Baliber 2007) have confirmed these results, which now appear to be placed on a firm footing.

2.2. *Main sequence angular momentum losses*

The mechanism for angular momentum loss on the main sequence, at least for solar mass stars, is thought to be a solar-type magnetised stellar wind. The time-dependence of rotation rates in this age range has been firmly established observationally for solar-type stars since Kraft (1967), and from the age of the Hyades to the Sun, obeys well the famous Skumanich (1972) $t^{1/2}$ law (i.e. $\omega \propto t^{-1/2}$). This can be reproduced on a more theoretically motivated framework from parameterised angular momentum loss laws (usually based on Kawaler 1988; the $t^{1/2}$ law can be reproduced by setting $n = 3/2$ and $a = 1$ in his model).

Although the Skumanich (1972) law provides a good description of the evolution of solar-type stars from the age of the Hyades to the Sun, and for the majority of stars in the Pleiades to the age of the Hyades, simply evolving the Pleiades distribution forward in time to the age of the Hyades using this law produces stars which are rotating much more rapidly than observed in the Hyades. Therefore, most modelers modify the Kawaler (1988) formalism to incorporate saturation of the angular momentum losses above a critical angular velocity ω_{sat} (Stauffer & Hartmann 1987; Barnes & Sofia 1996). The saturation is further assumed to be mass-dependent, to account for the mass-dependent spin-down timescales observed on the early-MS.

The resulting angular momentum loss law assumed in our models is:

$$\left(\frac{dJ}{dt}\right)_{\rm wind} = \begin{cases} -K\,\omega^3 \left(\frac{R}{R_\odot}\right)^{1/2} \left(\frac{M}{M_\odot}\right)^{-1/2} &, \omega < \omega_{\rm sat} \\ -K\,\omega\,\omega_{\rm sat}^2 \left(\frac{R}{R_\odot}\right)^{1/2} \left(\frac{M}{M_\odot}\right)^{-1/2} &, \omega \geqslant \omega_{\rm sat} \end{cases} \tag{2.1}$$

We note that although saturation is reasonably well-motivated observationally (e.g. by the saturation observed in the relationship between X-ray activity and rotation rate), this is a rather unsatisfactory feature of the present models, since there is a certain amount of arbitrariness in the way this parameter is introduced (for example, the choice of power of ω in Eq. 2.1 for the saturated angular velocity dependence of dJ/dt is arbitrary). An important area for future work is to develop a theoretical framework for this phenomenon, and for solar-type winds in general.

2.3. *Comparison of models to data*

The models of rotational evolution we use are described in detail in Bouvier *et al.* (1997), Allain (1998), Irwin *et al.* (2007) and Irwin (2007). We summarise only the most salient features in this contribution.

The methodology of Bouvier *et al.* (1997) which we adopt here essentially implements a rotational evolution model by using a standard non-rotating PMS stellar evolution track to compute the variation of the stellar parameters as a function of time, principally the moments of inertia of the radiative and convective regions of the star, which are then used as an input to determine the angular velocity as a function of time.

This model in the simple form described has four parameters (which could all be functions of mass): the initial angular velocity ω_0, normalisation of the solar-type angular momentum loss law K (see Eq. 2.1), saturation velocity $\omega_{\rm sat}$, and the lifetime of the circumstellar disc $\tau_{\rm disc}$. For solar-type stars, we can fix K by requiring that the model reproduces the observed rotation rate of the Sun.

Figure 4 shows the result of fitting this model to the rotation period data using a simple nonlinear least squares routine. For the moment, we have required the radiative core and convective envelope to have the same angular velocity, i.e. a "solid body" model.

The results show that the model does a reasonable job for the late-time evolution from the Hyades to the Sun, but this is largely by construction since we used the Skumanich (1972) law as an input! At earlier times, in general, we find that the rapid rotators are better-reproduced by the model, with the only major issue being at $\sim$ 150 Myr where the model underpredicts the upper bound to the observed rotation rates. For the slow rotators, the model rotates too rapidly given the ONC and NGC 2362 as an initial condition around the ZAMS and early-MS. The disc lifetimes behave as expected, being shorter for the rapid rotators.

The difficulty fitting the slow rotators motivates the introduction of an additional parameter into the model. In particular, we now relax the assumption of solid body rotation, and allow the core and envelope to have different angular velocities, coupling angular momentum between them on a timescale τ_c (this is done in a fashion which tries to equalise their angular velocities). $\tau_c = 0$ represents the solid body case already considered. The net effect of this "core-envelope decoupling" from the point of view of the observed rotation rate (which is that of the envelope) is to "hide" angular momentum in the core when it forms, which is then coupled back into the envelope gradually, providing a "late-time replenishment" of the surface angular velocity.

Figure 5 shows the result of applying this revised model to the solar-type stars. We can see that the fit is substantially improved. Importantly, we find that the value of τ_c for the

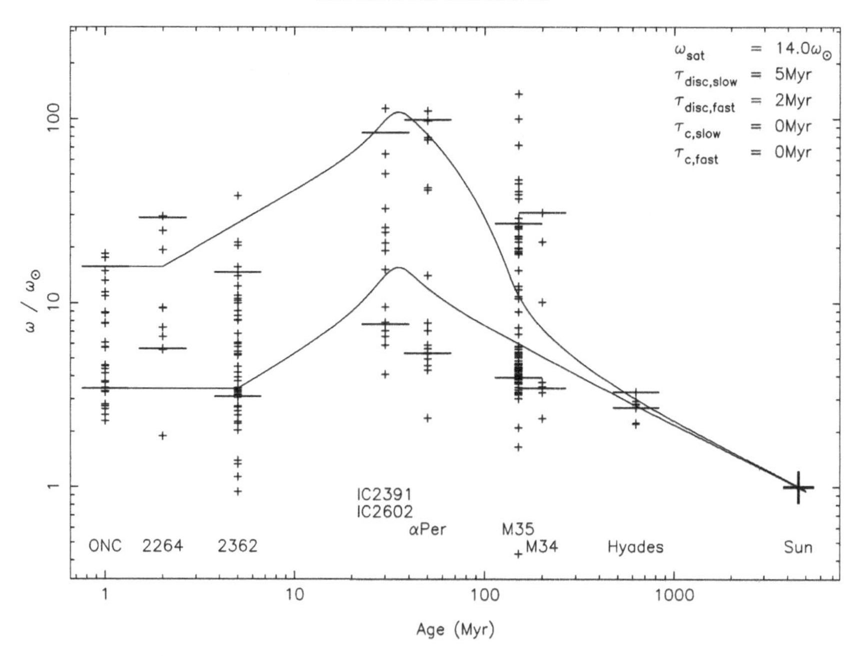

Figure 4. Rotational angular velocity ω plotted as a function of time for stars with masses $0.9 < M/\mathrm{M}_\odot \leqslant 1.1$. Crosses show the open cluster rotation period data such that each cluster collapses into a vertical stripe on the diagram, and short horizontal lines show the 25th and 90th percentiles of ω, used to characterise the slow and fast rotators respectively. The lines show rotational evolution models for 1.0 $\mathrm{M}_\odot$ stars, fit to the percentiles for each cluster using a simple unweighted least squares method. For this plot, we have assumed the stars rotate as solid bodies (i.e. constant ω as a function of radius inside the star). Plotted are the ONC, NGC 2264, NGC 2362, IC 2391, IC 2602, α Per, M35, M34, the Hyades, and the Sun.

slow rotators is relatively large, $\sim$ 110 Myr indicating inefficient core-envelope coupling, whereas for the fast rotators it is short, $\sim$ 6 Myr: these stars have efficient core-envelope coupling and rotate more like solid bodies. A corollary of this is that slow rotators develop a higher degree of rotational shear across the convective/radiative boundary.

This result may have important observationally-testable consequences. Bouvier (2008) discusses one of these: the impact on the surface abundance of lithium. In particular, the higher rotational shear in the slow rotators is likely to induce additional mixing in these stars, which may bring Li down into the star where it can be burnt. The net effect would be a higher level of lithium depletion in slow rotators than fast rotators, and these should therefore display a lower lithium abundance on the main sequence. Indeed, Soderblom *et al.* (1993) report that this appears to be the case in the Pleiades.

Having examined the models for solar-type stars, we now proceed to the fully-convective, very low-mass stars ($M \lesssim 0.4\ \mathrm{M}_\odot$). Figure 6 shows the results of applying the same models to this mass domain. Since these stars are fully-convective, there is no possibility for core-envelope decoupling, so they should rotate as solid bodies.

The model is clearly a poor match to the observations. In particular, the solar-type wind appears to loose too much angular momentum at late-times, resulting in a sharp drop in the predicted rotation rates starting at $\sim$ 200 Myr, which does not appear to be seen, although the available data are still very limited at present, with $v \sin i$ measurements giving only upper limits for the slowest rotating stars. This cannot be "fixed" by invoking saturation, since this has already been done for the model shown in Figure 6, which has $\omega_{\mathrm{sat}} = 1.8\ \omega_\odot$ (i.e. every star is saturated in the relevant age range).

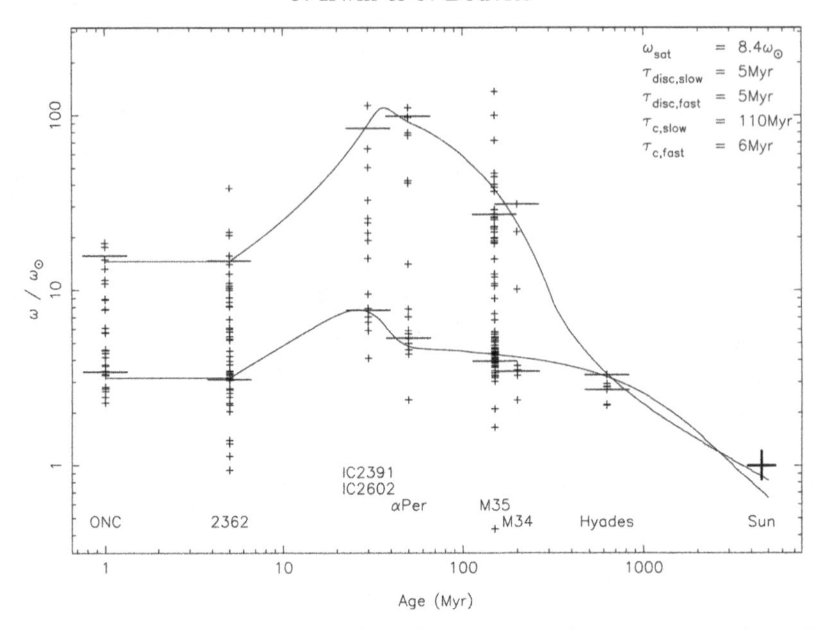

Figure 5. As Figure 4, but now allowing the radiative core and convective envelope of the stars to have different angular velocities, coupling angular momentum between the two on a timescale τ_c. We have also omitted NGC 2264, since the models presently have difficulty explaining the observations in this cluster.

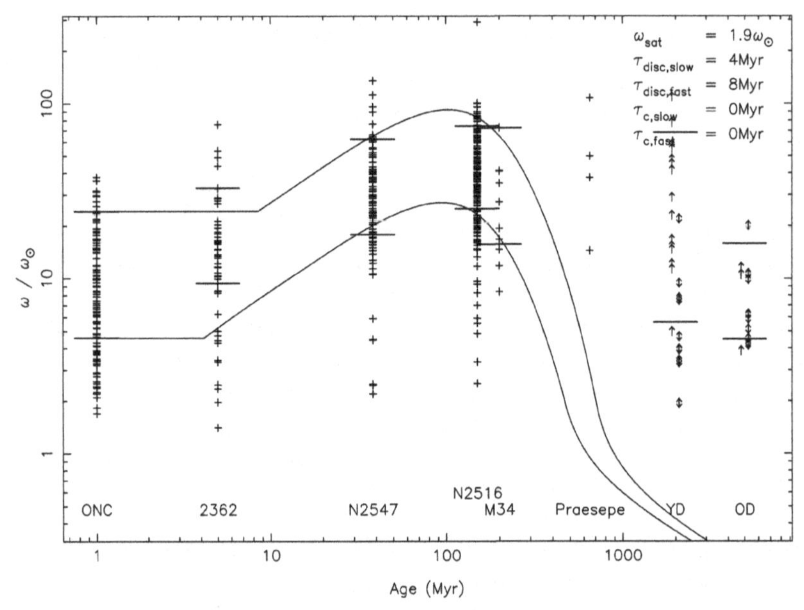

Figure 6. As Figure 5, except for stars with masses $0.1 < M/M_{\odot} \leqslant 0.4$, and models for $0.25\ M_{\odot}$. Data for NGC 2547, NGC 2516 and Praesepe are shown in addition to those clusters already included in Figure 5. We have also included lower limits from $v \sin i$ measurements for field M-dwarfs from Delfosse *et al.* (1998), placing the young disc population at 2 Gyr and the old disc at 5 Gyr (e.g. Feltzing & Bensby 2008; although note that in reality, these populations have a wide dispersion in age). Many of their points are only upper limits in $v \sin i$, and are shown as double-headed arrows in the figure, plotted offset slightly from the other data in age for clarity. Note the sparse coverage of cluster data, and particularly rotation periods, at the oldest ages. This is largely due to the intrinsic faintness of old, mid–late M-dwarfs. Several proposed surveys aim to fill in this region of parameter space within the next couple of years.

It would not be surprising that a solar-type wind does not reproduce the evolution of fully-convective stars, since this is thought to be launched at a tachocline of shear at the radiative/convective interface, and these stars posses none. However, as highlighted by J. Stauffer in the questions, we should be cautious in overinterpreting the rather sparse data presently available at these masses and ages, especially given the large number of $v \sin i$ upper limits and the natural bias of the rotation period method toward more active, and hence presumably more rapidly-rotating, stars, resulting from the extremely small photometric amplitudes of the rotational modulations at these old ages.

This domain of very low-masses and late ($\gtrsim$ 500 Myr) ages represents an important area for future observational studies to explore, and could provide us with a valuable insight into the winds of fully-convective stars.

Acknowledgments

This research has made use of the SIMBAD database, operated at CDS, Strasbourg, France, the WEBDA database, operated at the Institute for Astronomy of the University of Vienna, and the Open Cluster Database, as provided by C.F. Prosser and J.R. Stauffer, which may currently be accessed at `http://www.noao.edu/noao/staff/cprosser/`, or by anonymous ftp to `140.252.1.11`, `cd /pub/prosser/clusters/`. We thank Isabelle Baraffe for providing the stellar evolution tracks used to compute the models in §2.3.

References

Allain, S., Fernandez, M., Martín, E. L., & Bouvier, J. 1996, *A&A*, 314, 173

Allain, S. 1998, *A&A*, 333, 629

Baraffe, I., Chabrier, G., Allard, F., & Hauschildt, P. H. 1998, *A&A*, 337, 403

Barnes, S. A., Sofia, S., Prosser, C. F., & Stauffer, J. R. 1999, *ApJ*, 516, 263

Barnes, S. & Sofia, S. 1996, *ApJ*, 462, 746

Barnes, J. R., Collier Cameron, A., Unruh, Y. C., Donati, J. F., & Hussain, G. A. J. 1998, *MNRAS*, 299, 904

Bouvier, J., Forestini, M., & Allain, S. 1997, *A&A*, 326, 1023

Bouvier, J. 2008, *A&A*, 489, 53

Cieza, L. & Baliber, N. 2006, *ApJ*, 649, 862

Cieza, L. & Baliber, N. 2007, *ApJ*, 671, 605

Cohen, R. E., Herbst, W., & Williams, E. C. 2004, *AJ*, 127, 1602

Collier Cameron, A., Campbell, C. G., & Quaintrell, H. 1995, *A&A*, 298, 133

D'Antona, F. & Mazzitelli, I. 1998, in R. Rebolo, E. L. Martín, & M. R. Zapatero Osorio (eds.), *Brown dwarfs and extrasolar planets*, ASP Conf. Series 134, p. 442.

Delfosse, X., Forveille, T., Perrier, C., & Mayor, M. 1998, *A&A*, 331, 581

Feltzing, S. & Bensby, T. 2008, in P. Barklem, A. Korn, & B. Plez (eds.), *A stellar journey*, Physica Scripta, in press (`arXiv:0811.1777`)

Haisch, K. E., Lada, E. A., & Lada, C. J. 2001, *ApJ*, 553, 153

Hartman, J. D., *et al.* 2008, *ApJ*, in press (`arXiv:0803.1488`)

Herbst, W., Bailer-Jones, C. A. L. & Mundt, R. 2001, *ApJ*, 554, 197

Herbst, W., Bailer-Jones, C. A. L., Mundt, R., Meisenheimer, K., & Wackermann, R. 2002, *A&A*, 396, 513

Irwin, J., Aigrain, S., Hodgkin, S., Irwin, M., Bouvier, J., Clarke, C., Hebb, L., & Moraux, E. 2006, *MNRAS*, 370, 954

Irwin, J., Hodgkin, S., Aigrain, S., Hebb, L., Bouvier, J., Clarke, C., Moraux, E. & Bramich D. M. 2007, *MNRAS*, 377, 741

Irwin, J. 2007, Ph.D. thesis, University of Cambridge

Irwin, J., Hodgkin, S., Aigrain, S., Bouvier, J., Hebb, L., & Moraux, E. 2008a, *MNRAS*, 383, 1588

Irwin, J., Hodgkin, S., Aigrain, S., Bouvier, J., Hebb, L., Irwin, M., & Moraux, E. 2008b, *MNRAS*, 384, 675
Irwin, J., Aigrain, S., Bouvier, J., Hebb, L., Hodgkin, S., Irwin, M., & Moraux, E. 2009, *MNRAS*, in press (`arXiv:0810.5110`)
Jeffries, R. D., James, D. J., & Thurston, M. R. 1998, *MNRAS*, 300, 550
Jones, B. F., Fischer, D. A., & Stauffer, J. R. 1996, *AJ*, 112, 1562
Königl, A. 1991, *ApJ*, 370, L37
Kawaler, S. D. 1998, *ApJ*, 333, 236
Kraft, R. P. 1967, *ApJ*, 150, 551
Krishnamurthi, A., *et al.* 1998, *ApJ*, 493, 914
Lamm, M. H., Mundt, R., Bailer-Jones, C. A. L., & Herbst, W. 2005, *A&A*, 430, 1005
Littlefair, S. P., Naylor, T., Burningham, B., & Jeffries, R. D. 2005, *MNRAS*, 358, 341
Magnitskii, A. K. 1987, *Soviet Astron.* (Letters), 13, 451
Makidon, R. B., Rebull, L. M., Strom, S. E., Adams, M. T., & Patten, B. M. 2004, *AJ*, 127, 2228
Martín, E. L. & Zapatero Osorio, M. R. 1997, *MNRAS*, 286, L17
Matt, S. & Pudritz, R. E. 2005, *ApJ*, 632, 135
Meibom, S., Mathieu, R. D., & Stassun K. G. 2008, *ApJ*, in press (`arXiv:0805.1040`)
O'Dell, M. A. & Collier Cameron, A. 1993, *MNRAS*, 262, 521
O'Dell, M. A., Hendry, M. A., & Collier Cameron, A. 1994, *MNRAS*, 268, 181
O'Dell, M. A., Hilditch, R. W., Collier Cameron, A. & Bell, S. A. 1996, *MNRAS*, 284, 874
Patten, B. M. & Simon, T. 1996, *ApJS*, 106, 489
Prosser, C. F. 1991, Ph.D. Thesis, University of California, Santa Cruz
Prosser, C. F., Schild, R. E., Stauffer, J. R., Jones, B. F. 1993a, *PASP*, 105, 269
Prosser, C. F., *et al.* 1993b, *PASP*, 105, 1407
Prosser, C. F., *et al.* 1995, *PASP*, 107, 211
Prosser, C. F. & Randich, S. 1998, *AN*, 319, 210
Prosser, C. F., Randich, S., & Simon, T. 1998, *AN*, 319, 215
Queloz, D., Allain, S., Mermilliod, J.-C., Bouvier, J., & Mayor, M. 1998, *A&A*, 335, 183
Radick, R. R., Thompson, D. T., Lockwood, G. W., Duncan, D. K., & Baggett, W. E. 1987, *ApJ*, 321, 459
Rebull, L. M., Wolff S. C., & Strom S. E. 2004, *AJ*, 127, 1029
Rebull, L. M., Stauffer, J. R., Megeath, S. T., Hora, J. L., & Hartmann, L. 2006, *ApJ*, 646, 297
Scholz, A. & Eislöffel, J. 2004, *A&A*, 419, 249
Scholz, A. & Eislöffel, J. 2004, *A&A*, 421, 259
Scholz, A. & Eislöffel, J. 2005, *A&A*, 429, 1007
Scholz, A. & Eislöffel, J. 2007, *MNRAS*, 381, 1638
Skumanich, A. 1972, *ApJ*, 171, 565
Soderblom, D. R., Stauffer, J. R., Hudon, J. D., & Jones, B. F. 1993, *ApJS*, 85, 315
Stassun, K. G., Mathieu, R. D., Mazeh, T., & Vrba, F. J. 1999, *AJ*, 117, 2941
Stauffer, J. R., Hartmann, L., Soderblom, D. R., & Burnham, N. 1984, *ApJ*, 280, 202
Stauffer, J. R., Hartmann, L. W., Burnham, J. N., & Jones, B. F. 1985, *ApJ*, 289, 247
Stauffer, J. R. & Hartmann, L. W. 1987, *ApJ*, 318, 337
Stauffer, J. R., Schild, R. A., Baliunas, S. L., & Africano, J. L. 1987, *PASP*, 99, 471
Stauffer, J. R., Hartmann, L. W., & Jones, B. F. 1989, *ApJ*, 346, 160
Terndrup, D. M., Krishnamurthi, A., Pinsonneault, M. H., & Stauffer, J. R. 1999, *AJ*, 118, 1814
Terndrup, D. M., Stauffer, J. R., Pinsonneault, M. H., Sills, A., Yuan, Y., Jones, B. F., Fischer, D., & Krishamurthi, A. 2000, *AJ*, 119, 1303
Terndrup, D. M., Pinsonneault, M., Jeffries, R. D., Ford, A., & Sills, A. 2002, *ApJ*, 576, 950
Van Leeuwen, F., Alphenaar, P, & Meys, J. J. M. 1987, *A&AS*, 67, 483

Discussion

T. NAYLOR: 1) You mentioned that you had missed the young cluster NGC 2264 from your plots. Why was this missing, and what about IC348? 2) Could it be there is a second parameter for the young clusters which means that age ordering and gyrochronological orderings are different?

J. IRWIN: 1) NGC 2264 (and indeed IC 348) using the "literature" ages don't seem to fit in well with the models I showed. This is somewhat puzzling and I am not really sure of the reason yet. 2) Yes, absolutely, this is quite strongly possible especially given the difference in "environment" in these young clusters.

J. STAUFFER: The old (Praesepe/Hyades) M stars are an example of where it is useful to combine the information from spectroscopy ($v \sin i$) with the period data. Most of the Hyades and Praesepe M dwarfs are slow rotators: the small number of Praesepe stars with periods are very likely a biased set (but the $v \sin i$ values show most are slow rotators, and you also do see the fast rotators).

J. IRWIN: Absolutely, this is something that needs to be included in the models I showed. Indeed, the periods may well be an activity-biased sample at these old ages so we must be cautious in their interpretation.

G. MEYNET: What might be the physical cause of decoupling/non-decoupling behavior between the core/envelope. Can it be magnetic fields?

J. IRWIN: I am no expert but it seems that magnetic fields are often invoked to explain the core-envelope coupling in the literature.

D. SODERBLOM: You have presented a large quantity of high-quality observations of rotation periods in clusters, and have applied some physics to the problem as well. Converting color to a mass is the right thing to do, even given the differences in PMS mass tracks. My question is this: Your clusters range from the sparse to the very rich. Is there any evidence that rich clusters partition their angular momentum any differently from sparse clusters?

J. IRWIN: At the moment, I have not really seen evidence for differences. In Irwin (2008, arXiv:0810.5110), I show there is no statistically significant difference between the almost identically-aged clusters NGC 2516, M50, and M35, but more work is clearly needed. This may explain the differences between some of the earliest PMS clusters.

Jonathan Irwin

The Ages of Stars
Proceedings IAU Symposium No. 258, 2008
E.E. Mamajek, D.R. Soderblom & R.F.G. Wyse, eds.

doi:10.1017/S1743921309032037

How accurately can we age-date solar-type dwarfs using activity/rotation diagnostics?

Eric E. Mamajek

Department of Physics & Astronomy,
University of Rochester, Rochester, NY 14624 USA
email: emamajek@pas.rochester.edu

Abstract. It is well established that activity and rotation diminishes during the life of sun-like main sequence (~F7-K2V) stars. Indeed, the evolution of rotation and activity among these stars appears to be so deterministic that their rotation/activity diagnostics are often utilized as estimators of stellar age. A primary motivation for the recent interest in improving the ages of solar-type field dwarfs is in understanding the evolution of debris disks and planetary systems. Reliable isochronal age-dating for field, solar-type main sequence stars is very difficult given the observational uncertainties and multi-Gyr timescales for significant structural evolution. Observationally, significant databases of activity/rotation diagnostics exist for field solar-type field dwarfs (mainly from chromospheric and X-ray activity surveys). *But how well can we empirically age-date solar-type field stars using activity/rotation diagnostics?* Here I summarize some recent results for F7-K2 dwarfs from an analysis by Mamajek & Hillenbrand (2008), including an improved "gyrochronology" [Period(color, age)] calibration, improved chromospheric ($R'_{\rm HK}$) and X-ray ($\log(L_X/L_{bol})$) activity vs. rotation (via Rossby number) relations, and a chromospheric vs. X-ray activity relation that spans four orders of magnitude in $\log(L_X/L_{bol})$. Combining these relations, one can produce predicted chromospheric and X-ray activity isochrones as a function of color and age for solar type dwarfs.

Keywords. Sun: (activity, rotation), stars: (activity, chromospheres, coronae, fundamental parameters, late-type, low-mass, rotation), Galaxy: evolution

1. Introduction

Observational and theoretical studies regarding the evolution of circumstellar disks and planetary systems have fueled a renewed interest in assessing how accurately we can determine the ages for solar-type field dwarfs (Mamajek *et al.* 2008, ; Meyer, this volume). For field stars we can place reasonable constraints on their effective temperatures, luminosities, and metallicities from spectroscopic, photometric, and astrometric measurements. Plotting these observables against theoretical evolutionary tracks allows us to infer ages and masses. However, for main sequence solar-type stars, the observational uncertainties in a star's HRD diagram position can be large enough that they encompass a large fraction of the star's main sequence lifetime, even for stars with precise distances and metallicities (e.g. Nordström *et al.* 2004; Valenti & Fischer 2005; Takeda *et al.* 2007). The situation is worse for the hordes of stars lacking trigonometric parallaxes and metallicity estimates. For this reason, we are motivated to explore alternative age indicators beyond deriving individual isochronal ages.

It has been long appreciated that solar-type stars lose angular momentum via a magnetized wind, spin down, and become less active during their main sequence phase (e.g. Skumanich 1972; Soderblom *et al.* 1991). Here I discuss recent efforts by the author and collaborators to improve the estimation of ages for solar-type (~F7-K2) field dwarfs using rotation/activity diagnostics. By "activity diagnostics", I will discuss two common examples: the Ca II H & K chromospheric activity index $\log R'_{\rm HK}$, and the X-ray-to-bolometric luminosity ratio $\log(L_X/L_{bol})(=\log R_X)$ in the 0.2-2.4 keV band (ROSAT band). For more exhaustive and wavelength-balanced reviews, especially from the perspective of the Sun's evolution, I refer the reader to Güdel (2007), Ayres (1997), and Walter & Barry (1991).

For the Sun, radiometric dating of the oldest meteorites have converged on an age within a few Myr of 4.57 Gyr (e.g. Baker *et al.* 2005). Pleasingly, solar models which match the observed helioseismological constraints (sound speed profiles, acoustic modes) can produce the Sun with an age within a few percent of the meteoritic age (Houdek & Gough 2008). For members of nearby young open clusters (e.g. the Pleiades, Hyades, etc.), detailed modelling of the HR diagram positions of the high-mass members, and HRD positions and Li-depletion pattern of the low-mass members, has led to age-dating with claimed accuracy of ~5-15% (e.g. de Bruijne *et al.* 2001; Barrado y Navascués, *et al.* 2004). Recent results for the small sample of solar-type stars which have been asteroseismologically observed and modeled are also yielding age uncertainties of typically ~5-15% (Thévenin *et al.* 2002; Eggenberger *et al.* 2004), with some claims of even ~1% (Mosser *et al.* 2008). While the observational data for these well-studied clusters and asteroseismological target stars is impeccable, the accuracy of the inferred cluster ages hinge on the input physics (e.g. treatment of convection, opacities, etc.) and abundances of the stellar evolution models, both of which are intimately tied to solar modeling efforts. While the Sun provides us with a "gold" age standard (<few % accuracy) and open clusters and asteroseismological targets provide us "silver" age standards (~15%), how well can we estimate ages for solar-type field dwarfs using "bronze" indicators like rotation and activity?

2. Ages from activity and/or rotation

2.1. *The chromospheric activity-age correlation*

In solar-type stars, the majority of chromospheric and X-ray activity is believed to be generated as a result of the stellar magnetic dynamo. The strength of the dynamo and its ability to nonthermally heat the outer atmospheres of Sun-like stars is ultimately tied to stellar rotation – and more specifically – differential rotation (Noyes *et al.* 1984; Donahue *et al.* 1996). Both activity and rotation among Sun-like stars are observed to decay with isochronal age (e.g. Wilson 1963; Skumanich 1972; Soderblom *et al.* 1991). As an illustration of this, in Fig. 1 (left) we plot chromospheric activity $\log R'_{\rm HK}$ vs. color for the Sun and members of age-dated clusters (Mamajek & Hillenbrand 2008). Using our large modern database of activity and age estimates, we find shortcomings among all previous activity-age relations. The new fit to the cluster and field star activity-age data is shown as a solid line in Fig. 1 (right):

$$\log\tau = -38.053 - 17.912\,\log R'_{HK} - 1.6675\,\log(R'_{HK})^2 \tag{2.1}$$

As is obvious from the leftside figure of Fig. 1, there are color-dependent effects which force us to dismiss a simple activity-age relationship as an oversimplification. Simply using this activity-age polynomial can provide age estimates of ~±0.25 dex or 60% accuracy (1σ; uncertainties come from investigating the scatter in inferred ages among coeval cluster or binary samples), however color-dependent systematic effects will be present. Mamajek & Hillenbrand (2008) were unable to find a simple way to parameterize age as a function of activity and color which simultaneously satisfied the available cluster, binary,

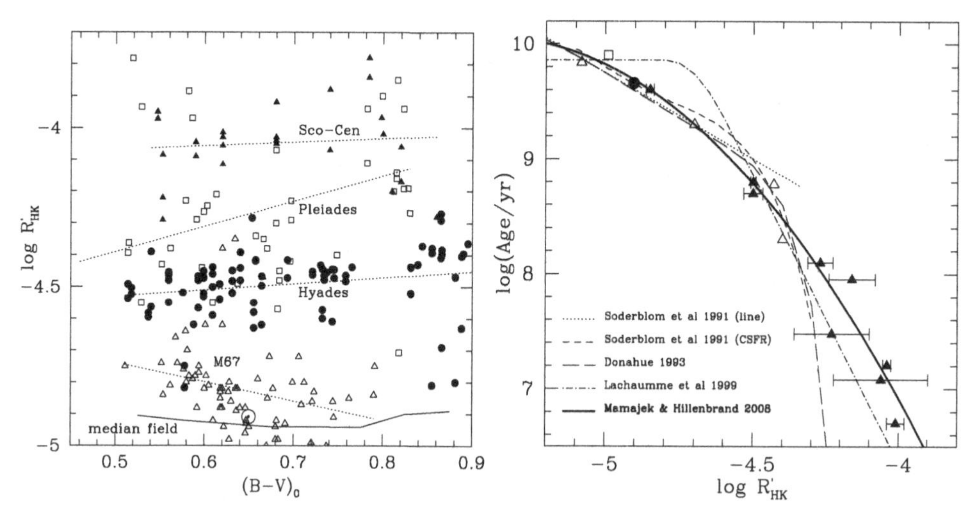

Figure 1. *Left:* Intrinsic B-V color vs. chromospheric activity index $\log R'_{\rm HK}$ for members of age-dated clusters and the Sun (from Mamajek & Hillenbrand 2008). Sco-Cen members are typically ~5-17 Myr old (filled triangles), the Pleiades are ~130 Myr old (open squares), the Hyades are ~625 Myr old (filled circles), and M67 is ~4 Gyr old (open triangles; see Mamajek & Hillenbrand 2008, and references therein). The median $\log R'_{\rm HK}$ value for field dwarfs as a function of $(B-V)_0$ is plotted. *Right:* Color-independent activity-age relationships from previous studies and Mamajek & Hillenbrand (2008). The best fit to the cluster data and a sample of isochronally-dated older field dwarfs is $\log(\tau/{\rm yr}) = -38.053 - 17.912\log R'_{\rm HK} - 1.6675\log R'^{\,2}_{\rm HK}$ (solid line). However cluster and binary $\log R'_{\rm HK}$ data suggests that one needs to allow for color-dependence. This is best taken into account by converting $\log R'_{\rm HK} \rightarrow$ rotation $\rightarrow$ age via the Rossby number and a gyrochronology relation.

and field star datasets. But there is hope in the form of the activity-rotation correlation (e.g. Noyes *et al.* 1984) and the gyrochronology relations (Barnes 2007).

2.2. *Ages from chromospheric activity via rotation*

Theoretical models exist for explaining the decay of rotation speeds among solar-type stars due to angular momentum loss via magnetized winds and changes in the moment of inertia of the star (Kawaler 1988). Some of the model parameters are poorly constrained, (e.g. mass loss, magnetic field geometry), but large rotation period datasets for clusters can be used to constrain the parameters (Irwin, this volume). Empirically, however, one can fit a series of simple curves in color-period-age space. These "gyrochronology" curves introduced by Barnes (2007) and improved upon by Mamajek & Hillenbrand (2008) can be used to derive ages from rotation rates with statistical accuracy of order ~15%. For the solar-type dwarfs, Mamajek & Hillenbrand (2008) fit a gyrochronology relation for period P in days and age t in Myr:

$$P(B-V,t) = (0.407\pm0.021)[(B-V)_o - 0.495\pm0.010]^{0.325\pm0.024}\,(t/{\rm Myr})^{0.566\pm0.008} \quad (2.2)$$

Rotation rate can be tied to dynamo strength via the Rossby number ($\dot{R}_o$), which is observationally defined as the rotation period divided by an estimate of the local convective turnover time just above the convective-radiative boundary (τ_c; e.g. Noyes *et al.* 1984). Using the best available data for solar-type dwarfs, Mamajek & Hillenbrand (2008) find the strong correlation between the Rossby number and chromospheric activity for "normal" and "inactive" ~F7-K2 dwarfs ($\log R'_{\rm HK} < -4.35$) to be:

$$\log R'_{HK} = -4.522 - 0.337(R_o - 0.814) \quad (2.3)$$

Through applying the conversion activity → rotation → age via the Rossby number and gyrochronology, an analysis of (presumably) coeval stars in resolved binaries and star clusters suggests that the derived ages have precision of $\sim\pm0.1$-0.2 dex ($\sim$25-50%; 1σ). By combining a gyrochronology relation with the $\log R'_{\rm HK}$-Rossby number correlation, one can predict chromospheric activity as a function of color and age for solar-type dwarfs ("gyrochromochrones"; Fig. 2). When combining the activity vs. rotation and rotation vs. age relations, it becomes apparent that for a given chromospheric activity level $\log R'_{\rm HK}$, the late F-type and early G-type stars are systematically younger than the late G-type and early K-type stars. In the future, it will be prudent to take into account the effects of metallicity in Fig. 2, and produce isochrones in mass-metallicity-activity.

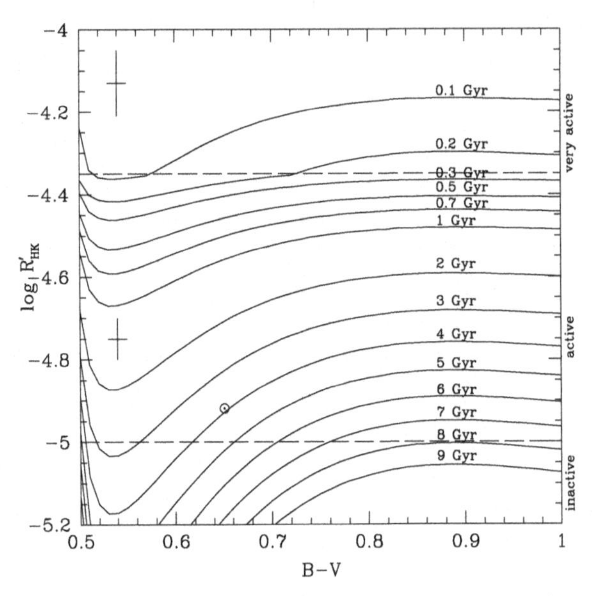

Figure 2. Predicted chromospheric activity levels as a function of age ("gyrochromochrones"), from combining the age-rotation relations (Eqn. 2.2) with the rotation-activity relations (Eqn. 2.3 for age >100 Myr). Typical uncertainty bars are shown in the very active and active regimes, reflecting the r.m.s. in the Rossby number-activity fits, and typical photometric errors. The behavior of the gyrochromochrones at the blue end (i.e. the obvious upturn) is not well-constrained, and is particularly sensitive to the c parameter in the gyrochronology fits. Figure from Mamajek & Hillenbrand (2008).

2.3. *Coronal X-rays as an age indicator*

It is clear that one can estimate useful stellar ages for solar-type dwarfs from rotation periods (e.g. Barnes 2007), however rotation periods are currently difficult to measure in large numbers for all but the most active, starspotted stars (although periods can be inferred from long-term monitoring of chromospheric emission; Donahue *et al.* 1996). For these older stars lacking rotation periods, of order $\sim10^{3.5}$ solar-type field dwarfs have chromospheric activity measurements (predominantly from large surveys by e.g. Henry *et al.* 1996; Wright *et al.* 2004; Gray *et al.* 2006) to help us estimate their ages. However, another larger, and mostly untapped, stellar activity database exists for age estimation: X-ray fluxes.

Using *ROSAT* soft X-ray (0.2-2.4 keV) fluxes, Sterzik & Schmitt (1997) demonstrated that coronal X-ray activity ($\log(L_{\rm X}/L_{bol}) = \log R_{\rm X}$) scales with chromospheric activity $\log R'_{\rm HK}$ over $\sim$4 orders of magnitude in $\log R_{\rm X}$ and $\sim$1 order of magnitude in $\log R'_{\rm HK}$.

Mamajek & Hillenbrand (2008) improved the correlation through including more high- and low-activity stars, and provided an improved quantification of this correlation for solar-type dwarfs:

$$\log R'_{HK} = (-4.54 \pm 0.01) + (0.289 \pm 0.015)\,(\log R_X + 4.92) \qquad (2.4)$$

with an r.m.s. scatter of 0.06 in $\log R'_{\rm HK}$. The inverse relation is:

$$\log R_X = (-4.90 \pm 0.04) + (3.46 \pm 0.18)\,(\log R'_{HK} + 4.53) \qquad (2.5)$$

with an r.m.s. of 0.19 dex (~55%) in $\log R_{\rm X}$. Equation 2.5 is statistically consistent with the relation found by Sterzik & Schmitt (1997), but our uncertainties are ~2× smaller.

As with the chromospheric activity, their is a strong correlation between coronal X-ray activity and rotation via the Rossby number:

$$R_o = (0.86 \pm 0.02) - (0.79 \pm 0.05)\,(\log R_X + 4.83) \qquad (2.6)$$

This best fit appears to be useful over ~4 orders of magnitude in $\log R_{\rm X}$ ($-7 < \log R_{\rm X} < -4$). The fit is similar to the linear-log fit quoted by Hempelmann *et al.* (1995), but is severely at odds with the oft-cited $\log R_{\rm X}$ vs. $\log R_o$ fit quoted by Randich *et al.* (1996). Hence, one can relate X-ray fluxes to rotation periods via the Rossby number (typically ~0.25 1σ accuracy in R_o), and estimate ages from the periods via a gyrochronology relation. The typical spread in $\log R_{\rm X}$ as a function of age is ± 0.4 dex (1σ) and should be factored into the age uncertainty. It appears that a few hundred second X-ray snapshot with an X-ray satellite can be used to predict the multi-decadal average value of $\log R'_{\rm HK}$ to within ± 0.1 (1σ) accuracy (minimum age precision ~30-50%). A star with X-ray emission similar to that of the Sun can be seen out to ~15 pc in the ROSAT All-Sky Survey, and younger, more active stars can be seen to larger distances. Using X-ray emission measured by the ROSAT All-Sky Survey, one should (in principle) be able to derive useful ages for $\sim 10^{2-3}$ solar-type field dwarfs in the solar neighborhood.

The very active stars ($\log R'_{\rm HK} > -4.35$; $\log R_{\rm X} > -4.0$) have negligible correlation between rotation and activity (i.e. the "saturated" regime). So while setting upper limits to ages might be fruitful in the saturated regime, quoting exact ages appears not to be.

3. Implications for nearby Sun-like field dwarfs

What we have constructed are useful empirical relations between stellar measurements, which when combined, yield a parameter more difficult to measure: age. In future work, we would like to understand the physics underlying these empirical relations in terms of how it constrains stellar dynamo theory (e.g. Montesinos *et al.* 2001) and the evolution of stellar angular momentum (e.g. Kawaler 1988). For the time being, let us use our new and improved rotation-activity-age tools to see what their implications are.

We have already seen that the evolution of activity appears to be fairly color/mass-dependent among solar-type dwarfs (Fig. 2), contrary to previous studies which employed a color/mass-independent activity-age correlation. As a first use of our activity → rotation → age calibrations, we constructed a histogram of the chromospheric activity-derived ages for a volume-limited ($d < 16$ pc) sample of the nearest 108 solar-type dwarfs to the Sun†. A table of the names, parallaxes, B-V colors, $\log R'_{\rm HK}$ values, absolute magnitudes, spectral types, and inferred ages for the sample stars is given in Table 13 of Mamajek

† The ages for these nearest solar-type dwarfs will be of astrobiological interest for proposed missions designed to image and take spectra of extrasolar terrestrial planets, like the TPF and Darwin (Kaltenegger *et al.* 2007) and New Worlds Observer (Cash *et al.* 2005)

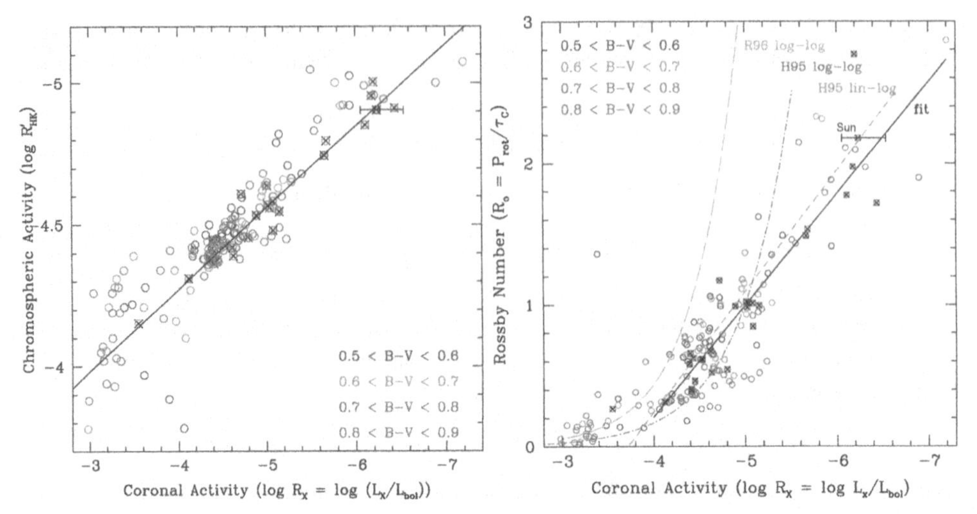

Figure 3. *Left:* $\log R_{\rm X}$ vs. $\log R'_{\rm HK}$ for solar-type dwarfs with known rotation periods and chromospheric and X-ray activity levels (from Mamajek & Hillenbrand 2008). Stars from Donahue *et al.* (1996) and Baliunas *et al.* (1996) with well-determined periods also have dark Xs and conveniently provide an X-ray unbiased sample via the *ROSAT* All-Sky Survey. Shaded color bins are illustrated in the legend. The Solar X-ray and $\log R'_{\rm HK}$ datum is described in Mamajek & Hillenbrand (2008). *Right:* $\log R_{\rm X}$ vs. Rossby number R_o for stars in our sample of solar-type stars with known rotation periods and chromospheric and X-ray activity levels (from Mamajek & Hillenbrand 2008). Donahue-Baliunas stars with well-determined periods also have dark Xs. Previously published $R_{\rm X}$ vs. R_o fits are drawn: *cyan long-dashed line* is a log-log fit from Randich *et al.* (1996), *magenta dot-dashed line* is a log-log fit from Hempelmann *et al.* (1995), and the *green dashed line* is a linear-log fit from Hempelmann *et al.* (1995). Our new log-linear fit for stars in the range $-7 < \log R_{\rm X} < -4$ is the *solid dark line*, consistent with the Hempelmann linear-log relation. Saturated X-ray emission ($\log R_{\rm X} > -4$) is consistent with $R_o < 0.5$.

& Hillenbrand (2008). We plot the fruits of this effort as a histogram of inferred ages in Figure 4. The histogram can not be directly interpreted as a "star-formation history" at this time, as we have not accounted for the effects of kinematic disk-heating, the loss of some older stars due to stellar evolution (given the constraint that the "dwarf" stars must lie within 1 mag of the main sequence), and the effects of metallicity on the sample. As these effects will mostly conspire to skew our conclusions regarding the old end of the histogram (i.e. evolved and/or metal-poor and/or high-velocity stars), we focus on the stars younger than the Sun. First, we note that when a simple activity → age relation is adopted, one sees a pronounced dip in the age histogram at age ~2-3 Gyr, right in the region of the "Vaughan-Preston gap" (Vaughan & Preston 1980). One one derives age using the recommended activity → rotation → age (via the Rossby number and revised gyrochronology relations), one gets a much flatter age distribution between 0-6 Gyr. As our color-magnitude selection biases should have negligible impact on the age distribution of these young to middle-aged dwarfs, it appears that the histogram is *consistent with a more-or-less flat star-formation history over the past ~5 Gyr or so.* The histogram is in disagreement with assertions in previous historical studies (e.g. Barry 1988, which used an activity-age relation) which concluded that there has been a recent enhancement of the stellar birth-rate in the past ~Gyr, which followed a lower birth-rate ~2-3 Gyr ago. Applying our age-estimation methods to a larger sample of the nearest solar-type stars out to ~25-40 pc should place our conclusions on firmer statistical footing.

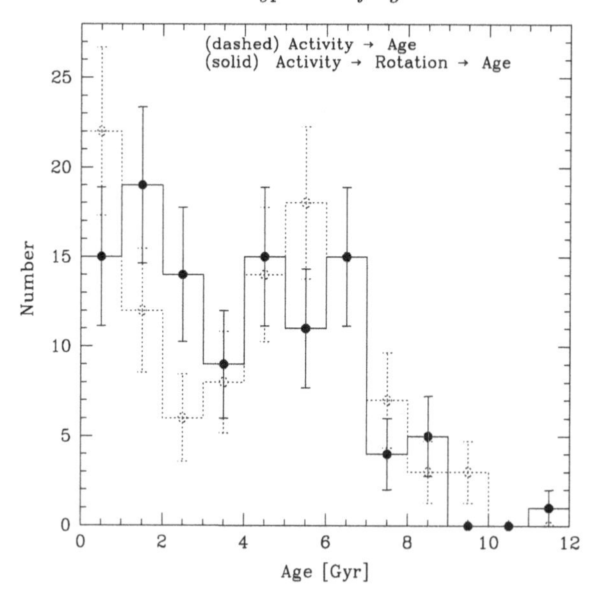

Figure 4. Histogram of inferred ages for the nearest 108 solar-type dwarfs (F7-K2V) within 16 pc (from Mamajek & Hillenbrand 2008). *Dashed histogram* is for ages inferred directly from chromospheric activity using equation 3. *Solid histogram* is for ages derived from converting activity $\log R'_{\rm HK}$ to rotation period, then converting rotation period and color to age via the Rossby number and revised gyro relation.

These techniques should also be prove valuable for more accurately assessing the ages of extrasolar planetary systems (Mamajek & Slipski, in prep.) and dusty debris disk systems (Hillenbrand *et al.*, in prep.).

Acknowledgements

The author collaborated with Lynne Hillenbrand on much of the material presented for this talk (presented in Mamajek & Hillenbrand 2008), and acknowledges helpful conversations with D. Soderblom, J. Stauffer, and M. R. Meyer.

References

Ayres, T. R. 1997, *J*nl. Geophys. Res., 102, 1641
Baker, J., Bizzarro, M., Wittig, N., Connelly, J., & Haack, H. 2005, *N*ature, 436, 1127
Baliunas, S., Sokoloff, D., & Soon, W. 1996, *A*pJ, 457, L99
Barnes, S. A. 2007, *A*pJ, 669, 1167
Barrado y Navascués, D., Stauffer, J. R., & Jayawardhana, R. 2004, *A*pJ, 614, 386
Barry, D. C. 1988, *A*pJ, 334, 436
de Bruijne, J. H. J., Hoogerwerf, R., & de Zeeuw, P. T. 2001, *A*&A, 367, 111
Cash, W., Kasdin, J., Seager, S., & Arenberg, J. 2005, *P*roc. SPIE, 5899, 274
Donahue, R. A., Saar, S. H., & Baliunas, S. L. 1996, *A*pJ, 466, 384
Eggenberger, P., *et al.* 2004, *A*&A, 417, 235
Gray, R. O., *et al.* 2006, *A*J, 132, 161
Güdel, M. 2007, Living Reviews in Solar Physics, 4, 3
Hempelmann, A., *et al.* 1995, *A*&A, 294, 515
Henry, T. J., Soderblom, D. R., Donahue, R. A., & Baliunas, S. L. 1996, *A*J, 111, 439
Houdek, G. & Gough, D. O. 2008, IAU Symposium, 252, 149
Kaltenegger, L., Traub, W. A., & Jucks, K. W. 2007, *A*pJ, 658, 598
Kawaler, S. D. 1988, *A*pJ, 333, 236

Mamajek, E. E., Barrado y Navascués, D., Randich, S., Jensen, E. L. N., Young, P. A., Miglio, A., & Barnes, S. A. 2008, 14th Cambridge Workshop on Cool Stars, Stellar Systems, and the Sun, 384, 374
Mamajek, E. E. & Hillenbrand, L. A. 2008, *ApJ*, 687, 1264
Montesinos, B., Thomas, J. H., Ventura, P., & Mazzitelli, I. 2001, *MNRAS*, 326, 877
Mosser, B., *et al.* 2008, *A&A*, 488, 635
Nordström, B., *et al.* 2004, *A&A*, 418, 989
Noyes, R. W., *et al.* 1984, *ApJ*, 279, 763
Randich, S., Schmitt, J. H. M. M., Prosser, C. F., & Stauffer, J. R. 1996, *A&A*, 305, 785
Skumanich, A. 1972, *ApJ*, 171, 565
Soderblom, D. R., Duncan, D. K., & Johnson, D. R. H. 1991, *ApJ*, 375, 722
Sterzik, M. F. & Schmitt, J. H. M. M. 1997, *AJ*, 114, 1673
Takeda, G., *et al.* 2007, *ApJS*, 168, 297
Thévenin, F., *et al.* 2002, *A&A*, 392, L9
Valenti, J. A. & Fischer, D. A. 2005, *ApJS*, 159, 141 (VF05)
Vaughan, A. H. & Preston, G. W. 1980, *PASP*, 92, 385
Walter, F. M. & Barry, D. C. 1991, The Sun in Time, 633
Wilson, O. C. 1963, *ApJ*, 138, 832
Wright, J. T., Marcy, G. W., Butler, R. P., & Vogt, S. S. 2004, *ApJS*, 152, 261

Discussion

D. SODERBLOM: First, maybe it's a coincidence, but the line for Rossby number = 1 runs right down the middle of the "Vaughan-Preston gap," and that may explain the build-up of stars at $\log R'_{\rm HK} = -4.5$. Second, have you compared activity ages to isochrones ages? When I do I see zero correlation.

E. MAMAJEK: Rossby number = 1 depends on the choice of convective overturn time, which differs by a factor of a few among the models. I adopted those from Noyes *et al.* (1984). The scatter between ages from activity or gyrochronology and those from isochrones (Valenti & Fischer 2005) is large; however, Valenti has suggested that Takeda *et al.* (2007) ages are to be preferred, but I have not yet compared to those.

F. WALTER: The Rossby number is a convenient way to sweep a lot of our ignorance about convection under the rug (or at least into a single parameter). How well do we really understand convective turnover timescales in convective stars, and how might this uncertainty affect the details of your activity-age relations?

E. MAMAJEK: Modelers have evaluated convective turnover times differently – primarily at different depths with respect to the base of the convection zone – but there appears to broad agreement in the *relative* turnover times for main sequence stars as a function of mass. Combining the theoretical turnover times with the observed rotation and activity data shows that a strong correlation between Rossby number and activity exists over a wide range of parameter space for solar-type dwarfs when one uses MLT models with $\alpha = 1.9$ (see Noyes *et al.* 1984 and Montesinos *et al.* 2001). Deriving rough ages for solar-type dwarfs using activity is then supported by two empirical correlations: the activity-rotation relation (via the Rossby number) and the rotation-age relation (gyrochronology; see talk by S. Barnes).

The Ages of Stars
Proceedings IAU Symposium No. 258, 2008
E.E. Mamajek, D.R. Soderblom & R.F.G. Wyse, eds.

doi:10.1017/S1743921309032049

Isochrones for late-type stars

Pierre Demarque
Department of Astronomy, Yale University
P.O. Box 208101, New Haven , CT 06520-8101, USA
email: pierre.demarque@yale.edu

Abstract. A brief summary of the history of stellar evolution theory and the use of isochrones is given. The present state of the subject is summarized. The major uncertainties in isochrone construction are considered: chemical abundances and color calibrations, and the treatment of turbulent convection in stellar interior and atmosphere models. The treatment of convection affects the modeling of stellar interiors principally in two ways: convective core overshoot which increases evolutionary lifetimes, and the depth of convection zones which determines theoretical radii. Turbulence also modifies atmospheric structure and dynamics, and the derivation of stellar abundances. The symbiosis of seismic techniques with increasingly more realistic three-dimensional radiation hydrodynamics simulations is transforming the study of late-type stars. The important case of very low mass stars, which are fully convective, is briefly visited.

Keywords. Convection, turbulence, stars: abundances, stars: evolution, stars: late-type, stars: low-mass; stars: oscillations

1. Introduction

The determination of stellar ages has been a primary motivation of the study of stellar structure beginning in the late 19th century. Early estimate of stellar ages had relied on gravitational contraction and the release of gravitational energy (the Kelvin-Helmholtz timescale), leading to an apparent conflict with the much longer geological timescale estimates. Although the possible role of nuclear fusion in stellar energy generation was recognized early in the 20th century, the development of wave mechanics and nuclear physics in the 1930's provided the solution to the origin of stellar energy, culminating in the work of Bethe (1939).

In this talk, I wish to focus on some of the uncertainties met in comparing the observational data to the models. There has been much effort in understanding source of uncertainties in fitting the data to isochrones that are considered more or less free of errors. I wish to emphasize the role played by present uncertainties in the underlying physics of stellar evolution. The story of stellar dating parallels that of our understanding the physics of stellar interiors and atmospheres, and future progress will depend on advances in modeling and in new observational tools, such as high precision abundance determinations, parallaxes, interferometric angular diameters and astero-seismology. The reliability of the next generations of isochrones will depend on how well we understand the physics of cool stars.

Early applications of stellar models to the Hertzsprung-Russell diagrams (HRD) of star clusters illustrate this point, drawing an example taken from Chapter 7 in Chandrasekhar's monograph (1939). The author describes a theory that explains misleadingly the differences in the HRD of star clusters of different ages to their internal helium abundance, assumed to be uniform within stars. The major breakthrough came from Öpik (1938) who realized that stars in general do not evolve mixed, and that they develop helium rich cores which make them evolve into giant stars. The concept of the main sequence turnoff was born, and with the subsequent discovery of the Schönberg-Chandrasekhar (1942) limit at core exhaustion. Stars like the Sun were then believed to have a convective core (Cowling 1936). Quantitative estimate of stellar ages at the main sequence turnoff became possible. The evolution of globular cluster stars was outlined in physical terms and sketched as a sequence of static models in the HRD by Hoyle & Schwarschild (1955). The authors derived a preliminary age of 6.2 Gyr for the globular cluster M3.

The first automatic calculations of stellar evolution are due to Haselgrove & Hoyle (1956a). They derived an approximate age of 6.5 Gyr for M3 on the basis of 16 evolutionary models (1956b). The construction of grids of stellar evolutionary tracks, from which isochrones are constructed, came in the early and mid-1960's in papers by (Demarque & Larson 1964; Kippenhahn *et al.* 1967; Iben 1967) and many others since then.

2. The last twenty years

As emphasized in the introduction, recent improvements are the result of advances in input physics motivated, for late-type stars, in major part by studies of the solar interior stimulated by solar neutrino research, and by helioseismology. Indeed helioseismology, by probing the run of sound speed in the core of the Sun, played a key role in the "resolution" of the classical solar neutrino problem (Bahcall *et al.* 1995, 2001, 2005).

2.1. *Input physics*

Giant strides have been made in the input physics for late-type stars in the last twenty years in the following areas, which are crucial in isochrone construction:

Nuclear reactions. Progress was motivated principally by the solar neutrino research.

Interior opacities. Due to the remarkable agreement of several independent projects (Los Alamos, OPAL, the OPACITY Project), using different conceptual and numerical approaches (Iglesias & Rogers 1996; Rogers & Iglesias 1994; Seaton & Badnell 2004; Seaton 2007), confidence in radiative opacities for the stellar interior has greatly increased.

Low temperature opacities. Low temparature opacities are particularly critical in cool stars. Calculations have been performed by Kurucz (1995), Alexander and Ferguson (1994), Ferguson *et al.* (2005) and OPAL, extending the range and density of the opacity grids.

Equation of state In both opacities and the equation of state (e.g. OPAL, Rogers *et al.* 1996), but primarily in the equation of state, because it affects the calculation of the speed of sound most directly, helioseismology has played a central role in stimulating progress.

2.2. *The role of element diffusion*

Helioseismology has also demonstrated the role played by element diffusion in the solar interior, both in the layers just below the convection zone, and in the solar core. As a result, the effects of diffusion are normally included in the construction of the so-called standard solar model (SSM) (Bahcall *et al.* 1995; Guenther & Demarque 1997).

Helium and heavy element diffusion are known to affect isochrone ages (Chaboyer *et al.* 1992), particularly in globular cluster stars which exhibit thin convection zone near the main sequence turnoff due to their low metallicities. The efficiency of diffusion can be inhibited by the presence of turbulence or other internal mixing processes (Chaboyer *et al.* 1995). Diffusion also plays an important role in the interpretation of lithium abundance data (Deliyannis *et al.* 1990). The effects of helium diffusion are included in the Y^2 isochrones (Yi *et al.* 2001; Kim *et al.* 2002; Demarque *et al.* 2004).

2.3. *Theoretical isochrones and their applications*

There are now many sources of theoretical isochrones. The reader is referred to proceedings of IAU Symposium 252, held earlier this year in Sanya, China, for a more extensive description of isochrone grids available in the literature. Increasingly, grids of isochrones are being developed for special research purposes. A recent example is the version version 3 of the Yonsei-Yale (Y^2) isochrones, described by Yi *et al.* (2008) and soon to be posted on the web, which includes a larger range of helium abundances and improved equation of state in the low mass stars.

Isochrones are used in the interpretation of color-magnitudes of resolved stellar systems and clusters (Gallart *et al.* 2008; Lee *et al.* 2005; Norris 2004; Piotto *et al.* 2005; Sarajedini *et al.* 2007; Stetson *et al.* 2004), unresolved systems (Bruzual & Charlot 2003; Yi 2003) and field stars (Pont & Eyer 2005).

3. Main uncertainties in isochrone construction

While enormous progress has been achieved in the constitutive microscopic physics of stellar models for late-type stars, serious uncertainties still remain in a number of areas central to the construction of isochrones. I discuss here briefly three main uncertainties. These uncertainties are (1) convective core overshoot, (2) the depth of convection zones and the mixing length theory, and (3) atmospheric structure and dynamics, and the effect on abundance determinations and color transformations, which are known to be particularly deficient for cool stars, where atmospheric convection is known to dominate. All three of these uncertainties are primarily affected by our inadequate knowledge of turbulent convection.

3.1. *Isochrones and convective core overshoot*

Convective core overshoot beyond the boundary of the convective core set by the Schwarzschild criterion (1906) lengthens the core hydrogen burning phase in stellar evolutionary tracks. On isochrones, the shape of the turnoff region is changed. Figure 1 shows the effect. The resulting turnoff luminosity function is very different. Evolutionary timescales both near the main sequence and on the giant branch are also modified. The consequence of core overshoot for the integrated spectral energy distribution of a simple stellar population (SSP) has been discussed by Yi (2003).

3.2. *Isochrones and the mixing length uncertainties*

On an isochrone, the radii of all stellar models with a convective envelope is affected by the choice of the mixing length parameter α, a free parameter usually calibrated on solar models. Primarily for simplicity and convenience, theoretical isochrones are constructed using a constant vale of α during evolution. There is really neither theoretical nor empirical justification for assuming that α should remain constant along an evolutionary track (Demarque *et al.* 1992; Robinson *et al.* 2004). In fact, there is evidence that the effective α is a function of both position in the HRD and depth within stars. The sensitivity of the giant branch position to the choice of α within the framework of the MLT is illustrated in Figure 2. The effect on the spectral energy distribution of the corresponding SSP has also been discussed by Yi (2003).

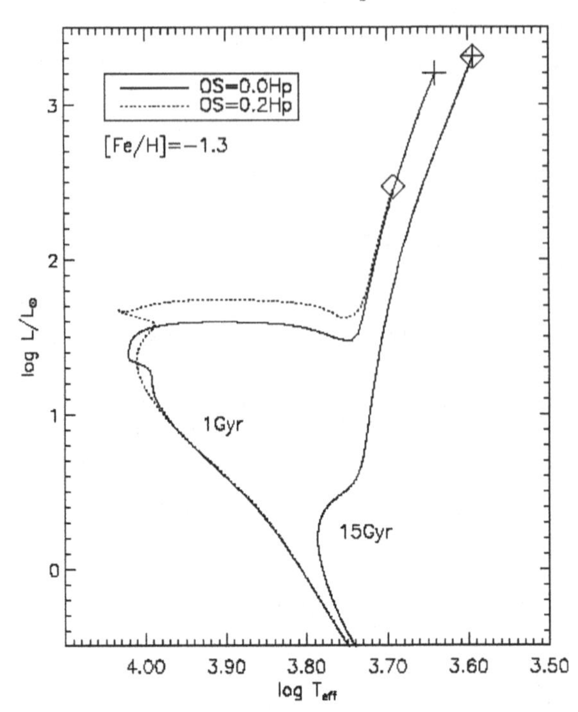

Figure 1. Convective core overshoot (OS) affects the shape of the isochrone and the hydrogen burning lifetime (Yi 2003).

3.3. *Modeling atmospheric structure and dynamics*

The crucial importance of a physically consistent treatment of turbulence in the atmospheres of late-type stars is illustrated by the recent revision of solar abundances, based on the use of a 3D simulation of the solar atmosphere (Asplund *et al.* 2005). A detailed discussion of the implications of this revision for stellar physics and helioseismology is given by Basu & Antia (2008).

Colors are also sensitive to atmospheric structure and dynamics; it is well known that present calibrations are very unreliable for cool stars (see e.g. Yi 2003). An extensive discussion of the differences between color calibrations by several authors can be found in VandenBerg & Clem (2003).

4. New research tools

Next generation isochrones will require improvements in the description of turbulent convection, and turbulence in regions which are formally in radiative equilibrium (according to the Schwarzschild criterion), such as overshoot regions.

More physically consistent turbulence in model stellar atmospheres is needed for high precision abundance estimates. A more realistic treatment of core overshoot is needed to derive reliable ages for intermediate mass and massive stars. In particular, reliable spectral dating of distant galaxies depends on this information (Yi *et al.* 2000). The depth of convection zones and the precise position of the giant branch for different chemical compositions depend on a description of convection to replace the MLT. In the case of fully convective stars at the low mass end, the derivation of the M-L-R relation is determined by physical conditions near the surface (Limber 1958; Allard *et al.* 1997; Chabrier & Baraffe 2000).

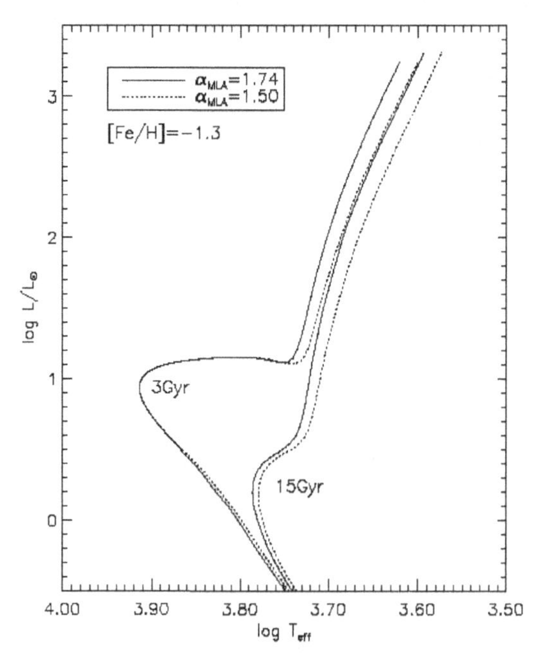

Figure 2. The mixing length approximation parameter α affects the shape of isochrones. A larger value of α makes the position of the giant branch bluer (Yi 2003).

Two new research tools are now available to study the effects of turbulent convection in stars. On the theoretical side, it is now possible to carry out physically realistic three-dimensional simulations of radiation hydrodynamics (3D RHD) (Chan & Sofia 1989; Nordlund & Dravins 1990; Kim *et al.* 1995; Stein & Nordlund 1998; Kim & Chan 1998; Robinson *et al.* 2003, 2004). More recently, simulations for metal-poor dwarf stars have also been carried out (Jung *et al.* 2007; Ludwig *et al.* 2008). On the observational side, seismology offers radically new ways to probe the deep interior of stars, by means of the WIRE, MOST, CoRoT and Kepler space missions, and of ground-based networks such as SONG. Seismic techniques are particularly well suited to determine precisely the position of radiative-convective interfaces and chemical composition discontinuities (Basu & Antia 1995).

4.1. *Some of what we can learn from 3D RHD*

There is a wide variety of basic stellar physics problems that can be addressed using three dimensional radiation hydrodynamics simulations. Some of the most relevant are listed below:

- Physically realistic surface boundary conditions for late-type stars (the entropy jump and the convection zone depth).
- Improved model atmospheres for abundance determinations and color calibrations.
- Effects of turbulence on p-mode frequency shifts in stellar outer layers
- The excitation of p-modes (Samadi *et al.* 2007).
- viscous dissipation due to turbulence in stellar convection zones(e.g. orbital circularization in binary stars and exoplanet orbital decay) (Zahn & Bouchet 1989; Penev *et al.* 2007).
- The properties of convection, which when combined with rotation, drive magnetic activity in late-type stars.
- The formation of shocks leading to mass loss.

4.2. *Parameterization of turbulent convection*

One of the challenge is the parameterization of 3D simulations to 1D and its implementation into 1D stellar models (Lydon *et al.* 1992). From the 3D simulations, one can derive the following pieces of information:

- The form of the $T(\tau)$ relation in the atmosphere.
- The specific entropy in the deep convection zone. This quantity determines the depth of the convection zone, and the position of the tachocline at the interface between the convection zone and the radiative envelope.
- The specific entropy is determined by the structure of the outer convection zone (CZ) and of the highly superadiabatic layer (SAL). In the language of the mixing length theory (MLT), α is the parameter that is used to tweak the specific entropy.
- The sound speed (and the p-mode frequencies) are sensitively affected by the outer envelope and atmospheric structure.
- The parameterization should describe as faithfully as possible the turbulent velocity field to which absorption line profiles are sensitive. The simulation is thus capable of providing a measure of the microturbulence, which is a free parameter in conventional 1D stellar atmosphere modeling.

An example of simple parameterization of turbulence is given in the next section.

4.3. *Including turbulence in stellar models*

A method to incorporate the effects of turbulence into the outer layers of one-dimensional (1D) stellar models has been implemented by Li *et al.* (2002) in the stellar evolution code YREC (Demarque *et al.* 2008). The approach is based on a formalism due to Lydon & Sofia (1995). It requires a detailed three-dimensional hydrodynamical simulation (3D RHD) of the atmosphere and highly superadiabatic layer of the star (Robinson *et al.* 2003). The basic idea is to extract from the velocity field of the 3D simulation three important quantities: the turbulent pressure, the turbulent kinetic energy and the anisotropy of the flow. Implementation into a 1D stellar model thus requires two additional parameters, i.e. χ, the specific turbulent kinetic energy, and γ, which reflects the flow anisotropy. These parameters, which modify the hydrostatic equilibrium equation and the internal energy equation, must be introduced in a thermodynamically self-consistent way. As a result, they also change the adiabatic and convective temperature gradients, as well as the energy conservation equation.

Figure 3 illustrates the effect on p-mode frequencies of incorporating parameterized turbulence into the standard solar model (SSM). An application to the subgiant η Boo is shown in Figure 4 where the resulting echelle diagram is shown (Straka *et al.* 2006).

4.4. *What we can learn from seismology*

Seismology provides uniquely sensitive probes of stellar interiors, primarily using the properties of acoustic (p-modes) known to be excited in the Sun and late-type stars. In particular, seismology provides:

- Age estimates from a consideration of **small** and **large spacings** of the observed p-modes (Christensen-Dalsgaard 1988)(the C-D diagram);
- Detailed information about **convective core** size and **core overshoot** (Straka *et al.* 2005; Mazumdar *et al.* 2006);
- A measure of the **convection zone depth** (Christensen-Dalsgaard *et al.* 1991);
- Sensitive tests of the **structure** of the outer convective layers (Rosenthal *et al.* 1999);
- Precise estimates of the **envelope helium abundance** (Basu & Antia 1995).

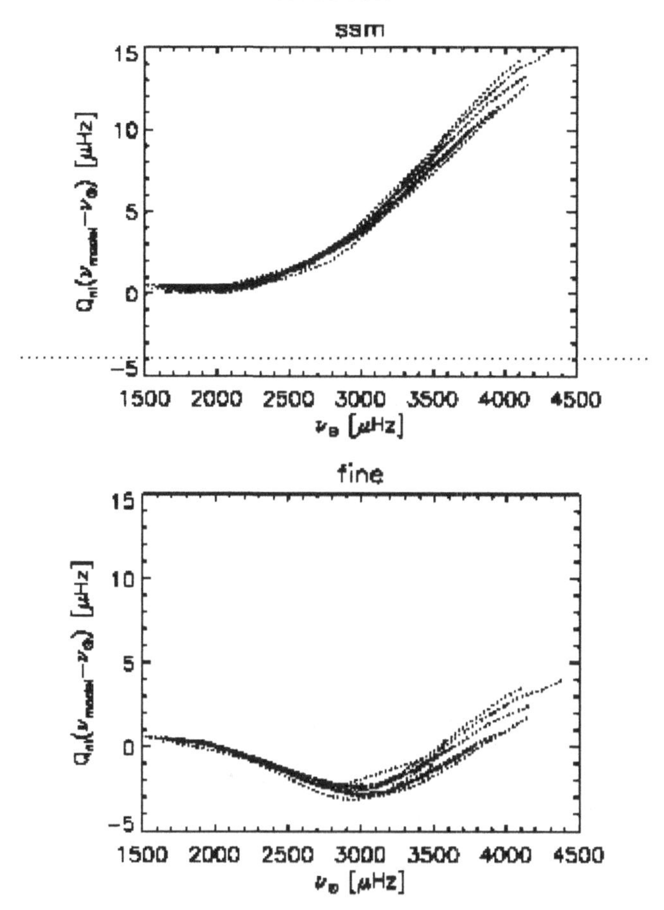

Figure 3. Difference between observed and computed p-mode frequencies for a SSM without (top) and with (bottom) turbulence.

One must emphasize here that seismic techniques are **complementary** to other observational techniques, and cannot replace them. For sun-like stars, seismic techniques are most powerful when used in conjunction with other observations, i.e. when estimates of mass, metallicity, parallax, angular diameter, are also available.

5. An extreme case: the very low mass stars

The very low mass stars present a challenging problem. Because they have fully convective interiors, they obey a mass-luminosity-radius (M-L-R) relation, and their central conditions depend sensitively on their surface boundary conditions Limber 1958). As a result, the important independent age indicator, the lithium depletion boundary (LDB) (Burke *et al.* 2004 is sensitive to details of the atmospheric structure. The 3D RHD simulations by Ludwig (2006) indicate that the chemical equilibrium is complex. Interior models yield masses that do not agree with observations of binary stars by Henry (2004) and Hillenbrand & White (2004).

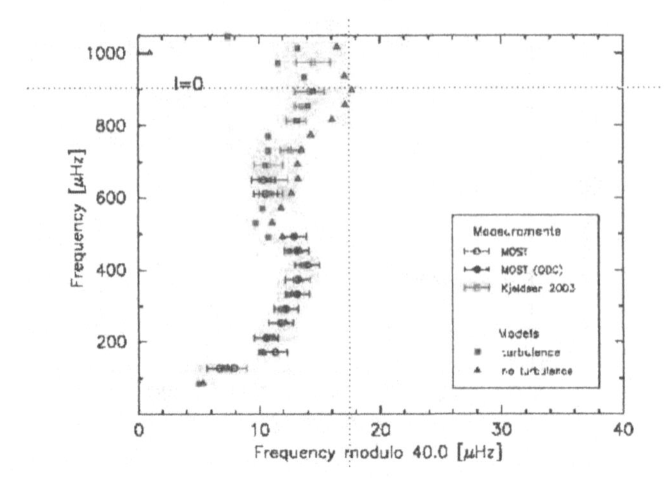

Figure 4. Echelle diagram for a calibrated model of the subgiant η Boo with and without the effects of turbulence. The data are taken from Kjeldsen *et al.* (2003) and the MOST satellite at the lowest frequencies (Guenther 2007).

6. Future generation isochrones

This review focuses on some major uncertainties in the construction of stellar evolutionary tracks and isochrones for late-type stars.

More specifically, these uncertainties are (1) the color transformations used to convert the position of stars in the theoretical HRD to their positions in the observational plane; (2) convective core overshoot which principally affects evolutionary lifetimes near the main sequence, but also advanced evolutionary phases; and (3) the treatment of convection in the outer layers of the convection zone. All three of these uncertainties have at their root our insufficient understanding of turbulent convection and its interaction with radiation.

In the extreme case of very low mass stars, which are fully convective, the treatment of convection in the outer layers affect sensitively the M-L-R relation. As a result, masses derived from theoretical isochrones are extremely uncertain. The position of the lithium depletion boundary is also affected.

The theoretical and observational tools are now available to address these problems. On the theoretical side, grids of physically realistic numerical 3D simulations of radiation hydrodynamics can now be constructed. From these, physically realistic model atmospheres will yield more accurate abundances and calibrated spectral energy distributions. At the same time appropriate parameterizations are being devised that will create the stellar evolutionary models for future generation isochrones. On the observational side, the advent of seismic techniques, both space based or in ground-based networks of telescopes distributed around the globe, are beginning to provide the necessary tests of the models. We note that in order to be effective, asteroseismic data will have to be analyzed in conjunction with the high precision astrometric observations and spectroscopic data.

Acknowledgements

I am especially grateful to S. Yi for his communications in the preparation of this paper.

References

Alexander, D. R. & Ferguson, J. W. 1994, *ApJ*, 437, 879

Allard, F., Hauschildt, P. H., Alexander, D. R., & Starrfield, S. 1997, *ARA&A*, 35, 137

Asplund, M., Grevesse, N., & Sauval, A. J. 2005, *ASPC*, 336, 25

Bahcall, J. N., Pinsonneault, M. H., & Wasserburg, G. J. 1995, *RvMP*, 67, 781

Bahcall, J. N., Pinsonneault, M., & Basu, S. 2001, *ApJ*, 555, 990

Bahcall, J. N., Basu, S., Pinsonneault, M., & Serenelli, A. M. 2005, *ApJ*, 618, 1049

Basu, S. & Antia, H. M. 1995, *MNRAS*, 276, 1402

Basu, S. & Antia, H. M. 2008, *Phys. Rep.*, 457, 217

Bethe, H. 1939, *Phys. Rev.*, 55, 434

Burke, C. J., Pinsonneault, M. J., & Sills, A. 2004, *ApJ*, 604, 272

Bruzual, A. G. & Charlot, S. 2003, *MNRAS*, 344, 1000

Chaboyer, B., Deliyannis, C. P., Demarque, P., Pinsonneault, M. H., & Sarajedini, A. 1992 *ApJ*, 388, 372

Chaboyer, B., Demarque, P., Guenther, D. B., & Pinsonneault, M.H. 1995, *ApJ*, 446, 435

Chabrier, G. & Baraffe, I. 2000, *ARA&A*, 38, 337

Chan, K. L. & Sofia, S. 1989, *ApJ*, 336, 1022

Chandrasekar, S. 1939, *An Introduction to the Study of Stellar Structure* (Chicago: U. of Chicago Press), Chap. 7

Christensen-Dalsgaard, J. 1988, *IAUS*, 123, 295

Christensen-Dalsgaard, J., Gough, D. O., & Thompson, M. J. 1991, *ApJ*, 378, 413

Cowling, T. G. 1936, *MNRAS*, 96, 42

Deliyannis, C. P., Demarque, P., & Kawaler, S. D. 1990, *ApJS*, 73, 21

Demarque, P. R. & Larson, R. B. 1964, *ApJ*, 140, 544

Demarque, P., Green, E. M., & Guenther, D. B. 1992, *AJ*, 103, 1501

Demarque, P., Woo, J.-H., Kim, Y.-C., & Yi, S. K. 2004, *ApJS* 155, 667

Demarque, P., Guenther, D. B., Li, L. H., Mazumdar, A., & Straka, C. W. 2008, *Ap&SS*, 316, 31

Ferguson, J. W. *et al.* 2005, *ApJ*, 623, 585

Gallart, C. *et al.* 2008, *ApJ*, 682, 89

Guenther, D. B. & Demarque, P. 1997, *ApJ*, 484, 937

Guenther *et al.* 2007, *CoAst*, 151, 5

Haselgrove, C. B. & Hoyle, F. 1956a, *MNRAS*, 116, 515

Haselgrove, C. B. & Hoyle, F. 1956b, *MNRAS*, 116, 527

Henry, T. J. 2004, *ASPC*, 348, 159

Hillenbrandt, L. A. & White, R. J. 2004, *ApJ*, 604, 741

Hoyle, F. & Schwarzschild, M. 1955, *ApJS*, 2, 1

Iben, I, Jr. 1967, *ARA&A*, 5, 57

Iglesias, C. A. & Rogers, F.G. 1996 *ApJ*, 464, 943

Jung, Y. K., Kim, Y.-C., Robinson, F. J., Demarque, P., & Chan, K. L. 2007, *ASPC*, 362, 306

Kim, Y.-C, Fox, P. A., Demarque,P., & Sofia, S. 1995, *ApJ*, 442, 422

Kim, Y.-C. & Chan, K. L. 1998, *ApJ*, 496, 121

Kim, Y. -C., Demarque, P., Yi, S. & Alexander, D. R. 2002, *ApJS*, 144, 259

Kippenhahn, R., Weigert, A., & Hofmeister, E. 1967, *Meth. Comp. Phys.*, 7, 129

Kjeldsen, H. *et al.* 2003, *AJ*, 126, 1483

Kurucz, R. L. 1995, *ASPC*, 81, 583

Lee, Y.-W. *et al.* 2005, *ApJ*, 621, L57

Li, L. H., Robinson, F. J., Demarque, P., Sofia, S., & Guenther, D. B. 2002, *ApJ*, 567, 1192

Limber, D. N. 1958, *ApJ*, 127, 387

Ludwig, H.-G., Allard, F., & Hauschildt, P. H. 2006, *A&A*, 459, 599

Ludwig, H.-G., González Hernández, J. I., Behara, N., Caffau, E., & Steffen, M. 2008, *AIPC*, 990, 268

Lydon, T. J., Fox, P. A., & Sofia, S. 1992, *ApJ*, 397, 701

Lydon, T. J. & Sofia, S. 1995, *ApJS*, 101, 357

Mazumdar, A., Basu, S., Collier, B. L., & Demarque, P. 2006, *MNRAS*, 372, 949

Nordlund, Å. & Dravins, D. 1990, *A&A*, 228, 155

Norris, J. E. 2004, *ApJ*, 612, 25

Öpik, E. 1938, *Pub. Obs. Astr. de l'Univ. de Tartu*, vol. 30, no. 3

Penev, K. *et al.* 2007, *ApJ*, 655, 1166

Piotto, G., Villanova, S., Bedin, L. G. *et al.* 2005, *ApJ*, 621, 777

Pont, F. & Eyer, L. 2005, *ESASP*, 576, 187

Robinson, F. J., Li, L. H., Demarque, P., Sofia, S., Kim, Y.-C., & Guenther, D. B. 2003, *MNRAS*, 340, 923

Robinson, F. J., Demarque, P., Li, L. H., Sofia, S., Kim, Y.-C., Chan, K. L., & Guenther, D. B. 2004, *MNRAS*, 347, 1204

Rogers, F. J. & Iglesias, C. A. 1994, *Science*, 263, 50

Rogers, F. J., Swenson, F. J., & Iglesias, C. A. 1996, *ApJ*, 456, 902

Rosenthal, C. S., Christensen-Dalsgaard, J., Nordlund, Å, Stein, R. F., & Trampedach, R. 1999, *A&A*, 351, 689

Samadi, R. *et al.* 2007, *A&A*, 463, 294

Sarajedini, A., *et al.* 2007, *AJ*, 133, 290

Schönberg, M., & Chandrasekar, S. 1942, *ApJ*, 96, 161

1906, *Nachr. Kön. Preus. Akad. Wissenschaften* 195, 41

Seaton, M. J. & Badnell, N. R. 2004, *MNRAS*, 354, 457

Seaton, M. J. 2007, *MNRAS*, 382, 245

Stein, R. F. & Nordlund, Å. 1998, *A&A*, 499, 914

Stetson, P. B. *et al.* 2004, *PASP*, 116, 1012

Straka, C. W., Demarque, P., & Guenther, D. B. 2005, *ApJ*, 629, 1075

Straka, C. W., Demarque, P., Guenther, D. B., Li, L. H., & Robinson, F. J. 2006, *ApJ*, 636, 1078

VandenBerg, D. A. & Clem, J. L. 2003, *AJ*, 126, 778

Yi, S. *et al.* 2000, *ApJ*, 533, 670

Yi, S., Demarque, P., Kim, Y. C., Lee, Y-W., Ree, C.-H., Lejeune, T., & Barnes, S. 2001, *(*ApJS), 136, 417

Yi, S. K. 2003, *ApJ*, 582, 202

Yi, S. K., Kim, Y.-C., Demarque, P., Lee, Y.-W., Han, S.-I., & Kim, D. G. 2008, *IAUS*, 252, 413

Zahn, J.-P. & Bouchet, L. 1989, *A&A*, 223, 112

Discussion

S. Yi: Where would the inadequateness of the mixing length convection approximation be more critically affecting? Would isochrone fits be affected much?

P. Demarque: The lowest masses would be most obviously affected because they have deep convective zones. In the more massive stars, core overshooting in the core-H-burning and core He- burning phases would affect evolutionary lifetimes most critically and would also have an impact on the advanced phases of evolution.

E. Jensen: Could you comment on any changes from your new simulations in terms of predictions of Li depletion for low-mass stars?

P. Demarque: I hope that we will have useful results in the next two or three years. We have completed the parallelization of the simulation code, and are improving the treatment of radiative opacities in the atmospheric layers. The improvements will be not only in predicting Li line strengths in the atmospheres, but also in calculating the depth of the convective overshoot in the interiors.

B. Weaver: Modeling microturbulence has been causing me problems in a stellar atmosphere synthetic spectrum simulation. Would the effect on spectra increase with increasing mass?

P. Demarque: Yes, it seems to be the case. The results we have show that the relative importance of turbulence increases with stellar mass on the main sequence and with decreasing gravity on the giant branch.

The Ages of Stars
Proceedings IAU Symposium No. 258, 2008
E.E. Mamajek, D.R. Soderblom, & R.F.G. Wyse, eds.

doi:10.1017/S1743921309032050

The Sun in time: age, rotation, and magnetic activity of the Sun and solar-type stars and effects on hosted planets

Edward F. Guinan[1] & Scott G. Engle[1,2]

[1]Department of Astronomy & Astrophysics, Villanova University,
Villanova, PA 19085, USA
email: edward.guinan@villanova.edu
[2]Centre for Astronomy, James Cook University,
Townsville, QLD 4811, Australia
email: scott.engle@villanova.edu

Abstract. Multi-wavelength studies of solar analogs (G0–5 V stars) with ages from ~50 Myr to 9 Gyr have been carried out as part of the "Sun in Time" program for nearly 20 yrs. From these studies it is inferred that the young (ZAMS) Sun was rotating more than 10× faster than today. As a consequence, young solar-type stars and the early Sun have vigorous magnetohydrodynamic (MHD) dynamos and correspondingly strong coronal X-ray and transition region/chromospheric FUV–UV emissions (up to several hundred times stronger than the present Sun). Also, rotational modulated, low amplitude light variations of young solar analogs indicate the presence of large starspot regions covering ~5–30% of their surfaces. To ensure continuity and homogeneity for this program, we use a restricted sample of G0–5 V stars with masses, radii, T_{eff}, and internal structure (i.e. outer convective zones) closely matching those of the Sun. From these analogs we have determined reliable rotation-age-activity relations and X-ray–UV (XUV) spectral irradiances for the Sun (or any solar-type star) over time. These XUV irradiance measures serve as input data for investigating the photo-ionization and photo-chemical effects of the young, active Sun on the paleo-planetary atmospheres and environments of solar system planets. These measures are also important to study the effects of these high energy emissions on the numerous exoplanets hosted by solar-type stars of different ages. Recently we have extended the study to include lower mass, main-sequence (dwarf) dK and dM stars to determine relationships among their rotation spin-down rates and coronal and chromospheric emissions as a function of mass and age. From rotation-age-activity relations we can determine reliable ages for main-sequence G, K, M field stars and, subsequently, their hosted planets. Also inferred are the present and the past XUV irradiance and plasma flux exposures that these planets have endured and the suitability of the hosted planets to develop and sustain life.

Keywords. planetary systems, stars: late-type, activity, evolution, magnetic fields, rotation; Sun: evolution, magnetic fields, corona, chromosphere

---

1. Introduction and background on the evolution of the Sun

Changes in the Sun's luminosity, temperature and radius over its ~10 Gyr main-sequence lifetime, arising from nuclear evolution, are explicable (except for "minor" details) by modern stellar evolution theory and models (Basu *et al.* 2000). However, an extrapolation of the Sun's magnetic dynamo-related properties back in time to study their manifestations – such as chromospheric and coronal emissions, flares and coronal mass ejections, winds, and sunspots – cannot be reliably modeled or predicted by theory alone. Even for the present Sun, the solar dynamo and magnetohydrodynamic (MHD) dynamo, and resulting coronal, models have problems explaining the observed magnetic-related phenomena – such as the ~11/22 year sunspot/magnetic cycle and the magnetic heating of the Sun's outer atmosphere. But much progress in solar and stellar dynamo theory is being made (see e.g. Dikpati 2005). However, we can trace the Sun's magnetic past by utilizing carefully selected solar analogs (with different ages and rotations) that serve as proxies for the Sun over its main-sequence lifetime. To address this problem, we established the "Sun in Time" program nearly 20 years ago. Unlike most other studies of stellar rotation-age-activity, this study focuses specifically on stars with nearly identical basic physical properties to the Sun. Moreover, the program aims at determining changes in the coronal/chromospheric X-ray to UV (XUV) fluxes over the Sun's main-sequence lifetime and the effects that these high energy emissions have on hosted planets. Related to the theme of this symposium, we also have developed rotation-age-activity relations for solar-type stars that permit the age of solar-like field stars to be determined from their rotation periods or from calibrated magnetic activity indicators such as Ca II *H&K* chromospheric emission (R'_{HK}), X-ray luminosity (L_x) and spot-modulated brightness variations.

Studies of single main-sequence G and K-type stars in open clusters and stellar moving groups with known ages show that stars rotate more slowly with age and have corresponding decreases in magnetic activity such as chromospheric Ca II *HK* emissions (e.g. Skumanich 1972; Soderblom 1982, 1983). The observed decrease in rotation with age (for solar-type and cooler stars) results from angular momentum loss due to magnetically threaded winds (see Schrijver & Zwann 2000 and references therein). The Sun in Time project was motivated, in part, by the early studies of rotation, age & activity of cool stars by Soderblom (see e.g. Soderblom 1982, 1983). Some recent examples of related studies are given in Barnes (2007), Mamajek & Hillenbrand (2008), several papers given in this volume (e.g. Barnes, Meiborn & Mamajek) and the numerous references given in these papers.

2. The Sun in Time Program

As an integral part of the Sun in Time program, studies of solar analogs (G0–5 V stars) have been carried out across the electromagnetic spectrum (see Dorren & Guinan 1994; Güdel *et al.* 1997; Guinan *et al.* 2003; Ribas *et al.* 2005). The study utilizes X-ray, EUV, FUV & UV data secured by us and others (including archival observations) from NASA, ESA and Japanese space missions as well as ground-based photometry and spectroscopy. These stars serve as proxies for the Sun (and other solar-mass stars) and their ages essentially cover the main-sequence lifetime of the Sun (from ~50 Myr to 9 Gyr). From these studies (and others) it is inferred that the young (ZAMS) Sun was rotating more than 10× faster than today. See Fig. 1a for a plot of 1/P (days^{-1}) versus age for representative program stars. This plot includes both G- and dM-type stars, so the behavior of the two stellar types can be seen and compared. As indicated in the figure (and discussed later in Section 7), dM stars rotate more slowly than corresponding G-type stars of similar ages. As a consequence of rapid rotation, young solar-type stars and the young Sun (as well as dM-type stars) have vigorous magnetic dynamos and correspondingly strong coronal (X-ray) and transition region/chromospheric (FUV–UV) emissions up to several hundred times stronger than the present Sun. Also, the rotationally modulated, low amplitude light variations of young solar analogs indicate the presence of large starspot regions that cover ~5–30% of their surfaces. As an example, the coronal X-ray (0.2–2.5 keV) luminosities (L_x) of the program stars are plotted versus age (Fig. 1b). The least-squares fits to these data are provided within the figure. Note the large decrease in L_x (by > 1000×) between young and older solar-type stars.

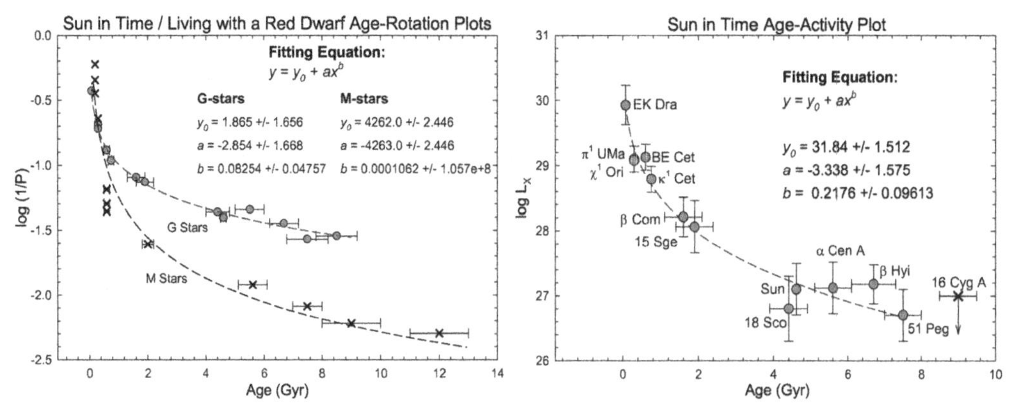

Figure 1. Left (a) – shows a plot of $1/P_{rot}$ (angular velocity $\Omega \propto 1/P_{rot}$) of the solar-type stars in units of 1/days plotted against stellar age. Also plotted for comparison are the same quantities for a representative sample of dM0–5 stars. As shown in the figure, both the G- and dM-stars decrease in angular velocity with age. For a given age the dM stars have longer rotation periods indicating that they spin down faster than the more massive G stars. As discussed in Section 7, the rapid magnetic braking of dM stars is most likely due to more efficient dynamos (from their deeper CZs) and their lower masses. **Right (b)** – shows a plot of the mean log L_x (ergs/s) versus age for the solar-type stars. As shown in the plot, the mean X-ray luminosity of the solar analogs decreases by 10^3 times from the youngest to solar-age stars in the sample. The least squares fitting equations for both plots are given within the figures.

To ensure consistency and homogeneity, the Sun in Time program uses a restricted sample of G0–5 V stars with masses, radii, T_{eff} and internal structure (i.e. outer convective zones) closely matching those of the Sun. The only major "free parameters" are age and rotation. From these analogs we have determined reliable rotation-age-activity relations and XUV spectral irradiances (flux/cm^2/sec at a reference distance of 1 AU) for the Sun over time. These XUV irradiance determinations serve as inputs to investigate the photo-ionization and photo-chemical effects of the young, active Sun on paleo-planetary atmospheres and environments. These measures are also important to study the numerous exoplanets hosted by solar-type stars of different ages. Recently we have extended the study to include lower mass, main-sequence (dwarf) dK and dM stars to determine their spin-down rates and coronal and chromospheric emissions as a function of mass and age. The present and past XUV irradiances that these planets experience can then be estimated and the suitability of the hosted planets to retain their atmospheres and develop/sustain life can be assessed. Also, from the studies by Wood *et al.* (2002, 2005) of solar-type, stellar astrospheres (equivalent to the solar heliosphere) using Hubble Space Telescope (HST) H I Lyman-α spectroscopy, it appears that the young Sun had more dense stellar winds and plasma outflows back to at least ~3.5 Gyr ago. As discussed by Lammer *et al.* (2003, 2008) and others, the interplay of strong XUV ionizing radiation and high plasma fluxes (winds), inferred for the young Sun, played a major role in the erosion of juvenile planetary atmospheres and loss of water for Venus and Mars. More details on this work are given later.

The Sun in Time program utilizes a relatively small number of stars (~25 primary stars and about ~30 secondary stars with less certain properties or measurements) that have similar basic physical properties such as mass (or mass parameterized by spectral types and colors). The homogeneity of the sample of ~25 stars (except for age and rotation) ensures that the depth of the outer convective zone (CZ) – which is important for a solar/stellar dynamo – is nearly identical for all stars. This is essential because the CZ, together with rotation, plays a crucial role in the magnetic dynamo, the generation of magnetic fields and the resulting outer atmospheric heating of solar-type (and cooler) stars – which, in turn, impacts the properties of the chromospheres, Transition Regions (TRs), coronae, flares and coronal winds. The program also utilizes rotation periods, instead of $v \sin i$ measures, to serve as the angular velocity ($\Omega \propto 1/\mathrm{P_{rot}}$). Most of the program stars are nearby, bright and have accurate distances. Thus they are accessible for securing measures of coronal X-ray & EUV fluxes, and accurate measures of transition region and chromospheric emission fluxes. Many of the target stars now have long-term photometry, permitting the investigation of activity cycles and differential rotation. Also, from the multi-band photometry, the properties of the starspots can be determined – such as spot temperatures, areal coverage, surface distributions and differential rotation (see Messina *et al.* 2006 and references therein). The primary science objectives are given below.

2.1. *Major goals*

• Study the solar (stellar) dynamo with rotation as the only important free parameter;
• investigate the energy transfer and heating of the stars' chromospheres, transition regions and coronae, as well as explosive events such as coronal mass ejections and flares;
• determine accurate rotation-age-activity relations for solar-type stars and, from these relations, the ages of G0–5 V stars can be estimated from knowing the rotation period, or values of coronal X-ray luminosity $\mathrm{L_X}$, Ca II *HK* emission (or $\mathrm{R'}_{HK}$), or other activity indicators;
• determine XUV irradiances of the young Sun to study the effects of its inferred high levels of XUV emissions and the resulting photo-ionization and heating of early planetary atmospheres;
• determine the ages and XUV flux histories of stars that host planets using rotation-age-activity relations and also to determine the effects of XUV radiation on these planets to determine if such planets could be suitable for life;
• continue to extend the sample to study the more common, lower mass and luminosity, cooler dK & dM stars. These stars are very common, comprising about 90% of all stars in the solar neighborhood. Also an increasing number of planets are being found hosted by dK/M stars. Since the CZs of cooler stars are deeper (becoming fully convective around dM 3.5+), we can investigate the dynamo as a function of CZ depth by comparing dG, dK and dM stars having similar rotation periods and/or ages.

3. Rotation periods of the program stars

Whenever possible, the rotation periods of the stars are directly measured from the rotation modulation in brightness arising from star spots (and bright faculae) and/or from rotational modulations of chromospheric Ca II *HK* line emissions. In some cases, UV chromospheric and transition region (TR) line emissions (such as Mg II *hk* (2800Å) and C IV (1550Å)) are used to determine the activity level and the rotation period – such as for the G2 V star α Cen A (see Carton *et al.* 2007). The determination of reliable rotation periods for younger stars (age $<$ 1.5 Gyr) is accomplished from conventional ($\pm \sim$5 mmag precision) ground-based photometry. However, for older solar-type stars (ages $>$ 2 Gyr), the star spots typically cover less than 1% of the stellar surface and rotational light modulations are difficult to detect. For example, the oldest solar-type star in our ongoing ground-based photometry program, found to show definite periodic light variations, is the G1 V star 15 Sge (age $\approx$ 1.9 Gyr; P $\approx$ 13.5 days; V-band light amplitude $\approx$ 0.008$\pm$0.003 mag). For comparison, the average rotationally-induced light modulation of the Sun, arising from bright plages and sunspots, is typically less than

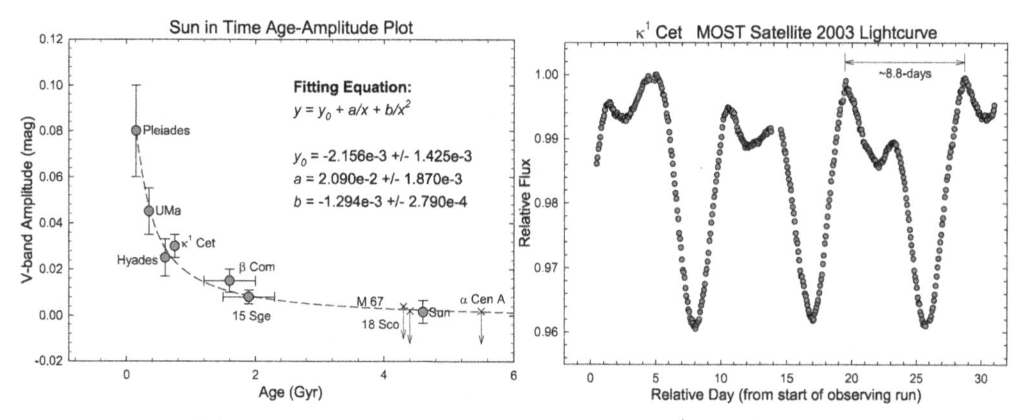

Figure 2. Left (a) – The V-band light amplitudes from starspot rotational modulations and cycles plotted against stellar age. Note that the younger stars have significant light variations due to the larger areal coverage of starspots. The least squares fitting equation for these data is given within the figure. **Right (b)** – High precision photometry carried by Walker *et al.* (2007) with the *MOST* satellite of the Hyades-age G5 V star κ^1 Cet is shown. The ~8.8 day brightness variations arise from the rotational modulation from at least two primary starspot regions with different areas on the surface of the star.

~0.15% (~0.0015 mag). Although, with high precision photometry from space missions such as *MOST* (www.astro.ubc.ca/MOST/) and *CoRot* (smsc.cnes.fr/COROT/) it is now possible to detect low amplitude (< 1 mmag) light variations. And soon, with *Kepler* (kepler.nasa.gov/), it will be easy to determine star spot modulated brightness variations as small as our Sun's, and thus rotation periods, and determine star spot areal coverage for solar-age and older stars. Fig. 2a shows the dependence of starspot-induced light variability (as a proxy for starspot areal coverage and magnetic activity) plotted against age. As shown, the young solar-type stars have significant light variations of a few percent up to 10% while the Sun and older stars have very small (<0.2%) light variations.

An excellent example of the high precision light curves that are now feasible from space missions is illustrated by the recent photometry of the young (age ≈ 0.7 Gyr) G5 V star κ^1 Cet using the Canadian microsatellite *MOST* – See Walker *et al.* (2007). From this study, Walker *et al.* were able to determine a precise rotation period of $P_{rot} = 8.77 \pm 0.04$ days (from the three MOST datasets combined: 2003, 2004 & 2005), along with starspot coverage and differential rotation. An example of the light curve obtained with *MOST* is shown in Fig. 2b. The light variations are modeled with the presence of two starspot regions very different in area and differential rotation. This high precision *MOST* light curve is illustrative of the thousands of even higher quality light curves expected from the *Kepler* mission.

Determinations of rotation periods also can be made from time-series measures of Ca II *HK* emission. Most of these Ca II emission studies have been made at Mt. Wilson and more recently from Lowell Observatory (e.g. see Baliunas *et al.* 1995; Hall *et al.* 2007). Although these programs are primarily focused on defining activity cycles, the period analyses also revealed, for some of the stars, apparent rotational modulations and rotation periods. A number of our program stars have rotation periods found from Ca II *HK* spectrophotometry. Also, from time series UV spectrophotometry of chromospheric and TR emissions, the rotation periods of several more stars have been secured (see Datin *et al.* 2009; Guinan *et al.* 2009).

As shown previously in Fig. 1a, the angular velocity ($\Omega \propto 1/P_{rot}$) of our program stars decreases with age. By solar age, the light variations are < 2-mmag and are barely detectable. The analytical expression giving the best fit to these data is provided with the plot. As shown in the figure, the rotation period of solar-type stars range from $<$ 3 days for the young (near-ZAMS) stars to $\gtrsim$ 35 days for the older 8–9 Gyr solar-type stars. Also shown is that, for a given age, the rotation period of a dM star is longer than the corresponding solar-type star. The more rapid magnetic braking of dM stars is most likely due to more efficient magnetic (and maybe turbulent) dynamos (from their deep CZs) as well as their lower masses. Not shown is that the initial rotation periods of very young near-ZAMS solar-type stars with ages less than 100 Myr (Pleiades age and younger) range from $\lesssim$ 1 day up to 3.5 days. Excellent discussions about rotation and activity for near-ZAMS and pre-main-sequence stars in nearby star forming regions are given elsewhere in this volume. The dispersion in initial rotation periods of these young stars arise from initial differences in the stars' angular velocities. Such differing conditions are – stars having different initial angular velocities, interactions with planetary accretion disks and possible undetected binary companions. However, by the age of about 300 Myr, the stars of a given age (for a set mass – in this case one solar mass) have similar rotation periods. For example, in the Hyades cluster (age = 650 Myr), most (if not all) of the single G0–5 V stars have rotation periods between 7.5 and 9 days. This rotation "funneling" effect with age (large dispersion in the rotations for very young stars to low dispersion in rotation for older stars) probably results because the loss of angular momentum from magnetic braking is strongly dependent on the rotation of the star. Stars that initially are rapid rotators spin-down more rapidly from stronger magnetic braking than those that are originally slower rotators. If this self-regulating process were not active, we would not have rotation-age-activity relations. Physically this is explicable by modern hydromagnetodynamo theory. Some in-depth recent reviews of solar and stellar dynamo theory are given by Weiss and Tobias (2000) and Dikpati (2005), and most recently by Brandenburg (2009).

4. Determining reliable ages for the program stars

As shown by the review & poster papers at this symposium, determining accurate stellar ages is crucial for astrophysics, but not an easy task. When possible, the ages of program stars are estimated from memberships in nearby open clusters – such as the Pleiades, Praesepe and Hyades clusters. Because there are few nearby solar-age stars with reliable ages, it would be important to use the ~14–15 mag solar-type stars in M67 (age $\approx$ 4.3 $\pm$ 0.5 Gyr) to calibrate the rotation-age-activity relations. Also these stars would be useful comparisons for the present Sun (i.e. serve as solar twins). At ~850-pc, though, M67 is a bit too distant (and stars too faint) to reliably determine rotational periods or measure magnetic activity indicators (such as L_X). However, Giampapa *et al.* (2006) have measured Ca II emissions for a number of the solar-type stars in M67 and find that the average Ca II emission level is similar to that of the Sun. More recently, Pasquini *et al.* (2008) have identified several G2 V stars in M67 that nearly perfectly match the Sun in age, $T_{\rm eff}$, [Fe/H], Lithium abundance and age. Unfortunately no time-series photometry or spectroscopic observations of the M67 cluster G0–5 V members have been carried out to determine their rotation periods. This is feasible with a 2-m class telescope and, as part of the Sun in Time, we hope to carry out the high precision photometry during 2009/10. These observations would have to be carried out over a few months to cover their expected 20-30 day rotation periods and expected $<$ 2 mmag light variations if indeed these stars behave like the Sun. Also, Wright (2009 – this volume) discusses the ~2 Gyr "benchmark" open cluster – Ruprecht 147. This nearby (~200 pc) cluster would fill in the age gap in rotation-age-activity studies between the Hyades (650 Myr) and the Sun (4.57 Gyr).

Stellar ages are also estimated from UVW space motions for members identified with stellar moving groups – such as the Pleiades, Castor, Ursa Major (UMa), Hyades and a few others. Soderblom *et al.* (1993) have intensively studied space motions and found many nearby dG and dK stars that are probable members of these moving groups. More recently, a few additional stars have been linked to the older (~2 Gyr) HR 1614 Moving Group (see Feltzing & Holmberg 2000). For solar-age stars (and older) with well determined values of T_{eff}, [Fe/H] and M_V, ages can be estimated from isochronal fits using modern stellar evolution codes. Several excellent codes (and isochrones) are currently available (e.g. VandenBerg *et al.* 2006; Yonsei-Yale Isochrones – Demarque *et al.* 2004). Unfortunately, isochronal fits for young stars (i.e. still close to the ZAMS) are not very reliable because of the sensitivity of age determinations to errors in mass, T_{eff}, log g, M_V and metal abundance. It should be noted that the stars are selected either from spectral types (G0–G5 V) or corresponding $B - V$ color indices (+0.60 – 0.70). However, it is preferable to use Strömgren photometry $(b - y)$ instead of Johnson/Bessell $(B - V)$ to estimate T_{eff} since the $(b - y)$ index is less sensitive to the metallicity of the star.

5. Magnetic activity indicators – coronal X-ray, transition region & chromospheric FUV–UV emissions and starspots

5.1. *Coronal X-ray and FUV/UV transition region & chromospheric emissions*

Almost all of the primary program stars have magnetic activity indicators since this was a major factor in the selection process. In particular, because this study was initiated when the International Ultraviolet Explorer (*IUE*) satellite was operating (1978–1995), most of the stars have *IUE* UV spectra. The wavelength interval covered by *IUE* (1160–3200Å) contains numerous important chromospheric and TR emission lines that include Ly-α 1215.6Å, Mg II h & k 2800Å and strong TR lines such as C III 1335Å, Si IV 1400Å, C IV 1550Å, He II 1640Å and others. Also, during the 1990s, *ROSAT* (0.1–2.5 keV) and *ASCA* (0.2–10 keV) were used to secure pointed X-ray observations of several young, active program stars. These observations are supplemented by archival *ROSAT* (both pointed and all sky survey data) and *ASCA* X-ray observations. Archival X-ray observations obtained more recently by *Chandra* and *XMM-Newton* are also used. Observations were also carried out in the 80–500Å wavelength range with the Extreme Ultraviolet Explorer (*EUVE*). The EUV region contains important diagnostic coronal and TR emissions lines. Because *EUVE* had a small effective collection area, only a few solar-type stars were observed. In addition to these data, FUV spectra of solar-type stars – particularly high dispersion spectrophotometry of the strong H I Ly-α feature – were also used in the FUV irradiance studies. Ly-α emission is the largest contributor to the X-FUV (1–1700Å) flux, comprising 70–90% of the total flux, in that wavelength range, for solar-type stars. Wood *et al.* (2002, 2005) use *HST* spectroscopy of the Ly-α feature to estimate the winds and mass loss rates of solar-type (and cooler) stars. They provide Ly-α emission fluxes for the stars in their sample, after the careful removal of interstellar absorption and geocoronal emission.

5.2. *Chromospheric Ca* II *H & K emissions*

Most of the program stars have measures of chromospheric Ca II *HK* emission. Almost all of these observations are from Mount Wilson or Lowell Observatory. The Ca II *HK* emissions are important, widely available and well-calibrated indicators of chromospheric activity and numerous papers have focused on these lines. (See e.g. Soderblom *et al.* 1993; Baliunas *et al.* 1995; Hall *et al.* 2007; Mamajek (in this volume); and references therein). One drawback with using Ca II emission is that it levels off for stars older than ~3–4 Gyr. Thus it becomes insensitive to rotation and age for older stars. Also, care must be taken in using Ca II *HK* emission measurements to determine age and rotation because chromospheric emission can vary from rotation modulation, flares, and also can have relatively large cyclic variations. Multiple measures should be used.

5.3. *Light modulations by starspots*

A usually overlooked indicator of activity is starspot areal coverage which is manifested by low amplitude brightness variations. Young, chromospherically-active stars have relatively large rotationally modulated, periodic brightness variations arising from an uneven distribution of cool starspots over the stars' surfaces. For example, Pleiades G0–5 V stars have light variations of typically ~0.06–0.12 mag whereas the Sun has light modulations of ~0.12% (0.0012 mag = 1.2 mmag). Recently, from long-term high precision photometry, Hall *et al.* (2007) report brightness variation as small as $\approx$ 0.1 percent (1 mmag) for the solar twin – 18 Sco (G2 V; age ~ 4.4 Gyr). Although the light amplitudes change from season to season (from activity cycles and differential rotation), the light modulations and resulting rotation periods could make nice age indicators, as shown in Fig. 2a. Excellent long-term high precision photometry and Ca II *HK* spectrophotometry of solar-type stars have been carried out at Lowell Observatory (See e.g. Lockwood *et al.* 2007). Results from our long-term photometry of some program stars are given by Messina *et al.* (2006) and references therein.

6. Effects of the young active Sun's XUV & plasma fluxes on terrestrial planets

The Sun in Time program also focuses on the crucial question of the influence of the young Sun's strong XUV emissions on the developing planetary system – in particular on the photo-chemical and photo-ionization evolution and possible erosion of their early atmospheres. The XUV spectral irradiance (1–1700Å) measures have been made for stars covering ages from ~100 Myr to 6.7 Gyr (Ribas *et al.* 2005). These data are of interest to researchers of young planetary atmospheres and for studies of the atmospheric evolution of the large number of extrasolar planets found orbiting other solar-type stars. As an illustration, Fig. 3a shows a plot of the FUV O VI irradiances (i.e.– flux densities at 1 AU) for solar-type stars of different ages. The emission fluxes decrease by nearly 50× over the stellar ages sampled. Fig. 3b shows the averaged XUV irradiance over time, inferred for the Sun and other solar-type (dG0–5) stars, from this study. The fluxes in the plot have been normalized to the corresponding mean flux values of the present Sun. As shown, the coronal X-ray/EUV emissions of the young main sequence Sun were ~ 100 – 1000× stronger than the present Sun. Similarly, the TR and chromospheric FUV & UV emissions of the young Sun (by inference) were 10 – 100× and 5 – 10× stronger, respectively, than at present. Also shown in the figure is the slow increase (~8%/Gyr) in the bolometric luminosity of the Sun over the past 4.6 Gyr. Over this time, the Sun's luminosity increased from ~0.7 $L_\odot$ to 1.0 $L_\odot$. So in the past, the young Sun was dimmer by 30% overall but, due to strong magnetic activity, had stronger XUV fluxes than today.

Since 2002 we have been collaborating with planetary scientists and astrobiology groups (in particular – Helmut Lammer and others at the Astrobiology Group at the Space Research Institute, Graz, Austria) to study the effects of the strong XUV radiation and inferred high plasma fluxes of the young Sun on hosted planets. Brief summaries

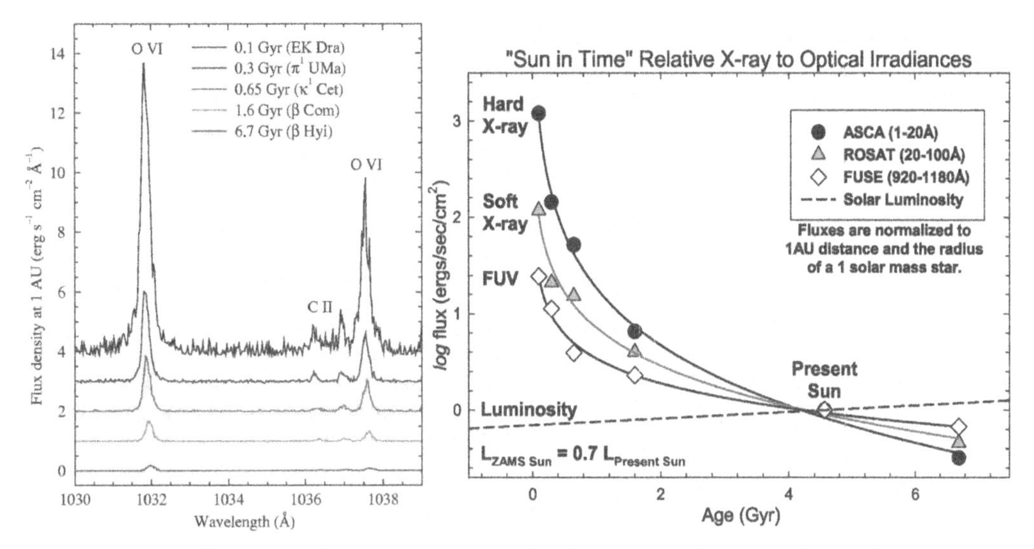

Figure 3. Left (a) – The FUV irradiance of the O VI spectral region of six "Sun in Time" program stars with ages from ~130 Myr to 9 Gyr. As shown, the emission fluxes decrease by nearly 50× from the youngest sun to the oldest sun in the sample. **Right (b)** – Irradiances (flux densities at a reference distance of 1 AU) of the Sun in Time program stars are plotted for hard (1–20Å) & soft (20–100Å) X-rays (from *ASCA* and *ROSAT*, respectively) along with the total FUV emission fluxes (from FUSE). The fluxes have been normalized to the corresponding mean fluxes of the Sun. These measures are based on the paper by Ribas *et al.* (2005). Also plotted is the change in the bolometric luminosity (L) of the Sun based on modern stellar evolution models. Note that the ZAMS Sun had a luminosity of 30% less than today.

of some of the results are discussed below for Mercury, Venus and Mars, as well as for exoplanets. Because of space constraints, we omit a discussion of the early Earth except to say that the Earth was shielded to a large extent from the sputtering/ion pickup processes discussed below because of the its protective magnetic field. A good review of the early evolution of the Earth's atmosphere (and the effects of the young Sun's high XUV and plasma fluxes) is given by Lammer *et al.* (2008).

6.1. *Possible erosion of Mercury by the active young Sun*

Mercury is the nearest planet to the Sun at a distance of 0.39 AU. Because of its proximity to the Sun, Mercury is subjected to over 6.5× more solar radiation and wind fluxes than the Earth. Thus, early Mercury was much more strongly exposed to the early Sun's very high particle fluxes and XUV radiation than any other planet. Even today, with a kinder, more gentle Sun, the solar wind and coronal & chromospheric XUV emissions (as well as flares and coronal mass ejection events) produce sputter/ion pickup erosion of its surface in which ions of Na^+, K^+, and O^+ and even OH^+ and H_2O^+ have been observed in its very thin, transient exosphere. (See e.g. Zurbuchen *et al.* 2008.)

As pointed out by Tehrany *et al.* (2002), some of the unusual features of Mercury – such as its high mean density (5.43 gm/cm^3) for its size – can be explained by the extreme erosive XUV/plasma actions of the early Sun. Mercury is an exceptionally dense planet for its size because of the large relative size of its core. Mercury's iron core extends out to 74% of its radius and it is estimated that the iron core occupies about 42% of its volume; for Earth this proportion is 17%. Because of this, Mercury is often referred to as the "Iron Planet". Preliminary modeling of early Mercury by Tehrany *et al.* (2002) was carried out using solar XUV irradiances from the Sun in Time program and a range of plasma fluxes. This study indicates that Mercury could have undergone significant erosion (by sputtering and ion pickup mechanisms) of its surface and mantle during the Sun's active phase – the first ~0.5–1.0 Gyr of its lifetime. This hypothesis is speculative and there are several other explanations; one of these is a major collision with large planetesimal that may have stripped away Mercury's outer mantle. Perhaps this question will be answered with measurements being carried out by the Mercury mission MESSENGER (www.nasa.gov/messenger) and also by BepiColumbo (www.esa.int/science/bepicolombo).

6.2. *The Loss of the water inventory from Venus*

Using the XUV irradiance appropriate for the young Sun (from Ribas *et al.* 2005) and solar wind flux estimates from Wood *et al.* (2002, 2005), Kulikov *et al.* (2006) have studied the atmosphere and water loss from early Venus. Venus has D/H ratios indicating that the juvenile planet once had water, possible even oceans. But Kulikov *et al.* show that Venus (at 0.71 AU from the Sun) has essentially lost all of its water reserves during the first ~0.5 Gyr after its formation from the vigorous action of strong (massive) winds and high XUV fluxes. However, as discussed recently by Lammer *et al.* (2008) and others, there are many uncertainties about early Venus that still leave many unanswered questions and problems. For example, Venus today rotates very slowly and does not have a protective magnetic field. But did early Venus rotate faster and have a strong dynamo with a resulting protective magnetosphere? Because of these and other uncertainties about Venus' early properties and initial conditions, many questions about its early history remain.

6.3. *Loss of water and soil oxidation of Mars*

It has been assumed, from topographic and geological studies, that Mars was originally warmer and much wetter than at present, and likely possessed a ~1 bar atmosphere (see discussions in Lammer *et al.* 2008). Lammer *et al.* (2003) considered ion pick-up sputtering as well as dissociative recombination processes to model the effects of the active young Sun's high XUV & plasma fluxes on the planet. The loss of H_2O from Mars over the last 3.5 Gyr was estimated to be equivalent to a global Martian H_2O ocean with a depth of $\gtrsim$12-m. If ion momentum transport is important on Mars, the water loss may be enhanced by a factor of ~2. For their study it has been assumed that, for the first billion years, Mars had a hot liquid iron-nickel core and, through rotation, possessed a significant magnetic field and resulting magnetosphere. This magnetosphere essentially shielded the early Martian environment from the combination of high levels of solar XUV radiation and plasma fluxes that would have otherwise removed its atmosphere. However, Mars is a smaller, less massive planet than the Earth [$M_{Mars} \approx 1/10\ M_{\oplus}$] with a smaller iron core which lost heat at a much faster rate. Thus its iron core is expected to have solidified ~1 billion years after the planet's formation. Without the protective magnetic field, the Martian exosphere was exposed to the ionizing effects and strong winds of the early Sun, and thus partially eroded. Photolysis of water ($H_2O \rightarrow 2H + O$) ensued, with a preferential loss of the lighter hydrogen over the (8×) heavier oxygen. The loss of water and water vapor from the atmosphere resulted in a greatly diminished greenhouse effect and subsequent rapid cooling of the lower Martian atmosphere. This rapid cooling permitted some water to remain behind, possibly as ice and permafrost trapped below the Martian surface.

6.4. *Exoplanets – possible evaporation of Hot Jupiters*

Our XUV irradiance data from Ribas *et al.* (2005) have been used by Grießmeier *et al.* (2004) to investigate the atmospheric loss of extrasolar planets from the XUV heating and winds of their host stars. This study indicates (among other interesting results) that strong planetary magnetic fields and resulting significant magnetospheres are critically important in shielding a planet from the atmospheric erosion and possible atmospheric loss (via sputtering and ion pickup mechanisms) expected from the strong winds and high XUV radiation of the active young Sun. This is particularly important for lower mass, terrestrial size planets with low gravities, located close to their host stars.

7. The next step: extension of the "Sun in Time" program to main-sequence K- and M- stars

The discovery of 330+ extrasolar planets orbiting mostly main-sequence dG, dK, and dM stars during the last decade has motivated the expansion of the "Sun in Time" program to include dK & dM stars. These cool, low mass and low luminosity stars comprise over 90% of all stars in the Galaxy. With the numerous ground-based extrasolar planet search programs and space missions such as *Kepler*, thousands of additional extrasolar planets are expected to be found in the near future. The goals of our dK–dM star program are to understand magnetic activity of red dwarf stars with deep CZs and to determine their XUV and plasma fluxes over a wide range of ages. This program is helping to identify and characterize dK & dM stars that have planets which may be suitable for life. Many dK & dM stars are members of the old disk and Pop II populations of our Galaxy. In particular, late dK & dM stars make interesting targets for exobiology and SETI programs because of their long lifetimes, exceeding 50 Gyr, and high space frequencies; See Tarter *et al.* (2007).

Because of the low luminosities of dM stars (e.g. $L \approx 0.06\ L_{\odot}$ for dM0 stars and $L \approx 0.008\ L_{\odot}$ for dM5 stars), their liquid water habitable zones (HZs) are located close to the host star ($\lesssim$0.4 AU). However, because of their deep CZs and efficient magnetic dynamos, late dK & dM red dwarf stars exhibit strong XUV coronal and chromospheric emissions and frequent flares. The high magnetic activity of dM stars causes a hypothetical hosted HZ planet to be strongly affected by stellar flares, winds and plasma ejection events that are frequent in young dM stars (e.g. Kasting *et al.* 1993; Lammer *et al.* 2006; Guinan & Engle 2007). Relevant to the "Ages of Stars" Symposium, we are developing rotation-age-activity relations for these ubiquitous little stars. Because of the very slow nuclear evolution of dK & dM stars, applying isochronal fits to determine age is not feasible. Reliable ages can be obtained from cluster and moving group memberships, as well as memberships in wide binary systems (such as Proxima Cen (M5.5 V) as an outlying member of the α Cen triple star system) in which the age of the companion is determinable from isochrones or, in the case of white dwarf companions, from the white dwarf cooling ages – See Catalan (2009) in this volume.

A description of this program – "Living with a Red Dwarf" – can be found on our website at (www.astronomy.villanova.edu/lward/). Also, some recent papers discussing initial results are given by Guinan and Engle (2009) & Engle *et al.* (2009).

Acknowledgements

This research is supported by grants from NSF and NASA and utilizes data from the IUE, ROSAT, ASCA, EUVE, FUSE, HST, XMM-Newton & the Chandra missions. The photometry used in the program is supported by a NSF/RUI grant. We are very grateful for this support. We also thank the Symposium organizer, David Soderblom, for the invitation to participate.

References

Baliunas, S. L., *et al.* 1995, *A*pJ, 438, 269

Barnes, S. A. 2007, *ApJ*, 669, 1167

Basu, S., Pinsonneault, M. H., & Bahcall, J. N. 2002, *A*pJ, 529, 1084

Brandenburg, A. 2009, arXiv:0901.3789

Carton, J. M., Dewarf, L. E., & Guinan, E. F. 2007, Bulletin of the American Astronomical Society, 38, 928

Datin, K., Dewarf, L. E., Guinan, E. F., & Carton, J. M. 2009, American Astronomical Society Meeting Abstracts, 213, #406.09

Demarque, P., Woo, J.-H., Kim, Y.-C., & Yi, S. K. 2004, *A*pJS, 155, 667

Dikpati, M., Gilman, P. A., & MacGregor, K. B. 2005, *A*pJ, 631, 647

Dorren, J. D. & Guinan, E. F. 1994, *A*pJ, 428, 805

Engle, S. G., Guinan, E. F., & Mizusawa, T. 2009, to appear in the proceedings of "Future Directions in Ultraviolet Spectroscopy"

Feltzing, S. & Holmberg, J. 2000, *A&A*, 357, 153

Giampapa, M. S., Hall, J. C., Radick, R. R., & Baliunas, S. L. 2006, *A*pJ, 651, 444

Grießmeier, J.-M., *et al.* 2004, *A&A*, 425, 753

Güdel, M., Guinan, E. F., & Skinner, S. L. 1997, *A*pJ, 483, 947

Guinan, E. F. & Engle, S. G. 2009, arXiv:0901.1860

Guinan, E. F. & Engle, S. G. 2007, arXiv:0711.1530

Guinan, E. F., Ribas, I., & Harper, G. M. 2003, *A*pJ, 594, 561

Hall, J. C., Lockwood, G. W., & Skiff, B. A. 2007, *A*J, 133, 862

Kasting, J. F., Whitmire, D. P., & Reynolds, R. T. 1993, *I*carus, 101, 108

Kulikov, Y. N., *et al.* 2006, *P*&SS, 54, 1425

Lammer, H., Lichtenegger, H. I. M., Kulikov, Y. N., Khodachenko, M. L., Griessmeier, J.-M., Terada, N., & Ribas, I. 2006, *E*uropean Planetary Science Congress 2006, 392

Lammer, H., Selsis, F., Ribas, I., Guinan, E. F., Bauer, S. J., & Weiss, W. W. 2003, *A*pJL, 598, L121

Lammer, H., Terada, N., Kulikov, Y. N., Lichtenegger, H. I. M., Khodachenko, M. L., & Penz, T. 2008, 14th Cambridge Workshop on Cool Stars, Stellar Systems, and the Sun, 384, 303

Lockwood, G. W., Skiff, B. A., Henry, G. W., Henry, S., Radick, R. R., Baliunas, S. L., Donahue, R. A., & Soon, W. 2007, *A*pJS, 171, 260

Mamajek, E. E. & Hillenbrand, L. A. 2008, *A*pJ, 687, 1264

Messina, S., Cutispoto, G., Guinan, E. F., Lanza, A. F., & Rodonò, M. 2006, *A&A*, 447, 293

Ribas, I., Guinan, E. F., Güdel, M., & Audard, M. 2005, *A*pJ, 622, 680

Skumanich, A. 1972, *A*pJ, 171, 565

Soderblom, D. R. 1983, *A*pJS, 53, 1

Soderblom, D. R. 1982, *A*pJ, 263, 239

Soderblom, D. R., Fedele, S. B., Jones, B. F., Stauffer, J. R., & Prosser, C. F. 1993, *A*J, 106, 1080

Soderblom, D. R. & Mayor, M. 1993, *A*J, 105, 226

Tarter, J. C., *et al.* 2007, *A*strobiology, 7, 30

Tehrany, M. G., Lammer, H., Hanslmeier, A., Ribas, I., Guinan, E. F., & Kolb, C. 2002, EGS XXVII General Assembly, Nice, 21-26 April 2002, abstract #1903

VandenBerg, D. A., Bergbusch, P. A., & Dowler, P. D. 2006, *A*pJS, 162, 375

Walker, G. A. H., *et al.* 2007, *A*pJ, 659, 1611

Weiss, N. O. & Tobias, S. M. 2000, *S*pace Science Reviews, 94, 99

Wood, B. E., Müller, H.-R., Zank, G. P., Linsky, J. L., & Redfield, S. 2005, *A*pJL, 628, L143

Wood, B. E., Müller, H.-R., Zank, G. P., & Linsky, J. L. 2002, *A*pJ, 574, 412

Zurbuchen, T. H., *et al.* 2008, *S*cience, 321, 90

Discussion

E. MAMAJEK: The latest published correlation between mass loss rate and X-ray luminosity from Wood (2005) showed that the correlation seemingly disappeared for the highest activity stars. How well is the mass loss rate constrained at young ages?

E. GUINAN: Only a small sample of stars have had their wind properties measured via the Lyman-profile/astrosphere method of Brian Wood. The estimates of winds (mass loss rate) is indeed uncertain for young, chromospherically-active stars. Also the sample of stars includes no stars closely similar to the Sun that could serve as solar analogs for wind studies. It's hoped that additional young G dwarfs could be observed by Wood with HST/STIS to clear up this problem. Otherwise we have to wait for ALMA to get reliable stellar wind estimates.

F. WALTER: What is the timescale for the crust/mantle of Mercury to be blown away? Was there any similar evaporation of the lunar crust?

E. GUINAN: 1) There are many uncertainties with Mercury's initial properties. If we assume it was an Earth-sized planet, then most of the erosion via strong XUV radiation and solar winds ($\sim$1000 times the present) would take place in the first few hundreds of million years of its life. 2) Even today the Moon's surface is affected by solar XUV radiation and winds but not in a significant way due to the present Sun's lower activity compared to its past. Also the Moon is further away from the Sun than Mercury so less affected by the young Sun's high winds and XUV radiation. Another major problem is the currently held theory that the Moon formed from a collision of another body with the young Earth. So if the Moon formed after the peak of Sun's early activity, it would suffer much less erosion.

Ed Guinan

Carla Cacciari

The Ages of Stars
Proceedings IAU Symposium No. 258, 2008
E.E. Mamajek, D.R. Soderblom & R.F.G. Wyse, eds.

doi:10.1017/S1743921309032062

The promise of Gaia and how it will influence stellar ages

Carla Cacciari
INAF, Osservatorio Astronomico di Bologna,
Via Ranzani 1, 40127 Bologna, Italy
email: carla.cacciari@oabo.inaf.it

Abstract. The Gaia space project, planned for launch in 2011, is one of the ESA cornerstone missions, and will provide astrometric, photometric and spectroscopic data of very high quality for about one billion stars brighter than V = 20. This will allow to reach an unprecedented level of information and knowledge on several of the most fundamental astrophysical issues, such as mapping of the Milky Way, stellar physics (classification and parameterization), Galactic kinematics and dynamics, study of the resolved stellar populations in the Local Group, distance scale and age of the Universe, dark matter distribution (potential tracers), reference frame (quasars, astrometry), planet detection, fundamental physics, Solar physics, Solar system science.

I will present a description of the instrument and its main characteristics, and discuss a few specific science cases where Gaia data promise to contribute fundamental improvement within the scope of this Symposium.

Keywords. space vehicles, astrometry, stars: distances, stars: kinematics, stars: fundamental parameters, Galaxy: stellar content, Galaxy: fundamental parameters, Galaxy: formation, Galaxy: evolution, galaxies: stellar content

1. Introduction

The idea of measuring the position of stars on the sky systematically and for a scientific purpose dates back to the 2nd Century BC, when the Greek astronomer Hipparchus measured about 1000 stars naked eye. This has been repeated several times during the following centuries, with steadily increasing power and accuracy, until the most recent Hipparcos satellite (1989-1993) that measured $\sim$ 120,000 sources with $\sim$ 1 mas accuracy and produced a Catalogue (Perryman *et al.* 1997) later revised by van Leeuwen (2007).

Gaia represents the natural follow up of the Hipparcos mission, with huge improvement in terms of: i) measurement accuracy, ii) limiting magnitude and hence number of observed objects, and iii) the combination of nearly simultaneous astrometric, photometric and spectroscopic observations.

Gaia is a cornerstone mission of the ESA Space Program, that will perform an all-sky survey and produce accurate astrometry and photometry for about 1.5×10^9 objects down to a limiting magnitude of 20 mag, and additional spectroscopy for objects brighter than V = 16-17. This will allow to obtain a stereoscopic and kinematic view of the Galaxy, and to address key questions of modern astrophysics regarding the formation and evolution of the Milky Way. Such an observational effort has been compared to the mapping of the human genome for the impact that it will have in Galactic astrophysics. In addition, Gaia will provide a fundamental contribution in a much broader range of scientific areas (see Sect. 3).

More information on the Gaia mission and its science can be found in the *Gaia Concept and Technology Study Report* (2000), the Proceedings of the Symposium "The 3-Dimensional Universe with Gaia" (2005), and at **http://www.rssd.esa.int/Gaia**.

Table 1. Comparison of Hipparcos and Gaia characteristics and performance

	Hipparcos	**Gaia**
Completeness to ...	V $\sim$ 9	V $\sim$ 20 (blue) - 22 (red)
Magnitude limit	V$\sim$12.4	$\sim$ 1 mag. fainter than completeness
N. Sources Quasars galaxies	$\sim$1.2 10^5 0 0	$\sim$1.5 10^9 $\sim$$10^6$ $\sim$$10^7$
Astrometric accuracy	$\sim$ 1 mas	$\leqslant$ 7 μas at V$\leqslant$10 12(red)-25(blue) μas at V = 15 100(red)-300(blue) μas at V = 20
Photometry	2 bands	low-res spectrophotometry, 330-1050 nm
Spectroscopy (R$\sim$ 11,000)	none	1-10 km s^{-1} at V $\leqslant$ 16(blue)-17(red)
Target selection	input catalogue	real-time onboard selection

In Table 1 Gaia and Hipparcos characteristics and performances are compared.

2. Overview of the Gaia mission

Gaia is scheduled for launch at the end of 2011 from Kourou by a Soyuz-Fregat launcher, that will put it in a Lissajous-type eclipse-free orbit around L2 point of the Sun-Earth system. The design lifetime is 5 years, with a possible extension of one year. The satellite will perform a continuous scanning of the sky at a rate of 60 arcsec s^{-1}, with a precession period of the spin axis of 70 days. As a result of this scanning law, at the end of the mission the whole sky will have been observed several times, from a few tens to more than 200 depending on the position (see Fig. 1), the average value being around 80.

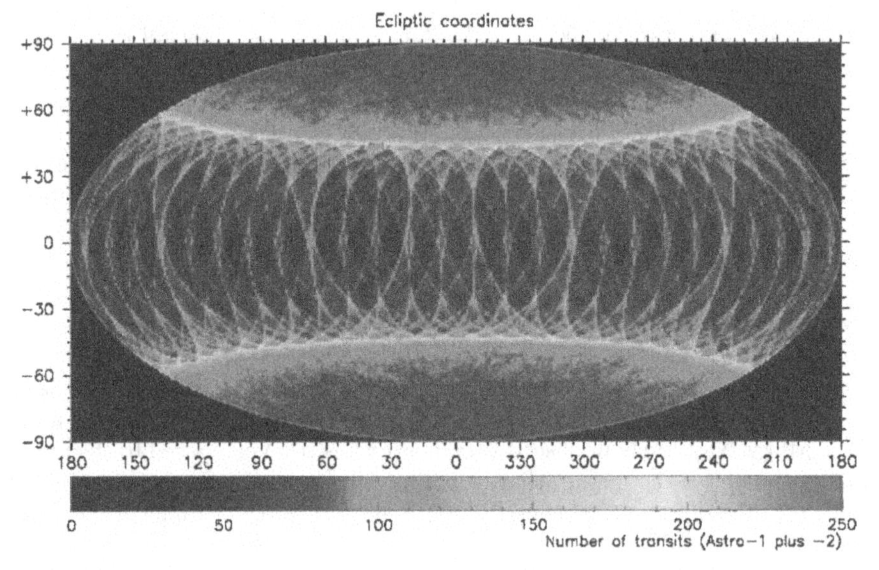

Figure 1. Dependence of the total end-of-mission number of focal plane transits on position on the sky. Shown is an all-sky equal-area Hammer projection in ecliptic coordinates. The maximum number of transits will occur in a $\sim$ 10-deg wide strip around ecliptic latitudes +/− 45 deg.

2.1. *Measurement Principle*

The mission is designed to perform global (wide field) astrometry as opposed to local (narrow field) astrometry. In local astrometry, the star position can only be measured with respect to a neighbouring star in the same field. Even with an accurate instrument, the propagation of errors is prohibitive when making a sky survey. The principle of global astrometry is to link stars with large angular distances in a network where each star is connected to a large number of other stars in every direction.

Global astrometry requires the simultaneous observation of two fields of view in which the star positions are measured and compared. Therefore, the payload will provide two lines of sight, obtained with two separate telescopes. Then, like Hipparcos, the two images will be combined, slightly spaced, on a unique focal plane assembly. Objects are matched in successive scans, attitude and calibrations are updated, and object positions are solved and fed back into the system. The procedure is iterated as more scans are added (Global Iterative Solution). In this way the system is self-calibrating by the use of isolated non variable point sources that will form a sufficiently large body of reference objects for most calibration purposes, including the definition of the celestial frame. Extragalactic objects (e.g. QSOs) will be used to attach this to the International Celestial Reference Frame.

2.2. *Instruments and data products*

The payload consists of a toroidal structure (optical bench) holding two primary mirrors whose viewing directions are separated by 106.5 deg (the Basic Angle). These two fields of view get superposed and combined on the same focal plane, that contains:

- the Sky Mapper: an array of 2×7 CCDs for on-board star detection and selection;
- the Astrometric Field (AF): an array of 9×7 CCDs for astrometric measurements and integrated white-light photometry;
- the Blue (BP) and Red (RP) Photometers: two columns of 7 CCDs each for prism spectrophotometry in the 330-680 nm wavelength range with dispersion 3-29 nm/px, and in the 640-1050 nm wavelength range with dispersion 7-15 nm/px, respectively;
- the Radial Velocity Spectrometer (RVS): an array of 3×4 CCDs for slitless spectroscopy (through grating and afocal field corrector) at the Ca II triplet (847-874 nm) with resolution R $\sim$ 11,000.

Therefore the data produced by Gaia will be of three types:

- Astrometry (parallaxes, proper motions);
- Photometry, both integrated (such as the white-light G-band from the Astrometric Field and the G_{BP} and G_{RP} from the blue and red photometers) and low-dispersion BP and RP spectra. We show in Figure 2 examples of simulated BP and RP spectra as will be observed by Gaia, for main sequence stars from O5 to M6, at G = 15 and zero reddening (Straižys *et al.* 2006).
- Spectroscopy (radial velocities; rotation, chemistry for the brighter sources).

2.3. *Performances*

2.3.1. *Astrometry*

Astrometric errors are dominated by photon statistics. The expected accuracies of parallax measures as a function of V are listed in Table 1, and are also reported as a function of M_V (i.e. distance) in Table 2 (left part), along with the corresponding number of objects that can be observed by Gaia. We consider V = 7 mag as the bright magnitude limit for astrometric observations, as saturation sets in at about V = 6 mag.
Accuracy on proper motions is about 20% better than on parallaxes.

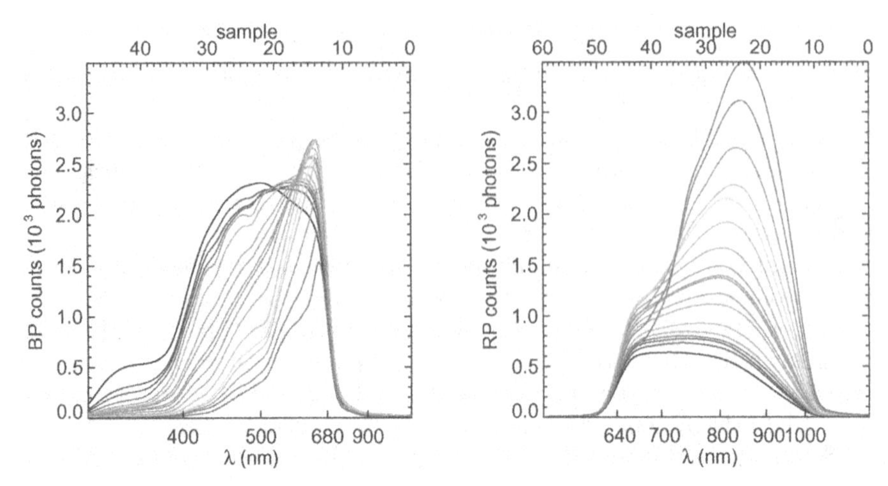

Figure 2. Examples of simulated BP and RP prism spectra for main sequence stars from O5 to M6, at G = 15 and $A_V = 0$ (Straižys *et al.* 2006).

Table 2. Accuracy of astrometric and photometric data. Left: parallax accuracy (and number of objects) as a function of M_V and V (and hence distance). Right: accuracy of internally calibrated integrated photometry as a function of *G* mag.

$\sigma(\pi)/\pi$ N. of objects	M_V	**-5**	**0**	**5**	**10**	**15**	*G* mag	**13**	**15**	**17**	**20**
0.1-0.3% $\sim 10^5$	V D(pc)		7 250	12 250	15 100	17 25	σG mmag	0.1	0.2	0.8	2.5
1-3% $\sim 10^7$	V D(pc)	7 2500	12 2500	15 1000	17 250	20 100	σG_{BP} mmag	0.3	0.8	6.5	37.8
10-30% $\geqslant 10^8$	V D(pc)	12 25000	15 10000	17 2500	20 1000		σG_{RP} mmag	0.3	0.7	6.1	35.2

We note that comparable astrometric accuracy is (will be) obtained from other (present) future space and ground-based facilities, but on **pencil-beam areas** of the sky.

2.3.2. *Photometry*

We show in Table 2 (right part) a summary of the accuracy that Gaia is expected to achieve for the internally calibrated integrated photometry in white light (*G*-band) and in the G_{BP} and G_{RP} bands, as a function of *G* mag. The quoted values are predicted end-of-mission (80 transits) mean values and include Poisson, background and readout noise. A systematic ~ 1 mmag calibration error should be considered in addition (Jordi *et al.* 2007). The accuracy per pixel of the BP/RP spectrophotometric data will be at least one order of magnitude worse, depending on the shape of the spectral energy distribution.

The external (absolute) flux calibration of the Gaia photometric system will be derived from constraining observations of spectrofotometric standard stars (SPSS). To this purpose a grid of SPSS is being built, based on CALSPEC stars and incremented by additional stars to cover adequately the widest possible spectral type range. An observing campaign has been started to ensure accurate and homogeneous data for a suitably large number of SPSS. The accuracy of the absolute calibrated data is expected to be $\sim 1\%$

to a few percent, depending on the number and spectral type of the SPSS and on the accuracy of their SEDs, and may vary with wavelength across the BP/RP spectra.

¿From the BP/RP spectral energy distributions it will be possible to estimate astrophysical parameters using pattern recognition techniques (Bailer-Jones 2008). For example, one expects to obtain (r.m.s. are internal uncertainties at $V = 15$):

- T_{eff} to 1-5 % for a wide range of T_{eff};
- log$g$ to 0.1-0.4 dex, < 0.1 dex for hot stars (SpT $\leqslant$ A);
- $[Fe/H]$ to < 0.2 dex for cool stars (SpT $>$ F) down to [Fe/H] = –2.0 dex;
- A_V to 0.05-0.1 mag for hot stars;

thus providing a complete characterisation of stellar populations.

2.3.3. *Spectroscopy*

The RVS provides the third component of the space velocity of each red (blue) source down to about 17th (16th) magnitude.

Radial velocities are the main product of the RVS, with typical accuracies of 1 to 10 km s^{-1} down to the limiting magnitude. For brighter sources (< 14 mag) the RVS spectra will provide information also on rotation and chemistry, and will allow to obtain more detailed and accurate astrophysical parameters than using the prism BP/RP spectra alone.

2.4. *Science data processing: the DPAC*

All aspects of the Gaia mission are charge and responsibility of ESA, except the processing and analysis of the science data that are assigned to the European astronomical community. To this purpose the community has formed the Data Processing and Analysis Consortium (DPAC), that collects the contribution of nearly 400 scientists from 24 Institutes of ESA member States, and is structured in 9 Coordination Units dealing with all aspects of the data processing. In addition, a number of Data Processing Centers (DPCs) are dedicated to the data handling and processing of specific parts of the pipeline, namely: ESAC (Spain), CNES (France) and the DPCs in Barcelona, Torino, Toulouse, Cambridge and Geneva.

A final data Catalogue will be produced around 2019-2020, containing the end-of-mission measurements for the complete sample of objects down to $V = 20$ mag. Intermediate catalogues might be released before the end of the mission, as appropriate. Science alerts data are released immediately.

No proprietary data rights are implemented.

3. Science with Gaia

The primary goal of the Gaia mission is to obtain data which allow for studying the structure, composition, formation and evolution of the Galaxy. The detailed knowledge of the Galaxy will provide a firm base for the analysis of other galaxies for which this level of accuracy cannot be achieved through direct observations. However, a large number of objects external to the Galaxy will be reached by the Gaia instruments, yielding results of no less interest and importance.

3.1. *Science products*

In the Galaxy: Gaia will provide a complete census of all stellar populations down to 20th magnitude. Based on the Besançon Galaxy model (Robin *et al.* 2003, 2004) Gaia is expected to measure about 9×10^8 stars belonging to the Disk, 4.3×10^8 Thick Disk stars, 2.1×10^7 Spheroid stars and 1.7×10^8 Bulge stars. Binaries, variable stars and rare stellar types (fast evolutionary phases) will be well sampled, as well as special objects such as Solar System bodies ($\sim 10^5$), extra-solar planets ($\sim 2 \times 10^4$), WDs ($\sim 2 \times 10^5$), BDs ($\sim 5 \times 10^4$).

One billion stars in 5-D (6-D if the radial velocity is available, and up to 9-D if the astrophysical parameters are know as well) will allow to derive the spatial and dynamical structure of the Milky Way, its formation and chemical history (e.g. by detecting evidence of accretion/merging events), and the star formation history throughout the Galaxy. The huge and accurate database will provide a powerful testbench for stellar structure and evolution models. The possibility to obtain clean Colour-Magnitude (and hence HR) diagrams throughout the Galaxy will lead to accurate mass and luminosity functions, as well as complete characterisation and dating of all spectral types and Galactic stellar populations. The distribution and rate of microlensing events will allow to map the dark matter distribution. The cosmic distance scale will get a definitive and robust definition (zero-point) thanks to the very accurate distance (i.e. luminosity calibration) of the primary standard candles, RR Lyraes and Cepheids.

Outside the Galaxy: the brightest stars in nearby (LG) galaxies will be observed by Gaia, as well as SNe and burst sources ($\sim 2 \times 10^4$), distant galaxies ($\sim 10^7$), QSOs ($\sim 10^6$), gravitational lensing events ($\leqslant 10^2$ photometric, a few 10^2 astrometric).

Fundamental physics and general relativity will also benefit from Gaia observations: as an example, the parameter γ, representing the deviation from Newtonian theory of the gravitational light bending, will be measured to $\sim 5 \times 10^{-7}$ as compared to the present accuracy of 10^{-4}-10^{-5}.

3.2. *The impact of Gaia on stellar ages: a few examples*

3.2.1. *Globular clusters*

Globular clusters (GCs) are among the densest fields in the sky, and considering that the maximum density that Gaia can handle is ~ 0.25 star arcsec^{-2} not all of the 150 GCs known in the Galaxy (Harris 1996) will be completely observable right to the centre. Simulations with King models and concentration parameter c = 0.5 to 2.5 have shown that 30% of them can be fully observed, most of the remaining ones can be observed at radial distances r > 1 arcmin, and only 5 will be observable at r > 3 arcmin. Therefore, for more than 100 GCs it will be possible to observe from 10^3 up to 10^6 stars each.

The availability of parallaxes and proper motions, as well as radial velocities for V < 16-17, allows to assess the membership and hence to derive clean CM diagrams. At the limiting magnitude V = 20, corresponding to $\geqslant 1$ mag below the main-sequence turn-off for the 30 GCs closer than 10 kpc, the accuracy on proper motions is expected to be ~ 0.08 mas (for red stars) to ~ 0.25 mas (for blue stars).

For comparison, we note that King *et al.* (1998), in their seminal work on NGC 6397, were able to perform a very good cleaning of the main-sequence using proper motions of accuracy ~ 10 mas obtained from WFPC2-WFC data over a 32 month time baseline. The most recent achievement by Anderson & King (2006) is ~ 0.5 mas astrometric accuracy from ACS/WFC data on well exposed images. This accuracy is indeed getting close to Gaia's, but on a field of view of only about 200×200 arcsec^2.

Clean CMD are the essential tool to derive the *absolute* (and relative) age of a GC, as we discuss in more detail in the following section using M3 as an example.

3.2.2. *The age of M3*

At a distance of about 10 kpc, the brightest features of M3 CM diagram, namely the upper RGB and the HB, are at V ~ 12.5 to 16 mag. Therefore these stars will be observed with *individual* accuracy $\sigma(\pi)/\pi \sim$ 7-30%. By averaging the results from 1000 such stars distributed according to the RGB luminosity function (Ferraro *et al.* 1997), the distance to M3 can be known to about 0.5% or better.

The most classical clock provided by stellar evolution theory for dating Population II stars is the luminosity of turn-off stars M_V(TO). This has been parameterised by Renzini (1993) as:

$$Log t_9 \propto 0.37 M_V(TO) - 0.43Y - 0.13[Fe/H]$$

M_V(TO) is sensitive to input physics and assumptions that affect the size and energy production of the radiative core. Comparison of the various most recent M_V(TO) vs. age relations shows that an intrinsic - and hence systematic - theoretical error in the age determination may be present. We refer to the presentation by B. Chaboyer (this conference) for more details on the intrinsic/systematic errors of theoretical models.

In addition to this, errors on the observable parameters entering the M_V(TO)-age relation must be considered. By assuming typical values currently obtained for these errors, the error budget can be summarised as:

- $0.85\sigma(V_{TO})$: error associated to the photometric determination of the TO. Extremely accurate and well defined main sequences can presently be obtained with instruments such as the HST and other large ground-based facilities, however the isochrones are nearly vertical at the TO, and V_{TO} is rather difficult to define. We assume $\sigma(V_{TO})$ ~0.10 mag;
- 0.85σ(mod): error associated to the distance determination, we assume ~0.10 mag;
- $0.85\sigma(A_V)$: error associated to the extinction determination, we assume ~0.06 mag;
- 0.99σ(Y): error associated to the Helium abundance determination, we assume $\geqslant$0.02 dex;
- 0.30σ[M/H]: error associated to the chemical abundance determination, we assume ~0.10 dex.

The final accuracy on age determinations is ~13%, as also estimated by A. Sarajedini (this conference). In a few particular cases and with especially accurate data and analysis, the accuracy on age determination has been claimed to be as low as $\leqslant$10% (Gratton *et al.* 2003). However, before further improvement can be achieved systematic errors need to be solved, for example on chemical abundance determination (by defining the calibrating Solar mixture and metallicity scale), on helium abundance determination (which is confused by the possible presence of multiple populations), on the definition of a homogeneous reddening scale.

Gaia will do its share to improve absolute age by acting on most of the above items:
i) clean CMDs and very accurate photometry at the level of the turn-off will allow to obtain a more precise definition and estimate of the observed V_{TO}. Accuracies of ~0.01-0.02 mag can be foreseen, and are within reach also of the best present and future observing facilities.
ii) The reddening will be monitored by Gaia for each object as part of the astrophysical parameter determination and may not be very accurate individually, but the statistical use of all cluster stars could lead to a rather accurate mean estimate. To be conservative, we assume a factor two improvement in the accuracy of the reddening values.
iii) Similarly, $[Fe/H]$ will be estimated as part of the astrophysical parameter determination, and the mean of hundreds/thousands stars will carry rather low internal errors.
iv) The most important contribution, however, will be on the distance determination, by greatly reducing the error on distance (e.g. by a factor 10 at 10 kpc).
Altogether, we expect that *absolute* ages can be known to ~ 5% or better.

3.2.3. *Open clusters*

Open clusters (OCs) are much looser than GCs and will be completely observable to the centre by Gaia. Therefore, the same type of analysis described for GCs in Sections 3.2.1 and 3.2.2 can be applied to the entire stellar population for each and all of the presently known OCs. In addition, Gaia data may well be able to identify new (faint/loose) clusters.

With Hipparcos, firm results were obtained only on 7 OCs, whereas the Pleiades still represent a controversial case.

The Pleiades have been studied extensively in the last decade, and very similar parallax values have been found by various authors: $\pi = 7.59 \pm 0.14$ mas from MS fitting (Pinsonneault *et al.* 1998), $\pi = 7.69$ mas from various methods (Kharchenko *et al.* 2005), $\pi = 7.49 \pm 0.07$ mas from HST-FGS parallaxes of three stars in the inner halo (Soderblom *et al.* 2005). However, from the new reduction of Hipparcos data van Leeuwen (2007) finds $\pi = 8.18 \pm 0.13$ mas. This discrepancy of about 8% in the distance determination will be resolved unambiguously by Gaia. Since all the stars of the Pleiades are brighter than V = 15, they will have *individual* parallaxes determined to better than 0.1-0.2%, and the distance and internal stellar distribution will be derived with extremely high precision.

3.2.4. *The distance scale: local calibrators*

- **RR Lyraes**

RR Lyrae variable stars are the most traditional standard candles, as their absolute magnitude can be expressed to a first approximation as $M_V = \alpha + \beta[Fe/H]$, with $\beta \sim 0.2$. However, the zero-point α of this relation is determined to somewhat lower accuracy than the slope β.

Hipparcos measured parallaxes for 126 RR Lyrae stars with $<V> =$ 10 to 12.5 mag (750-2500 pc, Fernley *et al.* 1998), but only one star, RR Lyr itself, had a parallax measured to better than 20%, $\pi = 3.46 \pm 0.64$ mas (van Leeuwen 2007). However, the parallax measured by Benedict *et al.* (2002) using HST data, $\pi = 3.82 \pm 0.20$ mas, leads to a shorter distance modulus by ~0.2 mag. This discrepancy is far too large and definitely not acceptable for what is supposed to be the basic luminosity/distance calibrator and the first step in the cosmic distance scale.

Gaia will obtain the parallax of RR Lyr to $< 0.1\%$ and the trigonometric distances to *all* the field RR Lyraes within 3 kpc with *individual* accuracy $\sigma(\pi)/\pi < 3\%$ (better than 30% for most galactic RR Lyraes). This will allow to calibrate the $M_V - [Fe/H]$ relation with very high accuracy, for application to all stellar systems where a good estimate of the RR Lyrae metallicity and mean V magnitude is possible.

- **Cepheids**

Cepheids, along with RR Lyrae stars, form the cornerstone of the extragalactic distance scale. Classical (Pop I) Cepheids are several magnitudes brighter than RR Lyraes, and can be observed in many spiral and irregular galaxies as far as 25 Mpc (thus reaching the Fornax and Virgo clusters) with the use of the HST and other large ground-based or space telescopes. The Hipparcos data provided the first opportunity to calibrate independently the critical parameters in the Period-Luminosity-Colour (PLC) relation for classical Cepheids in the Galaxy. Hipparcos measured parallaxes for about 250 Cepheids, ~ 100 of which with parallax accuracies of 1 mas or less. With the use of these data and additional HST parallax measures for 10 of these stars, van Leeuwen *et al.* (2007) derived a new calibration of the PLC relation leading to a distance modulus for the LMC of 18.48 ± 0.03 mag (no metallicity correction), and hence $H_0 = 70 \pm 5$ km s^{-1} Mpc^{-1}. This is certainly an excellent result, but is still affected by uncertainties due to the various parameters involved in the definition of the calibration itself.

Gaia is expected to measure distances to <4% for all galactic Cepheids (<1% up to 3 kpc), therefore will provide a definitive resolution of the controversy about the zero-point of the PLC relation, as well as about the dependence on period, colour and metallicity.

Cepheid parallaxes can also be measured by Gaia in extragalactic systems such as the Sagittarius dwarf galaxy with $\sigma(\pi)/(\pi) < 10\%$, and the Magellanic Clouds with $\sigma(\pi)/(\pi) \leqslant 50\%$ for all stars with period longer than ~10 days ($M_V \leqslant -4.2$ mag).
This will allow to reach a few fundamental goals:
i) define a very accurate PLC relation, including the possible dependence on metallicity;
ii) establish the distance to the LMC on a completely trigonometric basis, and improve its accuracy with the additional help of the Galactic calibration relation;
iii) establish the universality of the PLC relation, namely its applicability to all galaxies and hence the possibility to derive H_0 and the age of the Universe.

• **Metal-Poor Subdwarfs**

Within the context of distance determination, metal-poor subdwarfs are very important as they constitute the reference frame for GC main sequence stars of similar metallicity. The availability of the high precision Hipparcos parallaxes prompted numerous determinations of distances to several Galactic GCs using this Main Sequence Fitting method. We refer the reader to e.g. Gratton *et al.* (2003) for a detailed description and review.

However, Hipparcos provided precise enough parallaxes only for ~ 30 metal-poor subdwarfs with M_V ~ 5.5 to 7.5 mag (i.e. 2-4 mag below the TO). Since the required astrometric accuracy could only be obtained within limiting magnitude V ~ 10, this allowed to sample a rather small volume of the local neighborhood within 30-80 pc (and hence the small number of metal-poor stars).

Gaia's limiting magnitude to V ~ 15 will allow to measure metal-poor subdwarfs in the same range of absolute magnitude as far as ~ 800 pc with astrometric accuracy better than ~ 3%. Several thousands are expected, thus providing a much better statistics and finer sampling in metallicity for a more accurate fitting to any given GC main sequence.

Acknowledgements

This overview of the Gaia project borrows freely from previous scientific and technical publications, and from the information available on the Gaia website. The effort of the many people involved in the Gaia project is implicit in this synthesis.

This work was done under the financial support of ASI grant I/016/07/0 and PRIN-INAF grant CRA1.06.10.04. The author gratefully acknowledges the support of a IAU travel grant.

References

Anderson, J. & King, I. R. 2006, *STScI Instrument Science Report ACS 2006-01*
Bailer-Jones, C. A. L. 2008, *Gaia-C8-TN-MPIA-CBJ-040 in Gaia Livelink*
Benedict, G. F., McArthur, B. E., Fredrick, L. W. *et al.* 2002, *AJ*, 123, 473
Gaia Concept and Technology Study Report 2000, *ESA-SCI(2000)4*
Fernley, J., Barnes, T. G., Skillen, I., Hawley, S. L., Hanley, C. J., Evans, D. W., Solano, E., & Garrido, R. 1998, *A&A*, 330, 515
Ferraro, F. R., Carretta, E., Corsi, C. E. Fusi Pecci, F., Cacciari, C., Buonanno, R., Paltrinieri, B., & Hamilton, D. 1997, *A&A*, 320, 757
Gratton, R. G., Bragaglia, A., Carretta, E., Clementini, G., Desidera, S., Grundhal, F., & Lucatello, S. 2003, *A&A*, 408, 529
Harris, W. E. 1996, *AJ*, 112, 1487 *(updates in http://www.physics.mcmaster.ca/Globular.html)*

Jordi, C., Fabricius, C., Figueras, F., Voss, H., & Carrasco, J. M. 2007, *Gaia-C5-TN-UB-CJ-042 in Gaia Livelink*

Kharchenko, N. V., Piskuniv, A. E., Röser, S., Schilbach, E., & Scholz, R. D. 2005, *A&A*, 440, 403

King, I., Anderson, J., Cool, A., & Piotto, G. 1998, *ApJ*, 492, L37

van Leeuwen, F. 2007, *Hipparcos, the New Reduction of the Raw Data*, Springer, ASSL 350

van Leeuwen, F., Feast, M. W., Whitelock, P. A., & Laney, C. D. 2007, *MNRAS*, 379, 723

Perryman, M. A. C, Bernacca, P. L., & SOC 1997, *ESA SP-402*

Pinsonneault, M. H., Stauffer, J., Soderblom, D. R., King, J. R., & Hanson, R. B. 1998, *ApJ*, 504, 170

Renzini, A. 1993, *Ann. NY Acad. Sci.*, 688, 124

Robin, A. C., Reylé, C., Derrière, S., & Picaud, S. 2003, *A&A*, 409, 523

Robin, A. C., Reylé, C., Derrière, S., & Picaud, S. 2004, *A&A*, 416, 157

Soderblom, D. R., Nelan, E., Benedict, G. F., McArthur, B., Ramirez, I., Spiesman, W., & Jones, B. F. 2005, *AJ*, 129, 1616

Straižys, V., Lazauskaitė, R., Brown, A. G. A., & Zdanavičius, K. 2006, *Gaia-C8-TN-ITPA-VS-001 in Gaia Livelink*

Symposium "The 3-Dimensional Universe with Gaia" 2005, *ESA-SP-576*

Discussion

C. CORBALLY: What relation will Gaia have with Earth- based surveys such as Pan-STARRS and LSST? Will there be collaboration and/or overlap between the data processing and analysis teams?

C. CACCIARI: The Gaia Data Processing and Analysis Consortium is totally European and there is no official collaboration or overlap with other teams. But on some specific issues there is indeed collaboration with "external" advisors, for example on the selection and observation of standard stars for the absolute calibration of the spectrophotometric system.

M. ROBBERTO: How much degradation do you expect from radiation damage to the CCD detectors?

C. CACCIARI: This is presently being tested by Astrium and closely monitored by a dedicated DPAC team. Solutions have been proposed (e.g., charge injection), but a definitive answer will be available only when the test will be completed, in April 2009.

The Ages of Stars
Proceedings IAU Symposium No. 258, 2008
E.E. Mamajek, D.R. Soderblom & R.F.G. Wyse, eds.

doi:10.1017/S1743921309032074

Stellar ages from asteroseismology

Yveline Lebreton[1,2] and Josefina Montalbán[3]

[1]Observatoire de Paris, GEPI, UMR CNRS 8111,
Place J. Janssen, 92195 Meudon, France
email: Yveline.Lebreton@obspm.fr

[2]IPR, Université de Rennes 1, 35042 Rennes, France

[3]Institut d'Astrophysique et de Géophysique, Université de Liège, Belgium
email: j.montalban@ulg.ac.be

Abstract. Asteroseismology has been recognized for a long time as a very powerful mean to probe stellar interiors. The oscillations frequencies are closely related to stellar internal structure properties via the density and the sound speed profiles. Since these properties are in turn tightly linked with the mass and evolutionary state, we can expect to determine the age and mass of a star from the comparison of its oscillation spectrum with the predictions of stellar models. Such a comparison will of course suffer both from the problems we face when modeling a particular star (for instance the uncertainties on its global parameters and chemical composition) and from our general misunderstanding of the physical processes at work in stellar interiors (for instance the various transport processes that may lead to core mixing and affect the ages predicted by models). However for stars where observations have provided very precise and numerous oscillation frequencies together with accurate global parameters and additional information (as the radius or the mass of the star if it is member of a binary system, the radius if it observable in interferometry or the mean density if the star is an exoplanet host), we can also expect to better constrain the physical description of the stellar structure and transport processes and to finally get a more reliable age estimation.

After a brief survey of stellar pulsations, we present some general seismic diagnostics that can be used to infer the age of a pulsating star as well as their limitations. We then illustrate the ability of asteroseismology to scrutinize stellar interiors on the basis of a few examples. In the years to come, extended very precise asteroseismic observations are expected, either in photometry or in spectroscopy, from present and future ground-based (HARPS, CORALIE, ELODIE, UVES, UCLES, SIAMOIS, SONG) or spatial devices (MOST, CoRoT, WIRE, Kepler, PLATO). This will considerably enlarge the sample of stars eligible to asteroseismic age determination and should allow to estimate the age of individual stars with a 10-20% accuracy.

Keywords. stars: atmospheres, evolution, fundamental parameters, interiors, oscillations, techniques: spectroscopic

---

1. Introduction

As can be seen in Fig. 1, pulsating stars are presently observed in nearly each region of the HR diagram where stars are observed. Stellar pulsations may be excited by different mechanisms, in a large range of amplitudes. Self-excitation results from the κ-mechanism which drives pulsations either in the HeII ionisation zone (oscillations in δ Scuti, Cepheids, RR Lyrae or DB white dwarfs), in the HI and HeI ionisation zones (oscillations in roAp, Miras and irregular variables) or in the metal opacity bump (see e.g, Dziembowski *et al.* 1993; Pamiatnykh 1999 for oscillations in β Cephei and slowly pulsating -SPB- stars). In γ Doradus stars, excitations are interpreted as due to the convective blocking of the radiative flux (see e.g, Guzik *et al.* 2000; Warner *et al.* 2003; Dupret *et al.* 2004). Finally, in stars with significant convective envelopes, the so-called

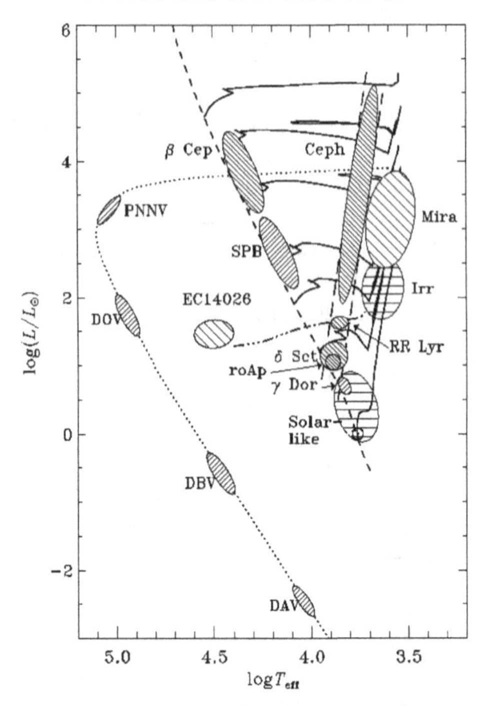

Figure 1. Pulsators in the HR diagram (courtesy J. Christensen-Dalsgaard).

solar-like oscillations are interpreted as resulting from stochastic excitation by turbulent convective motions. This can occur on the main sequence (Christensen-Dalsgaard 1982) and in subgiants and red giants (Dziembowski *et al.* 2001).

Oscillations are observed either by spectroscopic measurements (velocity variations) or by photometric ones (intensity variations). While only one or two modes but with large amplitudes are observed in Cepheids, W Virginis or RR Lyrae stars, a large number of modes with small amplitudes are observable in solar-like pulsators, δ Scuti, roAp, β Cephei, SPB, γ Doradus and white dwarfs pulsators. In solar-like pulsators the variations in velocity amplitudes range from less than 1.0 to about 50 m.s^{-1} (Sun: 0.2 m.s^{-1}) while relative intensity changes vary from 10^{-6} to 10^{-3} (Sun: 4. 10^{-6}) and the amplitude scales as $\left(\frac{L}{M}\right)^{\alpha}$ where α is in the range $0.7 - 1.0$ (Kjeldsen and Bedding 1995; Samadi *et al.* 2005). Amplitudes in δ Scuti, roAp, β Cephei, SPB, γ Doradus and white dwarfs pulsators range from 10^{-3} to a few 10^{-1} mag.

2. Characteristics of stellar oscillations

Each oscillation eigenmode of frequency $\nu_{n,\ell,m}$ consists in a radial part characterized by the radial order n (n is the number of nodes along the stellar radius) and in a surface pattern characterized by a spherical harmonic $Y_{\ell,m}$ where ℓ is the mode angular degree (ℓ is a measure of the number of wavelengths along the stellar circumference) and m is the azimuthal order (number of nodes along the equator).

In the resolution of the equations of stellar oscillations two characteristic frequencies appear, in addition to the cut-off frequency. First, the acoustic (Lamb) frequency $S_\ell = [\ell(\ell+1)]^{\frac{1}{2}}(c/r)$ which is a measure of the compressibility of the medium and depends on the adiabatic sound speed $c = (\Gamma_1 P/\rho)^{\frac{1}{2}}$ where $\Gamma_1 = (\partial \ln P/\partial \ln \rho)_{\rm ad}$ is the first adiabatic coefficient. Second, the buoyancy or Brunt-Väisälä frequency $N_{\rm BV}$, such that $N^2_{\rm BV} =$

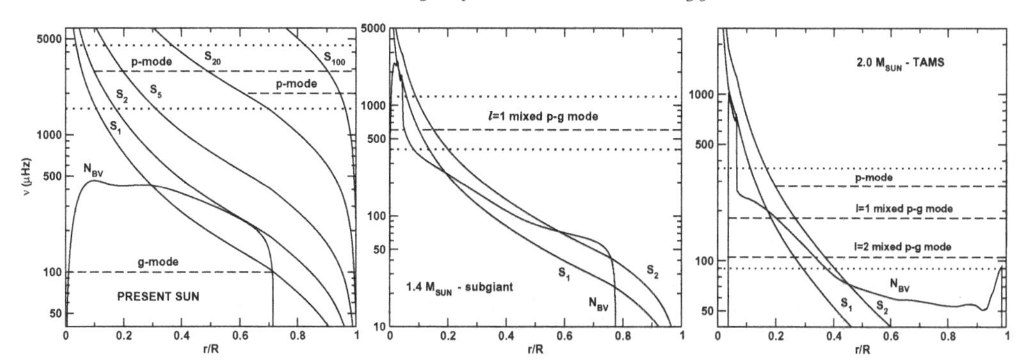

Figure 2. Propagation diagrams for a solar model (left), a 1.4 $M_\odot$ model on the subgiant branch (centre) and a 2 $M_\odot$ model (δ Scuti) at the end of the MS (right). The frequency domains of solar-like oscillations (high degree p-modes) is located in-between the dotted lines.

$g\left(\frac{1}{\Gamma_1}\frac{d\ln P}{dr} - \frac{d\ln\rho}{dr}\right)$. $N_{\rm BV}^2$ is positive in a radiative region where it corresponds to the oscillation frequency of a perturbed fluid element while it is negative in convective regions. It is worth noticing that for ideal gases, $c \propto (T/\mu)^{\frac{1}{2}}$ and $N_{\rm BV}^2 \simeq (g^2\rho/P)(\nabla_{\rm ad} - \nabla + \nabla_\mu)$ where μ is the mean molecular weight and ∇_μ is its gradient.

It can be shown that different waves can propagate inside a star: modes with frequencies higher than both S_ℓ and $N_{\rm BV}$ correspond to standing sound waves (acoustic pressure modes or p-modes) while modes with frequencies lower than both S_ℓ and $N_{\rm BV}$ correspond to standing gravity waves (g-modes). Otherwise modes are evanescent. In absence of rotation, the frequency of an oscillation mode only depends on two parameters (density ρ and Γ_1 or equivalently ρ and sound speed). Therefore frequencies change with mass, evolution (age) and chemical composition.

In Fig. 2, we show propagation diagrams. The left figure shows that in the present solar interior the domains of propagation of p and g-modes are well separated. The central figure shows a model of a star of 1.4 $M_\odot$ on the subgiant branch. In this star a μ-gradient has been built in the radiative core during the main sequence (MS) evolution and the core has become more and more dense on the subgiant branch. As a result, the Brunt-Väisälä frequency shows a central peak; this allows g-modes to propagate in a domain of frequency corresponding to the p-modes domain. Due to the closeness of the propagation regions (see the $\ell = 1$ mode on the figure), a g-mode can interact with a p-mode of the same frequency through the so-called avoided crossing and we therefore expect to observe p–g mixed modes in this kind of stars (Aizenman *et al.* 1977). This situation is also expected to happen in δ Scuti stars (right figure), on the MS, because of the building of a μ-gradient in the radiative regions just above the receding convective core. More generally, depending on the mass and evolutionary state, different kinds of modes are predicted to be observable: for instance in β Cephei and δ Scuti stars low order p and g modes are expected with periods in the range 2–8 hours for the former and 30 min–6 hours for the latter, in SPB stars and γ Doradus stars high order g-modes are expected with periods in the range 15 hours–5 days for the former and 8 hours–3 days for the latter while in solar-like pulsators high order p-modes are expected with periods in the range a few minutes (MS stars) to a few hours (red giants).

3. Age diagnostics through p-modes

In the case of the Sun, more than 10^5 modes can be observed, of all degrees, with an excellent precision. It is therefore possible to derive the density, the sound speed or the

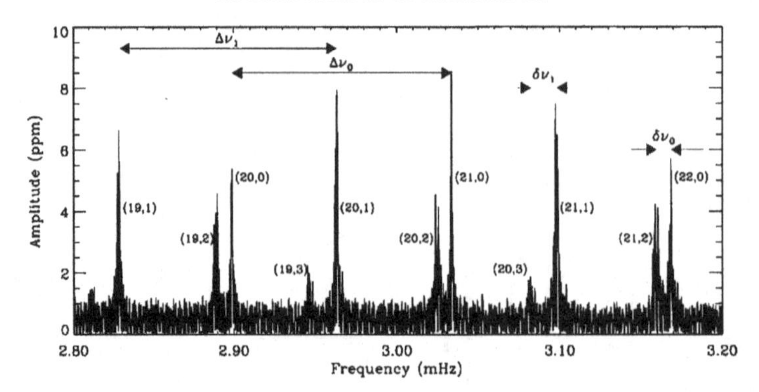

Figure 3. Detail of the solar oscillations amplitude spectrum, as observed by VIRGO on SOHO (Figure from Bedding & Kjeldsen 2003). Each frequency peak is labeled with the value of the mode order and degree and the large and small separations are indicated.

internal rotation rate by inversion of the frequencies which in turn gives access to very precise information on the solar properties such as the age, helium content and depth of the outer convective envelope (see e.g. Christensen-Dalsgaard, these proceedings). On the other hand, in the case of solar-like pulsators, only a few modes of low degree (mainly $\ell = 0, 1, 2, 3$) are accessible to observations. However, as discussed below, this can provide valuable information on the stellar interior.

3.1. *Asymptotic theory: frequency combinations and asteroseismic diagrams*

Figure 3 shows a detail of the solar power spectrum where the modes have been identified by their order and degree. The regular pattern seen is indeed predicted by the asymptotic theory of stellar oscillations (Vandakurov 1967, Tassoul 1980) from which the frequency of a mode of high order n and of degree $\ell \ll n$ is given by $\nu_{\mathrm{n},\ell} = \Delta\nu\left(n + \frac{1}{2}\ell + \epsilon_{n,\ell}\right) - \ell(\ell+1)D_0$ where $\epsilon_{n,\ell}$ is a function sensitive to surface physics, but only weakly sensitive to the order and degree of the mode, and $\Delta\nu$ and D_0 will be explained below. As a consequence the difference in frequency between two modes of consecutive orders and same degree is approximately constant and given by $\Delta\nu \simeq \nu_{\mathrm{n}+1,\ell} - \nu_{\mathrm{n},\ell} \equiv \Delta\nu_\ell$ while the difference in frequency between two modes of consecutive orders and degrees differing by two units is $\delta\nu_\ell \equiv \nu_{\mathrm{n},\ell} - \nu_{\mathrm{n}-1,\ell+2} = 4(\ell+6)D_0$. The differences $\Delta\nu$ and $\delta\nu$ are called the large and small frequency separations, respectively.

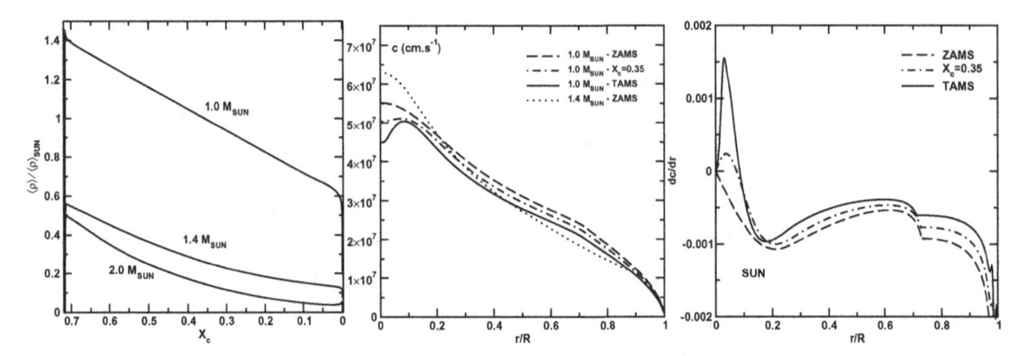

Figure 4. Left: the change of the mean density with evolution on the MS for 3 values of the stellar mass. Centre and right: the sound speed profile (centre) and the dc/dr profile (right) in 1 $M_\odot$ models, on the ZAMS, in the middle of the MS and at the end of the MS (TAMS)

It can be shown that the large frequency separation can be approximated by $\Delta\nu \sim (2\int_0^R \frac{dr}{c})^{-1}$. Therefore it is a measure of the inverse of the time it takes a sound wave to travel across the star and it scales as $(M/R^3)^{1/2}$, that is as the mean density. As shown in Fig. 4 left, for stars on the MS, the mean density increases as mass decreases and at given mass it decreases as evolution proceeds; we therefore expect the large separation to be quite sensitive to the mass and slightly to the evolutionary state on the MS. On the other hand, the small frequency separation depends on the value of $D_0 = \frac{\Delta\nu}{4\pi^2\nu_{n,\ell}}\left[\frac{c(R)}{R} - \int_0^R \frac{dc}{dr}\frac{dr}{r}\right]$ and is therefore sensitive to the integral over the star of the sound speed gradient weighted by $1/r$. During MS evolution, both the temperature T and the mean molecular weight μ increase in stellar cores but the relative increase in μ exceeds that in T so that the net effect is a decrease of the sound speed towards the center (see Figs. 4 centre and right). This in turns leads to a decrease in D_0 and therefore of the small separation as evolution proceeds making $\delta\nu$ a valuable age estimator.

Fig. 5 left recalls how the age of a star is poorly constrained in those regions of the HR diagram where the isochrones are degenerate (for instance in the regions close to the zero age main sequence of low mass stars and in the turnoff regions). On the contrary if we plot in a diagram the large separation versus the small one (Christensen-Dalsgaard 1988), or rather the ratio $r_{\ell,\ell+2} = \delta\nu_\ell/\Delta\nu_\ell$ of the small to the large separation which is less sensitive to surface effects (Roxburgh & Vorontsov 2003), we find that the degeneracy is removed

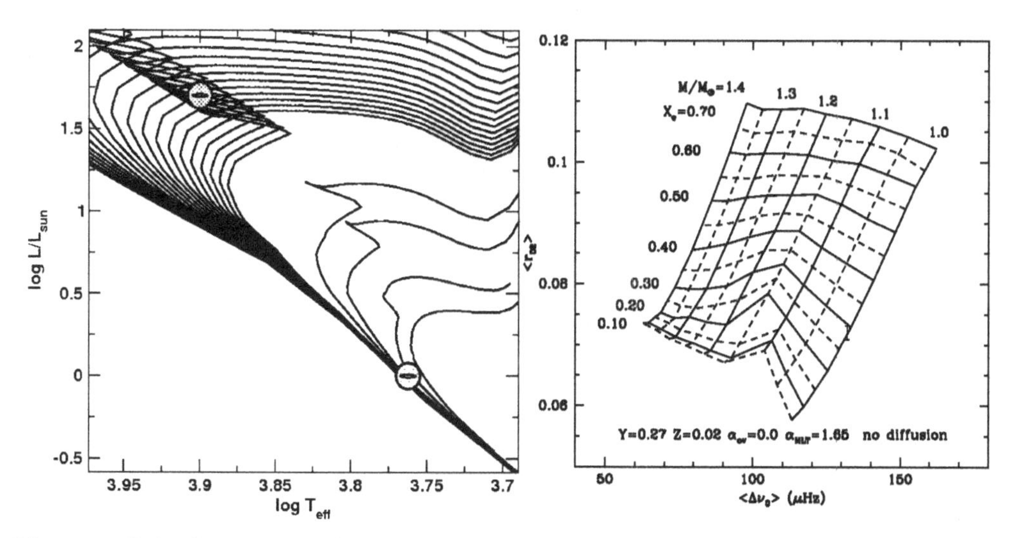

Figure 5. Left: the inversion of isochrones in the HR diagram doesn't allow to estimate the age of non evolved stars (the age of a G-star similar to the Sun is found to be in the range 0.5-10 Gyr!) while it gives poor accuracy on the age of stars close to the turnoff (the age of a typical A-star lying at the turnoff is found to be in the range 0.6-1 Gyr). Large ellipses correspond to the present uncertainty on the HR diagram position while small ones correspond to the accuracy achievable at GAIA time (see e.g. Lebreton 2005). Isochrones have been obtained with the CESAM code (Morel & Lebreton 2008). Right: Asteroseismic diagram in which the abscissa is the mean large frequency separation $\Delta\nu$ of $\ell = 0$ modes of different orders and the ordinate is the mean ratio of the small separation (between $\ell = 2$ and $\ell = 0$ modes) to the large separation. The frequencies have been calculated for standard models of different masses and evolution stages on the MS. Models have been obtained with either the CESAM or the CLES code (Morel & Lebreton 2008; Scuflaire *et al.* 2008a) and frequencies with the LOSC oscillation code (Scuflaire *et al.* 2008b). The frequency means have been estimated following Mazumdar (2005). The error bar located at the solar position corresponds to the one expected from CoRoT or Kepler missions measurements, i.e. to a relative accuracy on the individual frequency of 10^{-4}.

(see Fig. 5 right). Therefore a precise measure of p-modes frequencies should provide a valuable diagnostic of the age (and mass) of solar-like oscillators. Indeed Kjedsen *et al.* (2008) have estimated that with an accuracy on observed frequencies of a few 0.1 μHz (typical accuracy of the CoRoT and future Kepler missions), the age of a solar type oscillator can be determined to better than $\approx 10\%$ of its MS lifetime.

In the age estimation it is important to evaluate the uncertainties that result either from the uncertainty on the determination of the stellar fundamental parameters (as the chemical composition) or from the weak knowledge of several aspects of the physics entering stellar models calculation. For instance the description of convection, the occurrence of either microscopic or turbulent diffusion of chemical elements or the estimate of the size of the mixed core -which depends on processes like overshooting or rotational mixing- will all affect the calculated frequencies. We have estimated how the age estimated from an asteroseismic diagram ($\langle r_{0,2}\rangle$,$\langle\Delta\nu_0\rangle$) changes with stellar model inputs. We find that changing the mixing-length parameter from the value $\alpha_{\rm MLT} = 1.65$ to 1.80 changes the age at a given evolutionary stage on the MS (same central hydrogen content $X_{\rm c}$) by less than 1.5% while introducing microscopic diffusion in the model calculation changes the age by less than $\approx 5\%$. Changing the initial helium abundance in mass fraction from $Y = 0.27$ to 0.26 modifies the age at constant $X_{\rm c}$ by less than $\approx 10\%$. On the other hand, as illustrated in Fig. 6 (left) changing the initial metallicity in mass fraction from the value $Z = 0.02$ to 0.01 changes the age by 15 to 30% while changing the overshooting parameter from $\alpha_{\rm ov} = 0.0$ to 0.2 has huge effects on the age estimate for stars possessing convective cores in the late part of their MS evolution (see Fig. 6 right).

To improve the determination of the age based on asteroseismic diagrams, it is therefore crucial to improve the observational determination of their metal content (this requires high resolution spectroscopy and further improvements in model atmospheres). Also it is very interesting to get additional (preferably independent) information on their fundamental parameters. This is achievable for stars observable in interferometry (direct access to the radius), for members of binary systems (access to mass and/or radius), or for hosts of exoplanets (access to the mean density from the planetary transit). Finally, as discussed in the following, the full interpretation of seismic data may allow to get further information, such as the surface helium abundance or the size of the mixed core.

3.2. *Deviations from the asymptotic theory: signature of sharp features in the interior*

Rapid variations of physical quantities in the interior of a star can be detected by seismic analysis. Indeed, any sharp feature inside the star will produce an oscillatory signal in the frequencies of the form $\widetilde{\delta\nu} \sim A\cos(2\tau_m\omega + \phi)$ where $\tau_m = \int_{r_m}^{R} \frac{dr}{c}$ is the acoustic depth of the abrupt feature (see e.g. Gough 1990; Monteiro *et al.* 2000).

For instance, the second He ionisation in the outer convection zone of solar-like stars produces a depression in the profile of the adiabatic exponent Γ_1. The higher the helium abundance in the convection zone, the deeper the Γ_1 depression. The sound speed profile locally carries the signature of the Γ_1 depression which results in an oscillatory signal in the frequencies as a function of radial order. Basu *et al.* (2004) have shown that this signal, that can be extracted from low degree p modes frequencies, can be used to infer the helium abundance in the envelope of solar-like pulsators. They find that, with a relative accuracy of 10^{-4} on the frequencies (typical of CoRoT or Kepler missions), it should be possible to infer the envelope helium abundance (Y in mass fraction) of pulsators of mass in the range $0.8 - 1.4\ M_\odot$ with an accuracy of $\Delta Y \approx 0.02$ provided their radius or mass can be estimated independently. Also, it should be possible to infer the position of the basis of the convective envelope (see e.g. Ballot *et al.* 2004) or the boundary of the mixed core (see e.g. Mazumdar *et al.* 2006) in a similar way. The latter depends on the extra

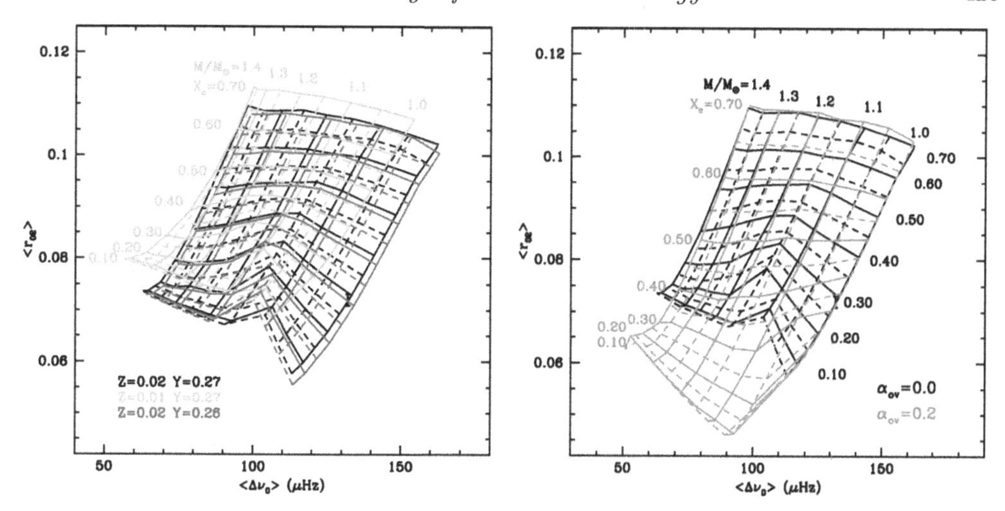

Figure 6. Left: effects on the asteroseismic diagram of a change of chemical composition. In black models, with $Y = 0.27$, $Z = 0.02$; in dark grey, models with $Y = 0.26$, $Z = 0.02$ and in light grey, models with $Y = 0.27$, $Z = 0.01$. Right: Effect of overshooting with in black the models without overshooting and, in grey, models calculated with $\alpha_{\rm ov} = 0.2$.

mixing processes that can take place at the border of the convective core (overshooting, rotationally induced mixing) and it is a crucial data for age estimation.

4. Age diagnostics through g-modes and p-g mixed modes

The properties of g-modes are determined by the behavior of the Brunt-Väisäla frequency in the stellar interior. These modes are sensitive probes of the chemical gradients resulting from the combined effect of nuclear burning and convective mixing, and hence they are potential diagnostics of the stellar evolutionary stage (and therefore of age).

In the case of high order low degree g-modes, such as those expected in SPB ($4-7\ M_\odot$) and γ Doradus stars ($1.4-1.8\ M_\odot$), the first order asymptotic theory predicts that the periods of the modes should be equally spaced (Tassoul 1980) with a period spacing for two modes of consecutive order k and $k+1$ and same degree ℓ given by $\Delta P_{k,\ell} = P_{k+1,\ell} - P_{k,\ell} = \frac{2\pi^2}{L}\Pi_0$, with the *buoyancy radius* $\Pi_0 = \left(\int_{r_0}^{R} \frac{|N_{BV}|}{r} dr\right)^{-1}$. This approximation is, however, no longer valid when variations of $N_{\rm BV}$ on a length scale smaller than the oscillation wavelength are present. That may occur in MS stars due to the building of a μ-gradient at the outer border of the receding convective core. The corresponding sharp feature in $N_{\rm BV}$ can lead to a resonant condition such that modes of different order k are periodically confined in the ∇_μ region. These modes have a different period and as a consequence the period spacing presents an oscillatory behavior rather than the constant value that would be expected in a model without sharp variations in $N_{\rm BV}$ (Fig. 7). By using a second order approximation, it is possible to derive an analytical expression relating the amplitude and periodicity of the oscillatory component to the location and sharpness of the feature in $N_{\rm BV}$ (Miglio *et al.* 2008 and references therein). This periodicity expressed in terms of radial order is given by the ratio between the buoyancy radius of the star, and the buoyancy radius of the ∇_μ region: $\Delta k = \Pi_\mu/\Pi_0$, and corresponds to the difference between radial orders of two consecutive trapped modes in the ∇_μ region. As shown in Figs. 7 and 8, the value of Δk is very sensitive to the evolutionary state of the star: for non evolved models ($X_c \sim X_0$) $\Delta k \to \infty$ and the period

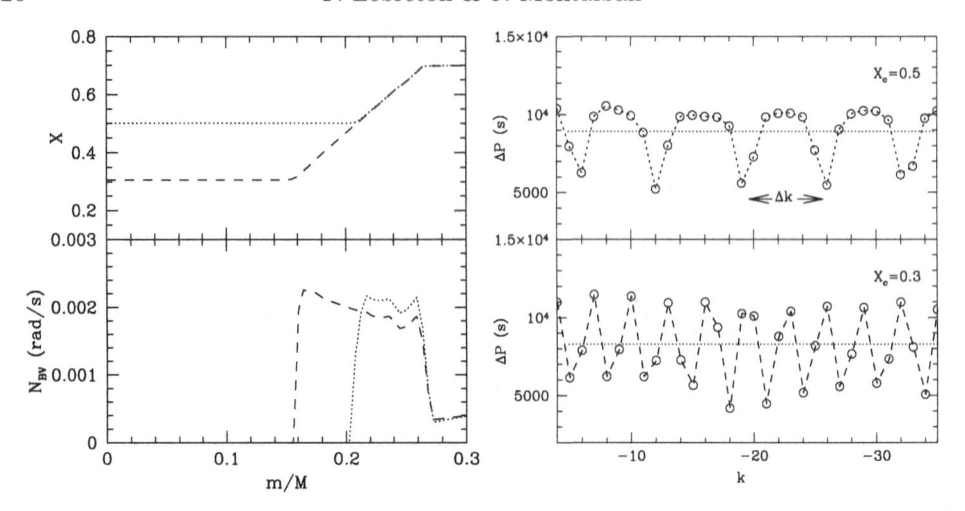

Figure 7. Left: Hydrogen abundance in the core of $6M_{\odot}$ models on the main sequence (upper panel) at $X_c \simeq 0.5$ (dotted line) and at $X_c \simeq 0.3$ (dashed line). The convective core recedes during the evolution leaving behind a chemical composition gradient and, a sharp feature in the Brunt-Väisäla frequency (lower panel). Right: Period spacing for the same models as a function of radial order k. Horizontal dotted lines represent constant period spacing as predicted by the asymptotic approximation, i.e. $\Delta P_{k,\ell} \propto \Pi_0$ (Miglio *et al.* 2008).

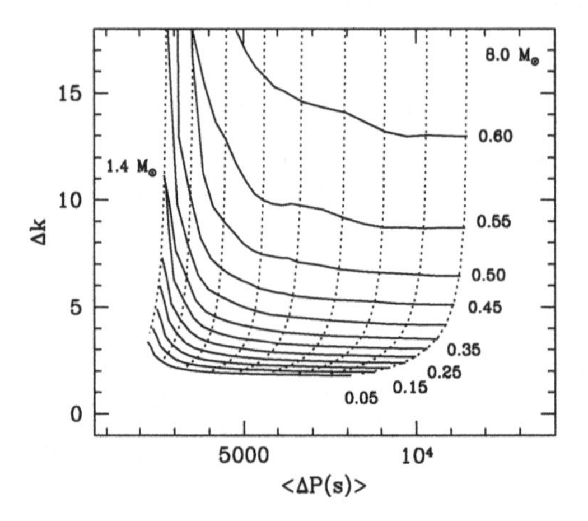

Figure 8. Diagram in which the abscissa is the main period spacing for $\ell = 1$, and the ordinate is the period of the oscillatory behavior of the period spacing (see Fig. 7, right) in terms of the difference in radial order between two consecutive trapped modes. Dotted lines represent the variation of these quantities along the MS evolution for models with masses between 1.4 and 8 $M_{\odot}$, initial chemical composition $X_0 = 0.70$, $Z_0 = 0.02$, and overshooting parameter $\alpha_{\rm OV} = 0.0$. Solid lines (grey) connect models at the same evolutionary state, i.e. with the same value of the central hydrogen mass fraction.

spacing is almost constant as predicted by the asymptotic theory but when the stars evolve the periodicity of the oscillatory component in ΔP decreases.

β Cephei and δ Scuti show p and g-modes of low radial order and, for stars evolved enough, also p-g mixed modes (see. Sect. 2). In this case, the quantitative predictions of the second order asymptotic approximation cannot be used, but they are still able to qualitatively describe the properties of g-modes. The periods of g-modes depend on the location and shape of the chemical composition gradient and therefore so do the frequencies of mixed modes. Since the frequency separation between consecutive modes varies rapidly during the avoided crossing phenomenon, the detection of mixed modes together with that of pure p-modes provides an important constraint to the stellar evolutionary state (see for instance Pamyatnykh *et al.* 2004).

Finally for low mass stars that leave the MS and are evolving as sub-giants, the increasing central condensation together with the chemical composition gradient lead to a large increase of N_{BV}. As a consequence, mixed modes may appear in the frequency domain of solar-like oscillations. Again, the frequencies of the modes undergoing an avoided crossing are direct probes of the evolutionary status: models on the MS do not present mixed modes and the large frequency separation is almost constant (as predicted by the asymptotic theory). On the contrary, as more evolved models are considered, the avoided crossing effects become more important and the large frequency spacing for non-radial modes becomes more irregular (see e.g. Di Mauro *et al.* 2004, Miglio *et al.* 2007).

In all the cases considered above, the frequencies of high-order g-modes and g-p mixed modes are sensitive to the location and shape of the chemical composition gradient. Therefore, these modes are sensitive probes of the evolutionary state and also of the properties of the transport processes acting in stellar cores, for instance, convective overshooting or diffusive mixing. Nevertheless, to clearly discriminate between different scenarios, high accuracy of the classical observables such as effective temperature, luminosity and chemical composition (and masses for binary stars) would also be required.

Age estimates of stars in the galactic disc are crucial for galactic evolution studies. Considering the progress expected on the determination of both the seismic and classical observables, Lebreton *et al.* (1995) (see also Lebreton 2005) find that we can reasonably expect to reduce the uncertainty on the age determination of disc A and F stars which is presently of about $35 - 40\%$ (of which $13 - 24\%$ result from the uncertainty on the size of the mixed cores) to a level of $\approx 10 - 15\%$.

5. Prospects and conclusions

We are presently in a very fruitful period where asteroseismic data are being obtained both from the ground and from space to an increasing level of precision for growing samples of stars in wide ranges of masses and evolutionary stages. The CoRoT satellite has observed for more than one year now. It will observe a total of more than 100 targets (solar-like stars, γ Doradus, δ Scuti, β Cephei, giants) in the so-called Seismo field and for stars observed during Long Runs of 150 days, a frequency accuracy as good as a few 10^{-7}Hz will be reached (see Michel *et al.* 2006). First CoRoT observations are currently under analysis and interpretation. Up to now, three solar-like pulsators have been observed (see e.g. Michel *et al.* 2008a) and for the first one HD49933, observed during the first 60 days Intermediate Run and then during a 150 days Long Run, the large frequency separation has been measured and a first frequency identification for modes of degrees $\ell = 0, 1, 2$ has been proposed (Appourchaux *et al.* 2008). Further very interesting results are being presently obtained related for instance to solar-like oscillations in giant F and G stars and to the very rich oscillation spectra of δ Scuti stars just revealed by CoRoT (see e.g. Michel *et al.* 2008b).

The Kepler photometry space mission to be launched in March 2009 will provide seismic data, at the same level of precision than CoRoT, for a large number of stars, hosts of exoplanets. The information on the stellar mean density, coming from the observation of the exoplanet transit, will add constraints for the age determination. Asteroseismic observations are presently obtained in spectroscopy with HARPS, CORALIE, ELODIE, UVES, UCLES and in photometry with MOST, CoRoT, WIRE while other projects are about to begin or are currently under study either in spectroscopy (SONG, SIAMOIS) or in photometry (Kepler, PLATO).

In parallel, much progress will be done in the determination of the stellar global parameters. For instance GAIA (ESA 2000; Perryman *et al.* 2001), to be launched in 2011, will make astrometric measurements, at the micro-arc second level together with photometric and spectroscopic observations of a huge number of stars covering the whole range of stellar masses, compositions and evolution stages. Important progress is also expected from interferometric and high resolution spectroscopy measurements (for instance with VLT-VLTI, CHARA, KECK, JWST etc.).

Asteroseismology provides diagnostics of stellar evolution all across the HR diagram. If we consider both the present and expected future accuracies on the observed frequencies and the future improvements expected in the determination of the stellar global parameters, it seems reasonable to forecast that asteroseismology will allow to determine the ages of individual stars with a 10–20% accuracy. This will certainly lead to significant progress in the understanding of the history and evolution of galaxies.

References

Aizenman, M., Smeyers, P., & Weigert, A. 1977, *A&A*, 58, 41

Appourchaux, T., *et al.* 2008, *A&A*, 488, 705

Ballot, J., Turck-Chize, S., & Garcia, R. A. 2004, *A&A* 423, 1051

Basu, S., Mazumdar, A., Antia, H. M., & Demarque, P. 2004, *MNRAS*, 350, 277

Bedding, T. R. & Kjeldsen, H. 2003, *PASA*, 20, 203

Christensen-Dalsgaard, J. 1982, *MNRAS*, 199, 735

Christensen-Dalsgaard, J. 1988, *Advances in Helio- and Asteroseismology*, Proc. IAU Symposium 123, 295

di Mauro, M. P., Christensen-Dalsgaard, J., Paternò, L., & D'Antona, F. 2004, *Solar Physics*, 220, 185

Dupret, M.-A., Grigahcène, A., Garrido, R., Gabriel, M., & Scuflaire, R. 2004, *A&A*, 414, L17

Dziembowski, W. A., Gough, D. O., Houdek, G., & Sienkiewicz, R. 2001, *MNRAS*, 328, 601

Dziembowski, W. A., Moskalik, P., & Pamyatnykh, A. A. 1993, *MNRAS*, 265, 588

Gough, D. O. 1990, *Lecture Notes in Physics* 367, 283

Guzik, J. A., Kaye, A. B., Bradley, P. A., Cox, A. N., & Neuforge, C. 2000, *ApJ* (Letters), 542, 57

Kjeldsen, H. & Bedding, T. R. 1995, *A&A*, 293, 87

Kjeldsen, H., Bedding, T. R., & Christensen-Dalsgaard, J. 2008, *American Institute of Physics Conference Series*, 1043, 365

Lebreton, Y., Michel, E., Goupil, M. J., Fernandes, J., & Baglin, A. 1995, *Proc. IAU Symp. 166*, Kluwer, p. 135

Lebreton, Y. 2005, *The Three-Dimensional Universe with Gaia* ESA Special Publication vol. 576, 493

Mazumdar, A. 2005, *A&A*, 441, 1079

Mazumdar, A., Basu, S., Collier, B. L., & Demarque, P. 2006, *MNRAS*, 372, 949

Michel, E., Baglin, A., & Auvergne, M. *et al.* 2006, *The CoRoT Mission* ESA Special Publication, 1306, 39

Michel, E., Baglin, A., Auvergne, M. *et al.* 2008, *Science*, 322, 558

Michel, E., Baglin, A., Weiss, W. W. *et al.* 2008, *Communications in Asteroseismology*, Proc. of the Wroclaw HELAS Workshop, in press

Miglio, A., Montalbán, J., & Maceroni, C. 2007 *MNRAS*, 377, 373

Miglio A., Montalbán J., Noels A., & Eggenberger P. 2008 *MNRAS*, 386, 1487

Monteiro, M. J. P. F. G., Christensen-Dalsgaard, J., & Thompson, M. J. 2000, *MNRAS*, 316, 165

Morel, P. & Lebreton, Y. 2008, *Ap&SS*, 316, 61

Pamyatnykh, A. A. 1999, *Acta Astronomica*, 49, 119

Pamyatnykh A. A., Handler G., & Dziembowski W. A. 2004, *MNRAS*, 350, 102

Perryman, M. A. C., de Boer, K. S., Gilmore, G., *et al.* 2001, *A&A*, 369, 339

Roxburgh, I. W. & Vorontsov, S. V. 2003, *A&A*, 411, 215

Samadi, R., Goupil, M.-J., Alecian, E., Baudin, F., Georgobiani, D., Trampedach, R., Stein, R., & Nordlund, Å. 2005, *Journal of Astrophysics and Astronomy*, 26, 171

Scuflaire, R., Montalbán, J., Théado, S. *et al.* *Ap&SS*, 316, 149

Scuflaire, R., Théado, S., Montalbán, J. *et al.* *Ap&SS*, 316, 83
Tassoul, M. 1980, *ApJ*, Suppl. 43, 469
Vandakurov, Y. V. 1967, *Astronomicheskii Zhurnal*, 44, 786
Warner, P. B., Kaye, A. B., & Guzik, J. A. 2003, *ApJ*, 593, 1049

Discussion

K. COVEY: Many of the examples and models shown in your talk were for solar mass or larger stars. Can you comment on how or if asteroseismology can help us understand the structure and age of the lowest mass ($\sim 0.3M_{\odot}$) stars?

Y. LEBRETON: Presently such low mass stars have not been observed in asteroseismology; Corot will go down to $\sim 0.9M_{\odot}$. For lower -mass stars, not a lot of diagnostics have been developed, however again we could expect to get information on the extent of the outer convective envelope and on transport of chemical elements in the radiative envelope (different terms in the microscopic diffusion process). However, the amplitudes of p-modes are expected to decrease as mass decreases (velocity amplitudes are proportional to $(L/M)^{0.7}$ according to the observational fit by Samadi *et al.* (2005, J. Ap. Astr., 26, 171)). This makes the mode detection quite hard in the very-low-mass stars.

J. STAUFFER: Will Corot tell us anything about pre- main sequence δ Scuti stars? There are a couple of them in NGC 2264, which was the target of a short run. Irrespective of Corot, what could one learn by doing asteroseismological observations of a PMS δ Scuti star?

Y. LEBRETON: For those stars, it would be very interesting to be able to discriminate between PMS δ Scuti stars close to their arrival on the MS and evolved MS δ Scuti stars. We expect that asteroseismology will give us some insight into the rotation profile inside the star and on the related angular momentum transport. For these stars there are also important clues related to the magnetic field. Corot will give us elements to explore those topics by observing a few δ Scuti stars in a young cluster and this should be improved by future missions.

J. CHRISTENSEN-DALSGAARD: Observing PMS stars may allow detection of frequency changes reflecting the evolution of the stars. In fact, with the expected precision of Kepler we may see the evolution on the main sequence of, e.g., δ Scuti stars.

LEBRETON: The variation of the oscillation periods of δ Scuti stars with evolution has been studied by Breger & Pamyatnykh (1998, A&A, 332, 958) who show that the fastest period changes, which are the easiest to detect, are found in PMS δ Scuti stars.

B. WEAVER: With mass and radius estimates, can you constrain [M/H] with your frequency data?

Y. LEBRETON: If enough modes are identified and additional precise information on the mass and radius of the star is available, it will be possible to probe the interior and possibly improve the determination of the chemical composition. However, for seismic analysis we still need to begin with an estimate of the value of [M/H] and ideally with individual elemental abundances which will provide a trial model of the star.

P. DEMARQUE: You showed tantalizing prospects for studying convective cores and core overshoot. What are the prospects for observing g-modes?

Y. LEBRETON: High-order g modes are observed in SBP and γ Doradus stars; moreover, the so-called mixed modes (p-g modes) which have kinetic energy both in interior and outer regions, have been detected in β Cephei and δ Scuti stars. In B supergiants on the post-main sequence theory predicts that g-modes can be excited (Saio *et al.* 2006) and they have indeed been observed in one star. For low-mass solar-type stars where their amplitudes are very weak, g-mode detection is still controversial.

S. VAUCLAIR: Most of your talk is based on the information we can obtain using this so-called "asymptotic theory". In real stars, deviations from this approximation are important. Can you comment on that?

Y. LEBRETON: I have just presented in my talk several examples of departures from the asymptotic theory which take the form of an oscillatory signal in the frequencies or in the frequency differences. Such signals may be (or are already) used to measure the He abundance in the convective envelopes of solar-like oscillators or to get information on core rotation and mixing in stars where mixed modes are excited. Another possibility is to look at features in echelle diagrams.

Yveline Lebreton

The Ages of Stars
Proceedings IAU Symposium No. 258, 2008
E.E. Mamajek, D.R. Soderblom & R.F.G. Wyse, eds.

doi:10.1017/S1743921309032086

The Sun as a fundamental calibrator of stellar evolution

Jørgen Christensen-Dalsgaard
Danish Asteroseismology Centre, and Department of Physics and Astronomy,
Aarhus University, DK 8000 Aarhus C, Denmark
email: jcd@phys.au.dk

Abstract. The Sun is unique amongst stars in having a precisely determined age which does not depend on the modelling of stellar evolution. Furthermore, other global properties of the Sun are known to much higher accuracy than for any other star. Also, helioseismology has provided detailed determination of the solar internal structure and rotation. As a result, the Sun plays a central role in the development and test of stellar modelling. Here I discuss solar modelling and its application to tests of asteroseismic techniques for stellar age determination.

Keywords. Sun: evolution, Sun: interior, Sun: helioseismology, Sun: fundamental parameters, stars: evolution, stars: interior, stars: oscillations, stars: fundamental parameters, asteroseismology

---

1. Introduction

Determining the age of a star from its observed properties requires a model describing how those properties change as the star evolves. Amongst the many properties discussed in this volume, those relevant to using the Sun as a calibrator depend on the changes in the internal structure of the star caused by evolution. Specifically, the change in the composition as hydrogen is fused to helium (with the additional effects of diffusion and settling) changes the internal structure and hence the observable properties. Evidently, both the details of the composition change, and the response of the structure and the observables, depend on the modelling of stellar interiors, and hence the age determination is sensitive to uncertainties in the modelling. The Sun provides a unique possibility for quantifying these uncertainties and attempting to reduce them.

In the present case the Sun has several major advantages: it is the only star for which the age can be inferred in a manner that is essentially independent of stellar modelling, through the radioactive dating of the solar system. Also, its proximity means that its mass, radius and luminosity are known to high accuracy. Finally, as a result of this proximity, helioseismology based on a broad range of modes has allowed inferences to be made of the detailed internal structure (and rotation) of the Sun. This allows a refined test of the modelling of stellar evolution. In addition, by applying techniques for age determination, based on stellar modelling, to the Sun we can test and possibly improve them. This, in particular, applies to the use of asteroseismic analyses for age determination; as discussed by Lebreton & Montalbán (this volume) these promise to be far more precise than other techniques based on the evolution of stellar structure.

2. Solar modelling

A summary of solar modelling and the helioseismic investigations of the Sun was given, for example, by Christensen-Dalsgaard (2002). Bahcall *et al.* (2006) made a detailed analysis of the sensitivity of the models to the choice of input parameters and physics. The solar models used here largely correspond to the so-called Model S of Christensen-Dalsgaard *et al.* (1996), although with some updates.

The mass of the Sun is known from planetary motion, with an accuracy limited by the accuracy of the determination of the gravitational constant G. For the modelling presented here I use $G = 6.67232 \times 10^{-8}$ in *cgs* units, and hence the solar mass is $M_\odot = 1.989 \times 10^{33}$ g. The solar radius $R_\odot$ is obtained from the solar angular diameter and the distance to the Sun; the radius should be defined in a manner that can be related precisely to the model, e.g., as the distance from the centre to the photosphere, defined by the location where the temperature is equal to the effective temperature. A commonly used value is $R_\odot = 6.9599 \times 10^{10}$ cm (Auwers 1891), and this value is used here. However, I note that Brown & Christensen-Dalsgaard (1998) obtained the value 6.9551×10^{10} cm which is probably more accurate. The solar luminosity $L_\odot$ is obtained from the solar constant, i.e., solar flux at the Earth, and the distance to the Sun; this assumes that the solar flux does not depend on latitude, a probably reasonable assumption although one that has never been verified; I use the value $L_\odot = 3.846 \times 10^{33}\,\mathrm{erg\,s^{-1}}$ (see also Fröhlich & Lean 2004).

The solar photospheric composition can be determined from spectral analysis. An important exception is the abundance of helium: the helium lines that led to the detection of helium in the Sun (and hence the name of the element) are formed in the chromosphere, under conditions such that the abundance determination is quite inaccurate. The abundances of other noble gases, amongst which neon is a relatively important constituent of the solar atmosphere, are similarly inaccurate. Thus only relative abundances, commonly defined relative to hydrogen, can be determined. In solar modelling this is typically characterized by the ratio $Z_\mathrm{s}/X_\mathrm{s}$ of the surface abundances by mass Z_s of elements heavier than helium and X_s of hydrogen. A commonly used value has been $Z_\mathrm{s}/X_\mathrm{s} = 0.0245$ (Grevesse & Noels 1993). However, redeterminations of the abundances, based on three-dimensional hydrodynamical models of the solar atmosphere and taking departures from local thermodynamical equilibrium into account, have had a major impact on the inferred abundances; in particular, the abundances of oxygen, carbon and nitrogen were substantially reduced (e.g., Asplund *et al.* 2004) (for a review, see also Asplund 2005). This resulted in $Z_\mathrm{s}/X_\mathrm{s} = 0.0165$; as discussed below, this has had drastic consequences for the comparison between solar models and the helioseismically inferred structure.

For the present discussion the solar age is of course of central importance. This can be inferred from radioactive dating of material from the early solar system, as represented by suitable meteorites. In a detailed discussion presented in the appendix to Bahcall & Pinsonneault (1995), G. J. Wasserburg concluded that the age of the Sun, since the beginning of its main-sequence evolution, is between 4.563×10^9 and 4.576×10^9 yr. The rounded value of 4.6×10^9 yr is often used, including for the reference Model S.

It is evident that computed solar models should match the observed quantities, at the age of the Sun. Models are typically computed without mass loss (see, however, Sackmann & Boothroyd 2003) and hence with the present mass of the Sun. A model with the correct radius, luminosity and $Z_\mathrm{s}/X_\mathrm{s}$ is obtained by adjusting three parameters that are *a priori* unknown: a parameter, such as the mixing-length parameter α_ML, † characterizing the properties of convection which largely determines the radius, the initial helium abundance Y_0 which mainly determines the luminosity, and the initial heavy-element abundance Z_0 which determines $Z_\mathrm{s}/X_\mathrm{s}$. This calibration provides a precise (although not necessarily

† i.e., the ratio of the mixing length to the pressure scale height in the mixing-length description of convection (e.g., Böhm-Vitense 1958; Gough & Weiss 1976)

accurate) determination of the initial solar helium abundance, of importance to studies of galactic chemical evolution. Also, the resulting value of $\alpha_{\rm ML}$ is often used for computations of other stellar models, although there is little justification for regarding $\alpha_{\rm ML}$ as being independent of stellar parameters. More significantly, the calibration of the properties of the solar convection zone can be used as a test of hydrodynamical simulations of convection (e.g., Demarque *et al.* 1999; Rosenthal *et al.* 1999), these may then be used to calibrate the dependence of convection-zone properties, e.g., characterized by $\alpha_{\rm LM}$, on stellar parameters (Ludwig *et al.* 1997, 1999; Trampedach *et al.* 1999).

An important solar observable is the neutrino flux. The discrepancy between the predicted and detected flux of electron neutrinos was long regarded as a potential problem of solar modelling, although even early helioseismic results strongly indicated that changes to the models designed to eliminate the discrepancy were inconsistent with the observed oscillation frequencies (e.g., Elsworth *et al.* 1990). However, it is now realized that the apparent discrepancy was caused by oscillations between different states of the neutrino; with the recent detection of neutrinos of other flavours the total observed flux of neutrinos agrees with predictions (Ahmad *et al.* 2002). Thus the emphasis in the study of solar neutrinos has shifted towards the investigation of the detailed properties of the neutrino oscillations (for recent reviews, see Bahcall *et al.* 2004; Robertson 2006; Haxton *et al.* 2006; Haxton 2008). Interestingly, the computed neutrino flux is not significantly affected by the recent revision in the solar composition (e.g. Bahcall & Serenelli 2005).

3. Helioseismic results on the solar interior

Very extensive data on solar oscillations have been obtained in last two decades (for a review, see Christensen-Dalsgaard 2002). Unlike any other pulsating star the availability of observations with high spatial resolution has provided accurate frequencies for modes over a broad range of spherical-harmonic degrees l, from 0 to more than 1000 (see also Lebreton & Montalbán, these proceedings, for an overview of the properties of stellar oscillations). Most of the observed modes are acoustic modes; these are essentially trapped between the solar surface and an inner *turning point* at a distance $r_{\rm t}$ from the centre given by $c(r_{\rm t})/r_{\rm t} = \omega/\sqrt{l(l+1)}$, where c is the adiabatic sound speed and ω is the angular frequency of the mode. Thus the broad range of l corresponds to a range of inner turning points varying from the centre to just beneath the solar surface. This availability of modes sensitive to very different parts of the Sun is essentially what allows inverse analyses to resolve the structure and rotation of the solar interior.

In the analysis of solar and solar-like pulsations an important issue is the effect of the near-surface layers: modelling of the structure of these layers and of their effect on the oscillation frequencies is highly uncertain, leading to systematic errors in the computed frequencies, which in many cases dominate the differences between observed and computed solar frequencies (e.g., Christensen-Dalsgaard *et al.* 1988). These errors depend essentially only on frequency (apart from a trivial dependence on the mode inertia) and furthermore are small at low frequency (e.g., Christensen-Dalsgaard & Thompson 1997). This allows their effect to be eliminated in the analysis of solar data but they should be kept in mind also in asteroseismic analyses (Kjeldsen *et al.* 2008a).

To illustrate the inference of solar structure, Fig. 1 shows the inferred difference in squared sound speed between the Sun and two solar models. One is what might be termed a 'standard' model, computed with the Grevesse & Noels (1993) composition. Here the relative differences are below 0.5%; although this is far more than the very small estimated errors in the difference it still indicates that the model provides a good representation of the solar interior. Also shown are the results for a corresponding model computed with the revised (Asplund 2005) composition. It is evident that this leads to a dramatic deterioration in the agreement between the model and the Sun (see also Basu & Antia 2008, for a review). Similar discrepancies are found for other helioseismically inferred quantities, such as the depth of the convection zone and the envelope helium abundance. Of particular relevance to the present discussion is the inconsistency found in a detailed analysis by Chaplin *et al.* (2007) in the small frequency separation $\delta\nu_{l\,l+2}(n) = \nu_{nl} - \nu_{n-1\,l+2}$, ν_{nl} being the cyclic frequency of a mode of degree l and radial order n; as discussed by Lebreton & Montalbán (this volume) this is a measure of stellar age. I return to the consequences of this below.

These discrepancies clearly indicate potential problems with solar modelling, if the revision of the solar abundances is accepted. Guzik (2006, 2008) reviewed the attempts to modify the model calculation to restore the agreement with helioseismology which so far have not led to any entirely satisfactory solution. A trivial modification is to postulate intrinsic errors in the opacity tables which compensate for the composition change (Bahcall *et al.* 2005; Christensen-Dalsgaard *et al.* 2009); however, as discussed in the latter reference the required change is as high as 30% at the base of the convection zone, at a temperature of 2×10^6 K, which may be unrealistic. A resolution of these issues is evidently of general importance to stellar modelling and hence to the determination of stellar ages from evolution calculations.

I finally recall that helioseismology has yielded detailed inferences of the solar internal rotation (see Thompson *et al.* 2003, for a review). This shows that the convection zone approximately shares the surface latitudinal differential rotation, while the radiative interior rotates at a nearly constant rate, somewhat smaller than the surface equatorial rotation rate. This is obviously relevant to the modelling of the, so far incompletely understood, evolution of stellar rotation and hence to the use of gyrochronology for stellar age determinations (Barnes 2007, Meibom, this volume).

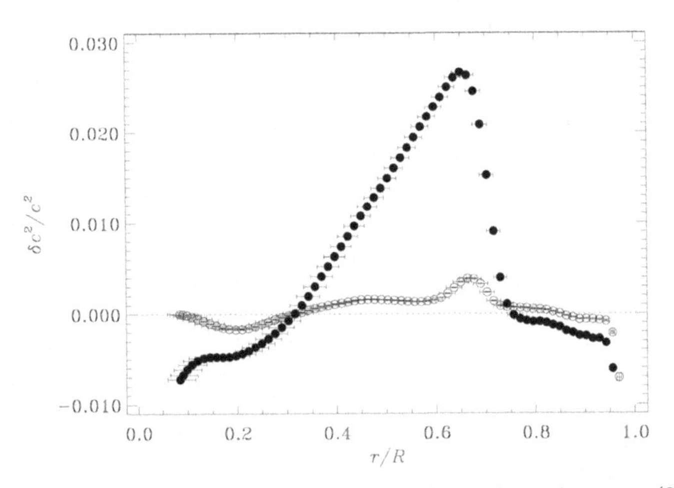

Figure 1. Inferred relative differences in squared sound speed, in the sense (Sun) – (model), from inversion of frequency differences between the Sun and two solar models. The open circles used Model S of Christensen-Dalsgaard *et al.* (1996), based on the Grevesse & Noels (1993) composition, while the filled circles are for a corresponding model but using the revised solar composition (Asplund 2005). The horizontal bars provide a measure of the resolution of the inversion. The standard error in the inferred differences are generally smaller than the size of the symbols. (From Christensen-Dalsgaard *et al.* 2009).

4. Asteroseismic age determination

As discussed by Lebreton & Montalbán (this volume), asteroseismology† provides sensitive diagnostics of stellar ages. The change in the internal structure of a star with evolution directly affects the oscillation frequencies, and hence the observed frequencies, when suitably analyzed, can be used to determine the age. Also, the computation of the relevant aspects of the frequencies from a given model structure is relatively insensitive to systematic errors; on the other hand, the dependence of the structure on age is clearly affected by uncertainties in the modelling.

Here I concentrate on acoustic modes in solar-like stars; these are typically of high radial order and hence their diagnostic potential can be investigated on the basis of asymptotic theory. The low-degree modes that are relevant to observations of distant stars penetrate to the core of the star; hence their frequencies are sensitive to the sound speed in the core and consequently to the composition change resulting from evolution. In the ideal-gas approximation $c^2 \propto T/\mu$, where T is temperature and μ is the mean molecular weight. Both increase as a result of evolution; however, since the temperature is strongly constrained by the high temperature sensitivity of the nuclear burning rates, the change in μ dominates, leading to a decrease in the sound speed in the core. This is illustrated in Fig. 2 for the evolution of a $1\,M_\odot$ star. The decrease in c affects most strongly the modes of the lowest degree which penetrate most deeply; consequently, the small frequency separations $\delta\nu_{02}$ and $\delta\nu_{13}$ decrease with increasing age. To characterize a star based on high-order acoustic-mode frequencies one can in addition use the *large frequency separation* $\Delta\nu_{nl} = \nu_{nl} - \nu_{n-1\,l}$ which essentially provides a measure of the mean density of the star. Thus the position of the star in a $(\langle\Delta\nu\rangle, \langle\delta\nu\rangle)$ diagram, based on suitable averages, provides an indication of the mass and age of the star (Christensen-Dalsgaard 1984, 1988; Ulrich 1986). Obviously, the calibration of the diagram depends on the physics and other parameters, such as the composition, of the stars (Gough 1987; Monteiro *et al.* 2002, see also Lebreton & Montalbán, these proceedings). These potential systematic errors must be taken into account in the interpretation of the results.

† For an illuminating and entertaining discussion of the etymology of *asteroseismology*, see Gough (1996).

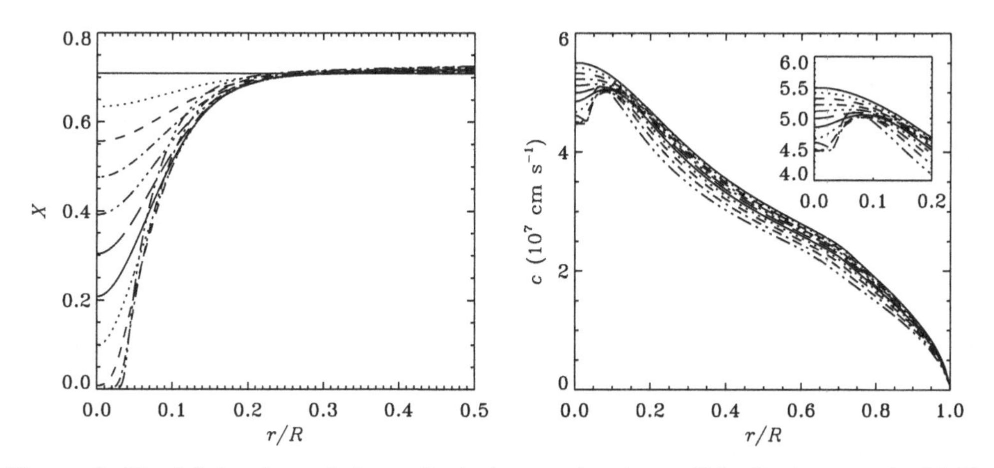

Figure 2. The left-hand panel shows the hydrogen abundance X in the inner part of $1\,M_\odot$ models of age $0 - 10\,\mathrm{Gyr}$, in steps of 1 Gyr. The right-hand panel shows the sound speed in these models, with an enlargement in the insert of the behaviour in the core. Except for the final model the central sound speed decreases with increasing age.

It was noted by Roxburgh & Vorontsov (2003) that the near-surface errors, although to a large extent canceling in the difference, has a significant effect on the small frequency separations. They showed that this effect is suppressed by evaluating separation ratios, such as

$$r_{02}(n) = \frac{\nu_{n\,0} - \nu_{n-1\,2}}{\nu_{n\,1} - \nu_{n-1\,1}} , \qquad r_{13}(n) = \frac{\nu_{n\,1} - \nu_{n-1\,3}}{\nu_{n+1\,0} - \nu_{n\,0}} , \tag{4.1}$$

and demonstrated that these ratios are directly related to the effect of the stellar core on the oscillation frequencies. This was further analyzed by Otí Floranes *et al.* (2005) who showed that, unlike $\delta\nu_{02}$, r_{02} is essentially insensitive to structure changes near the surface; they furthermore considered several examples of model modifications, including changes to the stellar radius, keeping the structure of the core unchanged and found that these did not affect the separation ratios. The use of $(\langle\Delta\nu\rangle, \langle r_{02}\rangle)$ diagrams to characterize stellar properties is discussed by Lebreton & Montalbán (this volume).

This correction for the near-surface effects assumes that they are independent of degree and hence essentially that the underlying physical cause is spherically symmetric. In fact, it is known from the case of the Sun that the magnetic activity causes frequency changes that are strongly related to the distribution in latitude of the magnetic field (e.g., Howe *et al.* 2002). As noted by Dziembowski & Goode (1997) the concentration of magnetic activity towards the equator causes a frequency shift for low-degree modes, observed with limited frequency resolution, that depends on degree and hence might corrupt the study of the solar core based, e.g., on the small frequency separations. Such degree-dependent frequency shifts were in fact observed by Chaplin *et al.* (2004) and Toutain & Kosovichev (2005). It was argued by Dziembowski & Goode (1997) that the effects could be eliminated in the solar case from observations of higher-degree modes; however, it is obvious that they are a significant concern in observations of distant stars where only low-degree data are available.

The independent radioactive age determination of the solar system provides an excellent test of the use of the oscillation frequencies of low-degree acoustic modes to determine stellar ages. Gough & Novotny (1990) made a careful analysis of the sensitivity of seismic age determinations to other aspects of the solar models. This was extended by Gough (2001), including also a determination of $Z_{\rm s}/X_{\rm s}$, from the analysis of $\delta\nu_{02}$ and $\delta\nu_{03}$; the results were consistent with the meteoritic ages although with a substantial uncertainty, owing to the strong sensitivity of the small frequency spacings to $Z_{\rm s}/X_{\rm s}$. Guenther & Demarque (1997) also estimated the solar age based on the small frequency separations. More systematic analyses, using a χ^2 fit to the observed values, fixing the value of $Z_{\rm s}/X_{\rm s}$, were carried out by Dziembowski *et al.* (1999) and Bonanno *et al.* (2002). These analyses showed that the seismically inferred age was in good agreement with the meteoritic age. Dziembowski *et al.* also found that the inferred age decreased with an increased $Z_{\rm s}/X_{\rm s}$, in accordance with the results of Gough (2001).

I have repeated this type of analysis, using the observed frequencies of Chaplin *et al.* (2007), based on 4572 days of observation with the BiSON network and corrected for frequency shifts caused by the solar magnetic activity. The models essentially corresponded to Model S of Christensen-Dalsgaard *et al.* (1996), except that more recent OPAL opacities (Iglesias & Rogers 1996) and OPAL equation of state (Rogers & Nayfonov 2002)†

† In particular, the equation of state tables included relativistic effects for the electrons; Bonanno *et al.* (2002) found that these had a noticeable effect on the age fit.

were used. The goodness of fit was determined by, for example,

$$\chi^2(\delta\nu_{02}) = \frac{1}{N-1}\sum_n \frac{[\delta\nu_{02}(n)^{(\mathrm{obs})} - \delta\nu_{02}(n)^{(\mathrm{mod})}]^2}{\sigma[\delta\nu_{02}(n)]^2}, \tag{4.2}$$

where N is the number of modes included, $\delta\nu_{02}(n)^{(\mathrm{obs})}$ and $\delta\nu_{02}(n)^{(\mathrm{mod})}$ are the observed and model values of the small separation, and $\sigma[\delta\nu_{02}(n)]^2$ is the variance of the observed small separation.

Preliminary results of this analysis are shown in Fig. 3. Fits have been made to both the unscaled small separations $\delta\nu_{02}$ and $\delta\nu_{13}$ and the separation ratios r_{02} and r_{13}. I have computed results as functions of age, in all cases calibrating the models to the solar radius and luminosity and a fixed value of $Z_\mathrm{s}/X_\mathrm{s}$. The left-hand curves assumed $Z_\mathrm{s}/X_\mathrm{s} = 0.0245$ (Grevesse & Noels 1993); here the best fits are evidently obtained close to the age interval obtained from the meteoritic analysis, indicated by the vertical lines. Interestingly, the fit for the separation ratios indicate a slightly lower age and, in the case of r_{13}, a substantially lower χ^2; in general, the values of χ^2 show that the models are not entirely consistent with the observations. On the other hand, the results in the right-hand curves for the Asplund (2005) composition, with $Z_\mathrm{s}/X_\mathrm{s} = 0.0165$, are clearly entirely inconsistent with the meteoritic age. This is in agreement with the analysis by Chaplin *et al.* (2007) of the small separations, similarly showing that they are incompatible with the revised composition. Also, the values of the minimal χ^2 are much larger than for the GN93 composition. The very large minimum χ^2 for r_{02} in this case clearly requires further investigation. Note that the shift in the inferred age with $Z_\mathrm{s}/X_\mathrm{s}$ is in accordance with the results obtained by Dziembowski *et al.* (1999) and Gough (2001).

As noted by Gough (2002) the systematic errors, arising from the other unknown parameters of the stars, far exceed the effects of the statistical errors in the oscillation frequencies. To constrain these parameters on the basis of asteroseismic data he proposed

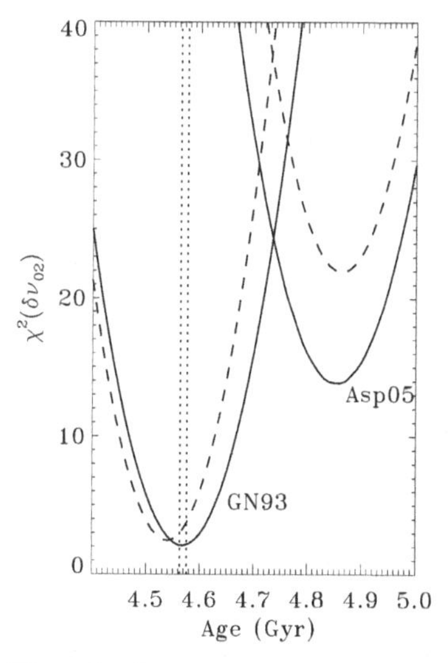

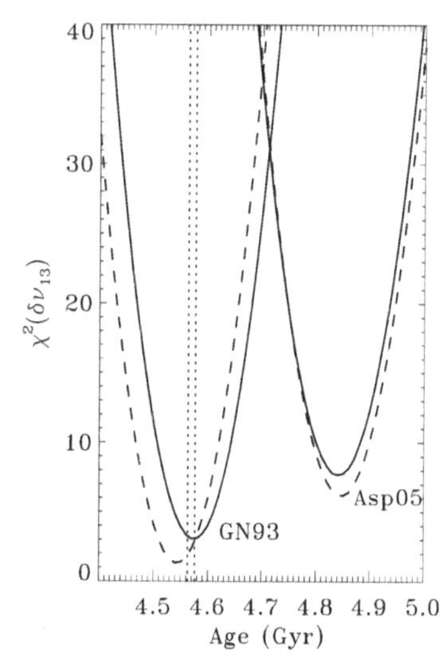

Figure 3. Fits of solar models, as a function of age, to the frequency separations obtained from BiSON observations (Chaplin *et al.* 2007); the left-hand panel shows results for the separations between $l = 0$ and 2, and the right-hand panel for separations between $l = 1$ and 3. The solid curves are for $\delta\nu_{l\,l+2}$ and the dashed curves for $r_{l\,l+2}$. Results are shown both for models computed with the Grevesse & Noels (1993) composition (GN93) and the Asplund (2005) composition (Asp05). The vertical dotted lines indicate the interval of solar age obtained by Wasserburg, in Bahcall & Pinsonneault (1995).

the analysis of other aspects of the frequencies, particularly the effects sharp features in the sound speed (see also Lebreton & Montalbán, this volume). Houdek & Gough (2007a) made a detailed analysis of the effects of such features, which they called 'acoustical glitches', in order to derive reliable diagnostics for the envelope helium abundance and the depth of the convection zone. This was applied by Houdek & Gough (2007b) and Houdek & Gough (2008) to the determination of the solar age, based solely on low-degree frequencies such as might be observed in other stars; they noted that removing the effects of the glitches from the frequencies resulted in a more robust calibration for the age. The resulting age, calibrating the radius and luminosity to solar values but determining the heavy-element abundance from the fit, is close to, but not entirely consistent with, the meteoritic age; preliminary results are an age $t_\odot = (4.68 \pm 0.02)\,\mathrm{Gyr}$ and an initial heavy-element abundance $Z_0 = 0.0169 \pm 0.0005$.

It is evident that processes modifying the core composition have a potentially serious effect on the age determination. This is particularly true for stars with convective cores where the uncertain extent of convective overshoot has a large effect on the relation between age and stellar structure, as illustrated by Lebreton & Montalbán (this volume) in a $(\langle\Delta\nu\rangle, \langle r_{02}\rangle)$ diagram. To detect such effects and correct the age determinations for them require asteroseismic analysis beyond the simple fits to the frequency separations. It is encouraging that it appears to be possible, with sufficiently good data, to resolve the structure of stellar cores in inverse analyses using just low-degree modes (e.g., Gough & Kosovichev 1993; Basu *et al.* 2002; Roxburgh 2002). This may allow determination of the extent of convective overshoot and other mixing processes that could affect the age determination. Also, based on asymptotic analysis Cunha & Metcalfe (2007) developed a diagnostic tool which may be used to characterize small convective cores and hence potentially eliminate the effects of additional mixing. A similar diagnostic was found by Mazumdar *et al.* (2006) on the basis of extensive model calculations.

5. Next steps

We have yet to see the full realization of the potential for age determination based on asteroseismology, but the observational prospects are excellent. The CoRoT mission has yielded the first results on solar-like stars (e.g., Appourchaux *et al.* 2008) and much more is expected in the next few years. The NASA Kepler mission, with planned launch in March 2009, will yield excellent asteroseismic data for a very large number of stars (e.g., Christensen-Dalsgaard *et al.* 2007; Kjeldsen *et al.* 2008b); an important aspect of the asteroseismic investigation based on Kepler data, given the main goal of the mission of characterizing extra-solar planetary systems, is to determine properties of the central stars in such systems, in particular their radius and age. In the longer term, ground-based projects for Doppler-velocity observations of stellar oscillations, such as the SONG (Grundahl *et al.* 2008) and SIAMOIS (Mosser *et al.* 2008) projects, are expected to yield exquisite data although for a smaller number of stars.

To utilize fully the data from these projects we need further development and tests of the asteroseismic diagnostic tools, taking into account also the additional unknown properties of the stars on the one hand, and other observed properties of the stars on the other. There is no doubt that, as in the past, asymptotic analyses will be extremely important guides in determining the optimal combinations of frequencies; however, extensive model calculations and analysis of artificial data, under the various relevant assumptions, will also be crucial. This clearly needs to take into account also the detailed properties of the oscillations and their effect on the inferred oscillation parameters (e.g., Chaplin *et al.* 2008a). A central effort in this regard is the asteroFLAG project (Chaplin *et al.* 2008b,c) to carry out blind tests on the analysis of artificial data, involving a substantial number of different techniques and groups.

As a result of these efforts, both observational and theoretical, we may hope to obtain reliable and precise age determinations for a number of stars of varying properties. These can then be used as calibrators for other, less direct, age diagnostics and thus extend the base for the general determination of stellar ages.

References

Ahmad, Q. R., Allen, R. C., Andersen, T. C., *et al.* 2002, *Phys. Rev. Lett.* 89, 011301

Appourchaux, T., Michel, E., Auvergne, M., *et al.* 2008, *A&A* 488, 705

Asplund, M. 2005, *ARAA* 43, 481

Asplund, M., Grevesse, N., Sauval, A. J., Allende Prieto, C., & Kiselman, D. 2004, *A&A* 417, 751 (Erratum: *A&A* 435, 339)

Auwers, A. 1891, *Astron. Nachr.* 128, 361

Böhm-Vitense, E. 1958, *ZfA* 46, 108

Bahcall, J. N. & Pinsonneault, M. H. 1995, (with an appendix by G. J. Wasserburg), *Rev. Mod. Phys.* 67, 781

Bahcall, J. N. & Serenelli, A. M. 2005, *ApJ* 626, 530

Bahcall, J. N., Gonzalez-Garcia, M. C., & Peña-Garay, C. 2004, *J. High Energy Phys.* 08, 016

Bahcall, J. N., Basu, S., Pinsonneault, M., & Serenelli, A. M. 2005, *ApJ* 618, 1049

Bahcall, J. N., Serenelli, A. M., & Basu, S. 2006, *ApJS* 165, 400

Barnes, S. A. 2007, *ApJ* 669, 1167

Basu, S. & Antia, H. M. 2008, *Phys. Rep.* 457, 217

Basu, S., Christensen-Dalsgaard, J., & Thompson, M. J. 2002, in: F. Favata, I. W. Roxburgh & D. Galadí-Enríquezi (eds), *Proc. 1st Eddington Workshop, 'Stellar Structure and Habitable Planet Finding'*, ESA SP-485 (Noordwijk, The Netherlands: ESA), p. 249

Bonanno, A., Schlattl, H., & Paternò, L. 2002, *A&A* 390, 1115

Brown, T. M. & Christensen-Dalsgaard, J. 1998, *ApJ* 500, L195

Chaplin, W. J., Elsworth, Y., Isaak, G. R., Miller, B. A., & New, R. 2004, *MNRAS* 352, 1102

Chaplin, W. J., Serenelli, A. M., Basu, S., Elsworth, Y., New, R., & Verner, G. A. 2007, *ApJ* 670, 872

Chaplin, W. J., Houdek, G., Appourchaux, T., Elsworth, Y., New, R., & Toutain, T. 2008a, *A&A* 485, 813

Chaplin, W. J., Appourchaux, T., Arentoft, T., *et al.* 2008b, *Astron. Nach.* 329, 549

Chaplin, W. J., Appourchaux, T., Arentoft, T., *et al.* 2008c, in: L. Gizon & M. Roth (eds), *Proc. HELAS II International Conference: Helioseismology, Asteroseismology and the MHD Connections*, *J. Phys.: Conf. Ser.* 118, 012048

Christensen-Dalsgaard, J. 1984, in: A. Mangeney & F. Praderie (eds), *Space Research Prospects in Stellar Activity and Variability* (Paris: Paris Observatory Press), p. 11

Christensen-Dalsgaard, J. 1988, in: J. Christensen-Dalsgaard & S. Frandsen (eds), *Advances in helio- and asteroseismology, Proc. IAU Symposium No 123* (Dordrecht: Reidel), p. 295

Christensen-Dalsgaard, J. 2002, *Rev. Mod. Phys.* 74, 1073

Christensen-Dalsgaard, J. & Thompson, M. J. 1997, *MNRAS* 284, 527

Christensen-Dalsgaard, J., Däppen, W., & Lebreton, Y. 1988, *Nature* 336, 634

Christensen-Dalsgaard, J., Däppen, W., Ajukov, S. V., *et al.* 1996, *Science* 272, 1286

Christensen-Dalsgaard, J., Arentoft, T., Brown, T. M., Gilliland, R. L., Kjeldsen, H., Borucki, W. J., & Koch, D. 2007, in: G. Handler & G. Houdek (eds), *Proc. Vienna Workshop on the Future of Asteroseismology*, *Comm. in Asteroseismology* 150, 350

Christensen-Dalsgaard, J., Di Mauro, M. P., Houdek, G., & Pijpers, F. 2009, *A&A*, in the press `[arXiv:0811.1001 [astro-ph]]`.

Cunha, M. S. & Metcalfe, T. S. 2007, *ApJ* 666, 413

Demarque, P., Guenther, D. B. & Kim, Y.-C. 1999, *ApJ* 517, 510

Dziembowski, W. A. & Goode, P. R. 1997, *A&A* 317, 919

Dziembowski, W. A., Fiorentini, G., Ricci, B., & Sienkiewicz, R. 1999, *A&A* 343, 990

Elsworth, Y., Howe, R., Isaak, G. R., McLeod, C. P., & New, R. 1990, *Nature* 347, 536

Fröhlich, C. & Lean, J. 2004, *A&AR* 12, 273

Gough, D. O. 1987, *Nature* 326, 257

Gough, D. O. 1996, *Observatory* 116, 313

Gough, D. O. 2001, in: T. von Hippel, C. Simpson & N. Manset (eds), *Astrophysical Ages and Time Scales*, ASP Conf. Ser. 245 (San Francisco: ASP), p. 31

Gough, D. O. 2002, in: F. Favata, I. W. Roxburgh & D. Galadí-Enríquezi (eds), *Proc. 1st Eddington Workshop: 'Stellar structure and habitable planet finding'*, ESA SP-485 (Noordwijk, The Netherlands: ESA), p. 65

Gough, D. O. & Kosovichev, A. G. 1993, in: T. M. Brown (ed.), *Proc. GONG 1992: Seismic investigation of the Sun and stars*, ASP Conf. Ser. 42, (San Francisco: ASP), p. 351

Gough, D. O. & Novotny, E. 1990, *Solar Phys.* 128, 143

Gough, D. O. & Weiss, N. O. 1976, *MNRAS* 176, 589

Grevesse, N. & Noels, A. 1993, in: N. Prantzos, E. Vangioni-Flam & M. Cassé (eds), *Origin and evolution of the Elements*, (Cambridge: Cambridge Univ. Press), p. 15

Grundahl, F., Arentoft, T., Christensen-Dalsgaard, J., Frandsen, S., Kjeldsen, H., & Rasmussen, P. K. 2008, in: L. Gizon & M. Roth (eds), *Proc. HELAS II International Conference: Helioseismology, Asteroseismology and the MHD Connections*, *J. Phys.: Conf. Ser.* 118, 012041

Guenther, D. B. & Demarque, P. 1997, *ApJ* 484, 937

Guzik, J. A. 2006, in: K. Fletcher (ed.), *Proc. SOHO 18 / GONG 2006 / HELAS I Conf. Beyond the spherical Sun*, ESA SP-624, (Noordwijk, The Netherlands: ESA).

Guzik, J. A. 2008, *MemSAI* 79, 481

Haxton, W. C. 2008, *PASA* 25, 44

Haxton, W. C., Parker, P. D., & Rolfs, C. E. 2006, *Nucl. Phys. A.* 777, 226

Houdek, G. & Gough, D. O. 2007a, *MNRAS* 375, 861

Houdek, G. & Gough, D. O. 2007b, in: R. J. Stancliffe, J. Dewi, G. Houdek, R. G. Martin & C. A Tout (eds), *Unsolved Problems in Stellar Physics*, AIP Conf. Proc. 948 (Melville: AIP), p. 219

Houdek, G. & Gough, D. O. 2008, in: L. Deng & K. L. Chan (eds), *The Art of Modelling Stars in the 21st Century*, Proc. IAU Symposium No 252 (Cambridge: Cambridge University Press), p. 149

Howe, R., Komm, R. W., & Hill, F. 2002, *ApJ* 580, 1172

Iglesias, C. A. & Rogers, F. J. 1996, *ApJ* 464, 943

Kjeldsen, H., Bedding, T. R., & Christensen-Dalsgaard, J. 2008a, *ApJ* 683, L175

Kjeldsen, H., Bedding, T. R., & Christensen-Dalsgaard, J. 2008b, in: F. Pont, D. Queloz & D. D. Sasselov (eds), *Transiting Planets*, Proc. IAU Symposium No 253 (Cambridge: Cambridge University Press), in the press [arXiv:0807.0508v1 [astro-ph]]

Ludwig, H.-G., Freytag, B., & Steffen, M. 1997, in: F. P. Pijpers, J. Christensen-Dalsgaard & C. S. Rosenthal, C. S. (eds), *SCORe'96: Solar Convection and Oscillations and their Relationship*, (Dordrecht: Kluwer), p. 59

Ludwig, H.-G., Freytag, B., & Steffen, M. 1999, *A&A* 346, 111

Mazumdar, A., Basu, S., Collier, B. L., & Demarque, P. 2006, *MNRAS* 372, 949

Monteiro, M. J. P. F. G., Christensen-Dalsgaard, J., & Thompson, M. J. 2002, in: F. Favata, I. W. Roxburgh & D. Galadí-Enríquezi (eds), *Proc. 1st Eddington Workshop: 'Stellar structure and habitable planet finding'*, ESA SP-485 (Noordwijk, The Netherlands: ESA), p. 291

Mosser, B., Appourchaux, T., Catala, C., Buey, J.-T. and the SIAMOIS team 2008, in: L. Gizon & M. Roth (eds), *Proc. HELAS II International Conference: Helioseismology, Asteroseismology and the MHD Connections*, *J. Phys.: Conf. Ser.* 118, 012042

Otí Floranes, H., Christensen-Dalsgaard, J., & Thompson, M. J. 2005, *MNRAS* 356, 671

Robertson, R. G. H. 2006, *Prog. Particle Nuclear Phys.* 57, 90

Rogers, F. J. & Nayfonov, A. 2002, *ApJ* 576, 1064

Rosenthal, C. S., Christensen-Dalsgaard, J., & Nordlund, Å., Stein, R. F. & Trampedach, R. 1999, *A&A* 351, 689

Roxburgh, I. W. & Vorontsov, S. V. 2003, *A&A* 411, 215

Roxburgh, I. W. 2002, in: F. Favata, I. W. Roxburgh & D. Galadí-Enríquezi (eds), *Proc.* 1st Eddington Workshop: 'Stellar structure and habitable planet finding', ESA SP-485 (Noordwijk, The Netherlands: ESA), p. 75
Sackmann, I.-Juliana & Boothroyd, A. I. 2003, *ApJ* 583, 1024
Thompson, M. J., Christensen-Dalsgaard, J., Miesch, M. S., & Toomre, J. 2003, *ARAA* 41, 599
Toutain, T. & Kosovichev, A. G. 2005, *ApJ* 622, 1314
Trampedach, R., Stein, R. F., Christensen-Dalsgaard, J., & Nordlund, *A&A.* 1999, in: A. Giménez, E.F. Guinan & B. Montesinos (eds), *Theory* and Tests of Convection in Stellar Structure, ASP Conf. Ser. 173 (San Francisco: ASP), p. 233
Ulrich, R. K. 1986, *ApJ* 306, L37

Discussion

G. MEYNET: Can you say a few words about the behavior of the angular velocity near the center of the Sun? Is it increasing or decreasing towards the center? What are the most recent results?

J. CHRISTENSEN-DALSGAARD: The data are consistent with constant rotation in the core. Unfortunately, improving the error bars with p-mode observations will require very extended observations and the g-mode claims, although very interesting, are so far tentative.

D. SODERBLOM: You mentioned that Kepler may be able to detect planets from phase shifts of oscillation frequencies. Has that effect been seen on the Sun?

J. CHRISTENSEN-DALSGAARD: I suppose that the effect of Jupiter might be visible, but it has not been seen, or looked for, as far as I know.

S. LEGGETT: Can you comment further on the Asplund abundances? Have they been revised upward?

J. CHRISTENSEN-DALSGAARD: There has been an independent analysis of a similar nature by Caffau *et al.* (2008, A &A, 488, 1031); preliminary results show an oxygen abundance halfway between the old and the Asplund values.

P. DEMARQUE: I draw your attention to a recent detailed review of the solar abundance problem by Basu & Antia (2008, Phys. Rep., 457, 217). There is also a poster downstairs in which my collaborators and I point out some internal inconsistencies in the Asplund *et al.* analysis. Having worked on both helioseismology and 3-D simulations, I must say that I consider the seismic results to be more trustworthy.

Eric Mamajek

Matteo Monelli

The Ages of Stars
Proceedings IAU Symposium No. 258, 2008
E.E. Mamajek, D.R. Soderblom & R.F.G. Wyse, eds.

doi:10.1017/S1743921309032098

Stellar ages from asteroseismology: a few examples

Sylvie Vauclair

Laboratoire d'Astronomie de Toulouse-Tarbes, Université de Toulouse, 14 Avenue Edouard Belin, 31400 Toulouse, France
email: sylvie.vauclair@ast.obs-mip.fr

Abstract. Asteroseismology is a powerful tool to derive stellar ages, masses, gravities, radii, etc. Precise determinations of these parameters need deep analyses for each individual stars. Approximate theories are not efficient enough. Here I present results for two stars, μ Arae and ι Hor, which have both been observed during eight nights with the HARPS spectrograph in La Silla. I also show that important constraints can be obtained on core overshooting using the same techniques.

Keywords. stars: atmospheres, interiors, fundamental parameters, oscillations, techniques: spectroscopic

1. Introduction

Asteroseismology can give much more precise values of the stellar parameters, including age, mass, radius, stellar gravity, effective temperature, metallicity, helium abundance value, than any other means. However such precise determinations need deep seismic analyses of individual stars, and cannot be obtained with approximate theories only.

The stellar oscillations have to be observed during a sufficiently long time, typically eight nights with the HARPS spectrograph, to allow precise comparisons with models. The Fourier analysis leads to frequency determinations and to mode identification, as discussed in Bouchy *et al.* 2005.

Meanwhile various evolutionary tracks are computed, using different input parameters (mass, chemical composition, presence or not of overshooting, etc.). The oscillation frequencies of the models are computed and compared with the observed ones. In this framework, departures from the "asymptotic" theory give fundamental information on the stellar parameters and internal structure.

For a given set of abundances (metallicity and helium value), only one model may reproduce the observed frequencies in a satisfying way (Soriano *et al.* 2007, Vauclair *et al.* 2008). We thus obtain a series of "best models" according to the abundance values. An important general result is that the various models obtained in this way have similar ages, gravities and radii. The other parameters are then constrained with the help of the spectroscopic observational boxes.

Asteroseismology can also give information about the internal structure of the stars, and more specifically about the regions where the sound velocity changes rapidly. This happens in various transitions layers, like the limits of convective regions, or layers with strong helium gradients. The transition layers in the stellar internal regions may be characterized using the "small separations", defined as $\delta\nu_{n,\ell} = \nu_{n-1} - \nu_{n-1,\ell+2}$ (Roxburgh & Vorontsov 1994).

In a previous paper dedicated to the study of the exoplanet-host star HD 52265 (Soriano *et al.* 2007), we showed that in some cases the small separations, which should be positive in first approximation, could become negative. We explained how this special behaviour was related to the presence of either a convective core or a helium core with abrupt frontiers, resulting from the presence of a convective core in the past history of the star.

I discuss below the examples of the stars μ Arae and ι Horologii

2. The case of μ Arae

The exoplanet-host star μ Arae (HD160691) is a G5V star with a visual magnitude V = 5.1 (SIMBAD Astronomical data base). The Hipparcos parallax was initially derived as $\pi = 65.5 \pm 0.8$ mas (Perryman *et al.* 1997). This was used in the paper Bazot *et al.* (2005) which first derived the stellar parameters of this star from seismology. More recently, a new analysis was carried out by Van Leeuwen (2007), who gives $\pi = 64.48 \pm 0.31$ mas. In a forthcoming paper (Soriano & Vauclair, 2009), this new value is used to obtain an absolute magnitude of $M_V = 4.20 \pm 0.04$ and a luminosity $\log(L/L_\odot) = 0.25 \pm 0.03$, lower than the previously derived one.

Spectroscopic observations by various authors gave five different effective temperatures and metallicities (see references in Bazot *et al.* 2005). The ages obtained for this star from chromospheric index vary between 1.45 Gyr (Rocha Pinto & Maciel 1998) and 6.41 Gyr (Donahue 1998)

This star was observed for seismology in August 2004 with HARPS. At that time, two planets were known. The observations aimed for seismology lead to the discovery of a third planet, μ Ara d, with period 9.5 days (santos *et al.* 2004). More recently, evidence for a fourth planet has been discovered (Pepe *et al.* 2007).

The HARPS seismic observations allowed to identify 43 oscillation modes of degrees $l = 0$ to $l = 3$ (Bouchy *et al.* 2005). In Figure 1, they are presented in the form of an echelle diagram and compared with one of our models. The best seismic fits lead to a mass of 1.10 ± 0.01 $M_\odot$, effective temperatures between $T_{eff} = 5620$ and 5750 K, gravities between log g = 4.211 and 4.215, and ages between 6 and 8 Gyr. The age of 1.45 Gyr given by Rocha Pinto & Maciel 1998 is definitively excluded. The values obtained from seismology, are much more precise than those obtained from spectroscopy alone, for masses, gravities and radii.

Computations of models including overshooting (modelled as an extension of the convective core) have also been done in (Soriano & Vauclair, 2009) for μ Arae. We showed how core mixing can be constraint from seismic observations. In this star, mixing by overshooting or any other means at the core limit cannot have an extent of more than 0.005 times the pressure scale height.

3. The case of ι Horologii

The case of the exoplanet-host star ι Hor (HD17051) has recently been discussed in detail in Laymand & Vauclair (2007) and Vauclair *et al.* 2008. Three different groups have given different stellar parameters for this star: Gonzalez *et al.* 2001, Santos *et al.* 2004 and Fischer & Valenti 2005. The derived values for T_{eff}, g and [Fe/H] can be quite different, according to the scale the authors use for the effective temperature determinations. The associated error bars are quite large (Figure 2). Meanwhile, Santos *et al.* 2004 suggested a mass of 1.32 $M_\odot$ while Fischer and Valenti 2005 gave 1.17 $M_\odot$. The ages given in the literature for ι Hor vary between 0.43 and 6.7 Gyr (Saffe *et al.* 2005).

Some authors (Chereul *et al.* 1999, Chereul and Grenon 2001, Montes *et al.* 2001) pointed out that this star belongs to the Hyades stream. There could be two different

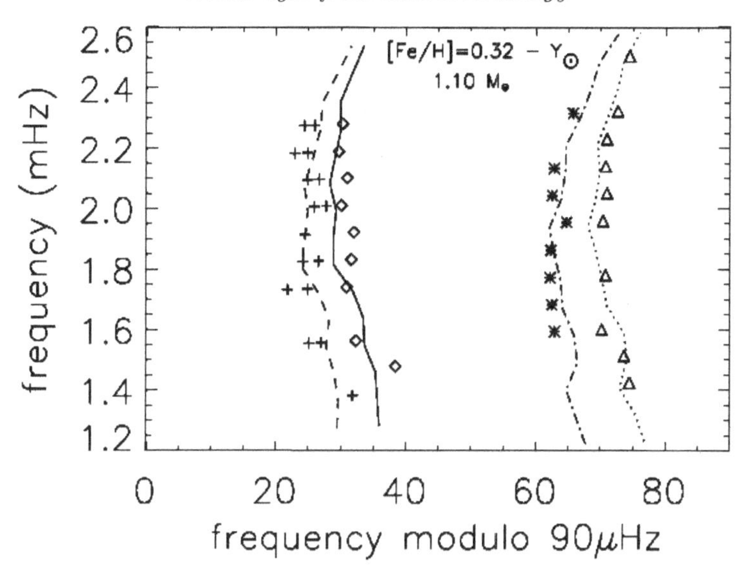

Figure 1. Echelle diagram for one of the best models obtained for the star μ Arae. The ordinates represent the frequencies of the modes and the absissae the same frequencies modulo the large separation, here 90 μHz. The lines represent the computational results for one of our models, respectively, from left to right : l = 2, 0, 3 and 1 and the symbols represent the observations (Soriano & Vauclair 2009).

reasons for this behaviour: either the star formed together with the Hyades, in a region between the Sun and the centre of the Galaxy, which would explain its overmetallicity compared to that of the Sun, or it was dynamically canalized by chance (see Famaey *et al.* 2007).

Solar-type oscillations of ι Hor were detected with HARPS in November 2006. Up to 25 oscillation modes could be identified and compared with stellar models. Our results first show that the error box given by Santos *et al.* lies below the zero age main sequence. No model can be computed in this box, while changing the effective temperature scale would move the box in the up right direction and become consistent with the other results (Figure 2).

We obtain the following general conclusions for this star: (Fe/H) is between 0.14 and 0.18; the helium abundance Y is small, 0.255 ± 0.015; The age of the star is 625 ± 5 Myr; the gravity is 4.40 ± 0.01 and its mass 1.25 ± 0.01 $M_{\odot}$. The values obtained for the metallicity, helium abundance and age of this star are those characteristic of the Hyades cluster (Lebreton *et al.* 2001). We proved from asteroseismology that ι Hor has been formed together with the Hyades cluster and that its metallicity is primordial, not due to planetary accretion.

4. Conclusion

From the few examples already available, asteroseismology has proved to be a powerful tool for determining stellar parameters and constraints on their internal structure. However, tests have to be done for individual stars, observed during long periods. Usual approximate theories are not precise enough to obtain such results. The scientific community is preparing for future observations with on-going or planned projects. Space missions like COROT, and later on KEPLER, are expected to give a large amount of new data for seismology, besides planet searches. Meanwhile ground based instruments

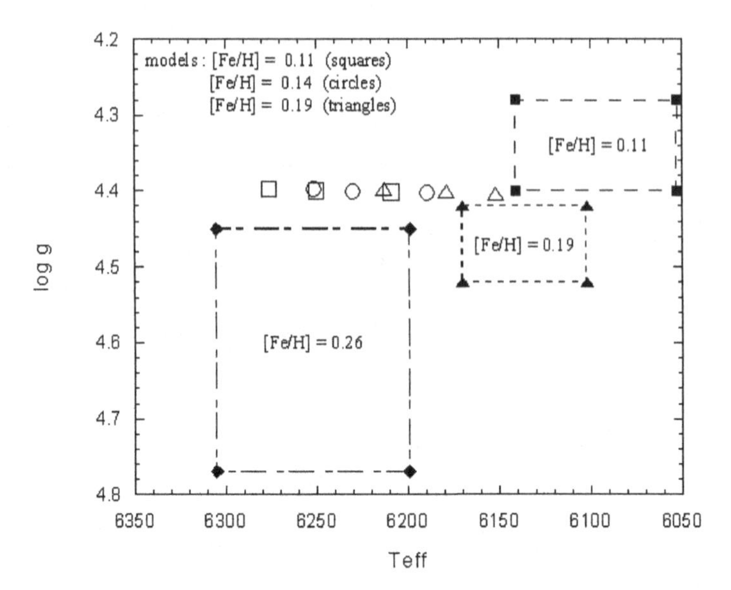

Figure 2. Observational boxes and models in the log g – log T_{eff} diagram for the star ι Hor. The three spectroscopic boxes correspond, from lower left to upper right, to Santos *et al.* 2004, Gonzalez *et al.* 2001 and Fischer & Valenti 2005. The open symbols correspond to the best models, which fit the observed echelle diagram. The squares are for [Fe/H] = 0.11, the circles for 0.14 and the triangles for 0.19. In each case three models are computed for three different helium abundances, decreasing from left to right (after Vauclair *et al.* 2008)

devoted to exoplanets like HARPS or SOPHIE can sometimes be deviated from their original purpose and be used for seismology. This represents a new dimension in the study of stellar structure and evolution.

References

Bazot, M., Vauclair, S., Bouchy, F., & Santos, N. C. 2005, *A&A*, 440, 615
Bouchy, F., Bazot, M., Santos, N. C., Vauclair, S., & Sosnowska, D. 2005, *A&A*, 440, 609
Chereul, E., Crézé, M., & Bienaymé, O., 1999, *A&AS*, 135, 5
Chereul, E. & Grenon, M., 2001, *ASP-CS*, 228, 398
Donahue, R. A., 1998, ASP Conf. Ser., 154, CD-834
Famaey, B., Pont, F., Luri, X., *et al.* 2007, *A&A*, 461, 957
Fischer,D. A. & Valenti, J. A. 2005, *ApJ*, 622, 1102
Gonzalez, G., Laws, C., Tyagi, S., & Reddy, B. E. 2001, *ApJ*, 121, 432
Laymand, M. & Vauclair, S. 2007, *A&A*, 463, 657
Lebreton, Y., Fernandes, J., & Lejeune, T., 2001, *A&A*, 374, 540
Montes, D., Lopez-Santiago, J., Galvez, M. C., *et al.*, 2001, *MNRAS*, 328, 45
Pepe, F., Correia, A. C. M., Mayor, M. *et al.* 2007, *A&A*, 462, 769
Perryman, M. A. C., Lindegren, L., Kovalevsky, J. *et al.*, 1997, *A&A*, 323, L49
Rocha Pinto, H. J. & Maciel, W. J. 1998, *MNRAS*, 298, 332
Roxburgh, I. W. & Vorontsov, S. V., 1994, *MNRAS*, 267, 297
Saffe, C., Gomez, M., & Chavero, C., 2005, *A&A*, 443, 609
Santos, N. C., Bouchy, F., Mayor, M. *et al.* 2004, *A&A*, 426, 19
Simbad Astronomical Database, http://simbad.u-strasbg.fr
Soriano, M., Vauclair, S., Vauclair, G., & Laymand, M. 2007, *A&A*, 471, 885
Soriano, M. & Vauclair, S. 2008, *Journal of Physics, C* 118, 12072
Soriano, M. & Vauclair, S. 2009, in preparation

Van Leeuwen, F., 2007, Hipparcos, the New Reduction of the Raw Data. By Floor van Leeuwen, Institute of Astronomy, Cambridge University, Cambridge, UK Series: Astrophysics and Space Science Library, Vol. 350 20 Springer Dordrecht

Vauclair, S., Laymand, M., Bouchy, F., Vauclair, G., Bon Hoa, A. Hui; Charpinet, S., & Bazot, M. 2008, *A&A*, 482, L5

Discussion

G. MEYNET: This is more a comment than a question. First, I think that these observations on the size of the convective core are exactly what is needed to make progress on convection. Second, I am not surprised that there is no overshooting in the 1.1. $M_\odot$ star you studied. Works done, for instance, by the Padova group, the Yale group, and that we did also in Geneva, showed that indeed in a small mass range, let's say for masses between $\sim 1.1 - 1.3M_\odot$, overshooting progressively increase from 0 to $\sim 0.1 - 0.2$.

S. VAUCLAIR: Yes, this is right. It would be interesting to check this behavior in the future.

R. WYSE: Could you comment on the large discrepancies in age determinations, particularly with the chromospheric activity ages?

S. VAUCLAIR: I am afraid this should be a question to the observers who derived these ages. I only think that this shows how poor these determinations are!

J. CHRISTENSEN-DALSGAARD: Have you measured the rotational splitting, and does this agree with other information about the rotation of the stars?

S. VAUCLAIR: Yes we did for μ Arae. For ι Hor we were not yet able to have convincing evidence of split modes. We intend to go on working on the data in the near future.

S. BARNES: In response to Joergen's question about gyrochoronology for any of these stars, ι Hor has a measured rotation period. Although I do not remember it right now, I remember calculating the gyro age when Sylvie's paper first came out on astro- ph and think that the gyro age worked out to be about 500 Myr, in rough agreement with Sylvie's age of $\sim$ 600 Myr.

Sylvie Vauclair

Anna Frebel

The Ages of Stars
Proceedings IAU Symposium No. 258, 2008
E.E. Mamajek, D.R. Soderblom & R.F.G. Wyse, eds.

doi:10.1017/S1743921309032104

Stellar age dating with thorium, uranium and lead

Anna Frebel[1] and Karl-Ludwig Kratz[2]

[1]McDonald Observatory and Department of Astronomy, University of Texas, 1 University Station, C1402, Austin TX, 78712
email: anna@astro.as.utexas.edu

[2]Max-Planck-Institut für Chemie (Otto-Hahn-Institut), D-55128 Mainz, Germany
email: klk@uni-mainz.de

Abstract. We present HE 1523−0901, a metal-poor star in which the radioactive elements Th and U could be detected. Only three stars have measured U abundances, of which HE 1523−0901 has the most confidently determined value. From comparing the stable Eu, Os, and Ir abundances with measurements of Th and U, stellar ages can be derived. Based on seven such chronometer abundance ratios, the age of HE 1523−0901 was found to be ~13 Gyr. Only an upper limit for Pb could be measured so far. Knowing all three abundances of Th, U, and Pb would provide a self-consistent test for r-process calculations. Pb is the beta- plus alpha-decay end-product of all decay chains in the mass region between Pb and the onset of dominant spontaneous fission above Th and U. Hence, in addition to Th/U also Th, U/Pb should be used to obtain a consistent picture for actinide chronometry. From recent r-process calculations within the classical "waiting-point" model, for a 13 Gyr old star we predict the respective abundance ratios of $\log\epsilon(\mathrm{Th/U}) = 0.84$, $\log\epsilon(\mathrm{Th/Pb}) = -1.32$ and $\log\epsilon(\mathrm{U/Pb}) = -2.16$. We compare these values with the measured abundance ratios in HE 1523−0901 of $\log\epsilon(\mathrm{Th/U}) = 0.86$, $\log\epsilon(\mathrm{Th/Pb}) > -1.0$ and $\log\epsilon(\mathrm{U/Pb}) > -1.9$. With this good level of agreement, HE 1523−0901 is already a vital probe for observational "near-field" cosmology by providing an independent lower limit for the age of the Universe.

Keywords. stars: abundances, stars: Population II, Galaxy: halo

1. Introduction

The first stars that formed from the pristine gas left after the Big Bang were very massive, of the order of $100\,\mathrm{M}_\odot$ (Bromm *et al.* 2002). After a very short life time these so-called Population III stars exploded as supernovae, which then provided the first metals to the interstellar medium. All subsequent generations of stars formed from chemically enriched material. Metal-poor stars are early Population II objects and belong to the stellar generations that formed from the non-zero metallicity gas left behind by the first stars. Due to their low masses ($\sim 0.8\,\mathrm{M}_\odot$) they have extremely long life times that exceed the current age of the Universe of $\sim 14\,\mathrm{Gyr}$. Hence, these stellar "fossils" of the early Universe are still observable today. In their atmospheres these old objects preserve information about the chemical composition of their birth cloud and thus provide archaeological evidence of the earliest times of the Universe. In particular, the chemical patterns provide detailed information about the formation and evolution of the elements and the involved nucleosynthesis processes. This knowledge is invaluable for our understanding of the cosmic chemical evolution and the onset of star- and galaxy formation. Galactic metal-poor stars are the local equivalent of the high-redshift Universe. They allow us to derive observational constraints on the nature of the first stars and supernovae, and on various theoretical works on the early Universe in general (Frebel *et al.* 2007b).

Focusing on long-lived low-mass metal-poor main-sequence and giant stars, we are observing stellar chemical abundances that reflect the composition of the interstellar medium during their star formation processes (Frebel *et al.* 2008). Stars spend $\sim$ 90% of their life time on the main sequence before they evolve to become red giants. Main-sequence stars only have a shallow convection zone that preserves the stars' birth composition over billions of years. Stars on the red giant branch have deeper convection zones that lead to a successive mixing of the surface layers with nuclear burning products from the stellar interior. In the lesser evolved giants the surface composition has not yet been significantly altered by any such mixing processes and are useful as tracers of the chemical evolution. Further evolved stars (e.g., asymptotic giant branch stars) usually have surface compositions that have been altered through repeated events that dredge up events of nucleosynthetic burning products.

The main indicator used to determine stellar metallicity is the iron abundance, [Fe/H], which is defined as $[\mathrm{A/B}] = \log_{10}(N_{\mathrm{A}}/N_{\mathrm{B}})_{\star} - \log_{10}(N_{\mathrm{A}}/N_{\mathrm{B}})_{\odot}$ for the number N of atoms of elements A and B, and $\odot$ refers to the Sun. With few exceptions, [Fe/H] traces the overall metallicity of the objects fairly well. This review focuses on metal-poor stars that have around 1/1,000 of the solar Fe abundances, and are thus able to probe the earliest epochs of nucleosynthesis processes. A summary of the history and the different "classes" of metal-poor stars and their role in the early Universe can be found in Frebel (2008) and an extensive review on the neutron-capture element abundances in (Sneden *et al.* 2008).

2. Observing the r-Process signature in the oldest stars

All elements except H and He are created in stars during stellar evolution and supernova explosions. About 5% of metal-poor stars with $[\mathrm{Fe/H}] < -2.5$ contain a strong enhancement of neutron-capture elements associated with the rapid (r-) nucleosynthesis process (Beers & Christlieb 2005) that is responsible for the production of the heaviest elements in the Universe. In those stars we can observe the majority (i.e., $\sim$ 70 of 94) of elements in the periodic table: the light, α, iron-peak, and light and heavy neutron-capture elements. These elements were not produced in the observed metal-poor star itself, but in a previous-generation supernova explosions. The so-called r-process metal-poor stars then formed from the material that was chemically enriched by such a supernova. This is schematically illustrated in Figure 1. We are thus able to study the "chemical fingerprint" of individual supernova explosions that occurred just prior to the formation of the observed star. So far, however, the nucleosynthesis site of the r-process has not yet unambiguously been identified, but supernovae with progenitor stars of $8 - 10\,\mathrm{M}_{\odot}$ are the most promising locations (Qian & Wasserburg 2003).

The giant HE 1523−0901 ($V = 11.1$) was found in a sample of bright metal-poor stars (Frebel *et al.* 2006) from the Hamburg/ESO Survey. It has the so far strongest enhancement in neutron-capture elements associated with the r-process†, $[\mathrm{r/Fe}] = 1.8$. Its metallicity is $[\mathrm{Fe/H}] = -3.0$ (Frebel *et al.* 2007a). The spectrum of HE 1523−0901 shows numerous strong lines of $\sim$ 25 neutron-capture elements, such as those of Sr, Ba, Eu, Os, and Ir. A full discussion of the complete abundance analysis will be given elsewhere (A. Frebel *et al.* 2009, in preparation). This makes possible a detailed study of the nucleosynthesis products of the r-process. This fortuitously also provides the opportunity of bringing together astrophysics and nuclear physics because these objects act as a "cosmic lab" for both fields of study.

† Stars with $[\mathrm{r/Fe}] > 1.0$; r represents the average abundance of elements from the r-process.

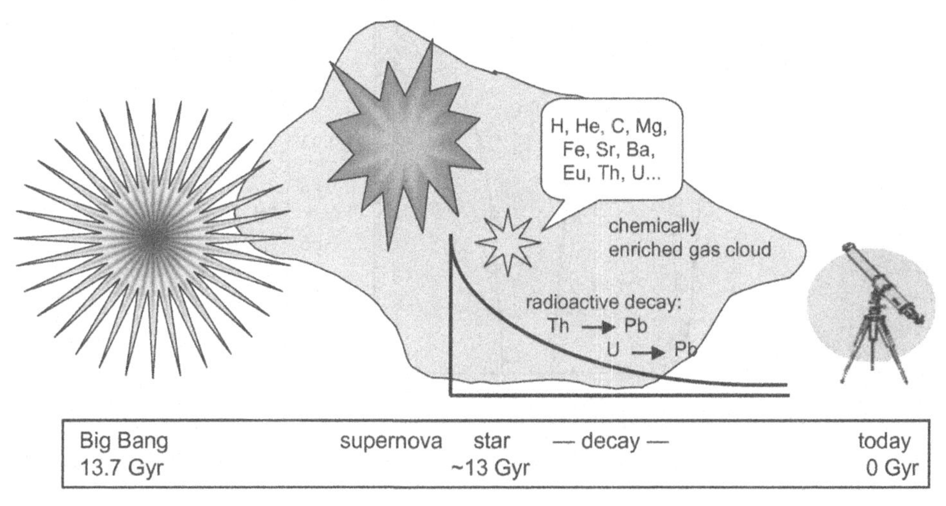

Figure 1. Formation process of r-process-enhanced metal-poor stars. They inherit the "chemical fingerprint" of a previous-generation supernova. Taken from Frebel (2009).

Although a rarity, HE 1523−0901 is not the only star that displays [r/Fe] > 1.5. In 1995, the first r-process star was discovered, CS 22892-052 (Sneden *et al.* 1996) with [r/Fe] = 1.6 and in 2001, CS 31082-001 (Cayrel *et al.* 2001) with the same overabundance in these elements. Their heavy neutron-capture elements follow the scaled *solar* r-process pattern, and offered the first vital clues to the universality of the r-process and the detailed study of the r-process by means of stars. As can be seen, in the mass range $56 < Z < 77$, the stellar abundances very closely follow the scaled solar r-process pattern Burris *et al.* (2000). This repeated behavior suggests that the r-process is universal – an important empirical finding that could not be obtained from any laboratory on earth. However, there are deviations among the lighter neutron-capture elements. It is not clear if the neutron-capture patterns are produced by a single r-process only, or if an additional new process might need to be invoked in order to explain all neutron-capture abundances.

3. Nucleo-chronometry

Among the heaviest elements are the long-lived radioactive isotopes ^{232}Th (half-life 14 Gyr) and ^{238}U (4.5 Gyr). While Th is often detectable in r-process stars, U poses a real challenge because *only one*, extremely weak, line is available in the optical spectrum. By comparing the abundances of the radioactive Th and/or U with those of stable r-process nuclei, such as Eu, stellar ages can be derived. Through individual age measurements, r-process objects become vital probes for observational "near-field" cosmology. Importantly, it also confirms that metal-poor stars with similarly low Fe abundances and no excess in neutron-capture elements are similarly old, and that the commonly made assumption about the low mass (0.6 to 0.8 $M_{\odot}$) of these survivors is well justified.

Most suitable for such age measurements are cool metal-poor giants that exhibit such strong overabundances of r-process elements. Since CS 22892-052 is very C-rich, however, the U line is blended and not detectable. Only the Th/Eu ratio could be employed, and an age of 14 Gyr was derived (Sneden *et al.* 2003). The U/Th chronometer was first measured in the giant CS 31082-001 (Cayrel *et al.* 2001) yielding an age of 14 Gyr. Since Eu and Th are much easier to detect than U, the Th/Eu chronometer is then used to derive stellar ages of r-process metal-poor stars. Compared to Th/Eu, the Th/U ratio is much more robust to uncertainties in the theoretically derived production ratio due to

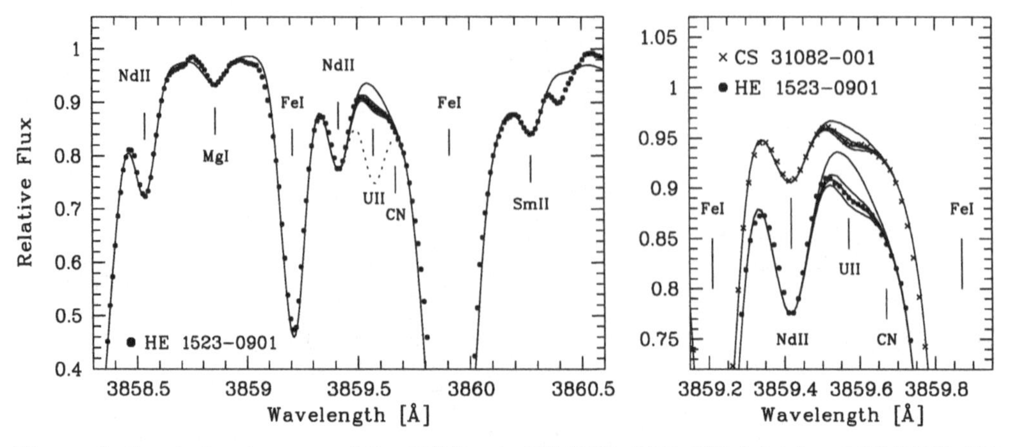

Figure 2. Spectral region around the U II line in HE 1523−0901 (*filled dots*) and CS 31082-001 (*crosses*; right panel only). Overplotted are synthetic spectra with different U abundances. The dotted line in the left panel corresponds to a scaled solar r-process U abundance present in the star if no U were decayed. Figure taken from (Frebel *et al.* 2007a).

the similar atomic masses of Th and U (Wanajo *et al.* 2002), (Kratz *et al.* 2004). Hence, stars displaying Th *and* U are the most valuable old stars.

In addition to the heaviest stable elements in HE 1523−0901, also the radioactive isotopes Th and U were measured. In fact, the U measurement in this star is the currently most reliable one of the only *three* stars with such detections. Figure 2 shows the spectral region around the only available optical U line from which the U abundance was deduced. For HE 1523−0901, the availability of both the Th and U opened up the possibility for the first time to use seven different chronometer pairs consisting of Eu, Os, Ir, Th, and U. It should be noted, that the other star with a reliable U abundance, CS 31082-001, suffers from what has been termed an "actinide boost" (Honda *et al.* 2004). Compared with the scaled solar r-process (i.e., with other stable r-process elements), it contains too much Th and U. Hence, its Th/Eu ratio yields a negative age. The origin of this issue has yet to be understood. As a result, however, it has become clear that this star likely has a different origin (Kratz *et al.* 2007).

The averaged stellar age of HE 1523−0901 derived from seven chronometers Th/r, U/r and Th/U involving combinations of Eu, Os, Ir, Th and U is $\sim$ 13 Gyr (Frebel *et al.* 2007a). Such an age provides a lower limit to the age of the Galaxy and hence, the Universe which is currently assumed to be 13.7 Gyr (Spergel *et al.* 2007). Unfortunately, realistic age uncertainties range from $\sim$2 to $\sim$ 5 Gyr. From a re-determination of the U/Th ratio in CS 31082-001 (Hill *et al.* 2002), a relative age of this star and HE 1523−0901 can be derived. HE 1523−0901 is found to be 1.5 Gyr younger than CS 31082-001, which is *independent* of the employed production ratio. This age difference is based on only a 0.07 dex difference in the observed U/Th ratios. Given that the observational uncertainties exceed that ratio difference, the derived ages of the two stars suggest that they formed at roughly the same time. This is also reflected in their almost identical metallicity of [Fe/H] ~ -3.0.

4. At the end of everything: lead

We also attempted to measure the Pb line at 4057 Å, the decay product of Th and U. However, it could not be detected in the current spectrum of HE 1523−0901 because the S/N is not high enough. The upper limit of $\log\epsilon(\mathrm{Pb}) < -0.2$ can be compared with predicted values for the total Pb abundance. Pb is produced through several decay channels. At the time when the r-process event stops, there is the direct (β- and β-delayed neutron) decay of very neutron-rich isobaric nuclei with $A = 206 - 208$ to ^{206}Pb, ^{207}Pb, and ^{206}Pb. Then there is α- and β-decay of nuclei with $A \geqslant 210$ back to Pb and finally the radioactive decay of the Th and U isotopes back to Pb over the course of the age of the Universe. The initial abundances of Th and U are driven in the same way as Pb, i.e., by a direct channel of nuclei with $A = 232$, 235 and 238, and an indirect way from the decay of r-process nuclei with heavier masses. Of course, the Th and U abundances determine, in part, the Pb abundance. It is highly debated what the abundances of individual nuclei directly after an r-process really are. Hence, different models yielding different abundance distributions can be self-consistently constrained by explaining the stellar triumvirate of Th, U and Pb abundances that are so intimately coupled not only with each other but also to the conditions (and potentially also the environment) of the r-process.

Following (Plez *et al.* 2004), we determine the Pb contribution from the decay of ^{238}U into ^{206}Pb, ^{232}U into ^{208}Pb, and ^{235}U into ^{207}Pb, whereby the last one is based on a theoretically derived ratio of $^{235}\mathrm{U}/^{238}\mathrm{U}$. The total abundance of these three decays amounts to $\log\epsilon(\mathrm{Pb}) = -0.72$. This is in agreement with our upper limit of $\log\epsilon(\mathrm{Pb}) < -0.2$. It also leaves "room" for the direct and indirect decay channels that likely produce the main portion of the Pb in the star. Using r-process model calculations, predictions can then be derived for the total Pb to be measured in HE 1523−0901. Our r-process calculations are based on a site-independent model of the classical "waiting-point" approximation (see e.g., (Kratz *et al.* 1993), (Kratz *et al.* 2007)) using as nuclear physics input the ETFSI-Q mass model (Pearson *et al.* 1996) overlaid by experimental masses Audi *et al.* (2003) Audi, Wapstra, & Thibault, and β-decay properties from QRPA calculations of Gamow-Teller and first-forbidden strength functions (Möller *et al.* 2003), again, where available, overlaid by experimental data (Pfeiffer *et al.* 2001).

After 13 Gyr of decay, the prediction for the Th/U ratio in an old metal-poor star is $\log\epsilon(\mathrm{Th/U}) = 0.84$, $\log\epsilon(\mathrm{Th/Pb}) = -1.32$ and $\log\epsilon(\mathrm{U/Pb}) = -2.16$. This agrees well with the values of $\log\epsilon(\mathrm{Th/U}) = 0.86$, $\log\epsilon(\mathrm{Th/Pb}) > -1.0$ and $\log\epsilon(\mathrm{U/Pb}) > -1.9$ in HE 1523−0901. The value for Th/U also agrees well with $\epsilon(\mathrm{Th/U}) = 0.89$ found in CS 31082-01. For Pb alone, our prediction of $\epsilon(\mathrm{Pb}) = -0.346$ is also consistent with our current upper limit of $\log\epsilon(\mathrm{Pb}) < -0.2$ in HE 1523−0901. With this good level of agreement, HE 1523-0901 is already a vital probe for observational "near-field" cosmology by providing an independent lower limit for the age of the Universe. However, our Pb prediction does not agree with the measured Pb abundance of $\log\epsilon(\mathrm{Pb}) = -0.55$ in CS31082-001. The exact reasons for this discrepancy, in having a lower Pb abundance despite the actinide boost, are currently unknown and subject to much debate. The only possibility of resolving this problem, wile still upholding the assumption of a universal (main) r-process pattern in extremely metal-poor stars, would be to assume that at least part of the heavy r-elements including Th and U in CS31082-001 were implanted long after the formation of this star (Kratz *et al.* 2004). Overall, our calculations suggest that the Pb abundance in HE 1523−0901 may be high enough so that with new, higher S/N data (500 or more at 4050 Å) a detection of the very weak Pb line at 4057 Å should become feasible, or at least provide a much tighter and more constraining upper limit.

5. Conclusion & outlook

Old metal-poor stars in our Galaxy have been shown to provide crucial observational clues to the nature of neutron-capture processes, in particular the r-process. However, even after dedicated searches, only about two dozens of these stars are known, and only three with any detection of U. Clearly, more such objects are needed to arrive at statistically meaningful abundance constraints for various r-process calculations. If more suitable candidates can be identified, and with large amounts of telescope time, more U as well as Pb abundances can be measured. Having all three measurements (Th, U and Pb) available provides an excellent test bed for different r-process models because all three abundances have to be reproduced in a self-consistent manner. Such constraints, in turn, should lead to a better understanding of how and where r-process nucleosynthesis can occur. Improved r-abundance calculations are crucial for reliably predicting the initial production ratios of Th/r, U/r and Th/U and are implicitly required for a better age dating of r-process enhanced stars. In the most optimistic (but probably not achievable) case of providing theoretical initial r-abundance ratios essentially free of systematic r-process model and nuclear-physics uncertainties, the deduced age uncertainties would then be reduced to the observational errors. In any case, better initial element production ratios will lead to improved age determinations, and with this will provide more meaningful lower limits of the age of the Universe, independent of and complementary to other methods. In this way, the possibility to carry out competitive "near-field" cosmology would strongly motivate our astronomer community to further search for search for the best and most suitable candidates for additional actinide measurements.

Acknowledgements

A. F. is supported through the W. J. McDonald Fellowship. K.-L. K. acknowledges partial support through the Deutsche Forschungsgemeinschaft under contract KR 806/13-1.

References

Audi, G., Wapstra, A. H., & Thibault, C. 2003, *N*uclear Physics A, 729, 337
Beers, T. C. & Christlieb, N. 2005, *A*RAA, 43, 531
Bromm, V., Coppi, P. S., & Larson, R. B. 2002, *A*pJ, 564, 23
Burris, D. L. *et al.* 2000, *A*pJ, 544, 302
Cayrel, R. *et al.* 2001, *N*ature, 409, 691
Frebel, A. 2009, subm. for PoS, Internatl. Symp. Nuclei in the Cosmos, Mackinaw Island, 2008
Frebel, A. 2008, in ASP Conference Series, Vol. 393, New Horizons in Astronomy, Frank N. Bash Symposium 2007, ed. A. Frebel, J. R. Maund, J. Shen, & M. H. Siegel, 63
Frebel, A. *et al.* 2006, *A*pJ, 652, 1585
Frebel, A. *et al.* 2007a, *A*pJ, 660, L117
Frebel, A., Johnson, J. L., & Bromm, V. 2007b, *M*NRAS, 380, L40
—. 2008, *M*NRAS, in press, astro-ph/0811.0020
Hill, V. *et al.* 2002, A&A, 387, 560
Honda, S. *et al.* 2004, *A*pJ, 607, 474
Kratz, K.-L., Bitouzet, J.-P., Thielemann, F.-K., Moeller, P., & Pfeiffer, B. 1993, *A*pJ, 403, 216
Kratz, K.-L. *et al.* 2007, *A*pJ, 662, 39
Kratz, K.-L., Pfeiffer, B., Cowan, J. J., & Sneden, C. 2004, *N*ew Astronomy Review, 48, 105
Möller, P., Pfeiffer, B., & Kratz, K.-L. 2003, *P*hysics Review C, 67, 055802
Pearson, J. M., Nayak, R. C., & Goriely, S. 1996, *P*hysics Letters B, 387, 455
Pfeiffer, B., Kratz, K. ., & Moeller, P. 2001, ArXiv Nuclear Experiment e-prints
Plez, B. *et al.* 2004, A&A, 428, L9
Qian, Y.-Z. & Wasserburg, G. J. 2003, *A*pJ, 588, 1099
Sneden, C., Cowan, J. J., & Gallino, R. 2008, *A*RA&A, 46, 241
Sneden, C. *et al.* 2003, *A*pJ, 591, 936
Sneden, C. *et al.* 1996, *A*pJ, 467, 819

Spergel, D. N., *et al.* 2007, *ApJS*, 170, 377
Wanajo, S., Itoh, N., Ishimaru, Y., Nozawa, S., & Beers, T. C. 2002, *ApJ*, 577, 853

Discussion

R. WYSE: As an advertisement for the RAVE survey, which is getting medium resolution spectra of $I < 12$ stars, we are obtaining follow-up high-resolution spectra for candidate extremely- metal-poor stars. They are bright enough that 4-8 m telescopes suffice.

A. FREBEL: Any bright metal-poor stars would of course be extremely valuable. It would also tell us more about the spatial and kinematic distributions in the Galaxy since, generally, the more metal-poor stars are fainter.

J. VALENTI: 1) How many stars are waiting to be discovered that could be age-dated using thorium or uranium diagnostics? 2) Is a high-resolution ultraviolet spectrograph required for this type of analysis?

A. FREBEL: 1) Most likely very few: bright r-process stars seem to be rare because otherwise we would have found them. Barklem *et al.* find 3-5% of stars with [Fe/H] <-2.0 to be r- process rich. We are finding those 3-5%, but they become fainter and fainter, making them inaccessible to high signal-to-noise spectroscopic observations.

I. ROEDERER: Regarding J. Valenti's question of needing STIS spectra, 1) all of the Pb-normal measurements in stars with [Fe/H] <-2.5 are from the Pb I resonance line at 2833 Å only. This is the only useful line in a number of stars in the regime of interest: old, metal-poor halo stars. 2) UV spectrographs are not required for identifying those stars, but they provide unprecedented information with regard to the overall chemical abundance pattern from which details about the r-process can be inferred.

I. ROEDERER: In the case where you have an ensemble of stars with measured thorium, and assuming that the stars are the same age and you just need to measure that age, this technique may hold great promise. It will be very difficult, I think, to improve upon the age uncertainties derived by Hill and Frebel for CS31082-001 and HE1523-0901, in the case of attempting to derive the age of a given field star using this technique.

Joergen Christensen-Dalsgaard

Jay Anderson

Author Index

Object Index

Subject Index